ATOMIC AND IONIC RADII OF SELECTED ELEMENTS[a]

Atomic number	Symbol	Atomic radius (nm)	Ion	Ionic radius (nm)
3	Li	0.152	Li^+	0.078
4	Be	0.114	Be^{2+}	0.054
5	B	0.097	B^{3+}	0.02
6	C	0.077	C^{4+}	<0.02
7	N	0.071	N^{5+}	0.01–0.02
8	O	0.060	O^{2-}	0.132
9	F	—	F^-	0.133
11	Na	0.186	Na^+	0.098
12	Mg	0.160	Mg^{2+}	0.078
13	Al	0.143	Al^{3+}	0.057
14	Si	0.117	Si^{4+}	0.039
15	P	0.109	P^{5+}	0.03–0.04
16	S	0.106	S^{2-}	0.174
17	Cl	0.107	Cl^-	0.181
19	K	0.231	K^+	0.133
20	Ca	0.197	Ca^{2+}	0.106
21	Sc	0.160	Sc^{2+}	0.083
22	Ti	0.147	Ti^{4+}	0.064
23	V	0.132	V^{4+}	0.061
24	Cr	0.125	Cr^{3+}	0.064
25	Mn	0.112	Mn^{2+}	0.091
26	Fe	0.124	Fe^{2+}	0.087
27	Co	0.125	Co^{2+}	0.082
28	Ni	0.125	Ni^{2+}	0.078
29	Cu	0.128	Cu^+	0.096
30	Zn	0.133	Zn^{2+}	0.083
31	Ga	0.135	Ga^{3+}	0.062
32	Ge	0.122	Ge^{4+}	0.044
35	Br	0.119	Br^-	0.196
39	Y	0.181	Y^{3+}	0.106
40	Zr	0.158	Zr^{4+}	0.087
41	Nb	0.143	Nb^{4+}	0.074
42	Mo	0.136	Mo^{4+}	0.068
46	Pd	0.137	Pd^{2+}	0.050
47	Ag	0.144	Ag^+	0.113
48	Cd	0.150	Cd^{2+}	0.103
50	Sn	0.158	Sn^{4+}	0.074
53	I	0.136	I^-	0.220
55	Cs	0.265	Cs^+	0.165
56	Ba	0.217	Ba^{2+}	0.143
74	W	0.137	W^{4+}	0.068
78	Pt	0.138	Pt^{2+}	0.052
79	Au	0.144	Au^+	0.137
80	Hg	0.150	Hg^{2+}	0.112
82	Pb	0.175	Pb^{2+}	0.132
92	U	0.138	U^{4+}	0.105

[a] For a complete listing, see Appendix 2.

INTRODUCTION TO MATERIALS SCIENCE FOR ENGINEERS,
Seventh Edition

Companion Website
Access Card

Thank you for purchasing a new copy of *Introduction to Materials Science for Engineers*, Seventh Edition, by James F. Shackelford. The information below provides instruction on how to access the Companion Student Website.

The Companion Website contains additional chapters which appeared in the sixth edition, student solutions to practice problems, as well as materials science programs, images, experiments, and more.

To access the Companion Website:

1. Go to www.prenhall.com/shackelford

2. From here you can register as a First-Time User or Returning User.

3. Use a coin to scratch off the coating below and reveal your student access code.
   **Do not use a knife or other sharp object as it may damage the code.*

4. On the registration page, enter your student access code. Do not type the dashes. You can use lower or uppercase letters.

5. Follow the on-screen instructions. If you need help during the online registration process, simply click on Need Help?

6. Once your personal Login Name and Password are confirmed, you can begin viewing the Companion Website.

To login to the website for the first time after you've registered:
Follow step 1 to return to the Companion Website. Then, follow the prompts for "Returning Users" to enter your Login Name and Password.

Note to Instructors: For access to the Instructor Resource Center, contact your Pearson Representative.

IMPORTANT: The access code on this page can only be used once to establish a subscription to the Shackelford, *Introduction to Materials Science for Engineers*, Seventh Edition Companion Website. If this access code has already been scratched off, it may no longer be valid. If this is the case, you can purchase a subscription by going to the *www.prenhall.com/shackelford* website and selecting "Get Access."

PEARSON
Prentice Hall

Upper Saddle River, NJ 07458
www.prenhall.com

To get help with registration, visit *http://247.prenhall.com*

SEVENTH EDITION

Introduction to MATERIALS SCIENCE FOR ENGINEERS

James F. Shackelford

University of California, Davis

PEARSON

Prentice Hall

Upper Saddle River, New Jersey 07458

Dedicated to Penelope and Scott

Library of Congress Cataloging-in-Publication Data
Shackelford, James F.
    Introduction to materials science for engineers / James F. Shackelford.–7th ed.
        p.   cm.
        Includes index.
    ISBN 0-13-601260-4
1. Materials I. Title
TA403.S515 2010
620.1′1–dc22   2008038698

Vice President and Editorial Director, ECS: Marcia J. Horton	**Marketing Assistant:** Mack Patterson
Executive Editor: Holly Stark	**Art Director:** Kenny Beck
Editorial Assistant: William Opaluch	**Manager, Rights and Permissions:** Zina Arabia
Director of Team-Based Project Management: Vince O'Brien	**Manager, Visual Research:** Beth Brenzel
Senior Managing Editor: Scott Disanno	**Manager, Cover Visual Research & Permissions:** Karen Sanatar
Production Manager: Clare Romeo	**Image Permission Coordinator:** Debbie Latronica
Marketing Manager: Tim Galligan	**Photo Researcher:** Marta Samsel

Cover Image: The molecular structure of nylon is superimposed on the surface of this nylon parachute, reminding us of the underlying principle of materials science: structure leads to properties. There is a ten order of magnitude difference in scale between this engineering design and the molecular dimensions responsible for the nature of nylon as an engineering material.

Cover photo courtesy of Stringer/Agence France Presse/Getty Images
Cover and text designer: Laura C. Ierardi

© 2009, 2005, 2000 Pearson Education, Inc.
Pearson Prentice Hall
Pearson Education, Inc.
Upper Saddle River, NJ 07458

Printed in the United States of America
10 9 8 7 6 5 4 3 2 1

ISBN-13: 978-0-13-601260-3
ISBN-10: 0-13-601260-4

Pearson Education LTD.
Pearson Education Australia PTY, Limited
Pearson Education Singapore, Pte. Ltd.
Pearson Education North Asia Ltd.
Pearson Education Canada, Ltd.
Pearson Educación de Mexico, S.A. de C.V.
Pearson Education–Japan
Pearson Education Malaysia, Pte. Ltd.

www.pearsonhighered.com

Contents

Preface

This book is designed for a first course in engineering materials. The field that covers this area of the engineering profession has come to be known as "materials science and engineering." To me, this label serves two important functions. First, it is an accurate description of the balance between scientific principles and practical engineering that is required in selecting the proper materials for modern technology. Second, it gives us a guide to organizing this book. After a short introductory chapter, "science" serves as a label for Part I on "The Fundamentals." Chapters 2 through 10 cover various topics in applied physics and chemistry. These are the foundation for understanding the principles of "materials science." I assume that some students take this course at the freshman or sophomore level and may not yet have taken their required coursework in chemistry and physics. As a result, Part I is intended to be self-contained. A previous course in chemistry or physics is certainly helpful, but should not be necessary. If an entire class has finished freshman chemistry, Chapter 2 (atomic bonding) could be left as optional reading, but it is important not to overlook the role of bonding in defining the fundamental types of engineering materials. The remaining chapters in Part I are less optional, as they describe the key topics of materials science. Chapter 3 outlines the ideal, crystalline structures of important materials. Chapter 4 introduces the structural imperfections found in real, engineering materials. These structural defects are the bases of solid-state diffusion (Chapter 5) and plastic deformation in metals (Chapter 6). Chapter 6 also includes a broad range of mechanical behavior for various engineering materials. Similarly, Chapter 7 covers the thermal behavior of these materials. Subjecting materials to various mechanical and thermal processes can lead to their failure, the subject of Chapter 8. In addition, the systematic analysis of material failures can lead to the prevention of future catastrophes. Chapters 9 and 10 are especially important in providing a bridge between "materials science" and "materials engineering." Phase diagrams (Chapter 9) are an effective tool for describing the equilibrium microstructures of practical engineering materials. Instructors will note that this topic is introduced in a descriptive and empirical way. Since some students in this course may not have taken a course in thermodynamics, I avoid the use of the free-energy property. (A chapter on thermodynamics is available on the related Web site for those instructors wishing to discuss phase diagrams with a complementary introduction to thermodynamics.) Kinetics (Chapter 10) is the foundation of the heat treatment of engineering materials.

The words "materials engineering" give us a label for Part II of the book that deals with "Materials and Their Applications." First, we discuss the five categories of *structural materials*: metals, ceramics, and glasses (Chapter 11) and polymers and composites (Chapter 12). In both chapters, we give examples of each type of structural material and describe their processing, the techniques used to produce the materials. In Chapter 13, we discuss *electronic materials* and discover a sixth category of materials, semiconductors, based on an electrical rather than bonding classification system. Metals are generally good electrical conductors, while ceramics, glasses, and polymers are generally good insulators, and semiconductors are intermediate. The exceptional discovery of superconductivity in certain ceramic materials at relatively high temperatures augments the long-standing use of superconductivity in certain metals at very low temperatures. Finally, in Chapter 14 (Materials in Engineering Design), we see that our previous discussions of properties have left us with "design parameters." Herein lies a final bridge between the principles of materials science and the use of those materials in modern engineering designs. We also must note that chemical degradation, radiation damage, wear, and recycling must be considered in making a final judgment on a materials application.

I hope that students and instructors alike will find what I have attempted to produce: a clear and readable textbook organized around the title of this important branch of engineering. It is also worth noting that materials play a central role across the broad spectrum of contemporary science and technology. In the report *Science: The End of the Frontier?* from the American Association for the Advancement of Science, 10 of the 26 technologies identified at the forefront of economic growth are various types of advanced materials.

In the presentation of this book I have attempted to be generous with examples and practice problems within each chapter, and I have tried to be even more generous with the end-of-chapter homework problems (with the level of difficulty for the homework problems clearly noted). Problems dealing with the role of materials in the engineering design process are noted with the use of a design icon **D**. One of the most enjoyable parts of writing the book was the preparation of biographical footnotes for those cases in which a person's name has become intimately associated with a basic concept in materials science and engineering. I suspect that most readers will share my fascination with these great contributors to science and engineering from the distant and not-so-distant past. In addition to a substantial set of useful data, the Appendixes provide convenient location of materials properties and key term definitions.

The various editions of this book have been produced during a period of fundamental change in the field of materials science and engineering. This change was exemplified by the change of name in the Fall of 1986 for the "American Society for Metals" to "ASM International"—a society for *materials*, as opposed to metals only. An adequate introduction to materials science can no longer be a traditional treatment of physical metallurgy with supplementary introductions to nonmetallic materials. The first edition was based on a balanced treatment of the full spectrum of engineering materials.

Subsequent editions have reinforced that balanced approach with the timely addition of new materials that are playing key roles in the economy of the twenty-first century: lightweight metal alloys, "high tech" ceramics for advanced structural applications, engineering polymers for metal substitution, advanced composites for aerospace applications, increasingly miniaturized semiconductor devices, high-temperature ceramic superconductors, fullerene carbons, and biomaterials. Since the debut of the first edition, we have also seen breakthroughs in materials characterization, such as the atomic force microscope (AFM), and in materials processing, such as self-propagating high temperature synthesis (SHS).

Finally, "feature boxes" have been introduced in recent editions. These one- or two-page case studies labeled "The Material World" are located in each chapter to provide a focus on some fascinating topics in the world of both engineered and natural materials.

Changes in the Seventh Edition

In addition to routine updates covering recent advances in materials science and engineering, the seventh edition has been condensed substantially in response to consistent recommendations from users, both students and faculty, over the years. Most instructors feel that the size of introductory textbooks has become excessive. These instructors have indicated that they cannot cover the entire book in a single term whether on the quarter or semester system.

The book is now divided into two basic parts, as described above. Part I (The Fundamentals) covering Chapters 2 through 10 has been condensed only slightly. Part II (Materials and Their Applications), however, has been condensed substantially, and ten chapters in the sixth edition are now covered in four chapters. Four chapters on structural materials are now covered in two (Chapters 11 and 12). Chapter 13 on electronic materials covers material previously dealt with in two chapters on electrical behavior (Chapter 15 of the sixth edition) and semiconductor materials (Chapter 17 of the sixth edition). Nonetheless, the full discussions of structural and electronic materials, as provided in the sixth edition, are available as supplementary chapters on the related Web site www.prenhall.com/shackelford. There, one will find Chapters 11A-B and 12A-B provide a more extensive treatment of structural materials for those desiring that, and similarly Chapters 13A and 13C contain the full discussions of electronic materials. Most instructors indicated that time did not permit coverage of previous chapters on optical behavior (Chapter 16 of the sixth edition) and magnetic materials (Chapter 18 of the sixth edition). As a result, these entire chapters are now available exclusively on the related Web site as Chapters 13B and 13D. Finally, Chapter 14 (Materials in Engineering Design) is a condensed combination of the final two chapters from the sixth edition. Again, a more extensive treatment of environmental degradation (Chapter 19 of the sixth edition) and materials selection (Chapter 20 of the sixth edition) is available on the related Web site as Chapters 14A and 14B, respectively.

A new feature of the Seventh Edition is to emphasize the concept of "Powers of Ten." In Chapter 1, we point out that an underlying principle of materials science is that understanding the behavior of materials in engineering designs (on the human scale) is obtained by looking at mechanisms that occur at various fine scales, such as the atomic-scale diffusion of carbon atoms involved in the heat treatment of steel. There is a full ten orders of magnitude difference between the size of typical engineered products and the size of typical atoms. Much of modern engineering practice has depended on engineering designs based on micrometer-scale structures, such as the transistors in an integrated circuit. Increasingly, engineers are designing systems involving the nanometer-scale. At various times throughout the text, a Powers of Ten icon will be used to highlight discussions that demonstrate this structure-property relationship.

Supplementary Material

An especially important addition to the sixth edition was the accompanying CD-ROMs. Primary credit for that supplement went to its senior author, Michael L. Meier. Mike did a tremendous amount of work in systematically collecting a vast amount of material: programs, data, articles, images, and other resources that complimented the book. All of that material is now available and updated on the related Web site.

In addition to the supplementary chapters providing the more extensive discussions in the sixth edition described in the previous section on seventh edition updates, there are other supplementary chapters on the related Web site. A chapter on "Advanced Structural Topics" can be used by instructors wishing to introduce the intriguing subjects of high-angle grain boundaries, quasicrystals and fractals. This optional discussion would naturally follow Chapters 3 and 4. Similarly, a chapter on "Materials Characterization" would complement the structural topics in Chapter 4 and the surface-related environmental issues in Chapter 14. The supplementary chapter "Thermodynamics" can be introduced just prior to Chapter 9 as a foundation for the discussions of phase diagrams and kinetics in Chapters 9 and 10.

A *Solutions Manual* is available to adopters of this textbook. The Solutions Manual contains fully worked-out solutions to the practice and homework problems **only**. The Solutions Manual is available from the publisher. The Web site contains the entire Solutions Manual in PDF format, as well as all the figures and tables from the textbook in PowerPoint® format. These materials can be very useful to faculty in preparing lectures.

Acknowledgments

Finally, I want to acknowledge a number of people who have been immensely helpful in making this book possible. My family has been more than the usual "patient and understanding." They are a constant reminder of the rich life beyond the material plane. Peter Gordon (first edition), David Johnstone (second and third editions), Bill Stenquist (fourth and fifth editions), Dorothy Marrero (sixth edition), and Holly Stark (seventh edition) are much appreciated in their roles as editors. Lilian Davila skillfully produced the computer-generated crystal structure images. A special appreciation is due to my colleagues at the University of California, Davis and to the many reviewers of all editions, especially D. J. Montgomery, John M. Roberts, D. R. Rossington, R. D. Daniels, R. A. Johnson, D. H. Morris, J. P. Mathers, Richard Fleming, Ralph Graff, Ian W. Hall, John J. Kramer, Enayat Mahajerin, Carolyn W. Meyers, Ernest F. Nippes, Richard L. Porter, Eric C. Skaar, E. G. Schwartz, William N. Weins, M. Robert Baren, John Botsis, D. L. Douglass, Robert W. Hendricks, J. J. Hren, Sam Hruska, I. W. Hull, David B. Knoor, Harold Koelling, John McLaughlin, Alvin H. Meyer, M. Natarajan, Jay Samuel, John R. Schlup, Theodore D. Taylor, Ronald Kander, Alan Lawley, Joanna McKittrick, Yu-Lin Shen, Kathleen R. Rohr, Jeffrey W. Fergus, James R. Chelikowsky, Christoph Steinbruchel, and James F. Fitz-Gerald.

I would especially like to thank the reviewers for the seventh edition: Blair London, Cal Poly State University San Luis Obispo; Valery Bliznyuk, Western

Michigan University; Patrick Ferro, Rose-Hulman Institute of Technology; David Bahr, Washington State University; K. Srinagesh, University of Massachusetts Dartmouth; Wayne L. Elban, Loyola College; Guanshui Xu, University of California at Riverside; Atin Sinha, Albany State University; Stacy Gleixner, San Jose University; Raj Vaidyanathan, University of Central Florida.

JAMES F. SHACKELFORD
Davis, California

About the Author

James F. Shackelford received the B.S. and M.S. degrees in ceramic engineering from the University of Washington and the Ph.D. degree in materials science and engineering from the University of California, Berkeley. He is currently a Professor with the Department of Chemical Engineering and Materials Science and the Director of the Integrated Studies Honors Program at the University of California, Davis. He teaches and conducts research in the areas of materials science, the structure of materials, nondestructive testing, and biomaterials. A member of ASM International and the American Ceramic Society, he was named a Fellow of the American Ceramic Society in 1992 and the Outstanding Educator of the American Ceramic Society in 1996. In 2003, he was given a Distinguished Teaching Award from the Academic Senate of the University of California, Davis. He has published over 100 archived papers and books, including *The CRC Materials Science and Engineering Handbook*, now in its third edition, and *Ceramic and Glass Materials: Structure, Properties and Processing*.

CHAPTER 1
Materials for Engineering

Modern sporting goods utilize numerous advances in materials science and engineering. Ski equipment is among the best examples. (Alamy Images)

1.1 | The Material World

We live in a world of material possessions that largely define our social relationships and economic quality of life. The material possessions of our earliest ancestors were probably their tools and weapons. In fact, the most popular way of naming the era of early human civilization is in terms of the materials from which these tools and weapons were made. The **Stone Age** has been traced as far back as 2.5 million years ago when human ancestors, or hominids, chipped stones to form weapons for hunting. The **Bronze Age** roughly spanned the period from 2000 B.C. to 1000 B.C. and represents the foundation of metallurgy, in which **alloys** of copper and tin were discovered to produce superior tools and weapons. (An *alloy* is a metal composed of more than one element.)

Contemporary archaeologists note that an earlier but less well known "Copper Age" existed between roughly 4000 B.C. and 3000 B.C. in Europe, in which relatively pure copper was used before tin became available. The limited utility of those copper products provided an early lesson in the importance of proper alloy additions. The **Iron Age** defines the period from 1000 B.C. to 1 B.C. By 500 B.C., iron alloys had largely replaced bronze for tool and weapon making in Europe.

Although archaeologists do not refer to a "pottery age," the presence of domestic vessels made from baked clay has provided some of the best descriptions

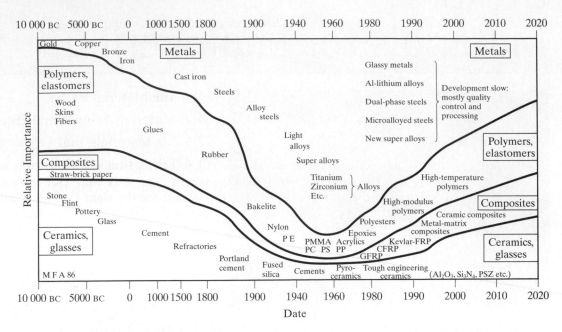

FIGURE 1.1 *The evolution of engineering materials with time. Note the highly nonlinear scale. (From M. F. Ashby,* Materials Selection in Mechanical Design, *2nd ed., Butterworth-Heinemann, Oxford, 1999.)*

of human cultures for thousands of years. Similarly, glass artifacts have been traced back to 4000 B.C. in Mesopotamia.

Modern culture in the second half of the 20th century is sometimes referred to as "plastic," a not entirely complimentary reference to the lightweight and economical polymeric materials from which so many products are made. Some observers have suggested instead that this same time frame should be labeled the "silicon age," given the pervasive impact of modern electronics largely based on silicon technology.

An intriguing visual summary of the relative importance of engineering materials over the course of human history is illustrated in Figure 1.1. Although the time scale is highly nonlinear due to the increasingly rapid evolution of technology in modern times, we can see that the increasingly dominant role of metal alloys reached a peak after World War II. Since the 1960s, pressures for weight and cost savings have led to an increasing demand for new, sophisticated nonmetallic materials. In Figure 1.1, the values of "relative importance" are necessarily subjective and come from a variety of sources. For example, for the Stone and Bronze Ages, the values are based on the assessment of archeologists. Values in the 1960s correspond to the allocated teaching hours in U.S. and U.K. universities. The values in 2020 are based on predictions made by automobile manufacturers.

1.2 | Materials Science and Engineering

Since the 1960s, the term that has come to label the general branch of engineering concerned with materials is *materials science and engineering*. This label is accurate in that this field is a true blend of fundamental scientific studies and practical engineering. It has grown to include contributions from many traditional fields, including metallurgy, ceramic engineering, polymer chemistry, condensed matter physics, and physical chemistry.

The term "materials science and engineering" will serve a special function in this introductory textbook. It will provide the basis for the text's organization. First, the word *science* describes the topics covered in Chapters 2 through 10, which deal with the fundamentals of structure, classification, and properties. Second, the word *materials* describes Chapters 11 through 13, which deal with the five types of *structural materials* (Chapters 11 and 12) and various electronic materials, especially *semiconductors* (Chapter 13). Finally, the word *engineering* describes Chapter 14, which puts the materials to work with discussions of key aspects of the selection of the right materials for the right job, along with some caution about the issue of environmental degradation in those real-world applications. Additional discussion relative to Chapters 11 through 14 is available on the Web site www.prenhall.com/shackelford.

1.3 | Six Materials That Changed Your World

The most obvious question to be addressed by the engineering student entering an introductory course on materials is, "What materials are available to me?" Various classification systems are possible for the wide-ranging answer to this question. In this book, we distinguish six categories that encompass the materials available to practicing engineers: metals, ceramics, glasses, polymers, composites, and semiconductors. We will introduce each of these categories with a single example.

STEEL BRIDGES—INTRODUCING METALS

If there is a "typical" material associated in the public's mind with modern engineering practice, it is structural *steel*. This versatile construction material has several properties that we consider **metallic**: First, it is strong and can be readily formed into practical shapes. Second, its extensive, permanent deformability, or **ductility**, is an important asset in permitting small amounts of yielding to sudden and severe loads. For example, many Californians have been able to observe moderate earthquake activity that leaves windows of glass, which is relatively **brittle** (i.e., lacking in ductility), cracked while steel-support framing still functions normally. Third, a freshly cut steel surface has a characteristic metallic luster, and fourth, a steel bar shares a fundamental characteristic with other metals: It is a good conductor of electrical current.

Among the most familiar uses of structural steel are bridges, and one of the most famous and beautiful examples is the Golden Gate Bridge connecting San Francisco, California with Marin County to the north (Figure 1.2). The opening on May 27, 1937, allowed 200,000 local residents to stroll across the impressive new structure. The following day, a ribbon cutting ceremony inaugurated automobile traffic that continues to be an important part of the fabric of life in the San Francisco Bay area. For many years, the Golden Gate held the title of "longest suspension bridge" in the world (2,737 meters). Although new bridge technologies have provided newer holders of that title, the Golden Gate is still, in the words of a local historian, a "symphony in steel."

FIGURE 1.2 *The Golden Gate Bridge north of San Francisco, California, is one of the most famous and most beautiful examples of a steel bridge. (Courtesy of Dr. Michael Meier.)*

Steel bridges continue to provide a combination of function and beauty with the Sundial Bridge in Redding, California being a stunning example (Figure 1.3). The Redding Bridge is a 66-meter pedestrian walkway designed by the famous Spanish architect Santiago Calatrava. It connects a walking trail system with the Turtle Bay Exploration Park. New bridges like this one are not merely serving as sculptural art projects. The aging infrastructure, including many bridges built as long as a century ago, also provides a challenge to engineers and the requirement for both maintenance and replacement of these important structures.

In Chapter 2, the nature of metals will be defined and placed in perspective relative to the other categories. It is useful to consider the extent of metallic behavior in the currently known range of chemical elements. Figure 1.4 highlights the chemical elements in the periodic table that are inherently metallic. This is a large family indeed. The shaded elements are the bases of the various engineering alloys, including the irons and steels (from Fe), aluminum alloys (Al), magnesium alloys (Mg), titanium alloys (Ti), nickel alloys (Ni), zinc alloys (Zn), and copper alloys (Cu) [including the brasses (Cu, Zn)].

FIGURE 1.3 *The Sundial Bridge in Redding, California is a modern masterpiece of bridge design. (Courtesy of Corbis.)*

I A								VIII				III A	IV A	V A	VI A	VII A	O
1 H	II A									I B	II B						2 He
3 Li	4 Be											5 B	6 C	7 N	8 O	9 F	10 Ne
11 Na	12 Mg	III B	IV B	V B	VI B	VII B						13 Al	14 Si	15 P	16 S	17 Cl	18 Ar
19 K	20 Ca	21 Sc	22 Ti	23 V	24 Cr	25 Mn	26 Fe	27 Co	28 Ni	29 Cu	30 Zn	31 Ga	32 Ge	33 As	34 Se	35 Br	36 Kr
37 Rb	38 Sr	39 Y	40 Zr	41 Nb	42 Mo	43 Tc	44 Ru	45 Rh	46 Pd	47 Ag	48 Cd	49 In	50 Sn	51 Sb	52 Te	53 I	54 Xe
55 Cs	56 Ba	57 La	72 Hf	73 Ta	74 W	75 Re	76 Os	77 Ir	78 Pt	79 Au	80 Hg	81 Tl	82 Pb	83 Bi	84 Po	85 At	86 Rn
87 Fr	88 Ra	89 Ac	104 Rf	105 Db	106 Sg												

58 Ce	59 Pr	60 Nd	61 Pm	62 Sm	63 Eu	64 Gd	65 Tb	66 Dy	67 Ho	68 Er	69 Tm	70 Yb	71 Lu
90 Th	91 Pa	92 U	93 Np	94 Pu	95 Am	96 Cm	97 Bk	98 Cf	99 Es	100 Fm	101 Md	102 No	103 Lw

FIGURE 1.4 *Periodic table of the elements. Those elements that are inherently metallic in nature are shown in color.*

LUCALOX LAMPS—INTRODUCING CERAMICS

Aluminum (Al) is a common metal, but aluminum *oxide*, a compound of aluminum and oxygen such as Al_2O_3, is typical of a fundamentally different family of engineering materials, **ceramics**. Aluminum oxide has two principal advantages over metallic aluminum. First, Al_2O_3 is chemically stable in a wide variety of severe

FIGURE 1.5 *Some common ceramics for traditional engineering applications. These miscellaneous parts with characteristic resistance to damage by high temperatures and corrosive environments are used in a variety of furnaces and chemical processing systems. (Courtesy of Duramic Products, Inc.)*

environments, whereas metallic aluminum would be oxidized (a term discussed further in Chapter 14). In fact, a common reaction product in the chemical degradation of aluminum is the more chemically stable oxide. Second, the ceramic Al_2O_3 has a significantly higher melting point (2020°C) than does the metallic Al (660°C), which makes Al_2O_3 a popular **refractory** (i.e., a high-temperature-resistant material of wide use in industrial furnace construction), with examples illustrated in Figure 1.5.

With its superior chemical and temperature-resistant properties, why isn't Al_2O_3 used for applications such as automotive engines in place of metallic aluminum? The answer to this question lies in the most limiting property of ceramics—brittleness. Aluminum and other metals have high ductility, a desirable property that permits them to undergo relatively severe impact loading without fracture, whereas aluminum oxide and other ceramics lack this property. Thus, ceramics are eliminated from many structural applications because they are brittle.

A significant achievement in materials technology is the development of transparent ceramics, which has made possible new products and substantial improvements in others (e.g., commercial lighting). To make traditionally opaque ceramics, such as aluminum oxide (Al_2O_3), into optically transparent materials required a fundamental change in manufacturing technology. Commercial ceramics are frequently produced by heating crystalline powders to high temperatures until a relatively strong and dense product results. Traditional ceramics made in this way contained a substantial amount of residual porosity (see also the Feature Box, "Structure Leads to Properties"), corresponding to the open space between the original powder particles prior to high-temperature processing. A significant reduction in porosity resulted from a relatively simple invention* that involved adding a small amount of impurity (0.1 wt % MgO), which caused the high-temperature densification process for the Al_2O_3 powder to go to completion. Cylinders of translucent Al_2O_3 became the heart of the design of high-temperature (1000°C) sodium vapor lamps, which provide substantially higher illumination than do conventional lightbulbs (100 lumens/W compared to 15 lumens/W). A commercial sodium vapor lamp is shown in Figure 1.6.

---

*R. L. Coble, U.S. Patent 3,026,210, March 20, 1962.

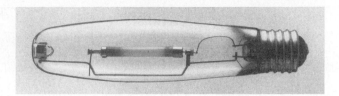

FIGURE 1.6 *High-temperature sodium vapor lamp made possible by use of a translucent Al_2O_3 cylinder for containing the sodium vapor. (Note that the Al_2O_3 cylinder is inside the exterior glass envelope.) (Courtesy of General Electric Company.)*

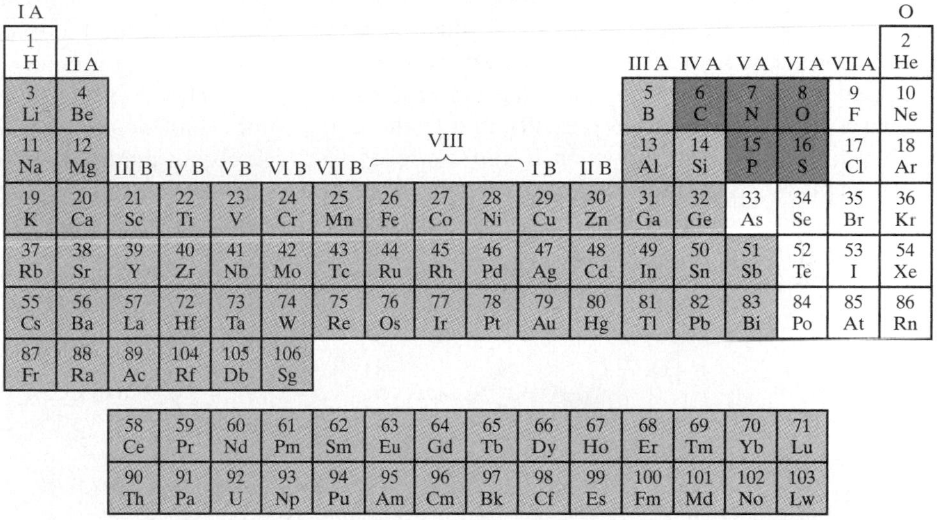

I A																	O
1 H	II A											III A	IV A	V A	VI A	VII A	2 He
3 Li	4 Be											5 B	6 C	7 N	8 O	9 F	10 Ne
11 Na	12 Mg	III B	IV B	V B	VI B	VII B	VIII			I B	II B	13 Al	14 Si	15 P	16 S	17 Cl	18 Ar
19 K	20 Ca	21 Sc	22 Ti	23 V	24 Cr	25 Mn	26 Fe	27 Co	28 Ni	29 Cu	30 Zn	31 Ga	32 Ge	33 As	34 Se	35 Br	36 Kr
37 Rb	38 Sr	39 Y	40 Zr	41 Nb	42 Mo	43 Tc	44 Ru	45 Rh	46 Pd	47 Ag	48 Cd	49 In	50 Sn	51 Sb	52 Te	53 I	54 Xe
55 Cs	56 Ba	57 La	72 Hf	73 Ta	74 W	75 Re	76 Os	77 Ir	78 Pt	79 Au	80 Hg	81 Tl	82 Pb	83 Bi	84 Po	85 At	86 Rn
87 Fr	88 Ra	89 Ac	104 Rf	105 Db	106 Sg												

58 Ce	59 Pr	60 Nd	61 Pm	62 Sm	63 Eu	64 Gd	65 Tb	66 Dy	67 Ho	68 Er	69 Tm	70 Yb	71 Lu
90 Th	91 Pa	92 U	93 Np	94 Pu	95 Am	96 Cm	97 Bk	98 Cf	99 Es	100 Fm	101 Md	102 No	103 Lw

FIGURE 1.7 *Periodic table with ceramic compounds indicated by a combination of one or more metallic elements (in light color) with one or more nonmetallic elements (in dark color). Note that elements silicon (Si) and germanium (Ge) are included with the metals in this figure but were not included in the periodic table shown in Figure 1.4. They are included here because, in elemental form, Si and Ge behave as semiconductors (Figure 1.16). Elemental tin (Sn) can be either a metal or a semiconductor, depending on its crystalline structure.*

Aluminum oxide is typical of the traditional ceramics, with magnesium oxide (MgO) and **silica** (SiO_2) being other good examples. In addition, SiO_2 is the basis of a large and complex family of **silicates**, which includes clays and claylike minerals. Silicon nitride (Si_3N_4) is an important nonoxide ceramic used in a variety of structural applications. The vast majority of commercially important ceramics are chemical compounds made up of at least one metallic element (see Figure 1.4) and one of five **nonmetallic** elements (C, N, O, P, or S). Figure 1.7 illustrates the various metals (in light color) and the five key nonmetals (in dark color) that can be combined to form an enormous range of ceramic materials. Bear in mind that many commercial ceramics include compounds and solutions of many more than two elements, just as commercial metal alloys are composed of many elements.

THE MATERIAL WORLD

Structure Leads to Properties

To understand the properties or observable characteristics of engineering materials, it is necessary to understand their structure. Virtually every major property of the six materials' categories outlined in this chapter will be shown to result directly from mechanisms occurring on a small scale (usually either the atomic or the microscopic level).

The dramatic effect that fine-scale structure has on large-scale properties is well illustrated by the development of transparent ceramics, just discussed in the introduction to ceramic materials. The microscopic-scale residual porosity in a traditional aluminum oxide (as shown in *a* and *b*), leads to loss of visible light transmission (i.e., a loss in transparency) by providing a light-scattering mechanism. Each Al_2O_3–air interface at a pore surface is a

source of light refraction (change of direction). Only about 0.3% porosity can cause Al_2O_3 to be translucent (capable of transmitting a diffuse image), and 3% porosity can cause the material to be completely opaque. The elimination of porosity provided by the Lucalox patent (adding 0.1 wt % MgO) produced a pore-free microstructure and a nearly transparent material (as shown in *c* and *d*) with an important additional property—excellent resistance to chemical attack by high-temperature sodium vapor.

The example just cited shows a typical and important demonstration of how properties of engineering materials follow directly from structure. Throughout this book, we shall be alert to the continuous demonstration of this interrelationship for all the materials of importance to engineers.

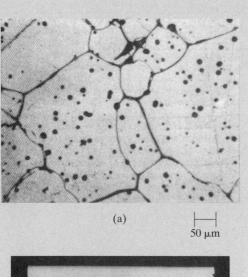

(a)

$\vdash\!\!-\!\!-\!\dashv$
50 μm

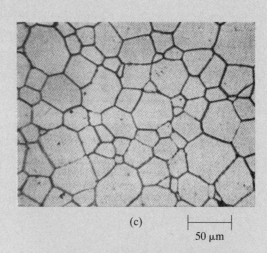

(c)

$\vdash\!\!-\!\!-\!\dashv$
50 μm

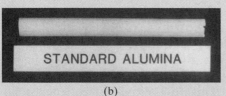

(b)

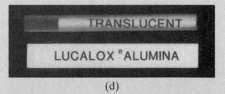

(d)

Porous microstructure in polycrystalline Al_2O_3 (a) leads to an opaque material (b). Nearly pore-free microstructure in polycrystalline Al_2O_3 (c) leads to a translucent material (d). (Courtesy of C. E. Scott, General Electric Company.)

OPTICAL FIBERS—INTRODUCING GLASSES

The metals and ceramics just introduced have a similar structural feature on the atomic scale: They are **crystalline**, which means that their constituent atoms are stacked together in a regular, repeating pattern. A distinction between metallic- and ceramic-type materials is that, by fairly simple processing techniques, many ceramics can be made in a **noncrystalline** form (i.e., their atoms are stacked in irregular, random patterns), which is illustrated in Figure 1.8. The general term for noncrystalline solids with compositions comparable to those of crystalline ceramics is **glass** (Figure 1.9). Most common glasses are silicates; ordinary window glass is approximately 72% silica (SiO_2) by weight, with the balance of the material

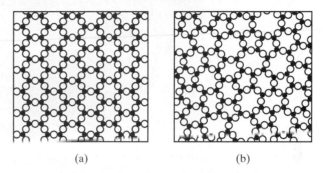

(a)   (b)

FIGURE 1.8 *Schematic comparison of the atomic-scale structure of (a) a ceramic (crystalline) and (b) a glass (noncrystalline). The open circles represent a nonmetallic atom, and the solid black circles represent a metal atom.*

FIGURE 1.9 *Some common silicate glasses for engineering applications. These materials combine the important qualities of transmitting clear visual images and resisting chemically aggressive environments. (Courtesy of Corning Glass Works.)*

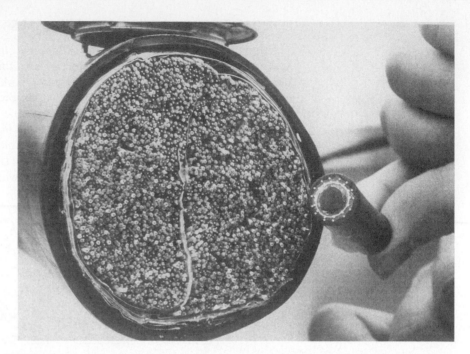

FIGURE 1.10 *The small cable on the right contains 144 glass fibers and can carry more than three times as many telephone conversations as the traditional (and much larger) copper-wire cable on the left. (Courtesy of the* San Francisco Examiner.*)*

being primarily sodium oxide (Na_2O) and calcium oxide (CaO). Glasses share the property of brittleness with crystalline ceramics. Glasses are important engineering materials because of other properties, such as their ability to transmit visible light (as well as ultraviolet and infrared radiation) and chemical inertness.

A major revolution in the field of telecommunications has occurred with the transition from traditional metal cable to optical glass fibers (Figure 1.10). Although Alexander Graham Bell had transmitted speech several hundred meters over a beam of light shortly after his invention of the telephone, technology did not permit the practical, large-scale application of this concept for nearly a century. The key to the rebirth of this approach was the invention of the laser in 1960. By 1970, researchers at Corning Glass Works had developed an **optical fiber** with a loss as low as 20 dB/km at a wavelength of 630 nm (within the visible range). By the mid-1980s, silica fibers had been developed with losses as low as 0.2 dB/km at 1.6 μm (in the infrared range). As a result, telephone conversations and any other form of digital data can be transmitted as laser light pulses rather than as the electrical signals used in copper cables. Glass fibers are excellent examples of **photonic materials**, in which signal transmission occurs by photons rather than by the electrons of electronic materials.

Glass-fiber bundles of the type illustrated in Figure 1.10 were put into commercial use by Bell Systems in the mid-1970s. The reduced expense and size, combined with an enormous capacity for data transmission, led to a rapid growth in the construction of optical communication systems. Now, virtually all telecommunications are transmitted in this way. Ten billion digital bits can be transmitted per second along an optical fiber in a contemporary system carrying tens of thousands of telephone calls.

FIGURE 1.11 *Miscellaneous internal parts of a parking meter are made of an acetal polymer. Engineered polymers are typically inexpensive and are characterized by ease of formation and adequate structural properties. (Courtesy of the Du Pont Company, Engineering Polymers Division.)*

I A	II A	III B	IV B	V B	VI B	VII B	VIII			I B	II B	III A	IV A	V A	VI A	VII A	O
1 H																	2 He
3 Li	4 Be											5 B	6 C	7 N	8 O	9 F	10 Ne
11 Na	12 Mg											13 Al	14 Si	15 P	16 S	17 Cl	18 Ar
19 K	20 Ca	21 Sc	22 Ti	23 V	24 Cr	25 Mn	26 Fe	27 Co	28 Ni	29 Cu	30 Zn	31 Ga	32 Ge	33 As	34 Se	35 Br	36 Kr
37 Rb	38 Sr	39 Y	40 Zr	41 Nb	42 Mo	43 Tc	44 Ru	45 Rh	46 Pd	47 Ag	48 Cd	49 In	50 Sn	51 Sb	52 Te	53 I	54 Xe
55 Cs	56 Ba	57 La	72 Hf	73 Ta	74 W	75 Re	76 Os	77 Ir	78 Pt	79 Au	80 Hg	81 Tl	82 Pb	83 Bi	84 Po	85 At	86 Rn
87 Fr	88 Ra	89 Ac	104 Rf	105 Db	106 Sg												

58 Ce	59 Pr	60 Nd	61 Pm	62 Sm	63 Eu	64 Gd	65 Tb	66 Dy	67 Ho	68 Er	69 Tm	70 Yb	71 Lu
90 Th	91 Pa	92 U	93 Np	94 Pu	95 Am	96 Cm	97 Bk	98 Cf	99 Es	100 Fm	101 Md	102 No	103 Lw

FIGURE 1.12 *Periodic table with the elements associated with commercial polymers in color.*

NYLON PARACHUTES—INTRODUCING POLYMERS

A major impact of modern engineering technology on everyday life has been made by the class of materials known as **polymers**. An alternative name for this category is **plastics**, which describes the extensive formability of many polymers during fabrication. These synthetic, or human-made, materials represent a special branch of organic chemistry. Examples of inexpensive, functional polymer products are readily available to each of us (Figure 1.11). The "mer" in a polymer is a single hydrocarbon molecule such as ethylene (C_2H_4). Polymers are long-chain molecules composed of many mers bonded together. The most common commercial polymer is **polyethylene** $+C_2H_4+_n$ where *n* can range from approximately 100 to 1,000. Figure 1.12 shows the relatively limited portion of the periodic table that is associated with commercial polymers. Many important polymers, including polyethylene, are simply compounds of hydrogen and carbon. Others contain oxygen (e.g., acrylics), nitrogen (nylons), fluorine (fluoroplastics), and silicon (silicones).

Nylon is an especially familiar example. Polyhexamethylene adipamide, or nylon, is a member of the family of synthetic polymers known as polyamides invented in 1935 at the DuPont Company. Nylon was the first commercially successful polymer and was initially used as bristles in toothbrushes (1938) followed by the highly popular use as an alternative to silk stockings (1940). Developed as a synthetic alternative to silk, nylon became the focus of an intensive effort during the early stages of World War II to replace the diminishing supply of Asian silk for parachutes and other military supplies. At the beginning of World War II, the fiber industry was dominated by the natural materials cotton and wool. By the end, synthetic fibers accounted for 25% of the market share. A contemporary example of a nylon parachute is shown in Figure 1.13. Today, nylon remains a popular fiber material, but it is also widely used in solid form for applications such as gears and bearings.

As the descriptive title implies, *plastics* commonly share with metals the desirable mechanical property of ductility. Unlike brittle ceramics, polymers are frequently lightweight, low-cost alternatives to metals in structural design applications. The nature of chemical bonding in polymeric materials will be explored in Chapter 2. Important bonding-related properties include lower strength compared

FIGURE 1.13 *Since its development during World War II, nylon fabric remains the most popular material of choice for parachute designs. (Courtesy of Stringer/Agence France Presse/Getty Images.)*

with metals and lower melting point and higher chemical reactivity compared with ceramics and glasses. In spite of their limitations, polymers are highly versatile and useful materials. Substantial progress has been made in recent decades in the development of engineering polymers with sufficiently high strength and stiffness to permit substitution for traditional structural metals.

KEVLAR®-REINFORCED TIRES—INTRODUCING COMPOSITES

The structural engineering materials we have discussed so far—metals, ceramics/ glasses, and polymers—contain various elements and compounds that can be classified by their chemical bonding. Metals are associated with metallic bonding, ceramics/glasses with ionic bonding, and polymers with covalent bonding. Such classifications are described further in Chapter 2. Another important set of materials is made up of some combinations of individual materials from the previous categories. This fourth group is **composites**, and an excellent example is **fiberglass**. This composite of glass fibers embedded in a polymer matrix is commonplace (Figure 1.14). Characteristic of good composites, fiberglass has the best properties of each component, producing a product that is superior to either of the components separately. The high strength of the small-diameter glass fibers is combined with the ductility of the polymer matrix to produce a strong material capable of withstanding the normal loading required of a structural material. There is no need to illustrate a region of the periodic table as characteristic of composites, since they involve virtually the entire table except for the noble gases (column O), equivalent to an overlay of the periodic table coverage for metals, ceramics, and polymers combined.

Kevlar fiber reinforcements provide significant advances over traditional glass fibers for **polymer–matrix composites**. Kevlar is a DuPont trade name for

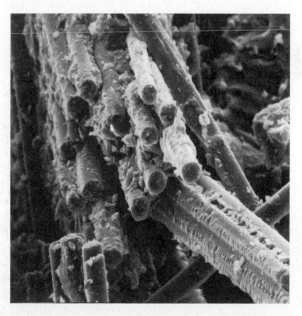

FIGURE 1.14 *Example of a fiberglass composite composed of microscopic-scale reinforcing glass fibers in a polymer matrix. (Courtesy of Owens-Corning Fiberglas Corporation.)*

FIGURE 1.15 *Kevlar reinforcement is a popular application in modern high-performance tires. In this case, the durability of sidewall reinforcement is tested along concrete ridges at a proving ground track. (Courtesy of the Goodyear Tire and Rubber Company.)*

poly *p*-phenyleneterephthalamide (PPD-T), a para-aramid. Substantial progress has been made in developing new polymer matrices, such as polyetheretherketone (PEEK) and polyphenylene sulfide (PPS). These materials have the advantages of increased toughness and recyclability. Kevlar-reinforced polymers are used in pressure vessels, and Kevlar reinforcement is widely used in tires (Figure 1.15). Kevlar was developed in 1965 and has been used commercially since the early 1970s. It is especially popular for demanding applications given that its strength-to-weight ratio is five times that of structural steel. The modern automobile tire is an especially good example.

SILICON CHIPS—INTRODUCING SEMICONDUCTORS

Whereas polymers are highly visible engineering materials that have had a major impact on contemporary society, semiconductors are relatively invisible but have had a comparable social impact. Technology has clearly revolutionized society, but solid-state electronics has revolutionized technology itself. A relatively small group of elements and compounds has an important electrical property, *semiconduction*, in which they are neither good electrical conductors nor good electrical insulators. Instead, their ability to conduct electricity is intermediate. These materials are called **semiconductors**, and in general they do not fit into any of the structural materials categories based on atomic bonding. As discussed earlier, metals are inherently good electrical conductors. Ceramics and polymers (nonmetals) are generally poor conductors, but good insulators. An important section of the periodic table is shown in dark color in Figure 1.16 These three

I A																		O
1 H	II A											III A	IV A	V A	VI A	VII A		2 He
3 Li	4 Be											5 B	6 C	7 N	8 O	9 F		10 Ne
11 Na	12 Mg	III B	IV B	V B	VI B	VII B		VIII			I B	II B	13 Al	14 Si	15 P	16 S	17 Cl	18 Ar
19 K	20 Ca	21 Sc	22 Ti	23 V	24 Cr	25 Mn	26 Fe	27 Co	28 Ni	29 Cu	30 Zn	31 Ga	32 Ge	33 As	34 Se	35 Br	36 Kr
37 Rb	38 Sr	39 Y	40 Zr	41 Nb	42 Mo	43 Tc	44 Ru	45 Rh	46 Pd	47 Ag	48 Cd	49 In	50 Sn	51 Sb	52 Te	53 I	54 Xe
55 Cs	56 Ba	57 La	72 Hf	73 Ta	74 W	75 Re	76 Os	77 Ir	78 Pt	79 Au	80 Hg	81 Tl	82 Pb	83 Bi	84 Po	85 At	86 Rn
87 Fr	88 Ra	89 Ac	104 Rf	105 Db	106 Sg												

58 Ce	59 Pr	60 Nd	61 Pm	62 Sm	63 Eu	64 Gd	65 Tb	66 Dy	67 Ho	68 Er	69 Tm	70 Yb	71 Lu
90 Th	91 Pa	92 U	93 Np	94 Pu	95 Am	96 Cm	97 Bk	98 Cf	99 Es	100 Fm	101 Md	102 No	103 Lw

FIGURE 1.16 *Periodic table with the elemental semiconductors in dark color and those elements that form semiconducting compounds in light color. The semiconducting compounds are composed of pairs of elements from columns III and V (e.g., GaAs) or from columns II and VI (e.g., CdS).*

semiconducting elements (Si, Ge, and Sn) from column IV A serve as a kind of boundary between metallic and nonmetallic elements. Silicon (Si) and germanium (Ge), widely used elemental semiconductors, are excellent examples of this class of materials. Precise control of chemical purity allows precise control of electronic properties. As techniques have been developed to produce variations in chemical purity over small regions, sophisticated electronic circuitry has been produced in exceptionally small areas (Figure 1.17). Such **microcircuitry** is the basis of the current revolution in technology.

The elements shaded in light color in Figure 1.16 form compounds that are semiconducting. Examples include gallium arsenide (GaAs), which is used as a high-temperature rectifier and a laser material, and cadmium sulfide (CdS), which is used as a relatively low-cost solar cell for conversion of solar energy to useful electrical energy. The various compounds formed by these elements show similarities to many of the ceramic compounds.

1.4 | Processing and Selecting Materials

Our use of materials in modern technology ultimately depends on our ability to make those materials. In Chapters 11 through 13, we will discuss how each of the six types of materials is produced. The topic of materials **processing** serves two functions. First, it provides a fuller understanding of the nature of each type of material. Second, and more importantly, it provides an appreciation of the effects of processing history on properties.

We shall find that processing technology ranges from traditional methods such as metal casting to the most contemporary techniques of electronic

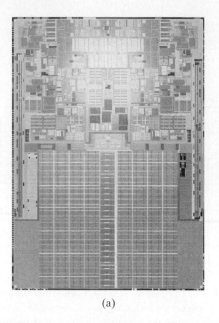

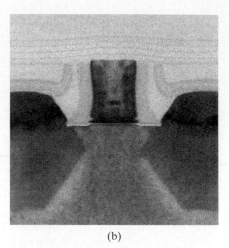

(a)

(b)

FIGURE 1.17 *(a) Typical microcircuit containing a complex array of semiconducting regions. (Photograph courtesy of Intel Corporation) (b) A microscopic cross section of a single circuit element in (a). The dark rectangular shape in the middle of the micrograph is a metal component less than 50 nm wide. (Micrograph courtesy of Intel Corporation)*

FIGURE 1.18 *The modern integrated circuit fabrication laboratory represents the state of the art in materials processing. (Courtesy of the College of Engineering, University of California, Davis.)*

microcircuit fabrication (Figure 1.18). In Section 1.3, we answered the question, "What materials are available to me?" We must next face a new and obvious question: "Which material do I now select for a particular application?" **Materials selection** is the final practical decision in the engineering design process and can determine that design's ultimate success or failure. Figure 1.19 illustrates the integral relationship among materials (and their properties), the processing of the materials, and the effective use of the materials in engineering design.

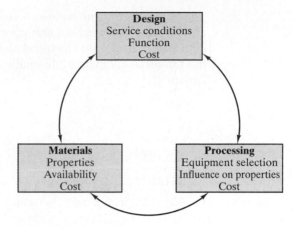

FIGURE 1.19 *Schematic illustration of the integral relationship among materials, the processing of those materials, and engineering design. (From G. E. Dieter, in* ASM Handbook, *Vol. 20:* Materials Selection and Design, *ASM International, Materials Park, OH, 1997, p. 243.)*

1.5 | Looking at Materials by Powers of Ten

In this chapter, we have seen that an underlying principle in materials science and engineering is that "structure leads to properties," that is, we explain the behavior of the materials that we use in engineering designs (on the human scale) by looking at mechanisms that involve the structure of the materials on some fine scale. This important concept is reminiscent of the delightful short documentary film *Powers of Ten* produced by the designers Charles and Ray Eames in 1977 and narrated by the physicist Philip Morrison. That film popularized the concept that human experience occurs several orders of magnitude below the scale of the universe and several orders of magnitude above the scale of the atom.

Throughout the next several chapters, we will see many examples of the relationship of materials properties on the human scale to fine-scale structural mechanisms. Some mechanisms involve the structure of the materials at the atomic scale (such as point defects explaining diffusion in Chapter 5), the microscopic scale (such as dislocations explaining plastic deformation in Chapter 6), or the millimeter scale (such as structural flaws that cause catastrophic failures as discussed in Chapter 8). In the past decade, the significance of the nanoscale has become widely emphasized. For example, some mechanical properties of metal alloys are improved when their polycrystalline grain size is reduced from the microscale to the nanoscale. So, the appropriate range of "powers of ten" that we discuss in the book is

This icon will appear throughout the book to highlight discussions involving the relationship between fine-scale structure and material properties.

the human scale: 1 meter
the milliscale: 1×10^{-3} meter
the microscale: 1×10^{-6} meter
the nanoscale: 1×10^{-9} meter
the atomic scale: 1×10^{-10} meter

The milli-, micro-, and nanoscales follow the SI unit system. The human scale and atomic scale are practical end points above and below the three SI scales. It is interesting to note that a typical atomic bond length (about 2×10^{-10} meter) is a full ten orders of magnitude smaller than the height of an average person (about 2 meters).

Summary

The wide range of materials available to engineers can be divided into six categories: metals, ceramics, glasses, polymers, composites, and semiconductors. The first four categories can be associated with three distinct types of atomic bonding. Composites involve combinations of two or more materials from the previous four categories. The first five categories comprise the structural materials. Semiconductors comprise a separate category of electronic materials that is distinguished by its unique, intermediate electrical conductivity. Understanding the human-scale properties of these various materials requires examination of structure at some fine scale. For example, the development of transparent ceramics has required careful control of a microscopic-scale architecture. Once the properties of materials are understood, the appropriate material for a given application can be processed and selected. We now move on to the body of the text, with the term *materials science and engineering* serving to define this branch of engineering. This term also provides the key words that describe the various parts of the text: "science" → the fundamentals covered in Chapters 2 though 10, "materials" → the structural and electronic materials of Chapters 11 through 13, and "engineering" → materials in engineering design as discussed in Chapter 14.

Key Terms

Many technical journals now include a set of key terms with each article. These words serve the practical purpose of information retrieval but also provide a convenient summary of important concepts in that publication. In this spirit, a list of key terms will be given at the end of each chapter. Students can use this list as a convenient guide to the major concepts that they should be learning from that chapter. A comprehensive glossary provided in Appendix 6 gives definitions of the key terms from all chapters.

alloy (1)
brittle (3)
Bronze Age (1)
ceramic (5)
composite (13)
crystalline (9)
ductility (3)
fiberglass (13)
glass (9)
Iron Age (1)

Kevlar (13)
materials selection (16)
metallic (3)
microcircuitry (15)
noncrystalline (9)
nonmetallic (7)
nylon (12)
optical fiber (10)
photonic material (10)
plastic (11)

polyethylene (11)
polymer (11)
polymer–matrix composite (13)
processing (15)
refractory (6)
semiconductor (14)
silica (7)
silicate (7)
Stone Age (1)

References

At the end of each chapter, a short list of selected references will be cited to indicate some primary sources of related information for the student who wishes to do outside reading. For Chapter 1, the references are some of the general textbooks in the field of materials science and engineering.

Askeland, **D. R.**, and **P. P. Phule**, *The Science and Engineering of Materials*, 5th ed., Thomson, London, 2006.

Callister, **W. D.**, *Materials Science and Engineering—An Introduction*, 7th ed., John Wiley & Sons, Inc., New York, 2007.

Schaffer, **J. P.**, **A. Saxena**, **S. D. Antolovich**, **T. H. Sanders**, **Jr.**, and **S. B. Warner**, *The Science and Design of Engineering Materials*, 2nd ed., McGraw-Hill Book Company, New York, 1999.

Smith, **W. F.**, and **J. Hashemi**, *Foundations of Materials Science and Engineering,* 4th ed., McGraw-Hill Higher Education, Boston, 2006.

PART I
The Fundamentals

A central principle of materials science is that the properties of materials that permit their engineering applications can be understood by examining the structure of those materials on a small scale. Shown on the facing page is a micrograph of a high carbon (1.4 wt % carbon) steel viewed under an optical microscope. The combination of phases present results in an alloy that exhibits high strength and low ductility, leading to applications such as cutting tools and truck springs. The instrument shown above is a scanning electron microscope that can provide higher magnifications with a greater depth of field than possible with optical microscopes. (Courtesy of the Department of Chemical Engineering and Materials Science, University of California, Davis.)

We begin our exploration of the field of *materials science and engineering* by focusing on *materials science*. Chapters 2 through 10 cover a variety of fundamental topics from physics and chemistry. A student may well have encountered many of the concepts in Chapter 2 (atomic bonding) in previous courses. Of special interest to the field of materials science is the role of atomic bonding in providing a classification scheme for materials. Metallic, ionic, and covalent bonding roughly correspond to the categories of structural materials: metals, ceramics/glasses, and polymers. Semiconductors, an important category of electronic materials, generally correspond to covalent bonding. Chapter 3 introduces the crystalline structures of many engineered materials and includes an introduction to x-ray diffraction, an important tool for determining crystal structure. Chapter 4 identifies various imperfections that can occur relative to the crystalline structures of Chapter 3. In Chapter 5, we see that some of these structural defects play a central role in solid-state diffusion, and, in Chapter 6, we find that other defects are responsible for some of the mechanical behavior of materials. Chapter 7 introduces the thermal behavior of materials, and, in Chapter 8, we see that certain mechanical and thermal processes (such as machining and welding) can lead to the failure of materials. In Chapter 9, we are introduced to phase diagrams that serve as useful tools for predicting the microscopic-scale structures of materials that are produced at a relatively slow rate, maintaining equilibrium along the way. In Chapter 10 on *kinetics*, we see the effect of more rapid heat treatments that lead to additional microstructures. Throughout Part I, we will find that fundamental principles from physics and chemistry underlie the practical behavior of engineered materials.

CHAPTER 2
Atomic Bonding

2.1 Atomic Structure
2.2 The Ionic Bond
2.3 The Covalent Bond
2.4 The Metallic Bond
2.5 The Secondary, or van der Waals, Bond
2.6 Materials—The Bonding Classification

Computer models of the structures of materials on the atomic scale require accurate knowledge of the bonding between adjacent atoms. In this model of a molecular network structure, atoms are shown as spheres joined by both single and double covalent bonds.

C hapter 1 introduced the basic types of materials available to engineers. One basis of that classification system is found in the nature of atomic bonding in materials. Atomic bonding falls into two general categories. *Primary bonding* involves the transfer or sharing of electrons and produces a relatively strong joining of adjacent atoms. Ionic, covalent, and metallic bonds are in this category. *Secondary bonding* involves a relatively weak attraction between atoms in which no electron transfer or sharing occurs. Van der Waals bonds are in this category. Each of the five fundamental types of engineering materials (metals, ceramics, glasses, polymers, and semiconductors) is associated with a certain type (or types) of atomic bonding. Composites, of course, are combinations of fundamental types.

2.1 | Atomic Structure

Outer orbital (with four sp^3 hybrid bonding electrons)

Inner orbital (with two $1s$ electrons)

Nucleus (with six protons and six neutrons)

FIGURE 2.1 *Schematic of the planetary model of a C^{12} atom.*

In order to understand bonding between atoms, we must appreciate the structure within the individual atoms. For this purpose, it is sufficient to use a relatively simple planetary model of atomic structure—that is, **electrons** (the planets) orbit about a **nucleus** (the sun).

It is not necessary to consider the detailed structure of the nucleus for which physicists have catalogued a vast number of elementary particles. We need consider only the number of **protons** and **neutrons** in the nucleus as the basis of the chemical identification of a given atom. Figure 2.1 is a planetary model of a carbon atom. This illustration is schematic and definitely not to scale. In reality, the nucleus is much smaller, even though it contains nearly all the mass of the atom. Each proton and neutron has a mass of approximately 1.66×10^{-24} g. This value is referred to as an **atomic mass unit (amu)**. It is convenient to express the mass of elemental materials in these units. For instance, the most common isotope of carbon, C^{12} (shown in Figure 2.1), contains in its nucleus six protons and six neutrons, for an **atomic mass** of 12 amu. It is also convenient to note that there are

FIGURE 2.2 *Periodic table of the elements indicating atomic number and atomic mass (in amu).*

I A	II A	III B	IV B	V B	VI B	VII B		VIII		I B	II B	III A	IV A	V A	VI A	VII A	0
1 H 1.008																	2 He 4.003
3 Li 6.941	4 Be 9.012											5 B 10.81	6 C 12.01	7 N 14.01	8 O 16.00	9 F 19.00	10 Ne 20.18
11 Na 22.99	12 Mg 24.31											13 Al 26.98	14 Si 28.09	15 P 30.97	16 S 32.06	17 Cl 35.45	18 Ar 39.95
19 K 39.10	20 Ca 40.08	21 Sc 44.96	22 Ti 47.90	23 V 50.94	24 Cr 52.00	25 Mn 54.94	26 Fe 55.85	27 Co 58.93	28 Ni 58.71	29 Cu 63.55	30 Zn 65.38	31 Ga 69.72	32 Ge 72.59	33 As 74.92	34 Se 78.96	35 Br 79.90	36 Kr 83.80
37 Rb 85.47	38 Sr 87.62	39 Y 88.91	40 Zr 91.22	41 Nb 92.91	42 Mo 95.94	43 Tc 98.91	44 Ru 101.07	45 Rh 102.91	46 Pd 106.4	47 Ag 107.87	48 Cd 112.4	49 In 114.82	50 Sn 118.69	51 Sb 121.75	52 Te 127.60	53 I 126.90	54 Xe 131.30
55 Cs 132.91	56 Ba 137.33	57 La 138.91	72 Hf 178.49	73 Ta 180.95	74 W 183.85	75 Re 186.2	76 Os 190.2	77 Ir 192.22	78 Pt 195.09	79 Au 196.97	80 Hg 200.59	81 Tl 204.37	82 Pb 207.2	83 Bi 208.98	84 Po (210)	85 At (210)	86 Rn (222)
87 Fr (223)	88 Ra 226.03	89 Ac (227)	104 Rf (261)	105 Db (262)	106 Sg (266)												

58 Ce 140.12	59 Pr 140.91	60 Nd 144.24	61 Pm (145)	62 Sm 150.4	63 Eu 151.96	64 Gd 157.25	65 Tb 158.93	66 Dy 162.50	67 Ho 164.93	68 Er 167.26	69 Tm 168.93	70 Yb 173.04	71 Lu 174.97
90 Th 232.04	91 Pa 231.04	92 U 238.03	93 Np 237.05	94 Pu (244)	95 Am (243)	96 Cm (247)	97 Bk (247)	98 Cf (251)	99 Es (254)	100 Fm (257)	101 Md (258)	102 No (259)	103 Lw (260)

0.6023×10^{24} amu per gram. This large value, known as **Avogadro's*** **number**, represents the number of protons or neutrons necessary to produce a mass of 1 g. Avogadro's number of atoms of a given element is termed a **gram-atom**. For a compound, the corresponding term is **mole**; that is, one mole of NaCl contains Avogadro's number of Na atoms *and* Avogadro's number of Cl atoms.

Avogadro's number of C^{12} atoms would have a mass of 12.00 g. Naturally occurring carbon actually has an atomic mass of 12.011 amu because not all carbon atoms contain six neutrons in their nuclei. Instead, some contain seven. Different numbers of neutrons (six or seven) identify different **isotopes**—various forms of an element that differ in the number of neutrons in the nucleus. In nature, 1.1% of the carbon atoms are the isotope C^{13}. However, the nuclei of all carbon atoms contain six protons. In general, the number of protons in the nucleus is known as the **atomic number** of the element. The well-known periodicity of chemical elements is based on this system of elemental atomic numbers and atomic masses arranged in chemically similar **groups** (vertical columns) in a **periodic table** (Figure 2.2).

While chemical identification is done relative to the nucleus, atomic bonding involves electrons and **electron orbitals**. The electron, with a mass of 0.911×10^{-27} g, makes a negligible contribution to the atomic mass of an element. However, this particle has a negative charge of 0.16×10^{-18} coulomb (C), equal in

---

*Amadeo Avogadro (1776–1856), Italian physicist, who, among other contributions, coined the word *molecule*. Unfortunately, his hypothesis that all gases (at a given temperature and pressure) contain the same number of molecules per unit volume was not generally acknowledged as correct until after his death.

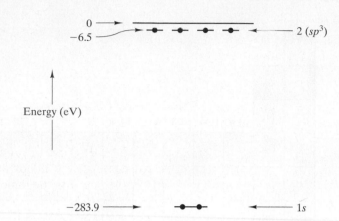

In Section 13.2, we shall see that energy level diagrams like this are central to the understanding of energy band gap structures that are, in turn, at the heart of semiconductor technology.

FIGURE 2.3 *Energy-level diagram for the orbital electrons in a C^{12} atom. Notice the sign convention. An attractive energy is negative. The 1s electrons are closer to the nucleus (see Figure 2.1) and more strongly bound (binding energy = -283.9 eV). The outer orbital electrons have a binding energy of only -6.5 eV. The zero level of binding energy corresponds to an electron completely removed from the attractive potential of the nucleus.*

magnitude to the $+0.16 \times 10^{-18}$ C charge of each proton. (The neutron is, of course, electrically neutral.)

Electrons are excellent examples of the wave-particle duality; that is, they are atomic-scale entities exhibiting both wavelike and particlelike behavior. It is beyond the scope of this book to deal with the principles of quantum mechanics that define the nature of electron orbitals (based on the wavelike character of electrons). However, a brief summary of the nature of electron orbitals is helpful. As shown schematically in Figure 2.1, electrons are grouped at fixed orbital positions about a nucleus. In addition, each orbital radius is characterized by an **energy level**, a fixed binding energy between the electron and its nucleus. Figure 2.3 shows an *energy-level diagram* for the electrons in a C^{12} atom. It is important to note that the electrons around a C^{12} nucleus occupy these specific energy levels, with intermediate energies forbidden. The forbidden energies correspond to unacceptable quantum mechanical conditions; that is, standing waves cannot be formed.

A detailed list of electronic configurations for the elements of the periodic table is given in Appendix 1, together with various useful data. The arrangement of the periodic table (Figure 2.2) is largely a manifestation of the systematic "filling" of the electron orbitals with electrons, as summarized in Appendix 1. The notation for labeling electron orbitals is derived from the quantum numbers of wave mechanics. These integers relate to solutions to the appropriate wave equations. We do not deal with this numbering system in detail in this book; instead, it is sufficient to appreciate the basic labeling system. For instance, Appendix 1 tells us that there are two electrons in the 1s orbital. The 1 is a principal quantum number, identifying this energy level as the first one which is closest to the atomic nucleus. There are also two electrons each associated with the 2s and 2p orbitals. The s, p, and so on, notation refers to an additional set of quantum numbers. The rather cumbersome letter notation is derived from the terminology of early spectrographers. The six electrons in the C^{12} atom are then described as a $1s^2 2s^2 2p^2$ distribution; that is, two electrons in the 1s orbital, two in 2s, and two in 2p. In fact, the four electrons in the outer orbital of C^{12} redistribute themselves in a more symmetrical fashion to produce the characteristic geometry of bonding between carbon atoms and adjacent atoms (generally described as $1s^2 2s^1 2p^3$). This sp^3 configuration in the second energy level of carbon, called **hybridization**, is indicated in Figures 2.1 and 2.3 and is discussed in further detail in Section 2.3. (Note especially Figure 2.19.)

The bonding of adjacent atoms is essentially an electronic process. Strong **primary bonds** are formed when outer orbital electrons are transferred or shared between atoms. Weaker **secondary bonds** result from a more subtle attraction between positive and negative charges with no actual transfer or sharing of electrons. In the next section, we will look at the various possibilities of bonding in a systematic way, beginning with the ionic bond.

THE MATERIAL WORLD

Naming a New Chemical Element

The periodic table is generally one of the first items to which we are introduced as we begin to seriously explore modern science. This systematic arrangement of the chemical elements is, of course, useful for providing a visual understanding of the similarities and differences of the various chemical elements. The periodic table's role as a permanent record of this important information sometimes overshadows the fact that, at some point in time, each element had to be given a name. Some names, such as *iron*, simply have evolved from earlier languages (the Old

High German *isarn* that led to the Old English *iren*, with the chemical symbol *Fe* coming from the Latin *ferrum*).

As some elements were discovered, they were given names in honor of the country in which they were discovered or synthesized (e.g., *germanium* for Germany). The advances in physics and chemistry in the 20th century made possible the synthesis of new elements that are not found in nature and that have atomic numbers greater than that of uranium (92). These transuranic elements were often named in

I A																		O
1 H	II A												III A	IV A	V A	VI A	VII A	2 He
3 Li	4 Be												5 B	6 C	7 N	8 O	9 F	10 Ne
11 Na	12 Mg	III B	IV B	V B	VI B	VII B		VIII			I B	II B	13 Al	14 Si	15 P	16 S	17 Cl	18 Ar
19 K	20 Ca	21 Sc	22 Ti	23 V	24 Cr	25 Mn	26 Fe	27 Co	28 Ni	29 Cu	30 Zn		31 Ga	32 Ge	33 As	34 Se	35 Br	36 Kr
37 Rb	38 Sr	39 Y	40 Zr	41 Nb	42 Mo	43 Tc	44 Ru	45 Rh	46 Pd	47 Ag	48 Cd		49 In	50 Sn	51 Sb	52 Te	53 I	54 Xe
55 Cs	56 Ba	57 La	72 Hf	73 Ta	74 W	75 Re	76 Os	77 Ir	78 Pt	79 Au	80 Hg		81 Tl	82 Pb	83 Bi	84 Po	85 At	86 Rn
87 Fr	88 Ra	89 Ac	104 Rf	105 Db														

| | | | | | 106 Sg | | | | | | | |

58 Ce	59 Pr	60 Nd	61 Pm	62 Sm	63 Eu	64 Gd	65 Tb	66 Dy	67 Ho	68 Er	69 Tm	70 Yb	71 Lu
90 Th	91 Pa	92 U	93 Np	94 Pu	95 Am	96 Cm	97 Bk	98 Cf	99 Es	100 Fm	101 Md	102 No	103 Lw

(Continued)

honor of great scientists of the past (e.g., *mendele-vium* for Dmitri Mendeleev, the 19th-century Russian chemist who devised the periodic table). The leading authority in synthesizing the transuranic elements was Dr. Glenn Seaborg (1913–1999), Professor of Chemistry at the University of California, Berkeley. (It was Seaborg's idea to reconfigure Mendeleev's original periodic table by breaking out the actinide series below the main table.) Seaborg and his team discovered plutonium and nine other transuranic elements, including element 106, which was named *seaborgium* after him.

Professor Seaborg received the singular honor of being the first person ever to have an element named after him while he was still alive. He rightfully viewed this honor as much greater than his Nobel Prize in Chemistry awarded in 1951. While seaborgium has been synthesized only in minute amounts and may not play a significant role in materials science and engineering, its namesake, Professor Seaborg, was a great advocate of the field. His enthusiasm for materials came in no small measure

from his long term of service as Chair of the Atomic Energy Commission (the predecessor of today's Department of Energy). He was quoted in the January 1980 issue of *ASM News* as saying that "materials science and engineering will be essential for the solution of the problems attendant with the energy sources of the future." Professor Seaborg's vision is as true today as it was three decades ago.

EXAMPLE 2.1

Chemical analysis in materials science laboratories is frequently done by means of the scanning electron microscope. In this instrument, an electron beam generates characteristic x-rays that can be used to identify chemical elements. This instrument samples a roughly cylindrical volume at the surface of a solid material. Calculate the number of atoms sampled in a 1-μm-diameter by 1-μm-deep cylinder in the surface of solid copper.

SOLUTION
From Appendix 1,

$$\text{density of copper} = 8.93 \text{ g/cm}^3$$

and

$$\text{atomic mass of copper} = 63.55 \text{ amu.}$$

The atomic mass indicates that there are

$$\frac{63.55 \text{ g Cu}}{\text{Avogadro's number of Cu atoms}}.$$

The volume sampled is

$$V_{sample} = \pi \left(\frac{1\ \mu m}{2} \right)^2 \times 1\ \mu m$$

$$= 0.785\ \mu m^3 \times \left(\frac{1\ cm}{10^4\ \mu m} \right)^3$$

$$= 0.785 \times 10^{-12}\ cm^3.$$

Thus, the number of atoms sampled is

$$N_{sample} = \frac{8.93\ g}{cm^3} \times 0.785 \times 10^{-12}\ cm^3 \times \frac{0.602 \times 10^{24}\ atoms}{63.55\ g}$$

$$= 6.64 \times 10^{10}\ atoms.$$

EXAMPLE 2.2

One mole of solid MgO occupies a cube 22.37 mm on a side. Calculate the density of MgO (in g/cm^3).

SOLUTION
From Appendix 1,

$$mass\ of\ 1\ mol\ of\ MgO = atomic\ mass\ of\ Mg\ (in\ g)$$
$$+ atomic\ mass\ of\ O\ (in\ g)$$
$$= 24.31\ g + 16.00\ g = 40.31\ g.$$
$$density = \frac{mass}{volume}$$
$$= \frac{40.31\ g}{(22.37\ mm)^3 \times 10^{-3}\ cm^3/mm^3}$$
$$= 3.60\ g/cm^3.$$

EXAMPLE 2.3

Calculate the dimensions of a cube containing 1 mol of solid magnesium.

SOLUTION
From Appendix 1,

$$density\ of\ Mg = 1.74\ g/cm^3.$$
$$atomic\ mass\ of\ Mg = 24.31\ amu.$$

$$\text{volume of 1 mol} = \frac{24.31 \text{ g/mol}}{1.74 \text{ g/cm}^3}$$

$$= 13.97 \text{ cm}^3/\text{mol.}$$

$$\text{edge of cube} = (13.97)^{1/3} \text{ cm}$$

$$= 2.408 \text{ cm} \times 10 \text{ mm/cm}$$

$$= 24.08 \text{ mm.}$$

Beginning at this point, a few elementary problems, called Practice Problems, will be provided immediately following the solved example. These exercises follow directly from the preceding solutions and are intended to provide a carefully guided journey into the first calculations in each new area. **(Note that the solutions to all practice problems are provided on the related Web site.)** More independent and challenging problems are provided at the conclusion of the chapter. Answers for nearly all of the practice problems are given following the appendixes.

PRACTICE PROBLEM 2.1

Calculate the number of atoms contained in a cylinder 1 μm in diameter by 1 μm deep of **(a)** magnesium and **(b)** lead. (See Example 2.1.)

PRACTICE PROBLEM 2.2

Using the density of MgO calculated in Example 2.2, calculate the mass of an MgO refractory (temperature-resistant) brick with dimensions 50 mm $\times$ 100 mm $\times$ 200 mm.

PRACTICE PROBLEM 2.3

Calculate the dimensions of **(a)** a cube containing 1 mol of copper and **(b)** a cube containing 1 mol of lead. (See Example 2.3.)

2.2 | The Ionic Bond

Calculations involving the ionic bond are included in a laboratory manual on the related Web site.

An **ionic bond** is the result of *electron transfer* from one atom to another. Figure 2.4 illustrates an ionic bond between sodium and chlorine. The transfer of an electron *from* sodium is favored because it produces a more stable electronic configuration; that is, the resulting Na^+ species has a full outer **orbital shell**, defined as a set of electrons in a given orbital. Similarly, the chlorine readily accepts the electron, producing a stable Cl^- species, also with a full outer orbital shell. The charged species (Na^+ and Cl^-) are termed **ions**, giving rise to the name *ionic bond*. The positive species (Na^+) is a **cation**, and the negative species (Cl^-) is an **anion**.

It is important to note that the ionic bond is *nondirectional*. A positively charged Na^+ will attract any adjacent Cl^- equally in all directions. Figure 2.5 shows how Na^+ and Cl^- ions are stacked together in solid sodium chloride (rock salt).

Electron transfer

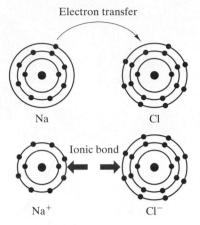

Na   Cl

Ionic bond

Na⁺   Cl⁻

FIGURE 2.4 *Ionic bonding between sodium and chlorine atoms. Electron transfer from Na to Cl creates a cation (Na⁺) and an anion (Cl⁻). The ionic bond is due to the coulombic attraction between the ions of opposite charge.*

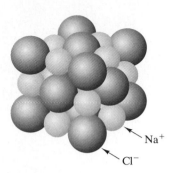

Na⁺

Cl⁻

FIGURE 2.5 *Regular stacking of Na⁺ and Cl⁻ ions in solid NaCl, which is indicative of the nondirectional nature of ionic bonding.*

Details about this structure will be discussed in Chapter 3. For now, it is sufficient to note that this structure is an excellent example of ionically bonded material, and the Na^+ and Cl^- ions are stacked together systematically to maximize the number of oppositely charged ions adjacent to any given ion. In NaCl, six Na^+ surround each Cl^-, and six Cl^- surround each Na^+.

The ionic bond is the result of the **coulombic*** **attraction** between the oppositely charged species. It is convenient to illustrate the nature of the bonding force for the ionic bond because the coulombic attraction force follows a simple, well-known relationship,

$$F_c = \frac{-K}{a^2},$$

(2.1)

where F_c is the coulombic force of attraction between two oppositely charged ions, a is the separation distance between the *centers* of the ions, and K is

$$K = k_0 (Z_1 q)(Z_2 q).$$

(2.2)

In the preceding equation, Z is the **valence** of the charged ion (e.g., $+1$ for Na^+ and -1 for Cl^-), q is the charge of a single electron (0.16×10^{-18} C), and k_0 is a proportionality constant (9×10^9 $V \cdot m/C$).

A plot of Equation 2.1, shown in Figure 2.6, demonstrates that the coulombic force of attraction increases dramatically as the separation distance between adjacent ion centers (a) decreases. This relationship, in turn, implies that the **bond**

---

*Charles Augustin de Coulomb (1736–1806), French physicist, was first to experimentally demonstrate the nature of Equations 2.1 and 2.2 (for large spheres, not ions). Beyond major contributions to the understanding of electricity and magnetism, Coulomb was an important pioneer in the field of applied mechanics (especially in the areas of friction and torsion).

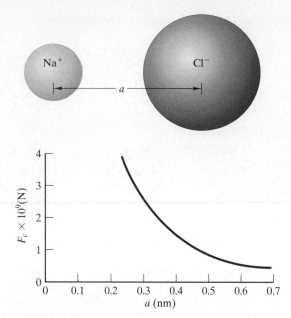

FIGURE 2.6 *Plot of the coulombic force (Equation 2.1) for a Na$^+$–Cl$^-$ pair.*

length (*a*) would ideally be zero. In fact, bond lengths are most definitely not zero because the attempt to move two oppositely charged ions closer together to increase coulombic attraction is counteracted by an opposing **repulsive force**, F_R, which is due to the overlapping of the similarly charged (negative) electric fields from each ion, as well as the attempt to bring the two positively charged nuclei closer together. The repulsive force as a function of *a* follows an exponential relationship

$$F_R = \lambda e^{-a/\rho}, \tag{2.3}$$

where λ and ρ are experimentally determined constants for a given ion pair. **Bonding force** is the net force of attraction (or repulsion) as a function of the separation distance between two atoms or ions. Figure 2.7 shows the *bonding force curve* for an ion pair in which the *net* bonding force, $F(= F_c + F_R)$, is plotted against *a*. The *equilibrium bond length*, a_0, occurs at the point where the forces of attraction and repulsion are precisely balanced ($F_c + F_R = 0$). It should be noted that the coulombic force (Equation 2.1) dominates for larger values of *a*, whereas the repulsive force (Equation 2.3) dominates for small values of *a*. Up to this point, we have concentrated on the attractive coulombic force between two ions of opposite charge. Of course, bringing two similarly charged ions together would produce a coulombic *repulsive force* (separate from the F_R term). In an ionic solid such as that shown in Figure 2.5, the similarly charged ions experience this "coulombic repulsion" force. The net cohesiveness of the solid is due to the fact that any given ion is immediately surrounded by ions of opposite sign for which the coulombic term (Equations 2.1 and 2.2) is positive. This overcomes the smaller, repulsive term due to more distant ions of like sign.

It should also be noted that an externally applied compressive force is required to push the ions closer together (i.e., closer than a_0). Similarly, an externally applied tensile force is required to pull the ions farther apart. This requirement

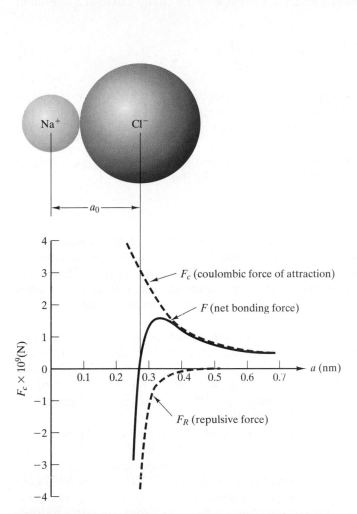

FIGURE 2.7 *Net bonding force curve for a Na^+−Cl^- pair showing an equilibrium bond length of $a_0 = 0.28$ nm.*

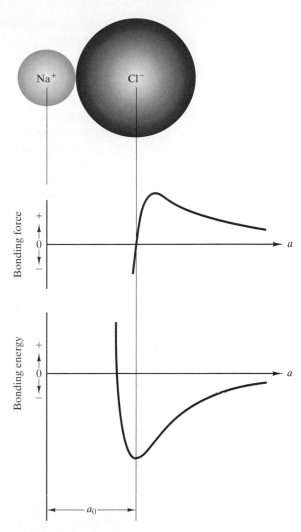

FIGURE 2.8 *Comparison of the bonding force curve and the bonding energy curve for a Na^+−Cl^- pair. Since $F = dE/da$, the equilibrium bond length (a_0) occurs where $F = 0$ and E is a minimum (see Equation 2.5).*

The atomic scale bonding force and energy curves shown on this page will help us understand the nature of elastic deformation in Section 6.2.

has implications for the mechanical behavior of solids, which is discussed in detail later (especially in Chapter 6).

Bonding energy, E, is related to bonding force through the differential expression

$$F = \frac{dE}{da}. \tag{2.4}$$

In this way, the net bonding force curve shown in Figure 2.7 is the derivative of the bonding energy curve. This relationship is shown in Figure 2.8. The relationship demonstrates that the equilibrium bond length, a_0, which corresponds to $F = 0$, also corresponds to a minimum in the energy curve. This correspondence is a consequence of Equation 2.5; that is, the slope of the energy curve at a minimum equals zero:

$$F = 0 = \left(\frac{dE}{da} \right)_{a=a_0}. \tag{2.5}$$

This important concept in materials science will be seen again many times throughout the book. The stable ion positions correspond to an energy minimum. To move the ions from their equilibrium spacing, energy must be supplied to this system (e.g., by compressive or tensile loading).

Having established that there is an equilibrium bond length, a_0, it follows that this bond length is the sum of two ionic radii; that is, for NaCl,

$$a_0 = r_{Na^+} + r_{Cl^-}. \tag{2.6}$$

This equation implies that the two ions are hard spheres touching at a single point. In Section 2.1 it was noted that, while electron orbitals are represented as particles orbiting at a fixed radius, electron charge is found in a range of radii. This is true for ions as well as for neutral atoms. An **ionic**, or **atomic**, **radius** is, then, the radius corresponding to the average electron density in the outermost electron orbital. Figure 2.9 compares three models of an Na^+–Cl^- ion pair: (a) shows a simple planetary model of the two ions, (b) shows a **hard-sphere** model of the pair, and (c) shows the **soft-sphere** model in which the actual electron density in the outer orbitals of Na^+ and Cl^- extends farther out than is shown for the hard sphere. The precise nature of actual bond lengths, a_0, allows us to use the hard-sphere model almost exclusively throughout the remainder of the book. Appendix 2 provides a detailed list of calculated ionic radii for a large number of ionic species.

Ionization has a significant effect on the effective (hard-sphere) radii for the atomic species involved. Although Figure 2.4 did not indicate this factor, the loss or gain of an electron by a neutral atom changes its radius. Figure 2.10 illustrates again the formation of an ionic bond between Na^+ and Cl^-. (Compare this figure with Figure 2.4.) In this case, atomic and ionic sizes are shown to correct scale. The loss of an electron by the sodium atom leaves 10 electrons to be drawn closer around the nucleus that still contains 11 protons. Conversely, a gain of 1 electron by the chlorine atom gives 18 electrons around a nucleus with 17 protons and, therefore, a larger effective radius.

COORDINATION NUMBER

Earlier in this section, the nondirectional nature of the ionic bond was introduced. Figure 2.5 shows a structure for NaCl in which six Na^+ surround each Cl^-, and vice versa. The **coordination number** (CN) is the number of adjacent ions (or atoms) surrounding a reference ion (or atom). For each ion shown in Figure 2.5, the CN is 6; that is, each ion has six nearest neighbors.

For ionic compounds, the coordination number of the smaller ion can be calculated in a systematic way by considering the greatest number of larger ions (of opposite charge) that can be in contact with, or coordinate with, the smaller one. This number (CN) depends directly on the relative sizes of the oppositely charged ions. This relative size is characterized by the **radius ratio** (r/R), where r is the radius of the smaller ion and R is the radius of the larger one.

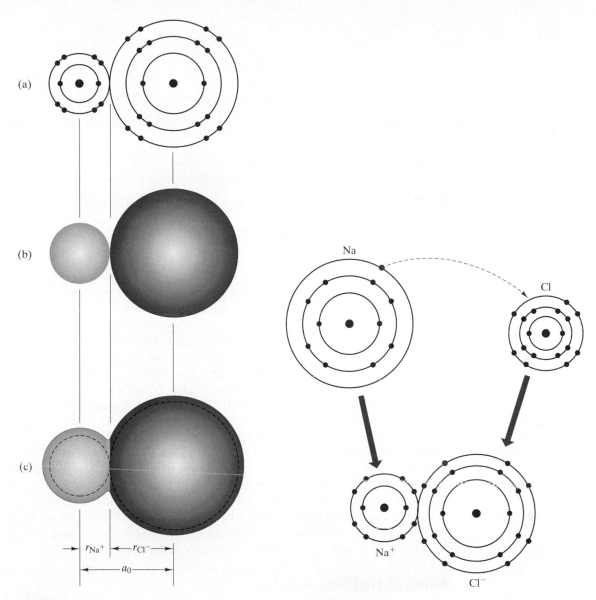

(a)

(b)

(c)

r_{Na^+} r_{Cl^-}

a_0

Na

Cl

Na$^+$

Cl$^-$

FIGURE 2.9 *Comparison of (a) a planetary model of a Na$^+$ −Cl$^-$ pair with (b) a hard-sphere model and (c) a soft-sphere model.*

FIGURE 2.10 *Formation of an ionic bond between sodium and chlorine in which the effect of ionization on atomic radius is illustrated. The cation (Na$^+$) becomes smaller than the neutral atom (Na), while the anion (Cl$^-$) becomes larger than the neutral atom (Cl).*

To illustrate the dependence of CN on radius ratio, consider the case of $r/R = 0.20$. Figure 2.11 shows how the greatest number of larger ions that can coordinate the smaller one is three. Any attempt to place four larger ions in contact with the smaller one requires the larger ions to overlap, which is a condition of great instability because of high repulsive forces. The minimum value of r/R that can produce threefold coordination ($r/R = 0.155$) is shown in Figure 2.12; that is, the larger ions are just touching the smaller ion as well as just touching each other. In the same way that fourfold coordination was unstable in Figure 2.11, an

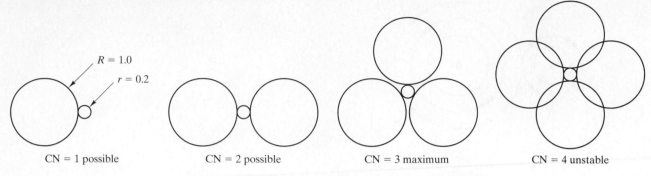

$R = 1.0$

$r = 0.2$

| CN = 1 possible | CN = 2 possible | CN = 3 maximum | CN = 4 unstable |

FIGURE 2.11 *The largest number of ions of radius R that can coordinate an atom of radius r is 3 when the radius ratio r/R = 0.2. (Note: The instability for CN = 4 can be reduced, but* not *eliminated, by allowing a three-dimensional, rather than a coplanar, stacking of the larger ions.)*

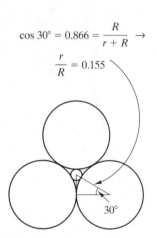

$$\cos 30° = 0.866 = \frac{R}{r + R} \rightarrow$$

$$\frac{r}{R} = 0.155$$

$30°$

FIGURE 2.12 *The minimum radius ratio, r/R, that can produce threefold coordination is 0.155.*

r/R value of *less* than 0.155 cannot allow threefold coordination. As r/R increases above 0.155, threefold coordination is stable (e.g., Figure 2.11 for $r/R = 0.20$) until fourfold coordination becomes possible at $r/R = 0.225$. Table 2.1 summarizes the relationship between coordination number and radius ratio. As r/R increases to 1.0, a coordination number as high as 12 is possible. As will be noted in Example 2.8, calculations based on Table 2.1 serve as guides, not as absolute predictors.

An obvious question is "Why doesn't Table 2.1 include radius ratios greater than 1?" Certainly, more than 12 small ions could simultaneously touch a single larger one. However, there are practical constraints in connecting the coordination groups of Table 2.1 into a periodic, three-dimensional structure, and the coordination number for the larger ions tends to be less than 12. A good example is again shown in Figure 2.5, in which the coordination number of Na^+ is 6, as predicted by the r/R value (= 0.098 nm/0.181 nm = 0.54), and the regular stacking of the six coordinated sodiums, in turn, gives Cl^- a coordination number of 6. These structural details will be discussed further in Chapter 3. One might also inquire why coordination numbers of 5, 7, 9, 10, and 11 are absent. These numbers cannot be integrated into the repetitive crystalline structures described in Chapter 3.

EXAMPLE 2.4

(a) Compare the electronic configurations for the atoms and ions shown in Figure 2.4.

(b) Which noble gas atoms have electronic configurations equivalent to those for the ions shown in Figure 2.4?

SOLUTION

(a) From Appendix 1,

$$Na: \quad 1s^2 2s^2 2p^6 3s^1$$

and

$$Cl: \quad 1s^2 2s^2 2p^6 3s^2 3p^5.$$

TABLE 2.1

Coordination Numbers for Ionic Bonding		
Coordination number	Radius ratio, r/R	Coordination geometry
2	$0 < \dfrac{r}{R} < 0.155$	
3	$0.155 \le \dfrac{r}{R} < 0.225$	
4	$0.225 \le \dfrac{r}{R} < 0.414$	
6	$0.414 \le \dfrac{r}{R} < 0.732$	
8	$0.732 \le \dfrac{r}{R} < 1$	
12	1	or[a]

[a]The geometry on the left is for the hexagonal close-packed (hcp) structure, and the geometry on the right is for the face-centered cubic (fcc) structure. These crystal structures are discussed in Chapter 3.

Because Na loses its outer orbital ($3s$) electron in becoming Na^+,

$$Na^+: \quad 1s^2 2s^2 2p^6.$$

Because Cl gains an outer orbital electron in becoming Cl^-, its $3p$ shell becomes filled:

$$Cl^-: \quad 1s^2 2s^2 2p^6 3s^2 3p^6.$$

(b) From Appendix 1,

$$Ne: \quad 1s^2 2s^2 2p^6,$$

which is equivalent to Na^+ (of course, the nuclei of Ne and Na^+ differ), and

$$Ar: \quad 1s^2 2s^2 2p^6 3s^2 3p^6,$$

which is equivalent to Cl^- (again, the nuclei differ).

EXAMPLE 2.5

(a) Using the ionic radii data in Appendix 2, calculate the coulombic force of attraction between Na^+ and Cl^- in NaCl.

(b) What is the repulsive force in this case?

SOLUTION

(a) From Appendix 2,

$$r_{Na^+} = 0.098 \text{ nm}$$

and

$$r_{Cl^-} = 0.181 \text{ nm}.$$

Then,

$$a_0 = r_{Na^+} + r_{Cl^-} = 0.098 \text{ nm} + 0.181 \text{ nm}$$

$$= 0.278 \text{ nm}.$$

From Equations 2.1 and 2.2,

$$F_c = -\frac{k_0(Z_1 q)(Z_2 q)}{a_0^2},$$

where the equilibrium bond length is used. Substituting the numerical data into the coulombic force equation, we get

$$F_c = -\frac{(9 \times 10^9 \text{ V} \cdot \text{m/C})(+1)(0.16 \times 10^{-18} \text{ C})(-1)(0.16 \times 10^{-18} \text{ C})}{(0.278 \times 10^{-9} \text{ m})^2}.$$

Noting that $1 \text{ V} \cdot \text{C} = 1 \text{ J}$, we obtain

$$F_c = 2.98 \times 10^{-9} \text{ N}.$$

Note. This result can be compared to the data shown in Figures 2.6 and 2.7.

(b) Because $F_c + F_R = 0$,

$$F_R = -F_c = -2.98 \times 10^{-9} \text{ N}.$$

EXAMPLE 2.6

Repeat Example 2.5 for Na_2O, an oxide component in many ceramics and glasses.

SOLUTION

(a) From Appendix 2,

$$r_{Na^+} = 0.098 \text{ nm}$$

and

$$r_{O^{2-}} = 0.132 \text{ nm}.$$

Then,

$$a_0 = r_{Na^+} + r_{O^{2-}} = 0.098 \text{ nm} + 0.132 \text{ nm}$$
$$= 0.231 \text{ nm}.$$

Again,

$$F_c = -\frac{k_0(Z_1q)(Z_2q)}{a_0^2}$$

$$= -\frac{(9 \times 10^9 \text{ V} \cdot \text{m/C})(+1)(0.16 \times 10^{-18} \text{ C})(-2)(0.16 \times 10^{-18} \text{ C})}{(0.231 \times 10^{-9} \text{ m})^2}$$

$$= 8.64 \times 10^{-9} \text{ N}.$$

(b) $F_R = -F_c = -8.64 \times 10^{-9}$ N.

EXAMPLE 2.7

Calculate the minimum radius ratio for a coordination number of 8.

SOLUTION
From Table 2.1, it is apparent that the ions are touching along a body diagonal. If the cube edge length is termed l, then

$$2R + 2r = \sqrt{3}l.$$

For the minimum radius ratio coordination, the large ions are also touching each other (along a cube edge), giving

$$2R = l.$$

Combining the two equations gives us

$$2R + 2r = \sqrt{3}(2R).$$

Then

$$2r = 2R(\sqrt{3} - 1)$$

and

$$\frac{r}{R} = \sqrt{3} - 1 = 1.732 - 1$$
$$= 0.732.$$

Note. There is no shortcut to visualizing three-dimensional structures of this type. It might be helpful to sketch slices through the cube of Table 2.1 with the ions drawn full scale. Many more exercises of this type will be given in Chapter 3.

EXAMPLE 2.8

Estimate the coordination number for the cation in each of these ceramic oxides: Al_2O_3, B_2O_3, CaO, MgO, SiO_2, and TiO_2.

SOLUTION

From Appendix 2, $r_{Al^{3+}} = 0.057$ nm, $r_{B^{3+}} = 0.02$ nm, $r_{Ca^{2+}} = 0.106$ nm, $r_{Mg^{2+}} = 0.078$ nm, $r_{Si^{4+}} = 0.039$ nm, $r_{Ti^{4+}} = 0.064$ nm, and $r_{O^{2-}} = 0.132$ nm.

For Al_2O_3,

$$\frac{r}{R} = \frac{0.057 \text{ nm}}{0.132 \text{ nm}} = 0.43,$$

for which Table 2.1 gives

$$CN = 6.$$

For B_2O_3,

$$\frac{r}{R} = \frac{0.02 \text{ nm}}{0.132 \text{ nm}} = 0.15, \quad \text{giving } CN = 2.^*$$

For CaO,

$$\frac{r}{R} = \frac{0.106 \text{ nm}}{0.132 \text{ nm}} = 0.80, \quad \text{giving } CN = 8.$$

For MgO,

$$\frac{r}{R} = \frac{0.078 \text{ nm}}{0.132 \text{ nm}} = 0.59, \quad \text{giving } CN = 6.$$

For SiO_2,

$$\frac{r}{R} = \frac{0.039 \text{ nm}}{0.132 \text{ nm}} = 0.30, \quad \text{giving } CN = 4.$$

For TiO_2,

$$\frac{r}{R} = \frac{0.064 \text{ nm}}{0.132 \text{ nm}} = 0.48, \quad \text{giving } CN = 6.$$

PRACTICE PROBLEM 2.4

(a) Make a sketch similar to that shown in Figure 2.4 illustrating Mg and O atoms and ions in MgO. **(b)** Compare the electronic configurations for the atoms and ions illustrated in part (a). **(c)** Show which noble gas atoms have electronic configurations equivalent to those illustrated in part (a). (See Example 2.4.)

---

*The actual CN for B_2O_3 is 3 and for CaO is 6. Discrepancies are due to a combinatio~~n~~ ~~of unce~~ in the estimation of ionic radii and bond directionality due to partially covalent char~~

PRACTICE PROBLEM 2.5

(a) Using the ionic radii data in Appendix 2, calculate the coulombic force of attraction between the $Mg^{2+}-O^{2-}$ ion pair. **(b)** What is the repulsive force in this case? (See Examples 2.5 and 2.6.)

PRACTICE PROBLEM 2.6

Calculate the minimum radius ratio for a coordination number of **(a)** 4 and **(b)** 6. (See Example 2.7.)

PRACTICE PROBLEM 2.7

In the next chapter, we will see that MgO, CaO, FeO, and NiO all share the NaCl crystal structure. As a result, in each case the metal ions will have the same coordination number (6). The case of MgO and CaO is treated in Example 2.8. Use the radius ratio calculation to see if it estimates CN = 6 for FeO and NiO.

2.3 | The Covalent Bond

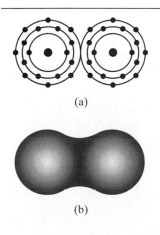

(a)

(b)

$: \overset{..}{Cl} : \overset{..}{Cl} :$

(c)

Cl ⎯ Cl

(d)

FIGURE 2.13 *The covalent bond in a molecule of chlorine gas, Cl_2, is illustrated with (a) a planetary model compared with (b) the actual electron density, (c) an electron-dot schematic, and (d) a bond-line schematic.*

While the ionic bond is nondirectional, the **covalent bond** is highly directional. The name *covalent* derives from the cooperative sharing of valence electrons between two adjacent atoms. **Valence electrons** are those outer orbital electrons that take part in bonding.* Figure 2.13 illustrates the covalent bond in a molecule of chlorine gas (Cl_2) with (a) a planetary model compared with (b) the actual **electron density**, which is clearly concentrated along a straight line between the two Cl nuclei. Common shorthand notations of electron dots and a bond line are shown in parts (c) and (d), respectively.

Figure 2.14a shows a bond-line representation of another covalent molecule, ethylene (C_2H_4). The double line between the two carbons signifies a **double bond**, or covalent sharing of two pairs of valence electrons. By converting the double bond to two single bonds, adjacent ethylene molecules can be covalently bonded together, leading to a long-chain molecule of *polyethylene* (Figure 2.14b). Such **polymeric molecules** (each C_2H_4 unit is a *mer*) are the structural basis of polymers. In Chapter 11, these materials will be discussed in detail. For now, it is sufficient to realize that long-chain molecules of this type have sufficient flexibility to fill three-dimensional space by a complex coiling structure. Figure 2.15 is a two-dimensional schematic of such a "spaghetti-like" structure. The straight lines between C and C and between C and H represent strong, covalent bonds. Only weak, secondary bonding occurs between adjacent sections of the long molecular chains. It is this secondary bonding that is the "weak link" that leads to the low strengths and low melting points for traditional polymers. By contrast, diamond, with exceptionally high hardness and a melting point of greater than 3,500°C, has covalent bonding between each adjacent pair of C atoms (Figure 2.16).

---

*Remember that, in ionic bonding, the valence of Na^+ was +1 because one electron had been transferred to an anion.

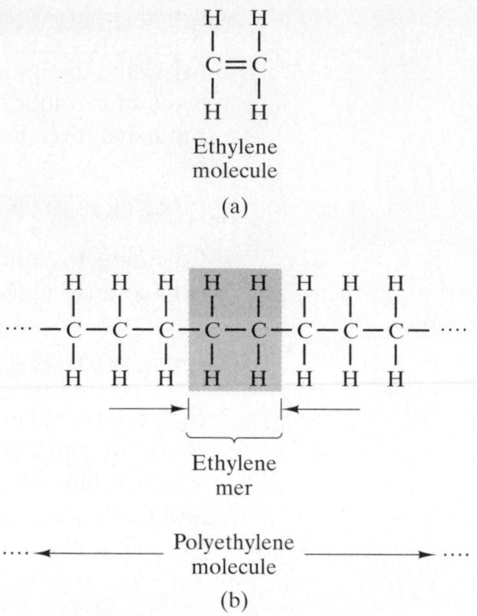

FIGURE 2.14 *(a) An ethylene molecule (C_2H_4) is compared with (b) a polyethylene molecule $+(C_2H_4)_n$ that results from the conversion of the C=C double bond into two C–C single bonds.*

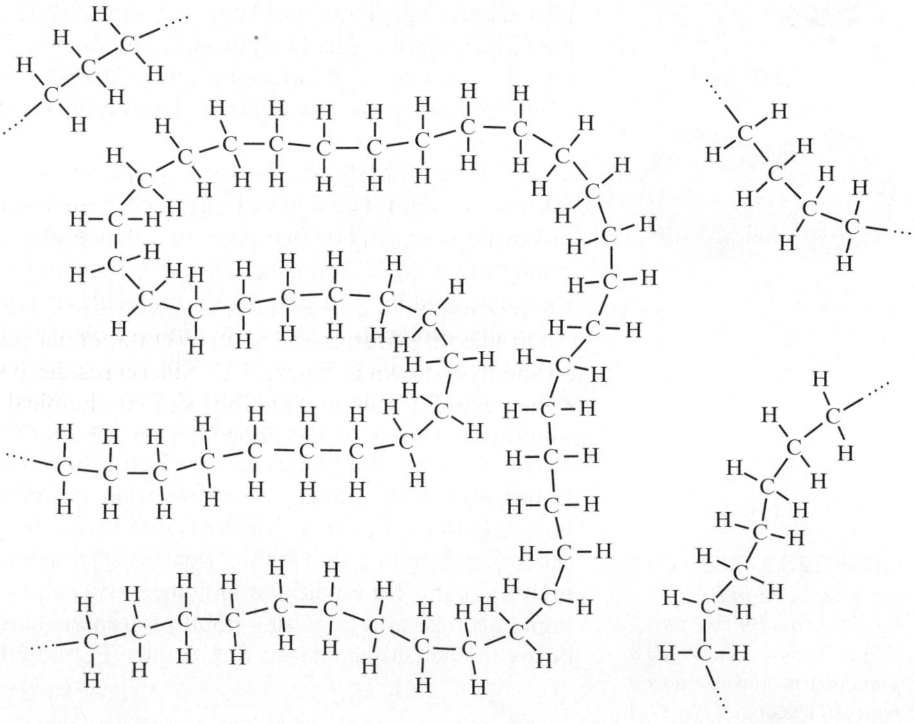

FIGURE 2.15 *Two-dimensional schematic representation of the "spaghetti-like" structure of solid polyethylene.*

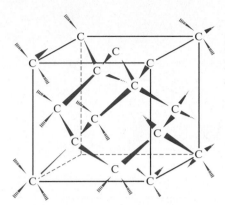

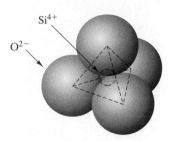

FIGURE 2.16 *Three-dimensional structure of bonding in the covalent solid, carbon (diamond). Each carbon atom (C) has four covalent bonds to four other carbon atoms. (This geometry can be compared with the diamond-cubic structure shown in Figure 3.20.) In this illustration, the bond-line schematic of covalent bonding is given a perspective view to emphasize the spatial arrangement of bonded carbon atoms.*

FIGURE 2.17 *The SiO_4^{4-} tetrahedron represented as a cluster of ions. In fact, the Si–O bond exhibits both ionic and covalent character.*

It is important to note that covalent bonding can produce coordination numbers substantially smaller than predicted by the radius ratio considerations of ionic bonding. For diamond, the radius ratio for the equally sized carbon atoms is $r/R = 1.0$, but Figure 2.16 shows that the coordination number is 4 rather than 12, as predicted in Table 2.1. In this case, the coordination number for carbon is determined by its characteristic sp^3 hybridization bonding in which the four outer-shell electrons of the carbon atom are shared with adjacent atoms in equally spaced directions (see Section 2.1).

In some cases, the efficient packing considerations shown in Table 2.1 are in agreement with covalent bonding geometry. For example, the basic structural unit in silicate minerals and in many commercial ceramics and glasses is the SiO_4^{4-} tetrahedron shown in Figure 2.17. Silicon resides just below carbon in group IV A of the periodic table and exhibits similar chemical behavior. Silicon forms many compounds with fourfold coordination. The SiO_4^{4-} unit maintains this bonding configuration, but simultaneously has strong ionic character, including agreement with Table 2.1. The radius ratio ($r_{Si^{4+}}/r_{O^{2-}} = 0.039$ nm/0.132 nm $= 0.295$) is in the correct range ($0.225 < r/R < 0.414$) to produce maximum efficiency of ionic coordination with CN $= 4$. In fact, the Si–O bond is roughly one-half ionic (electron transfer) and one-half covalent (electron sharing) in nature.

The bonding force and bonding energy curves for covalent bonding look similar to those shown in Figure 2.8 for ionic bonding. The different nature of the two types of bonding implies, of course, that the ionic force equations (2.1 and 2.2) do not apply. Nonetheless, the general terminology of bond energy and bond length apply in both cases (Figure 2.18). Table 2.2 summarizes values of bond energy and bond length for major covalent bonds.

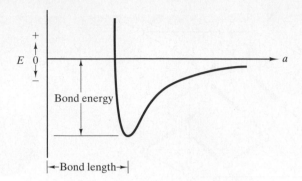

FIGURE 2.18 *The general shape of the bond-energy curve and associated terminology apply to covalent as well as ionic bonding. (The same is true of metallic and secondary bonding.)*

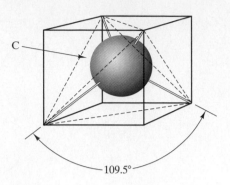

FIGURE 2.19 *Tetrahedral configuration of covalent bonds with carbon. The bond angle is 109.5°.*

Another important characteristic of covalent solids is the **bond angle**, which is determined by the directional nature of valence electron sharing. Figure 2.19 illustrates the bond angle for a typical carbon atom, which tends to form four equally spaced bonds. This tetrahedral configuration (see Figure 2.17) gives a bond angle of 109.5°. The bond angle can vary slightly depending on the species to which the bond is linked, double bonds, and so on. In general, bond angles involving carbon are close to the ideal 109.5°.

EXAMPLE 2.9

Sketch the polymerization process for polyvinyl chloride (PVC). The vinyl chloride molecule is C_2H_3Cl.

SOLUTION
Similar to the schematic shown in Figure 2.14, the vinyl chloride molecule would appear as

$$
\begin{array}{cc}
H & H \\
| & | \\
C & = C \\
| & | \\
H & Cl
\end{array}
$$

Polymerization would occur when several adjacent vinyl chloride molecules connect, transforming their double bonds into single bonds:

$$
\cdots -\overset{\displaystyle H}{\underset{\displaystyle H}{\overset{|}{\underset{|}{C}}}} - \overset{\displaystyle H}{\underset{\displaystyle Cl}{\overset{|}{\underset{|}{C}}}} - \overset{\displaystyle H}{\underset{\displaystyle H}{\overset{|}{\underset{|}{C}}}} - \overset{\displaystyle H}{\underset{\displaystyle Cl}{\overset{|}{\underset{|}{C}}}} - \overset{\displaystyle H}{\underset{\displaystyle H}{\overset{|}{\underset{|}{C}}}} - \overset{\displaystyle H}{\underset{\displaystyle Cl}{\overset{|}{\underset{|}{C}}}} - \overset{\displaystyle H}{\underset{\displaystyle H}{\overset{|}{\underset{|}{C}}}} - \overset{\displaystyle H}{\underset{\displaystyle Cl}{\overset{|}{\underset{|}{C}}}} - \overset{\displaystyle H}{\underset{\displaystyle H}{\overset{|}{\underset{|}{C}}}} - \overset{\displaystyle H}{\underset{\displaystyle Cl}{\overset{|}{\underset{|}{C}}}} - \cdots
$$

|→ mer ←|

TABLE **2.2**

Bond Energies and Bond Lengths for Representative Covalent Bonds			

Bond	Bond energy[a] kcal/mol	kJ/mol	Bond length, nm
C–C	88[b]	370	0.154
C=C	162	680	0.130
C≡C	213	890	0.120
C–H	104	435	0.110
C–N	73	305	0.150
C–O	86	360	0.140
C=O	128	535	0.120
C–F	108	450	0.140
C–Cl	81	340	0.180
O–H	119	500	0.100
O–O	52	220	0.150
O–Si	90	375	0.160
N–H	103	430	0.100
N–O	60	250	0.120
F–F	38	160	0.140
H–H	104	435	0.074

Source: L. H. Van Vlack, *Elements of Materials Science and Engineering,* 4th ed., Addison-Wesley Publishing Co., Inc., Reading, MA, 1980.

[a]Approximate. The values vary with the type of neighboring bonds. For example, methane (CH_4) has the value shown for its C–H bond; however, the C–H bond energy is about 5% less in CH_3Cl and 15% less in $CHCl_3$.

[b]All values are negative for forming bonds (energy is released) and positive for breaking bonds (energy is required).

EXAMPLE 2.10

Calculate the reaction energy for the polymerization of polyvinyl chloride in Example 2.9.

SOLUTION
In general, each C–C bond is broken to form two single C–C bonds:

$$C = C \rightarrow 2C - C.$$

By using data from Table 2.2, the energy associated with this reaction is

$$680 \text{ kJ/mol} \rightarrow 2(370 \text{ kJ/mol}) = 740 \text{ kJ/mol}.$$

The reaction energy is then

$$(740 - 680) \text{ kJ/mol} = 60 \text{ kJ/mol}.$$

Note. As stated in the footnote in Table 2.2, the reaction energy is released during polymerization, making this a spontaneous reaction in which the product, polyvinyl chloride, is stable relative to individual vinyl chloride molecules. Since carbon atoms in the backbone of the polymeric molecule are involved rather than side members, this reaction energy also applies for polyethylene (Figure 2.14) and other "vinyl"-type polymers.

EXAMPLE 2.11

Calculate the length of a polyethylene molecule, $+C_2H_4+_n$, where $n = 500$.

SOLUTION

Looking only at the carbon atoms in the backbone of the polymeric chain, we must acknowledge the characteristic bond angle of $109.5°$:

This angle produces an effective bond length, l, of

$$l = (\text{C–C bond length}) \times \sin 54.75°.$$

Using Table 2.2, we obtain

$$l = (0.154 \text{ nm})(\sin 54.75°)$$

$$= 0.126 \text{ nm}.$$

With two bond lengths per mer and 500 mers, the total molecule length, L, is

$$L = 500 \times 2 \times 0.126 \text{ nm}$$

$$= 126 \text{ nm}$$

$$= 0.126 \text{ } \mu\text{m}.$$

Note. In Chapter 12, we will calculate the degree of coiling of these long, linear molecules.

PRACTICE PROBLEM 2.8

In Figure 2.14, we see the polymerization of polyethylene $+C_2H_4+_n$ illustrated. Example 2.9 illustrates polymerization for polyvinyl chloride $+C_2H_3Cl+_n$. Make a similar sketch to illustrate the polymerization of polypropylene $+C_2H_3R+_n$, where R is a CH_3 group.

2.4 | The Metallic Bond

The ionic bond involves electron transfer and is nondirectional. The covalent bond involves electron sharing and is directional. The third type of primary bond, the **metallic bond**, involves electron sharing and is nondirectional. In this case, the valence electrons are said to be **delocalized electrons**; that is, they have an equal probability of being associated with any of a large number of adjacent atoms. In typical metals, this delocalization is associated with the entire material, leading to an electron cloud, or electron gas (Figure 2.20). This mobile "gas" is the basis for the high electrical conductivity in metals. (The role of electronic structure in producing conduction electrons in metals is discussed in Chapter 13.)

Again, the concept of a bonding **energy well**, or **trough**, as shown in Figure 2.18, applies. As with ionic bonding, bond angles and coordination numbers are

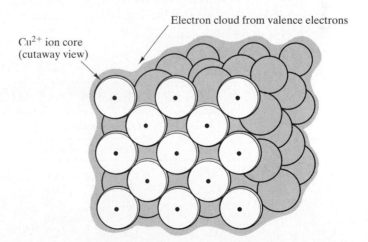

Electron cloud from valence electrons

Cu^{2+} ion core
(cutaway view)

FIGURE 2.20 *Metallic bond consisting of an electron cloud, or gas. An imaginary slice is shown through the front face of the crystal structure of copper, revealing Cu^{2+} ion cores bonded by the delocalized valence electrons.*

determined primarily by efficient packing considerations, so coordination numbers tend to be high (8 and 12). Relative to the bonding energy curve, a detailed list of atomic radii for the elements is given in Appendix 2, which includes the important elemental metals. Appendix 2 also includes a list of ionic radii. Some of these ionic species are found in the important ceramics and glasses. Inspection of Appendix 2 shows that the radius of the metal ion core involved in metallic bonding (Figure 2.20) differs substantially from the radius of a metal ion from which valence electrons have been transferred.

Instead of a list of bond energies for metals and ceramics, similar to those included for covalent bonds shown in Table 2.2, data that represent the energetics associated with the bulk solid rather than isolated atom (or ion) pairs are more useful. For example, Table 2.3 lists the heats of sublimation of some common metals and their oxides (some of the common ceramic compounds). The heat of sublimation represents the amount of thermal energy necessary to turn 1 mol of solid directly into vapor at a fixed temperature. It is a good indication of the relative strength of bonding in the solid. However, caution must be used in making direct comparisons to the bond energies shown in Table 2.2, which correspond to specific atom pairs. Nonetheless, the magnitudes of energies shown in Tables 2.2 and 2.3 are comparable in range.

In this chapter, we have seen that the nature of the chemical bonds between atoms of the same element and atoms of different elements depends on the transfer or sharing of electrons between adjacent atoms. American chemist Linus Pauling systematically defined **electronegativity** as the ability of an atom to attract electrons to itself. Figure 2.21 summarizes Pauling's electronegativity values for the elements in the periodic table. We can recall from Chapter 1 that the majority of elements in the periodic table are metallic in nature (Figure 1.4). In general, the values of electronegativities increase from the left to the right side of the periodic table, with cesium and francium (in group I A) having the lowest value (0.7) and fluorine (in group VII A) having the highest value (4.0). Clearly, the metallic elements tend to have the lower values of electronegativity, and the nonmetallic elements have the higher values. Although Pauling specifically based his electronegativities on thermochemical data for molecules, we shall see in Section 4.1 that the data of Figure 2.21 are useful for predicting the nature of metallic alloys.

TABLE 2.3

Heats of Sublimation (at 25°C) of Some Metals and Their Oxides					
Metal	**Heat of sublimation**		**Metal oxide**	**Heat of sublimation**	
	kcal/mol	**kJ/mol**		**kcal/mol**	**kJ/mol**
Al	78	326			
Cu	81	338			
Fe	100	416	FeO	122	509
Mg	35	148	MgO	145	605
Ti	113	473	α-TiO	143	597
			TiO_2 (rutile)	153	639

Source: Data from *JANAF Thermochemical Tables*, 2nd ed., National Standard Reference Data Series, Natl. Bur. Std. (U.S.), *37* (1971), and Supplement in *J. Phys. Chem. Ref. Data 4* (1), 1–175 (1975).

I A																		0
1 H 2.1	II A											III A	IV A	V A	VI A	VII A		2 He –
3 Li 1.0	4 Be 1.5											5 B 2.0	6 C 2.5	7 N 3.0	8 O 3.5	9 F 4.0		10 Ne –
11 Na 0.9	12 Mg 1.2	III B	IV B	V B	VI B	VII B		VIII		I B	II B	13 Al 1.5	14 Si 1.8	15 P 2.1	16 S 2.5	17 Cl 3.0		18 Ar –
19 K 0.8	20 Ca 1.0	21 Sc 1.3	22 Ti 1.5	23 V 1.6	24 Cr 1.6	25 Mn 1.5	26 Fe 1.8	27 Co 1.8	28 Ni 1.8	29 Cu 1.9	30 Zn 1.6	31 Ga 1.6	32 Ge 1.8	33 As 2.0	34 Se 2.4	35 Br 2.8		36 Kr –
37 Rb 0.8	38 Sr 1.0	39 Y 1.2	40 Zr 1.4	41 Nb 1.6	42 Mo 1.8	43 Tc 1.9	44 Ru 2.2	45 Rh 2.2	46 Pd 2.2	47 Ag 1.9	48 Cd 1.7	49 In 1.7	50 Sn 1.8	51 Sb 1.9	52 Te 2.1	53 I 2.5		54 Xe –
55 Cs 0.7	56 Ba 0.9	57-71 La-Lu 1.1-1.2	72 Hf 1.3	73 Ta 1.5	74 W 1.7	75 Re 1.9	76 Os 2.2	77 Ir 2.2	78 Pt 2.2	79 Au 2.4	80 Hg 1.9	81 Tl 1.8	82 Pb 1.8	83 Bi 1.9	84 Po 2.0	85 At 2.2		86 Rn –
87 Fr 0.7	88 Ra 0.9	89-102 Ac-No 1.1-1.7																

FIGURE 2.21 *The electronegativities of the elements. (From Linus Pauling,* The Nature of the Chemical Bond and the Structure of Molecules and Crystals; An Introduction to Modern Structural Chemistry, *3rd ed., Cornell University Press, Ithaca, NY, 1960.)*

EXAMPLE 2.12

Several metals, such as α-Fe, have a body-centered cubic crystal structure in which the atoms have a coordination number of 8. Discuss this structure in light of the prediction of Table 2.1 that nondirectional bonding of equal-sized spheres should have a coordination number of 12.

SOLUTION

The presence of some covalent character in these predominantly metallic materials can reduce the coordination number below the predicted value. (See Example 2.8.)

PRACTICE PROBLEM 2.12

Discuss the low coordination number (CN = 4) for the diamond cubic structure found for some elemental solids, such as silicon. (See Example 2.12.)

2.5 | The Secondary, or van der Waals, Bond

The major source of cohesion in a given engineering material is one or more of the three primary bonds just covered. As seen in Table 2.2, typical primary bond energies range from 200 to 700 kJ/mol ($\approx$50 to 170 kcal/mol). It is possible to obtain some atomic bonding (with substantially smaller bonding energies) without electron transfer or sharing. This bonding is known as *secondary bonding,* or

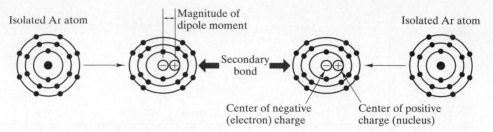

FIGURE 2.22 *Development of induced dipoles in adjacent argon atoms leading to a weak, secondary bond. The degree of charge distortion shown here is greatly exaggerated.*

van der Waals* bonding. The mechanism of secondary bonding is somewhat similar to ionic bonding (i.e., the attraction of opposite charges). The key difference is that no electrons are transferred.[†] Attraction depends on asymmetrical distributions of positive and negative charges within each atom or molecular unit being bonded. Such charge asymmetry is referred to as a **dipole**. Secondary bonding can be of two types, depending on whether the dipoles are (1) temporary or (2) permanent.

Figure 2.22 illustrates how two neutral atoms can develop a weak bonding force between them by a slight distortion of their charge distributions. The example is argon, a noble gas, which does not tend to form primary bonds because it has a stable, filled outer orbital shell. An isolated argon atom has a perfectly spherical distribution of negative electrical charge surrounding its positive nucleus. However, when another argon atom is brought nearby, the negative charge is drawn slightly toward the positive nucleus of the adjacent atom. This slight distortion of charge distribution occurs simultaneously in both atoms. The result is an *induced dipole*. Because the degree of charge distortion related to an induced dipole is small, the magnitude of the resulting dipole is small, leading to a relatively small bond energy (0.99 kJ/mol or 0.24 kcal/mol).

Secondary bonding energies are somewhat greater when molecular units containing *permanent dipoles* are involved. Perhaps the best example of this is the **hydrogen bridge**, which connects adjacent molecules of water, H_2O (Figure 2.23). Because of the directional nature of electron sharing in the covalent O–H bonds, the H atoms become positive centers and the O atoms become negative centers for the H_2O molecules. The greater charge separation possible in such a **polar molecule**, a molecule with a permanent separation of charge, gives a larger **dipole moment** (product of charge and separation distance between centers of positive and negative charge) and therefore a greater bond energy (21 kJ/mol or 5 kcal/mol). The secondary bonding between adjacent polymeric chains in polymers such as polyethylene is of this type.

Note that one of the important properties of water derives from the hydrogen bridge. The expansion of water upon freezing is due to the regular and

In Section 6.6, we shall see that secondary bonding between adjacent polymer molecules plays a critical role in understanding the nature of viscoelastic deformation in organic polymers.

---

*Johannes Diderik van der Waals (1837–1923), Dutch physicist, improved the equations of state for gases by taking into account the effect of secondary bonding forces. His brilliant research was first published as a thesis dissertation arising from his part-time studies of physics. The immediate acclaim for the work led to his transition from a job as headmaster of a secondary school to a professorship at the University of Amsterdam.

[†]Primary bonds are sometimes referred to as *chemical bonds*, with secondary bonds being *physical bonds*.

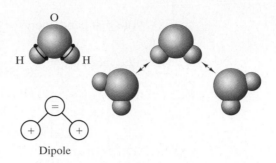

FIGURE 2.23 *Hydrogen bridge. This secondary bond is formed between two permanent dipoles in adjacent water molecules. (From W. G. Moffatt, G. W. Pearsall, and J. Wulff,* The Structure and Properties of Materials, *Vol. 1:* Structures, *John Wiley & Sons, Inc., New York, 1964.)*

repeating alignment of adjacent H_2O molecules, as seen in Figure 2.23, which leads to a relatively open structure. Upon melting, the adjacent H_2O molecules, while retaining the hydrogen bridge, pack together in a more random and more dense arrangement.

EXAMPLE 2.13

A common way to describe the bonding energy curve (Figure 2.18) for secondary bonding is the "6–12" potential, which states that

$$E = -\frac{K_A}{a^6} + \frac{K_R}{a^{12}},$$

where K_A and K_R are constants for attraction and repulsion, respectively. This relatively simple form is a quantum mechanical result for this relatively simple bond type. Given $K_A = 10.37 \times 10^{-78}$ J $\cdot$ m^6 and $K_R = 16.16 \times 10^{-135}$ J $\cdot$ m^{12}, calculate the bond energy and bond length for argon.

SOLUTION
The (equilibrium) bond length occurs at $dE/da = 0$:

$$\left(\frac{dE}{da}\right)_{a=a_0} = 0 = \frac{6K_A}{a_0^7} - \frac{12K_R}{a_0^{13}}.$$

Rearranging gives us

$$a_0 = \left(2\frac{K_R}{K_A}\right)^{1/6}$$

$$= \left(2 \times \frac{16.16 \times 10^{-135}}{10.37 \times 10^{-78}}\right)^{1/6} \text{m}$$

$$= 0.382 \times 10^{-9} \text{ m} = 0.382 \text{ nm}.$$

Note that bond energy $= E(a_0)$ yields

$$E(0.382 \text{ nm}) = -\frac{K_A}{(0.382 \text{ nm})^6} + \frac{K_R}{(0.382 \text{ nm})^{12}}$$

$$= -\frac{(10.37 \times 10^{-78} \text{ J} \cdot \text{m}^6)}{(0.382 \times 10^{-9} \text{ m})^6} + \frac{(16.16 \times 10^{-135} \text{ J} \cdot \text{m}^{12})}{(0.382 \times 10^{-9} \text{ m})^{12}}$$

$$= -1.66 \times 10^{-21} \text{ J}.$$

For 1 mol of Ar,

$$E_{\text{bonding}} = -1.66 \times 10^{-21} \text{ J/bond} \times 0.602 \times 10^{24} \frac{\text{bonds}}{\text{mole}}$$

$$= -0.999 \times 10^3 \text{ J/mol}$$

$$= -0.999 \text{ kJ/mol}.$$

Note. This bond energy is less than 1% of the magnitude of any of the primary (covalent) bonds listed in Table 2.2. It should also be noted that the footnote in Table 2.2 indicates a consistent sign convention (bond energy is negative).

PRACTICE PROBLEM 2.13

The bond energy and bond length for argon are calculated (assuming a "6–12" potential) in Example 2.13. Plot E as a function of a over the range 0.33 to 0.80 nm.

PRACTICE PROBLEM 2.14

Using the information from Example 2.13, plot the van der Waals bonding force curve for argon (i.e., F versus a over the same range covered in Practice Problem 2.13).

2.6 | Materials–The Bonding Classification

A dramatic representation of the relative bond energies of the various bond types of this chapter is obtained by comparison of melting points. The **melting point** of a solid indicates the temperature to which the material must be subjected to provide sufficient thermal energy to break its cohesive bonds. Table 2.4 shows representative examples used in this chapter. A special note must be made for polyethylene, which is of mixed-bond character. As discussed in Section 2.3, the secondary bonding is a weak link that causes the material to lose structural rigidity above approximately 120°C. This is not a precise melting point, but instead is a temperature above which the material softens rapidly with increasing

TABLE 2.4

Comparison of Melting Points for Some of the Representative Materials of Chapter 2		
Material	**Bonding type**	**Melting point ($°C$)**
NaCl	Ionic	801
C (diamond)	Covalent	~3,550
$+C_2H_4\}_n$	Covalent and secondary	~120[a]
Cu	Metallic	1,084.87
Ar	Secondary (induced dipole)	−189
H_2O	Secondary (permanent dipole)	0

[a]Because of the irregularity of the polymeric structure of polyethylene, it does not have a precise melting point. Instead, it softens with increasing temperature above 120°C. In this case, the 120°C value is a "service temperature" rather than a true melting point.

temperature. The irregularity of the polymeric structure (Figure 2.15) produces variable secondary bond lengths and, therefore, variable bond energies. More important than the variation in bond energy is the average magnitude, which is relatively small. Even though polyethylene and diamond each have similar C–C covalent bonds, the absence of secondary-bond weak links allows diamond to retain its structural rigidity more than 3,000°C beyond polyethylene.

We have now seen four major types of atomic bonding consisting of three primary bonds (ionic, covalent, and metallic) and secondary bonding. It has been traditional to distinguish the three fundamental structural materials (metals, ceramics/glasses, and polymers) as being directly associated with the three types of primary bonds (metallic, ionic, and covalent, respectively). This is a useful concept, but we have already seen in Sections 2.3 and 2.5 that polymers owe their behavior to both covalent and secondary bonding. We also noted in Section 2.3 that some of the most important ceramics and glasses have strong covalent as well as ionic character. Table 2.5 summarizes the bonding character associated with the five fundamental types of engineering materials together with some representative examples. Remember that the mixed-bond character for ceramics and glasses referred to both ionic and covalent nature for a given bond (e.g., Si–O), whereas the mixed-bond character for polymers referred to different bonds being

TABLE 2.5

Bonding Character of the Five Fundamental Types of Engineering Materials		
Material type	**Bonding character**	**Example**
Metal	Metallic	Iron (Fe) and the ferrous alloys
Ceramics and glasses	Ionic/covalent	Silica (SiO_2): crystalline and noncrystalline
Polymers	Covalent and secondary	Polyethylene $+C_2H_4\}_n$
Semiconductors	Covalent or covalent/ionic	Silicon (Si) or cadmium sulfide (CdS)

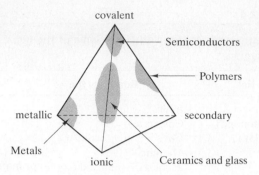

FIGURE 2.24 *Tetrahedron representing the relative contribution of different bond types to the five fundamental categories of engineering materials (the four structural types plus semiconductors).*

covalent (e.g., C–H) and secondary (e.g., between chains). The relative contribution of different bond types can be graphically displayed in the form of a tetrahedron of bond types (Figure 2.24) in which each apex of the tetrahedron represents a pure bonding type. In Chapter 13, we shall add another perspective on materials classification, electrical conductivity, which will follow directly from the nature of bonding and is especially helpful in defining the unique character of semiconductors.

Summary

One basis for the classification of engineering materials is atomic bonding. While the chemical identity of each atom is determined by the number of protons and neutrons within its nucleus, the nature of atomic bonding is determined by the behavior of the electrons that orbit the nucleus.

There are three kinds of strong, or primary, bonds responsible for the cohesion of solids. First, the ionic bond involves electron transfer and is nondirectional. The electron transfer creates a pair of ions with opposite charge. The attractive force between ions is coulombic in nature. An equilibrium ionic spacing is established due to the strong repulsive forces associated with attempting to overlap the two atomic cores. The nondirectional nature of the ionic bond allows ionic coordination numbers to be determined strictly by geometrical packing efficiency (as indicated by the radius ratio). Second, the covalent bond involves electron sharing and is highly directional, which can lead to relatively low coordination numbers and more open atomic structures. Third, the metallic bond involves sharing of delocalized electrons, producing a nondirectional bond. The resulting electron cloud or gas results in high electrical conductivity. The nondirec-tional nature results in relatively high coordination numbers, as in ionic bonding. In the absence of electron transfer or sharing, a weaker form of bonding is possible. This secondary bonding is the result of attraction between either temporary or permanent electrical dipoles.

The classification of engineering materials acknowledges a particular bonding type or combination of types for each category. Metals involve metallic bonding. Ceramics and glasses involve ionic bonding, but usually in conjunction with a strong covalent character. Polymers typically involve strong covalent bonds along polymeric chains, but have weaker secondary bonding between adjacent chains. Secondary bonding acts as a weak link in the structure, giving characteristically low strengths and melting points. Semiconductors are predominantly covalent in nature, with some semiconducting compounds having a significant ionic character. These five categories of engineering materials are, then, the fundamental types. Composites are a sixth category and represent combinations of the first four fundamental types and have bonding characteristics appropriate to their constituents.

Key Terms

anion (28)
atomic mass (22)
atomic mass unit (22)
atomic number (23)
atomic radius (32)
Avogadro's number (23)
bond angle (42)
bonding energy (31)
bonding force (30)
bond length (29)
cation (28)
coordination number (32)
coulombic attraction (29)
covalent bond (39)
delocalized electron (45)
dipole (48)
dipole moment (48)
double bond (39)

electron (22)
electron density (39)
electronegativity (46)
electron orbital (23)
energy level (24)
energy trough (45)
energy well (45)
gram-atom (23)
group (23)
hard sphere (32)
hybridization (25)
hydrogen bridge (48)
ion (28)
ionic bond (28)
ionic radius (32)
isotope (23)
melting point (50)
metallic bond (45)

mole (23)
neutron (22)
nucleus (22)
orbital shell (28)
periodic table (23)
polar molecule (48)
polymeric molecule (39)
primary bond (25)
proton (22)
radius ratio (32)
repulsive force (30)
secondary bond (25)
soft sphere (32)
valence (29)
valence electron (39)
van der Waals bond (48)

References

Virtually any introductory textbook on college-level chemistry will be useful background for this chapter. Good examples are the following:

Brown, **T. L.**, **H. E. LeMay**, **Jr.**, and **B. E. Bursten**, *Chemistry—The Central Science*, 10th ed., Prentice Hall, Upper Saddle River, NJ, 2006.

Oxtoby, **D. W.**, **H. P. Gillis**, and **A. Campion**, *Principles of Modern Chemistry*, 6th ed., Thomson Brooks/Cole, Belmont, CA, 2008.

Petrucci, **R. H.**, **W. S. Harwood**, **G. E. Herring**, and **J. Madura**, *General Chemistry—Principles and Modern Applications*, 9th ed., Prentice Hall, Upper Saddle River, NJ, 2007.

Problems

Beginning with this chapter, a set of problems will be provided at the conclusion of each chapter of the book. Instructors may note that there are few of the subjective, discussion-type problems that are so often used in materials textbooks. I strongly feel that such problems are generally frustrating to students who are being introduced to materials science and engineering. As such, I shall concentrate on objective problems. For this reason, no problems were given in the general, introductory chapter, Chapter 1.

A few points about the organization of problems should be noted. All problems are clearly related to the appropriate chapter section. Also, some Practice Problems for each section were already given following the solved examples within that section. These problems are intended to provide a carefully guided journey into the first calculations in each new area and could be used by students for self-study.

Answers are given for nearly all of the Practice Problems following the appendixes. **As noted on page 28, the solutions to all practice problems are provided on the related Web site.** The following problems are increasingly challenging. Problems not marked with a bullet are relatively straightforward but are not explicitly connected to an example. Those problems marked with a bullet (•) are intended to be relatively challenging. Answers to odd-numbered problems are given following the appendixes.

Section 2.1 • Atomic Structure

2.1. A gold O-ring is used to form a gastight seal in a high-vacuum chamber. The ring is formed from a 80-mm length of 1.5-mm-diameter wire. Calculate the number of gold atoms in the O-ring.

2.2. Common aluminum foil for household use is nearly pure aluminum. A box of this product at a local supermarket is advertised as giving 75 ft^2 of material (in a roll 304 mm wide by 22.8 m long). If the foil is 0.5 mil (12.7 μm) thick, calculate the number of atoms of aluminum in the roll.

2.3. In a metal-oxide-semiconductor (MOS) device, a thin layer of SiO_2 (density = 2.20 Mg/m^3) is grown on a single crystal chip of silicon. How many Si atoms and how many O atoms are present per square millimeter of the oxide layer? Assume that the layer thickness is 100 nm.

2.4. A box of clear plastic wrap for household use is polyethylene, $+C_2H_4 +_n$, with density = 0.910 Mg/m^3. A box of this product contains 100 ft^2 of material (in a roll 304 mm wide by 30.5 m long). If the wrap is 0.5 mil (12.7 μm) thick, calculate the number of carbon atoms and the number of hydrogen atoms in this roll.

2.5. An Al_2O_3 *whisker* is a small single crystal used to reinforce metal-matrix composites. Given a cylindrical shape, calculate the number of Al atoms and the number of O atoms in a whisker with a diameter of 1 μm and a length of 30 μm. (The density of Al_2O_3 is 3.97 Mg/m^3.)

2.6. An optical fiber for telecommunication is made of SiO_2 glass (density = 2.20 Mg/m^3). How many Si atoms and how many O atoms are present per millimeter of length of a fiber 10 μm in diameter?

2.7. Twenty-five grams of magnesium filings are to be oxidized in a laboratory demonstration. **(a)** How many O_2 molecules would be consumed in this demonstration? **(b)** How many moles of O_2 does this represent?

2.8. Naturally occurring copper has an atomic weight of 63.55. Its principal isotopes are Cu^{63} and Cu^{65}. What is the abundance (in atomic percent) of each isotope?

2.9. A copper penny has a mass of 2.60 g. Assuming pure copper, how much of this mass is contributed by **(a)** the neutrons in the copper nuclei and **(b)** electrons?

2.10. The orbital electrons of an atom can be ejected by exposure to a beam of electromagnetic radiation. Specifically, an electron can be ejected by a photon with energy greater than or equal to the electron's binding energy. Given that the photon energy (E) is equal to hc/λ, where h is Planck's constant, c the speed of light, and λ the wavelength, calculate the maximum wavelength of radiation (corresponding to the minimum energy) necessary to eject a 1s electron from a C^{12} atom. (See Figure 2.3.)

2.11. Once the 1s electron is ejected from a C^{12} atom, as described in Problem 2.10, there is a tendency for one of the 2(sp^3) electrons to drop into the 1s level. The result is the emission of a photon with an energy precisely equal to the energy change associated with the electron transition. Calculate the wavelength of the photon that would be emitted from a C^{12} atom. (You will note various examples of this concept throughout the text in relation to the chemical analysis of engineering materials.)

2.12. The mechanism for producing a photon of specific energy is outlined in Problem 2.11. The magnitude of photon energy increases with the atomic number of the atom from which emission occurs. (This increase is due to the stronger binding forces between the negative electrons and the positive nucleus as the numbers of protons and electrons increase with atomic number.) As noted in Problem 2.10, $E = hc/\lambda$, which means that a higher-energy photon will have a shorter wavelength. Verify that higher atomic number materials will emit higher-energy, shorter-wavelength photons by calculating E and λ for emission from iron (atomic number 26 compared to 6 for carbon), given that the energy levels for the first two electron orbitals in iron are at $-7{,}112$ eV and -708 eV.

Section 2.2 • The Ionic Bond

2.13. Make an accurate plot of F_c versus a (comparable to that shown in Figure 2.6) for an Mg^{2+}–O^{2-} pair. Consider the range of a from 0.2 to 0.7 nm.

2.14. Make an accurate plot of F_c versus a for an Na^+–O^{2-} pair.

2.15. So far, we have concentrated on the coulombic force of attraction between ions. But like ions repel each other. A nearest neighbor pair of Na^+ ions in Figure 2.5 are separated by a distance of $\sqrt{2}a_0$, where a_0 is defined in Figure 2.7. Calculate the coulombic force of *repulsion* between such a pair of like ions.

2.16. Calculate the coulombic force of attraction between Ca^{2+} and O^{2-} in CaO, which has the NaCl-type structure.

2.17. Calculate the coulombic force of repulsion between nearest-neighbor Ca^{2+} ions in CaO. (Note Problems 2.15 and 2.16.)

2.18. Calculate the coulombic force of repulsion between nearest-neighbor O^{2-} ions in CaO. (Note Problems 2.15, 2.16, and 2.17.)

2.19. Calculate the coulombic force of repulsion between nearest-neighbor Ni^{2+} ions in NiO, which has the NaCl-type structure. (Note Problem 2.17.)

2.20. Calculate the coulombic force of repulsion between nearest-neighbor O^{2-} ions in NiO. (Note Problems 2.18 and 2.19.)

2.21. SiO_2 is known as a *glass former* because of the tendency of SiO_4^{4-} tetrahedra (Figure 2.17) to link together in a noncrystalline network. Al_2O_3 is known as an intermediate glass former due to the ability of Al^{3+} to substitute for Si^{4+} in the glass network, although Al_2O_3 does not by itself tend to be noncrystalline. Discuss the substitution of Al^{3+} for Si^{4+} in terms of the radius ratio.

2.22. Repeat Problem 2.21 for TiO_2, which, like Al_2O_3, is an intermediate glass former.

2.23. The coloration of glass by certain ions is often sensitive to the coordination of the cation by oxygen ions. For example, Co^{2+} gives a blue-purple color when in the fourfold coordination characteristic of the silica network (see Problem 2.21) and gives a pink color when in a sixfold coordination. Which color from Co^{2+} is predicted by the radius ratio?

2.24. One of the first nonoxide materials to be produced as a glass was BeF_2. As such, it was found to be similar to SiO_2 in many ways. Calculate the radius ratio for Be^{2+} and F^-, and comment.

2.25. A common feature in high-temperature ceramic superconductors is a Cu–O sheet that serves as a superconducting plane. Calculate the coulombic force of attraction between a Cu^{2+} and an O^{2-} within one of these sheets.

2.26. In contrast to the calculation for the superconducting Cu–O sheets discussed in Problem 2.25, calculate the coulombic force of attraction between a Cu^+ and an O^{2-}.

*2.27. For an ionic crystal, such as NaCl, the net coulombic bonding force is a simple multiple of the force of attraction between an adjacent ion pair. To demonstrate this concept, consider the hypothetical, one-dimensional "crystal" shown:

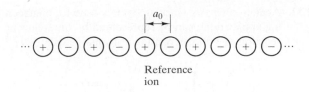

Reference ion

(a) Show that the net coulombic force of attraction between the reference ion and all other ions in the crystal is

$$F = AF_c,$$

where F_c is the force of attraction between an adjacent ion pair (see Equation 2.1) and A is a series expansion.

(b) Determine the value of A.

2.28. In Problem 2.27, a value for A was calculated for the simple one-dimensional case. For the three-dimensional NaCl structure, A has been calculated to be 1.748. Calculate the net coulombic force of attraction, F, for this case.

Section 2.3 • The Covalent Bond

2.29. Calculate the total reaction energy for polymerization required to produce the roll of clear plastic wrap described in Problem 2.4.

2.30. Natural rubber is polyisoprene. The polymerization reaction can be illustrated as

$$n\begin{pmatrix} H & H & CH_3 & H \\ | & | & | & | \\ C{=}C{-}C{=}C \\ | & & & | \\ H & & & H \end{pmatrix} \rightarrow \begin{pmatrix} H & H & CH_3 & H \\ | & | & | & | \\ {-}C{-}C{=}C{-}C{-} \\ | & & & | \\ H & & & H \end{pmatrix}_n$$

Calculate the reaction energy (per mole) for polymerization.

2.31. Neoprene is a synthetic rubber, polychloroprene, with a chemical structure similar to that of natural rubber (see Problem 2.30), except that it contains a Cl atom in place of the CH_3 group of the isoprene molecule. **(a)** Sketch the polymerization reaction for neoprene, and **(b)** calculate the reaction energy (per mole) for this polymerization. **(c)** Calculate the total energy released during the polymerization of 1 kg of chloroprene.

2.32. Acetal polymers, which are widely used for engineering applications, can be represented by the following reaction, the polymerization of formaldehyde:

$$n \left(\begin{matrix} H \\ \diagdown \\ \diagup \\ H \end{matrix} C=O \right) \longrightarrow \left(\begin{matrix} H \\ | \\ -C=O- \\ | \\ H \end{matrix} \right)_n$$

Calculate the reaction energy for this polymerization.

2.33. The first step in the formation of phenolformaldehyde, a common phenolic polymer, is shown in Figure 12.6. Calculate the net reaction energy (per mole) for this step in the overall polymerization reaction.

2.34. Calculate the molecular weight of a polyethylene molecule with $n = 500$.

2.35. The monomer upon which a common acrylic polymer, polymethyl methacrylate, is based is

$$\begin{matrix} H & CH_3 \\ | & | \\ C & = & C \\ | & | \\ H & C=O \\ & | \\ & O \\ & | \\ & CH_3 \end{matrix}$$

Calculate the molecular weight of a polymethyl methacrylate molecule with $n = 500$.

2.36. Bone "cement," used by orthopedic surgeons to set artificial hip implants in place, is methyl methacrylate polymerized during the surgery. The resulting polymer has a relatively wide range of molecular weights. Calculate the resulting range of molecular weights if $200 < n < 700$.

2.37. Orthopedic surgeons notice a substantial amount of heat evolution from polymethyl methacrylate bone cement during surgery. Calculate the reaction energy if a surgeon uses 15 g of polymethyl methacrylate to set a given hip implant.

2.38. The monomer for the common fluoroplastic, polytetrafluoroethylene, is

$$\begin{matrix} F & F \\ | & | \\ C & = & C \\ | & | \\ F & F \end{matrix}$$

(a) Sketch the polymerization of polytetrafluoroethylene.

(b) Calculate the reaction energy (per mole) for this polymerization.

(c) Calculate the molecular weight of a molecule with $n = 500$.

2.39. Repeat Problem 2.38 for polyvinylidene fluoride, an ingredient in various commercial fluoroplastics, that has the monomer

$$\begin{matrix} F & H \\ | & | \\ C & = & C \\ | & | \\ F & H \end{matrix}$$

2.40. Repeat Problem 2.38 for polyhexafluoropropylene, an ingredient in various commercial fluoroplastics, that has the monomer

$$\begin{matrix} F & F \\ | & | \\ C & = & C \\ | & | \\ & & F \\ & | \\ F-C-F \\ | \\ F \end{matrix}$$

Section 2.4 • The Metallic Bond

2.41. In Table 2.3, the heat of sublimation was used to indicate the magnitude of the energy of the metallic bond. A significant range of energy values is indicated by the data. The melting point data in Appendix 1 are another, more indirect indication of bond strength. Plot heat of sublimation versus melting point for the five metals of Table 2.3, and comment on the correlation.

2.42. In order to explore a trend within the periodic table, plot the bond length of the group II A metals (Be to Ba) as a function of atomic number. (Refer to Appendix 2 for necessary data.)

2.43. Superimpose on the plot generated for Problem 2.42 the metal-oxide bond lengths for the same range of elements.

2.44. To explore another trend within the periodic table, plot the bond length of the metals in the row Na to Si as a function of atomic numbers. (For this purpose, Si is treated as a semimetal.)

2.45. Superimpose on the plot generated for Problem 2.44 the metal-oxide bond lengths for the same range of elements.

2.46. Plot the bond length of the metals in the long row of metallic elements (K to Ga).

2.47. Superimpose on the plot generated for Problem 2.46 the metal–oxide bond lengths for the same range of elements.

*2.48. The heat of sublimation of a metal, introduced in Table 2.3, is related to the ionic bonding energy of a metallic compound discussed in Section 2.2. Specifically, these and related reaction energies are summarized in the Born–Haber cycle, illustrated next. For the simple example of NaCl,

$$\text{Na (solid)} + \tfrac{1}{2}\,\text{Cl}_2 \text{ (g)} \longrightarrow \text{Na (g)} + \text{Cl (g)}$$

$$\Big\downarrow \Delta H_f^{\circ} \qquad\qquad \Big\downarrow \qquad \Big\downarrow$$

$$\text{NaCl (solid)} \longleftarrow \text{Na}^+ \text{ (g)} + \text{Cl}^- \text{ (g)}$$

Given the heat of sublimation to be 100 kJ/mol for sodium, calculate the ionic bonding energy of sodium chloride. (Additional data: ionization energies for sodium and chlorine = 496 kJ/mol and −361 kJ/mol, respectively; dissociation energy for diatomic chlorine gas = 243 kJ/mol; and heat of formation, ΔH_f°, of NaCl = −411 kJ/mol.)

Section 2.5 • The Secondary, or van der Waals, Bond

2.49. The secondary bonding of gas molecules to a solid surface is a common mechanism for measuring the surface area of porous materials. By lowering the temperature of a solid well below room temperature, a measured volume of gas will condense to form a monolayer coating of molecules on the porous surface. For a 100-g sample of fused copper catalyst, a volume of 9×10^3 mm^3 of nitrogen (measured at standard temperature and pressure, 0°C and 1 atm) is required to form a monolayer upon condensation. Calculate the surface area of the catalyst in units of m^2/kg. (Take the area covered by a nitrogen molecule as 0.162 nm^2 and recall that, for an ideal gas, $pV = nRT$, where n is the number of moles of the gas.)

2.50. Repeat Problem 2.49 for a highly porous silica gel that has a volume of 1.16×10^7 mm^3 of N$_2$ gas [at standard temperature and pressure (STP)] condensed to form a monolayer.

2.51. Small-diameter noble gas atoms, such as helium, can dissolve in the relatively open network structure of silicate glasses. (See Figure 1.8b for a schematic of glass structure.) The secondary bonding of helium in vitreous silica is represented by a heat of solution, ΔH_s, of −3.96 kJ/mol. The relationship between solubility, S, and the heat of solution is

$$S = S_0 e^{-\Delta H_s/(RT)},$$

where S_0 is a constant, R is the gas constant, and T is the absolute temperature (in K). If the solubility of helium in vitreous silica is 5.51×10^{23} atoms/(m$^3 \cdot$ atm) at 25°C, calculate the solubility at 200°C.

2.52. Due to its larger atomic diameter, neon has a higher heat of solution in vitreous silica than helium. If the heat of solution of neon in vitreous silica is −6.70 kJ/mol and the solubility at 25°C is 9.07×10^{23} atoms/(m$^3 \cdot$ atm), calculate the solubility at 200°C. (See Problem 2.51.)

CHAPTER 3
Crystalline Structure—Perfection

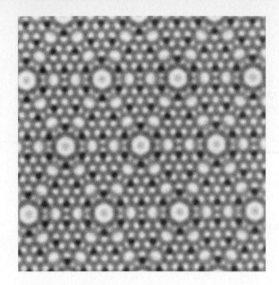

This high-resolution transmission electron micrograph illustrates the elegant beauty of the arrangement of atoms in a silicon crystal.

With the categories of engineering materials firmly established, we can now begin characterizing these materials. We will begin with atomic-scale structure, which for most engineering materials is crystalline; that is, the atoms of the material are arranged in a regular and repeating manner.

Common to all crystalline materials are the fundamentals of crystal geometry. We must identify the seven crystal systems and the 14 crystal lattices. Each of the thousands of crystal structures found in natural and synthetic materials can be placed within these few systems and lattices.

The crystalline structures of most metals belong to one of three relatively simple types. Ceramic compounds, which have a wide variety of chemical compositions, exhibit a similarly wide variety of crystalline structures. Some are relatively simple, but many, such as the silicates, are quite complex. Glass is noncrystalline, and its structure and the nature of noncrystalline materials are discussed in Chapter 4. Polymers share two features with ceramics and glasses. First, their crystalline structures are relatively complex. Second, because of this complexity, the material is not easily crystallized, and common polymers may have as much as 50% to 100% of their volume noncrystalline. Elemental semiconductors, such as silicon, exhibit a characteristic structure (diamond cubic), whereas semiconducting compounds have structures similar to some of the simpler ceramic compounds.

Within a given structure, we must know how to describe atom positions, crystal directions, and crystal planes. With these quantitative ground rules in hand, we conclude this chapter with a brief introduction to x-ray diffraction, the standard experimental tool for determining crystal structure.

3.1 | Seven Systems and Fourteen Lattices

The central feature of crystalline structure is that it is regular and repeating. In order to quantify this repetition, we must determine which structural unit is being repeated. Actually, any crystalline structure could be described as a pattern formed

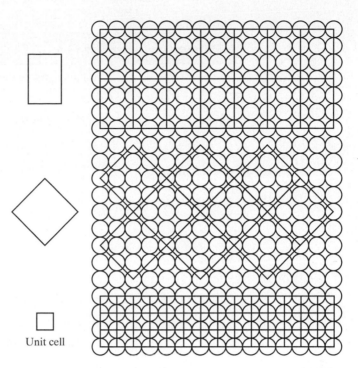

FIGURE 3.1 *Various structural units that describe the schematic crystalline structure. The simplest structural unit is the unit cell.*

Unit cell

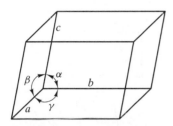

FIGURE 3.2 *Geometry of a general unit cell.*

by repeating various structural units (Figure 3.1). As a practical matter, there will generally be a simplest choice to serve as a representative structural unit. Such a choice is referred to as a **unit cell**. The geometry of a general unit cell is shown in Figure 3.2. The length of unit-cell edges and the angles between crystallographic axes are referred to as **lattice constants**, or **lattice parameters**. The key feature of the unit cell is that it contains a full description of the structure as a whole because the complete structure can be generated by the repeated stacking of adjacent unit cells face to face throughout three-dimensional space.

The description of crystal structures by means of unit cells has an important advantage. All possible structures reduce to a small number of basic unit-cell geometries, which is demonstrated in two ways. First, there are only seven unique unit-cell shapes that can be stacked together to fill three-dimensional space. These are the seven **crystal systems** defined and illustrated in Table 3.1. Second, we must consider how atoms (viewed as hard spheres) can be stacked together within a given unit cell. To do this in a general way, we begin by considering **lattice points**, theoretical points arranged periodically in three-dimensional space, rather than actual atoms or spheres. Again, there are a limited number of possibilities, referred to as the 14 **Bravais*** **lattices**, defined in Table 3.2. Periodic stacking of unit cells from Table 3.2 generates **point lattices**, arrays of points with identical

---

*Auguste Bravais (1811–1863), French crystallographer, was productive in an unusually broad range of areas, including botany, astronomy, and physics. However, it is his derivation of the 14 possible arrangements of points in space that is best remembered. This achievement provided the foundation for our current understanding of the atomic structure of crystals.

TABLE 3.1

The Seven Crystal Systems		
System	**Axial lengths and angles[a]**	**Unit cell geometry**
Cubic	$a = b = c, \alpha = \beta = \gamma = 90°$	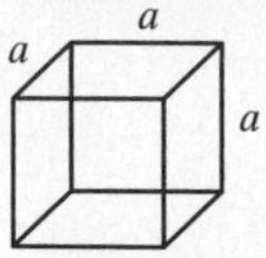
Tetragonal	$a = b \neq c, \alpha = \beta = \gamma = 90°$	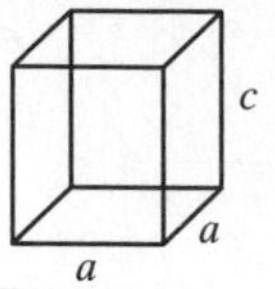
Orthorhombic	$a \neq b \neq c, \alpha = \beta = \gamma = 90°$	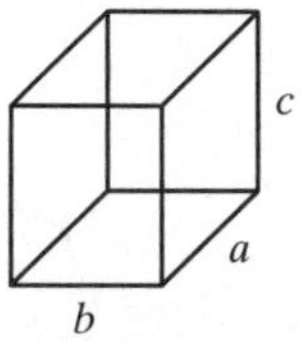
Rhombohedral	$a = b = c, \alpha = \beta = \gamma \neq 90°$	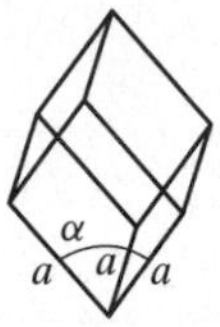
Hexagonal	$a = b \neq c, \alpha = \beta = 90°, \gamma = 120°$	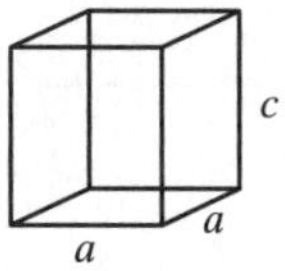
Monoclinic	$a \neq b \neq c, \alpha = \gamma = 90° \neq \beta$	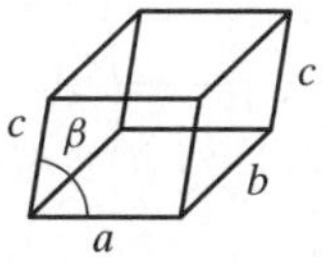
Triclinic	$a \neq b \neq c, \alpha \neq \beta \neq \gamma \neq 90°$	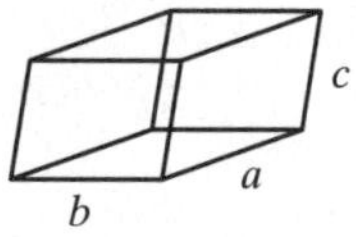

[a]The lattice parameters a, b, and c are unit-cell edge lengths. The lattice parameters α, β, and γ are angles between adjacent unit-cell axes, where α is the angle viewed along the a axis (i.e., the angle between the b and c axes). The inequality sign ($\neq$) means that equality is not required. Accidental equality occasionally occurs in some structures.

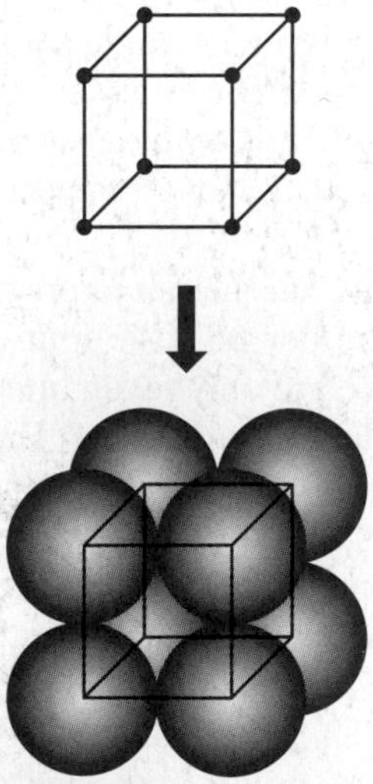

FIGURE 3.3 *The simple cubic lattice becomes the simple cubic crystal structure when an atom is placed on each lattice point.*

surroundings in three-dimensional space. These lattices are skeletons upon which crystal structures are built by placing atoms or groups of atoms on or near the lattice points. Figure 3.3 shows the simplest possibility, with one atom centered on each lattice point. Some of the simple metal structures are of this type. However, a very large number of actual crystal structures is known to exist. Most of these

TABLE 3.2

The 14 Crystal (Bravais) Lattices

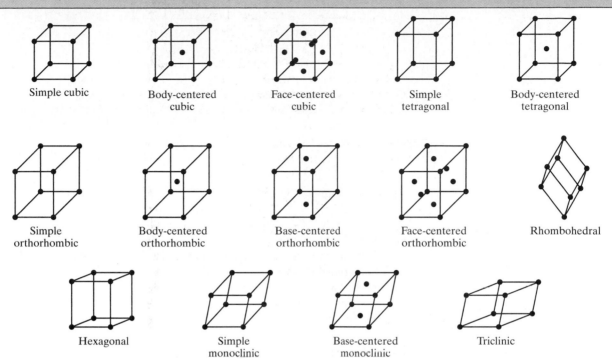

Simple cubic   Body-centered cubic   Face-centered cubic   Simple tetragonal   Body-centered tetragonal

Simple orthorhombic   Body-centered orthorhombic   Base-centered orthorhombic   Face-centered orthorhombic   Rhombohedral

Hexagonal   Simple monoclinic   Base-centered monoclinic   Triclinic

structures result from having more than one atom associated with a given lattice point. We shall find many examples in the crystal structures of common ceramics and polymers (Sections 3.3 and 3.4).

EXAMPLE 3.1

Sketch the five point lattices for two-dimensional crystal structures.

SOLUTION
Unit-cell geometries are

   i. Simple square
  ii. Simple rectangle
 iii. Area-centered rectangle (or rhombus)
 iv. Parallelogram
  v. Area-centered hexagon

Note. It is a useful exercise to construct other possible geometries that must be equivalent to these five basic types. For example, an area-centered square can be resolved into a simple square lattice (inclined at 45°).

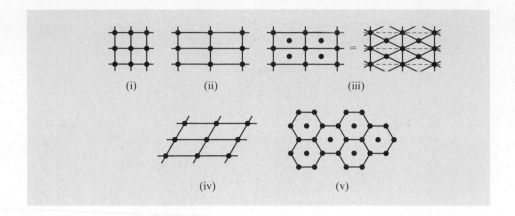

(i)   (ii)   (iii)

(iv)   (v)

PRACTICE PROBLEM 3.1

The note in Example 3.1 states that an area-centered square lattice can be resolved into a simple square lattice. Sketch this equivalence. **(Note that the solutions to all practice problems are provided on the related Web site.)**

3.2 | Metal Structures

With the structural ground rules behind us, we can now list the main crystal structures associated with important engineering materials. For our first group, the metals, this list is fairly simple. As we see from an inspection of Appendix 1, most elemental metals at room temperature are found in one of three crystal structures.

Figure 3.4 shows the **body-centered cubic** (bcc) structure, which is the body-centered cubic Bravais lattice with one atom centered on each lattice point. There is one atom at the center of the unit cell and one-eighth atom at each of eight unit-cell corners. (Each corner atom is shared by eight adjacent unit cells.) Thus, there are two atoms in each bcc unit cell. The **atomic packing factor** (APF) for this structure is 0.68 and represents the fraction of the unit-cell volume occupied by the two atoms. Typical metals with this structure include α-Fe (the form stable at room temperature), V, Cr, Mo, and W. An alloy in which one of these metals is the predominant constituent will tend to have this structure also. However, the presence of alloying elements diminishes crystalline perfection, a topic that will be discussed in Chapter 4.

Figure 3.5 shows the **face-centered cubic** (fcc) structure, which is the fcc Bravais lattice with one atom per lattice point. There is one-half atom (i.e., one atom shared between two unit cells) in the center of each unit-cell face and one-eighth atom at each unit-cell corner, for a total of four atoms in each fcc unit cell. The atomic packing factor for this structure is 0.74, a value slightly higher than the 0.68 found for bcc metals. In fact, an APF of 0.74 is the highest value possible for filling space by stacking equal-sized hard spheres. For this reason, the fcc structure is sometimes referred to as **cubic close packed** (ccp). Typical metals with the fcc structure include γ-Fe (stable from 912 to 1,394°C), Al, Ni, Cu, Ag, Pt, and Au.

A complete gallery of the computer-generated images used in this chapter is available on the related Web site.

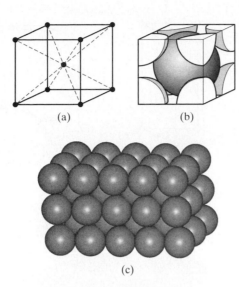

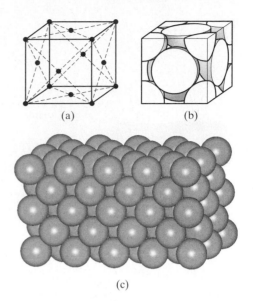

Structure: body-centered cubic (bcc)
Bravais lattice: bcc
Atoms/unit cell: $1 + 8 \times \frac{1}{8} = 2$
Typical metals: α-Fe, V, Cr, Mo, and W

Structure: face-centered cubic (fcc)
Bravais lattice: fcc
Atoms/unit cell: $6 \times \frac{1}{2} + 8 \times \frac{1}{8} = 4$
Typical metals: γ-Fe, Al, Ni, Cu, Ag, Pt, and Au

FIGURE 3.4 *Body-centered cubic (bcc) structure for metals showing (a) the arrangement of lattice points for a unit cell, (b) the actual packing of atoms (represented as hard spheres) within the unit cell, and (c) the repeating bcc structure, equivalent to many adjacent unit cells. [Part (c) courtesy of Accelrys, Inc.]*

FIGURE 3.5 *Face-centered cubic (fcc) structure for metals showing (a) the arrangement of lattice points for a unit cell, (b) the actual packing of atoms within the unit cell, and (c) the repeating fcc structure, equivalent to many adjacent unit cells. [Part (c) courtesy of Accelrys, Inc.]*

The **hexagonal close-packed** (hcp) structure (Figure 3.6) is our first encounter with a structure more complicated than its Bravais lattice (hexagonal). There are two atoms associated with each Bravais lattice point. There is one atom centered within the unit cell and various fractional atoms at the unit-cell corners (four $\frac{1}{6}$ atoms and four $\frac{1}{12}$ atoms), for a total of two atoms per unit cell. As the *close-packed* name implies, this structure is as efficient in packing spheres as is the fcc structure. Both hcp and fcc structures have atomic packing factors of 0.74, which raises two questions: (1) In what other ways are the fcc and hcp structures alike? and (2) How do they differ? The answers to both questions can be found in Figure 3.7. The two structures are each regular stackings of close-packed planes. The difference lies in the sequence of packing of these layers. The fcc arrangement is such that the fourth close-packed layer lies precisely above the first one. In the hcp structure, the third close-packed layer lies precisely above the first. The fcc stacking is re-ferred to as an *ABCABC . . . sequence*, and the hcp stacking is referred to as an *ABAB . . . sequence*. This subtle difference can lead to significant differences in material properties, as we shall see in Section 6.3. Typical metals with the hcp structure include Be, Mg, α-Ti, Zn, and Zr.

Although the majority of elemental metals falls within one of the three structural groups just discussed, several display less common structures. We shall

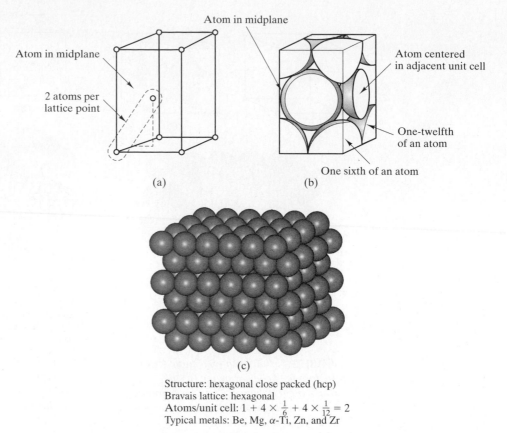

Structure: hexagonal close packed (hcp)
Bravais lattice: hexagonal
Atoms/unit cell: $1 + 4 \times \frac{1}{6} + 4 \times \frac{1}{12} = 2$
Typical metals: Be, Mg, α-Ti, Zn, and Zr

FIGURE 3.6 *Hexagonal close-packed (hcp) structure for metals showing (a) the arrangement of atom centers relative to lattice points for a unit cell. There are two atoms per lattice point (note the outlined example). (b) The actual packing of atoms within the unit cell. Note that the atom in the midplane extends beyond the unit-cell boundaries. (c) The repeating hcp structure, equivalent to many adjacent unit cells. [Part (c) courtesy of Accelrys, Inc.]*

not dwell on these cases, which can be found from a careful inspection of Appendix 1.

In the course of analyzing the metallic structures introduced in this section, we shall frequently encounter the useful relationships between unit-cell size and atomic radius given in Table 3.3. Our initial discovery of the utility of these relationships is found in the following examples and practice problems.

TABLE 3.3

Relationship between Unit-Cell Size (Edge Length) and Atomic Radius for the Common Metallic Structures	
Crystal structure	**Relationship between edge length, a, and atomic radius, r**
Body-centered cubic (bcc)	$a = 4r/\sqrt{3}$
Face-centered cubic (fcc)	$a = 4r/\sqrt{2}$
Hexagonal close packed (hcp)	$a = 2r$

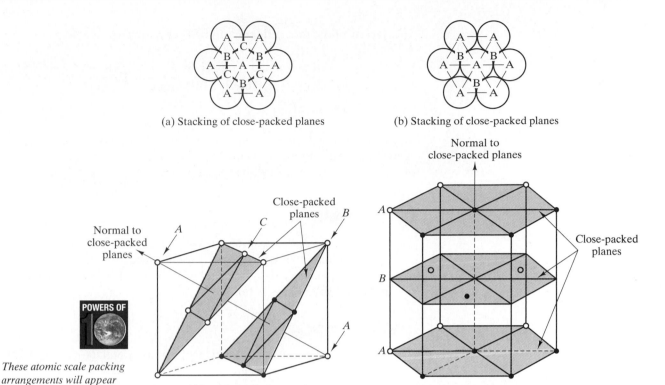

(a) Stacking of close-packed planes   (b) Stacking of close-packed planes

(c) Face-centered cubic   (d) Hexagonal close packed

These atomic scale packing arrangements will appear again in Table 6.8, central to our understanding of the plastic deformation of metals.

FIGURE 3.7 *Comparison of the fcc and hcp structures. They are each efficient stackings of close-packed planes. The difference between the two structures is the different stacking sequences. (From B. D. Cullity and S. R. Stock,* Elements of X-Ray Diffraction, *3rd ed., Prentice Hall, Upper Saddle River, NJ, 2001.)*

EXAMPLE 3.2

Using the data of Appendixes 1 and 2, calculate the density of copper.

SOLUTION

Appendix 1 shows copper to be an fcc metal. The length, l, of a face diagonal in the unit cell (Figure 3.5) is

$$l = 4r_{\text{Cu atom}} = \sqrt{2}a$$

or

$$a = \frac{4}{\sqrt{2}} r_{\text{Cu atom}},$$

as given in Table 3.3. From the data of Appendix 2,

$$a = \frac{4}{\sqrt{2}}(0.128 \text{ nm}) = 0.362 \text{ nm}.$$

The density of the unit cell (containing four atoms) is

$$\rho = \frac{4 \text{ atoms}}{(0.362 \text{ nm})^3} \times \frac{63.55 \text{ g}}{0.6023 \times 10^{24} \text{ atoms}} \times \left(\frac{10^7 \text{ nm}}{\text{cm}}\right)^3$$

$$= 8.89 \text{ g/cm}^3.$$

This result can be compared with the tabulated value of 8.93 g/cm^3 in Appendix 1. The difference would be eliminated if a more precise value of $r_{Cu \text{ atom}}$ were used (i.e., with at least one more significant figure).

PRACTICE PROBLEM 3.2

In Example 3.2, the relationship between lattice parameter, a, and atomic radius, r, for an fcc metal is found to be $a = (4/\sqrt{2})r$, as given in Table 3.3. Derive the similar relationships in Table 3.3 for **(a)** a bcc metal and **(b)** an hcp metal.

PRACTICE PROBLEM 3.3

Calculate the density of α-Fe, which is a bcc metal. (*Caution:* A different relationship between lattice parameter, a, and atomic radius, r, applies to this different crystal structure. See Practice Problem 3.2 and Table 3.3.)

3.3 | Ceramic Structures

The wide variety of chemical compositions of ceramics is reflected in their crystalline structures. We cannot begin to give an exhaustive list of ceramic structures, but we can give a systematic list of some of the most important and representative ones. Even this list becomes rather long, so most structures will be described briefly. It is worth noting that many of these ceramic structures also describe intermetallic compounds. Also, we can define an **ionic packing factor** (IPF) for these ceramic structures, similar to our definition of the APF for metallic structures. The IPF is the fraction of the unit-cell volume occupied by the various cations and anions.

Let us begin with the ceramics with the simplest chemical formula, MX, where M is a metallic element and X is a nonmetallic element. Our first example is the **cesium chloride** (CsCl) structure shown in Figure 3.8. At first glance, we might want to call this a body-centered structure because of its similarity in appearance to the structure shown in Figure 3.4. In fact, the CsCl structure is built on the simple cubic Bravais lattice with two ions (one Cs^+ and one Cl^-) associated with each lattice point. There are two ions (one Cs^+ and one Cl^-) per unit cell.

Although CsCl is a useful example of a compound structure, it does not represent any commercially important ceramics. By contrast, the **sodium chloride** (NaCl) structure shown in Figure 3.9 is shared by many important ceramic materials. This structure can be viewed as the intertwining of two fcc structures, one of sodium ions and one of chlorine ions. Consistent with our treatment of the hcp and CsCl structures, the NaCl structure can be described as having an fcc Bravais

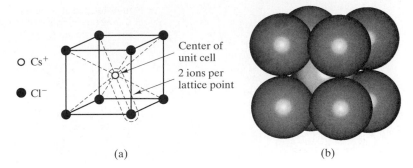

Structure: CsCl type
Bravais lattice: simple cubic
Ions/unit cell: $1Cs^+ + 1Cl^-$

FIGURE 3.8 *Cesium chloride (CsCl) unit cell showing (a) ion positions and the two ions per lattice point and (b) full-size ions. Note that the Cs^+-Cl^- pair associated with a given lattice point is not a molecule because the ionic bonding is nondirectional and because a given Cs^+ is equally bonded to eight adjacent Cl^-, and vice versa. [Part (b) courtesy of Accelrys, Inc.]*

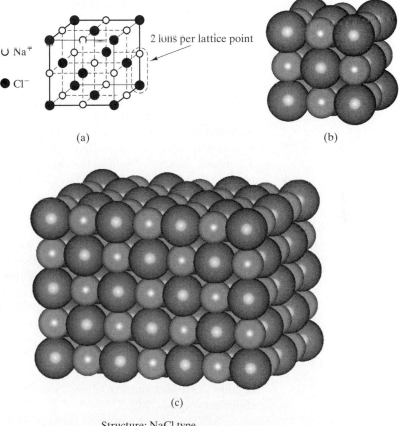

Structure: NaCl type
Bravais lattice: fcc
Ions/unit cell: $4Na^+ + 4Cl^-$
Typical ceramics: MgO, CaO, FeO, and NiO

FIGURE 3.9 *Sodium chloride (NaCl) structure showing (a) ion positions in a unit cell, (b) full-size ions, and (c) many adjacent unit cells. [Parts (b) and (c) courtesy of Accelrys, Inc.]*

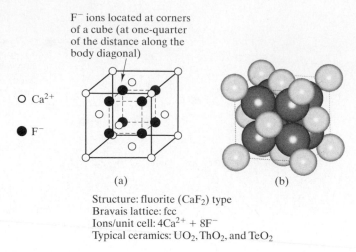

F⁻ ions located at corners
of a cube (at one-quarter
of the distance along the
body diagonal)

○ Ca^{2+}

● F^-

(a)   (b)

Structure: fluorite (CaF_2) type
Bravais lattice: fcc
Ions/unit cell: $4Ca^{2+} + 8F^-$
Typical ceramics: UO_2, ThO_2, and TeO_2

FIGURE 3.10 *Fluorite (CaF_2) unit cell showing (a) ion positions and (b) full-size ions. [Part (b) courtesy of Accelrys, Inc.]*

lattice with two ions (1 Na^+ and 1 Cl^-) associated with each lattice point. There are eight ions (4 Na^+ plus 4 Cl^-) per unit cell. Some of the important ceramic oxides with this structure are MgO, CaO, FeO, and NiO.

The chemical formula MX_2 includes a number of important ceramic structures. Figure 3.10 shows the **fluorite** (CaF_2) structure, which is built on an fcc Bravais lattice with three ions (1 Ca^{2+} and 2 F^-) associated with each lattice point. There are 12 ions (4 Ca^{2+} and 8 F^-) per unit cell. Typical ceramics with this structure are UO_2, ThO_2, and TeO_2. There is an unoccupied volume near the center of the fluorite unit cell that plays an important role in nuclear-materials technology. Uranium dioxide (UO_2) is a reactor fuel that can accommodate fission products such as helium gas without troublesome swelling. The helium atoms are accommodated in the open regions of the fluorite unit cells.

Included in the MX_2 category is perhaps the most important ceramic compound, **silica** (SiO_2), which is widely available in raw materials in the earth's crust. Silica, alone and in chemical combination with other ceramic oxides (forming silicates), represents a large fraction of the ceramic materials available to engineers. For this reason, the structure of SiO_2 is important. Unfortunately, this structure is not simple. In fact, there is not a single structure to describe it, but instead there are many (under different conditions of temperature and pressure). For a representative example, Figure 3.11 shows the **cristobalite** (SiO_2) structure. Cristobalite is built on an fcc Bravais lattice with six ions (2 Si^{4+} and 4 O^{2-}) associated with each lattice point. There are 24 ions (8 Si^{4+} plus 16 O^{2-}) per unit cell. In spite of the large unit cell needed to describe this structure, it is perhaps the simplest of the various crystallographic forms of SiO_2. The general feature of all SiO_2 structures is the same—a continuously connected network of SiO_4^{4-} tetrahedra (see Section 2.3). The sharing of O^{2-} ions by adjacent tetrahedra gives the overall chemical formula SiO_2.

We have already noticed (in Section 3.2) that iron, Fe, had different crystal structures stable in different temperature ranges. The same is true for silica, SiO_2. Although the basic SiO_4^{4-} tetrahedra are present in all SiO_2 crystal structures, the arrangement of connected tetrahedra changes. The equilibrium structures of SiO_2 from room temperature to its melting point are summarized in Figure 3.12. Caution must always be exercised in using materials with transformations of these types.

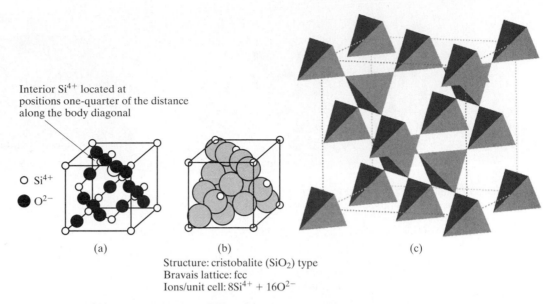

Interior Si^{4+} located at
positions one-quarter of the distance
along the body diagonal

○ Si^{4+}
● O^{2-}

(a)   (b)   (c)

Structure: cristobalite (SiO_2) type
Bravais lattice: fcc
Ions/unit cell: $8Si^{4+} + 16O^{2-}$

FIGURE 3.11 *The cristobalite (SiO_2) unit cell showing (a) ion positions, (b) full-size ions, and (c) the connectivity of SiO_4^{4-} tetrahedra. In the schematic, each tetrahedron has a Si^{4+} at its center. In addition, an O^{2-} would be at each corner of each tetrahedron and is shared with an adjacent tetrahedron. [Part (c) courtesy of Accelrys, Inc.]*

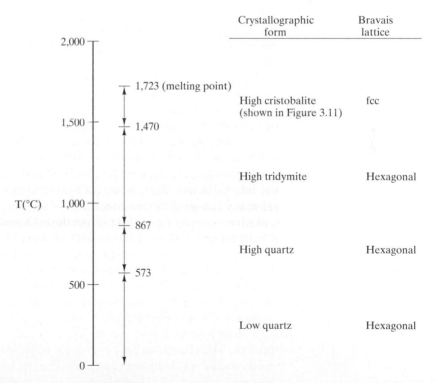

FIGURE 3.12 *Many crystallographic forms of SiO_2 are stable as they are heated from room temperature to melting temperature. Each form represents a different way to connect adjacent SiO_4^{4-} tetrahedra.*

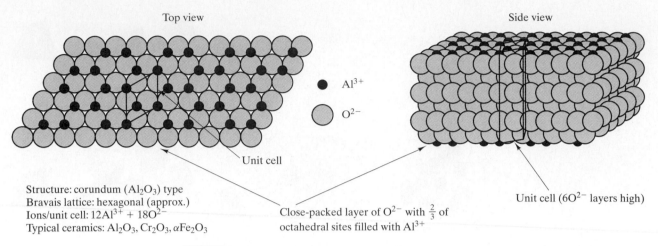

Top view   Side view

● Al^{3+}

○ O^{2-}

Unit cell

Structure: corundum (Al$_2$O$_3$) type
Bravais lattice: hexagonal (approx.)
Ions/unit cell: 12Al^{3+} + 18O^{2-}
Typical ceramics: Al$_2$O$_3$, Cr$_2$O$_3$, αFe$_2$O$_3$

Close-packed layer of O^{2-} with $\frac{2}{3}$ of
octahedral sites filled with Al^{3+}

Unit cell (6O^{2-} layers high)

FIGURE 3.13 *The corundum (Al$_2$O$_3$) unit cell is shown superimposed on the repeated stacking of layers of close-packed O^{2-} ions. The Al^{3+} ions fill two-thirds of the small (octahedral) interstices between adjacent layers.*

Even the relatively subtle low to high quartz transformation can cause catastrophic structural damage when a silica ceramic is heated or cooled through the vicinity of 573°C.

The chemical formula M$_2$X$_3$ includes the important **corundum** (Al$_2$O$_3$) structure shown in Figure 3.13. This structure is in a rhombohedral Bravais lattice, but it closely approximates a hexagonal lattice. There are 30 ions per lattice site (and per unit cell). The Al$_2$O$_3$ formula requires that these 30 ions be divided as 12 Al^{3+} and 18 O^{2-}. One can visualize this seemingly complicated structure as being similar to the hcp structure described in Section 3.2. The Al$_2$O$_3$ structure closely approximates close-packed O^{2-} sheets with two-thirds of the small interstices between sheets filled with Al^{3+}. Both Cr$_2$O$_3$ and α-Fe$_2$O$_3$ have the corundum structure.

In discussing the complexity of the SiO$_2$ structures, we mentioned the importance of the many silicate materials resulting from the chemical reaction of SiO$_2$ with other ceramic oxides. The general nature of silicate structures is that the additional oxides tend to break up the continuity of the SiO$_4^{4-}$ tetrahedra connections. The remaining connectedness of tetrahedra may be in the form of silicate chains or sheets. One relatively simple example is illustrated in Figure 3.14, which shows the **kaolinite** structure. Kaolinite [2(OH)$_4$Al$_2$Si$_2$O$_5$] is a hydrated aluminosilicate and a good example of a clay mineral. The structure is typical of sheet silicates. It is built on the triclinic Bravais lattice with two kaolinite "molecules" per unit cell. On a microscopic scale, we observe many clay minerals to have a platelike or flaky structure (see Figure 3.15), a direct manifestation of crystal structures such as those shown in Figure 3.14.

In this section, we have surveyed crystal structures for various ceramic compounds. The structures have generally been increasingly complex as we considered increasingly complex chemistry. The contrast between CsCl (Figure 3.8) and kaolinite (Figure 3.14) is striking.

Before leaving ceramics, it is appropriate to look at some important materials that are exceptions to our general description of ceramics as compounds.

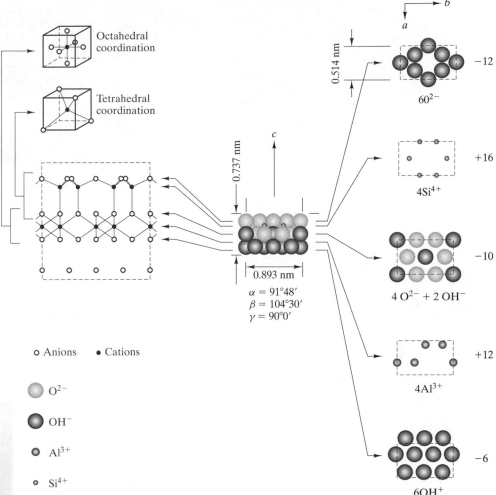

FIGURE 3.14 *Exploded view of the kaolinite unit cell, $2(OH)_4Al_2Si_2O_5$. (From F. H. Norton,* Elements of Ceramics, *2nd ed., Addison-Wesley Publishing Co., Inc., Reading, MA, 1974.)*

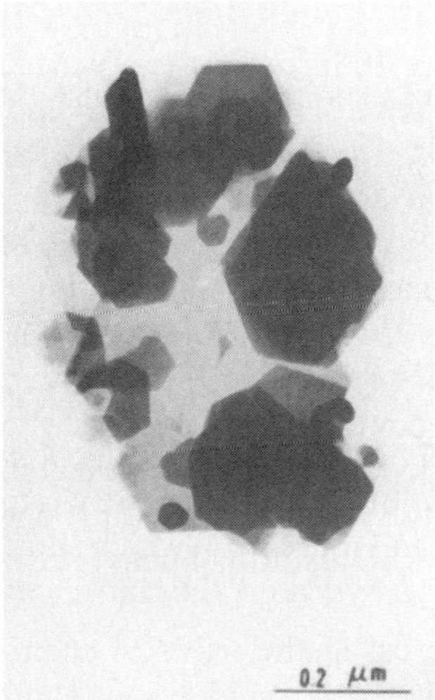

FIGURE 3.15 *Transmission electron micrograph of the structure of clay platelets. This microscopic-scale structure is a manifestation of the layered crystal structure shown in Figure 3.14. (Courtesy of I. A. Aksay.)*

First, Figure 3.16 shows the layered crystal structure of graphite, the stable room-temperature form of carbon. Although monatomic, graphite is much more ceramiclike than metallic. The hexagonal rings of carbon atoms are strongly bonded by covalent bonds. The bonds between layers are, however, of the van der Waals type (Section 2.5), accounting for graphite's friable nature and application as a useful "dry" lubricant. It is interesting to contrast the graphite structure with the high-pressure stabilized form, diamond cubic, which plays such an important role in solid-state technology because semiconductor silicon has this structure (see Figure 3.20).

Even more intriguing is a comparison of both graphite and diamond structures with an alternate form of carbon that was recently discovered as a byproduct of research in astrochemistry. Figure 3.17a illustrates the structure of a C_{60} molecule. This unique structure was discovered during experiments on the laser

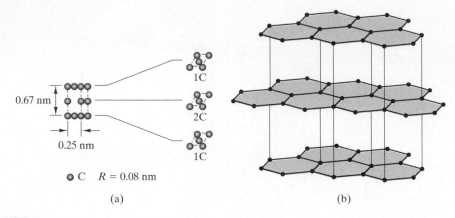

(a) 0.67 nm 0.25 nm 1C 2C 1C ● C $R = 0.08$ nm

(b)

FIGURE 3.16 *(a) An exploded view of the graphite (C) unit cell. (From F. H. Norton,* Elements of Ceramics, *2nd ed., Addison-Wesley Publishing Co., Inc., Reading, MA, 1974.) (b) A schematic of the nature of graphite's layered structure. (From W. D. Kingery, H. K. Bowen, and D. R. Uhlmann,* Introduction to Ceramics, *2nd ed., John Wiley & Sons, Inc., NY, 1976.)*

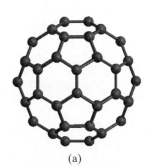

(a)

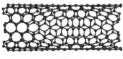

(b)

FIGURE 3.17 *(a) C_{60} molecule, or buckyball. (b) Cylindrical array of hexagonal rings of carbon atoms, or buckytube. (Courtesy of Accelrys, Inc.)*

POWERS OF

The buckyball is helpful in appreciating the size of the "nanoscale." The ratio of the diameter of the earth to a soccer ball (that has the same geometrical distribution of pentagons and hexagons as the buckyball) is the same as the ratio of the diameter of the soccer ball to the buckyball (about 10^8 in both cases!).

vaporization of carbon in a carrier gas such as helium. The experiments had intended to simulate the synthesis of carbon chains in carbon stars. The result, however, was a molecular-scale version of the geodesic dome, leading to this material's being named **buckminsterfullerene**,* or **fullerene**, in honor of the inventor of that architectural structure. A close inspection of the structure shown in Figure 3.17a indicates that the nearly spherical molecule is, in fact, a polyhedron composed of 5- and 6-sided faces.

The uniform distribution of 12 pentagons among 20 hexagons is precisely the form of a soccer ball, leading to the nickname for the structure as a **buckyball**. It is the presence of 5-membered rings that gives the positive curvature to the surface of the buckyball, in contrast to the flat, sheetlike structure of 6-membered rings in graphite (Figure 3.16b). Subsequent research has led to the synthesis of a wide variety of structures for a wide range of fullerenes. Buckyballs have been synthesized with the formula C_n, where n can take on various large, even values, such as 240 and 540. In each case, the structure consists of 12 uniformly distributed pentagons connecting an array of hexagons. Although pentagons are necessary to give the approximately spherical curvature of the buckyballs, extensive research on these unique materials led to the realization that cylindrical curvature can result from simply rolling the hexagonal graphite sheets. A resulting **buckytube** is shown in Figure 3.17b.

These materials have stimulated enormous interest in the fields of chemistry and physics, as well as the field of materials science and engineering. Their molecular structure is clearly interesting, but in addition they have unique chemical and physical properties (e.g., individual C_n buckyballs are unique, passive

---

*Richard Buckminster Fuller (1895–1983), American architect and inventor, was one of the most colorful and famous personalities of the 20th century. His creative discussions of a wide range of topics from the arts to the sciences (including frequent references to future trends) helped to establish his fame. In fact, his charismatic personality became as celebrated as his unique inventions of various architectural forms and engineering designs.

SECTION 3.3 Ceramic Structures **73**

surfaces on an nm-scale). Similarly, buckytubes hold the theoretical promise of being the highest-strength reinforcing fibers available for the advanced composites discussed in Chapter 12. Finally, we must acknowledge that the discrete molecular structures shown in Figure 3.17 are interesting but are not associated with a long-range crystallographic structure. Nonetheless, buckyballs and buckytubes have an intriguing set of atomic-scale structures that could lead to potentially important applications in materials technology.

EXAMPLE 3.3

Calculate the IPF of MgO, which has the NaCl structure (Figure 3.9).

SOLUTION
Taking $a = 2r_{Mg^{2+}} + 2r_{O^{2-}}$ and the data of Appendix 2, we have

$$a = 2(0.078 \text{ nm}) + 2(0.132 \text{ nm}) = 0.420 \text{ nm}.$$

Then,

$$V_{unit\ cell} = a^3 = (0.420 \text{ nm})^3 = 0.0741 \text{ nm}^3.$$

There are four Mg^{2+} ions and four O^{2-} ions per unit cell, giving a total ionic volume of

$$4 \times \frac{4}{3}\pi r_{Mg^{2+}}^3 + 4 \times \frac{4}{3}\pi r_{O^{2-}}^3$$

$$= \frac{16\pi}{3}[(0.078 \text{ nm})^3 + (0.132 \text{ nm})^3]$$

$$= 0.0465 \text{ nm}^3.$$

The ionic packing factor is then

$$IPF = \frac{0.0465 \text{ nm}^3}{0.0741 \text{ nm}^3} = 0.627.$$

EXAMPLE 3.4

Using data from Appendixes 1 and 2, calculate the density of MgO.

SOLUTION
From Example 3.3, $a = 0.420$ nm, which gave a unit-cell volume of 0.0741 nm^3. The density of the unit cell is

$$\rho = \frac{[4(24.31 \text{ g}) + 4(16.00 \text{ g})]/(0.6023 \times 10^{24})}{0.0741 \text{ nm}^3} \times \left(\frac{10^7 \text{ nm}}{\text{cm}}\right)^3$$

$$= 3.61 \text{ g/cm}^3.$$

> **PRACTICE PROBLEM 3.4**
>
> Calculate the IPF of **(a)** CaO, **(b)** FeO, and **(c)** NiO. All of these compounds share the NaCl-type structure. **(d)** Is there a unique IPF value for the NaCl-type structure? Explain. (See Example 3.3.)

> **PRACTICE PROBLEM 3.5**
>
> Calculate the density of CaO. (See Example 3.4.)

3.4 | Polymeric Structures

In Chapters 1 and 2, we defined the polymers category of materials by the chain-like structure of long polymeric molecules (e.g., Figure 2.15). Compared with the stacking of individual atoms and ions in metals and ceramics, the arrangement of these long molecules into a regular and repeating pattern is difficult. As a result, most commercial plastics are to a large degree noncrystalline. In those regions of the microstructure that are crystalline, the structure tends to be quite complex. The complexity of the unit cells of common polymers is generally beyond the scope of this text, but one relatively simple example will be shown.

Polyethylene, $+C_2H_4+_n$, is chemically quite simple. However, the relatively elaborate way in which the long-chain molecule folds back and forth on itself is illustrated in Figures 3.18 and 3.19. Figure 3.18 shows an orthorhombic unit cell, a common crystal system for polymeric crystals. For metals and ceramics, knowledge of the unit-cell structure implies knowledge of the crystal structure over a large volume. For polymers, we must be more cautious. Single crystals of polyethylene are difficult to grow. When produced (by cooling a dilute solution), they tend

FIGURE 3.18 *Arrangement of polymeric chains in the unit cell of polyethylene. The dark spheres are carbon atoms, and the light spheres are hydrogen atoms. The unit-cell dimensions are 0.255 nm × 0.494 nm × 0.741 nm. (Courtesy of Accelrys, Inc.)*

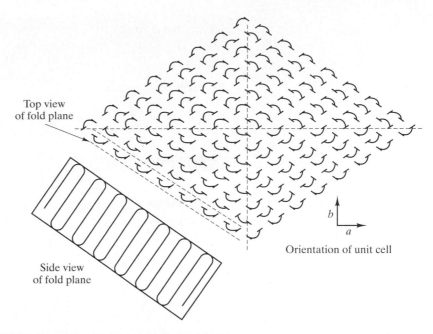

Top view
of fold plane

Side view
of fold plane

Orientation of unit cell

FIGURE 3.19 *Weaving-like pattern of folded polymeric chains that occurs in thin crystal platelets of polyethylene. (From D. J. Williams,* Polymer Science and Engineering, *Prentice Hall, Inc., Englewood Cliffs, NJ, 1971.)*

to be thin platelets, about 10 nm thick. Since polymer chains are generally several hundred nanometers long, the chains must be folded back and forth in a sort of atomic-scale weaving (as illustrated in Figure 3.19).

EXAMPLE 3.5

Calculate the number of C and H atoms in the polyethylene unit cell (Figure 3.18), given a density of 0.9979 g/cm^3.

SOLUTION

Figure 3.18 gives unit-cell dimensions that allow calculation of volume:

$$V = (0.741 \text{ nm})(0.494 \text{ nm})(0.255 \text{ nm}) = 0.0933 \text{ nm}^3.$$

There will be some multiple (n) of C_2H_4 units in the unit cell with atomic mass:

$$m = \frac{n[2(12.01) + 4(1.008)] \text{ g}}{0.6023 \times 10^{24}} = (4.66 \times 10^{-23} n) \text{ g}.$$

Therefore, the unit-cell density is

$$\rho = \frac{(4.66 \times 10^{-23} n) \text{ g}}{0.0933 \text{ nm}^3} \times \left(\frac{10^7 \text{ nm}}{\text{cm}}\right)^3 = 0.9979 \frac{\text{g}}{\text{cm}^3}.$$

Solving for n gives

$$n = 2.00.$$

As a result, there are

$$4 \ (= 2n) \ \text{C atoms} + 8 \ (= 4n) \ \text{H atoms per unit cell.}$$

PRACTICE PROBLEM 3.6

How many unit cells are contained in 1 kg of commercial polyethylene that is 50 vol % crystalline (balance amorphous) and that has an overall product density of 0.940 Mg/m^3? (See Example 3.5.)

3.5 | Semiconductor Structures

The technology developed by the semiconductor industry for growing single crystals has led to crystals of phenomenally high degrees of perfection. All crystal structures shown in this chapter imply structural perfection. However, all structures are subject to various imperfections, which will be discussed in Chapter 4. The "perfect" structures described in this section are approached in real materials more closely than in any other category.

A single structure dominates the semiconductor industry. The elemental semiconductors (Si, Ge, and gray Sn) share the **diamond cubic** structure shown in Figure 3.20. This structure is built on an fcc Bravais lattice with two atoms associated with each lattice point and eight atoms per unit cell. A key feature of this structure is that it accommodates the tetrahedral bonding configuration of these group IV A elements.

THE MATERIAL WORLD

Growing a (Nearly) Perfect Crystal

The technological and cultural revolution created by the many products based on the modern integrated circuit begins with single crystals of exceptionally high chemical purity and structural perfection. More than any other commercially produced materials, these crystals represent the ideal described in this chapter. The vast majority of integrated circuits are produced on thin slices (wafers) of **silicon** single crystals [see photo, which shows a state-of-the-art 300 mm (12 in.) diameter wafer]. The systematic procedures for producing fine-scale electrical circuits

on these wafers are discussed in some detail in Chapter 13.

As shown in the drawing, a large crystal of silicon is produced by pulling a small "seed" crystal up from a crucible containing molten silicon. The high melting point of silicon ($T_m = 1{,}414°C$) calls for a crucible made of high-purity SiO_2 glass. Heat is provided by radio-frequency (RF) inductive heating coils. The seed crystal is inserted into the melt and slowly withdrawn. Crystal growth occurs as the liquid silicon freezes next to the seed crystal, with

(Continued)

individual atoms stacking against those in the seed. Successive layers of atomic planes are added at the liquid–solid interface. The overall growth rate is approximately 10 μm/s. The resulting large crystals are sometimes called *ingots* or *boules*. This overall process is generally called the *Czochralski* or *Teal–Little technique*. As noted in Chapter 13, the economics of a circuit fabrication on single crystal wafers drives crystal growers to produce as large a diameter crystal as possible. The industry standard for crystal and corresponding wafer diameters has increased over the years to 200 mm and now to 300 mm.

The basic science and technology involved in growing single crystals as illustrated here have been joined by considerable "art" in adjusting the various specifics of the crystal-growing apparatus and procedures. The highly specialized process of growing these crystals is largely the focus of companies separate from those that fabricate circuits. The structural perfection provided by the Czochralski technique is joined with the ability to chemically purify the resulting crystal by the zone refining technique, as discussed in the feature box for Chapter 9 and in Chapter 13.

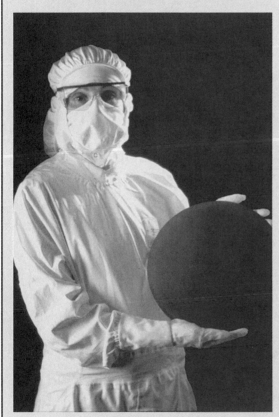

(Courtesy of SEMATECH.)

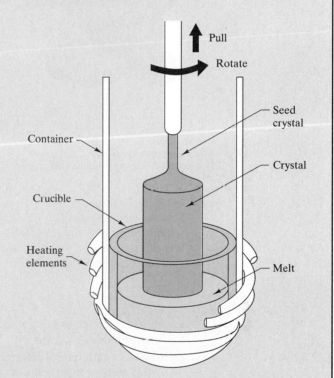

Schematic of growth of single crystals using the Czochralski technique. (From J. W. Mayer and S. S. Lau, Electronic Materials Science: For Integrated Circuits in Si and GaAs, *Macmillan Publishing Company, New York, 1990.)*

A small cluster of elements adjacent to group IV A forms semiconducting compounds, which tend to be MX-type compounds with combinations of atoms having an average valence of 4+. For example, GaAs combines the 3+ valence of gallium with the 5+ valence of arsenic, and CdS combines the 2+ valence of cadmium with the 6+ valence of sulfur. GaAs and CdS are examples of a **III–V compound** and a **II–VI compound**, respectively. Many of these simple MX compounds crystallize in a structure closely related to the diamond cubic.

Interior atoms located at
positions one-quarter of the
distance along the body diagonal

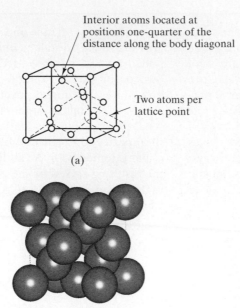

Two atoms per
lattice point

(a)

(b)

○ Zn^{2+} ● S^{2-}

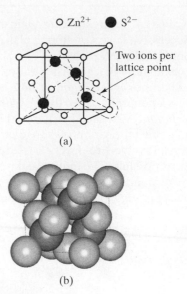

Two ions per
lattice point

(a)

(b)

Structure: diamond cubic
Bravais lattice: fcc
Atoms/unit cell: $4 + 6 \times \frac{1}{2} + 8 \times \frac{1}{8} = 8$
Typical semiconductors: Si, Ge, and gray Sn

FIGURE 3.20 *Diamond cubic unit cell showing (a) atom positions. There are two atoms per lattice point (note the outlined example). Each atom is tetrahedrally coordinated. (b) The actual packing of full-size atoms associated with the unit cell. [Part (b) courtesy of Accelrys, Inc.]*

Structure: Zinc blende (ZnS) type
Bravais lattice: fcc
Ions/unit cell: $4Zn^{2+} + 4S^{2-}$
Typical semiconductors:
  GaAs, AlP, InSb (III–V compounds),
  ZnS, ZnSe, CdS, HgTe (II–VI compounds)

FIGURE 3.21 *Zinc blende (ZnS) unit cell showing (a) ion positions. There are two ions per lattice point (note the outlined example). Compare this structure with the diamond cubic structure (Figure 3.20a). (b) The actual packing of full-size ions associated with the unit cell. [Part (b) courtesy of Accelrys, Inc.]*

Figure 3.21 shows the **zinc blende** (ZnS) structure, which is essentially the diamond cubic structure with Zn^{2+} and S^{2-} ions alternating in the atom positions. This is again the fcc Bravais lattice, but with two oppositely charged ions associated with each lattice site rather than two like atoms. There are eight ions (four Zn^{2+} and four S^{2-}) per unit cell. This structure is shared by both III–V compounds (e.g., GaAs, AlP, and InSb) and II–VI compounds (e.g., ZnSe, CdS, and HgTe).

EXAMPLE 3.6

Calculate the APF for the diamond cubic structure (Figure 3.20).

SOLUTION

Because of the tetrahedral bonding geometry of the diamond cubic structure, the atoms lie along body diagonals. Inspection of Figure 3.20 indicates that this orientation of atoms leads to the equality

$$2r_{Si} = \frac{1}{4}(\text{body diagonal}) = \frac{\sqrt{3}}{4}a$$

or

$$a = \frac{8}{\sqrt{3}} r_{Si}.$$

The unit-cell volume then, is,

$$V_{\text{unit cell}} = a^3 = (4.62)^3 r_{Si}^3 = 98.5 r_{Si}^3.$$

The volume of the eight Si atoms in the unit cell is

$$V_{\text{atoms}} = 8 \times \frac{4}{3}\pi r_{Si}^3 = 33.5 r_{Si}^3,$$

which gives an atomic packing factor of

$$APF = \frac{33.5 r_{Si}^3}{98.5 r_{Si}^3} = 0.340.$$

Note. This result represents a very open structure compared with the tightly packed metals' structures described in Section 3.2 (e.g., APF = 0.74 for fcc and hcp metals).

EXAMPLE 3.7

Using the data of Appendixes 1 and 2, calculate the density of silicon.

SOLUTION
From Example 3.6,

$$V_{\text{unit cell}} = 98.5 r_{Si}^3 = 98.5\,(0.117\text{ nm})^3$$
$$= 0.158\text{ nm}^3,$$

giving a density of

$$\rho = \frac{8\text{ atoms}}{0.158\text{ nm}^3} \times \frac{28.09\text{ g}}{0.6023 \times 10^{24}\text{ atoms}} \times \left(\frac{10^7\text{ nm}}{\text{cm}}\right)^3$$
$$= 2.36\text{ g/cm}^3.$$

As for previous calculations, a slight discrepancy between this result and Appendix 1 data (e.g., $\rho_{Si} = 2.33$ g/cm^3) is the result of not having another significant figure with the atomic radius data of Appendix 2.

PRACTICE PROBLEM 3.7

In Example 3.6, we find the atomic packing factor for silicon to be quite low compared with that of the common metal structures. Comment on the relationship between this characteristic and the nature of bonding in semiconductor silicon.

PRACTICE PROBLEM 3.8

Calculate the density of germanium using data from Appendixes 1 and 2. (See Example 3.7.)

3.6 | Lattice Positions, Directions, and Planes

There are a few basic rules for describing geometry in and around a unit cell. These rules and associated notations are used uniformly by crystallographers, geologists, materials scientists, and others who must deal with crystalline materials. What we are about to learn is, then, a vocabulary that allows us to communicate efficiently about crystalline structure. This vocabulary will prove to be most useful when we begin to deal with structure-sensitive properties later in the book.

Figure 3.22 illustrates the notation for describing **lattice positions** expressed as fractions or multiples of unit-cell dimensions. For example, the body-centered position in the unit cell projects midway along each of the three unit-cell edges and is designated the $\frac{1}{2}\frac{1}{2}\frac{1}{2}$ position. One aspect of the nature of crystalline structure is that a given lattice position in a given unit cell is structurally equivalent to the same position in any other unit cell of the same structure. These equivalent positions are connected by **lattice translations** consisting of integral multiples of lattice constants along directions parallel to crystallographic axes (Figure 3.23).

Figure 3.24 illustrates the notation for describing **lattice directions**. These directions are always expressed as sets of integers, which are obtained by identifying the *smallest integer positions* intercepted by the line from the origin of the crystallographic axes. To distinguish the notation for a direction from that of a position, the direction integers are enclosed in square brackets. The use of square brackets is important and is the standard designation for specific lattice directions. Other symbols are used to designate other geometrical features. Returning to Figure 3.24, we note that the line from the origin of the crystallographic axes through the $\frac{1}{2}\frac{1}{2}\frac{1}{2}$ body-centered position can be extended to intercept the 111 unit-cell corner position. Although further extension of the line will lead to interception of other integer sets (e.g., 222 or 333), the 111 set is the smallest. As a result, that direction is referred to as the [111].

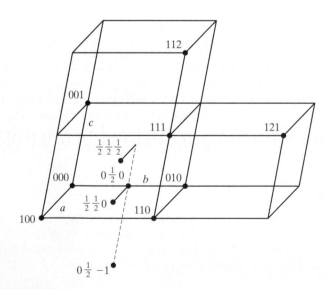

FIGURE 3.22 *Notation for lattice positions.*

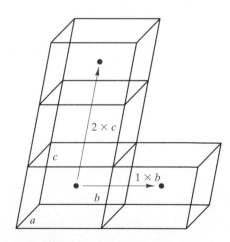

FIGURE 3.23 *Lattice translations connect structurally equivalent positions (e.g., the body center) in various unit cells.*

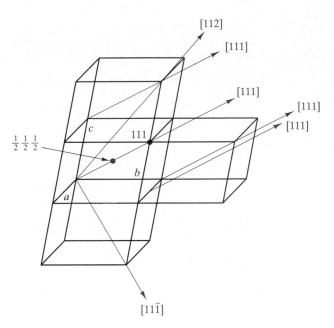

FIGURE 3.24 *Notation for lattice directions. Note that parallel [uvw] directions (e.g., [111]) share the same notation because only the origin is shifted.*

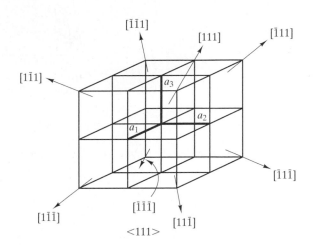

FIGURE 3.25 *Family of directions, ⟨111⟩, representing all body diagonals for adjacent unit cells in the cubic system.*

When a direction moves along a negative axis, the notation must indicate this movement. For example, the bar above the final integer in the [11$\bar{1}$] direction in Figure 3.24 designates that the line from the origin has penetrated the 11–1 position. Note that the directions [111] and [11$\bar{1}$] are structurally very similar. Both are body diagonals through identical unit cells. In fact, if you look at all body diagonals associated with the cubic crystal system, it is apparent that they are structurally identical, differing only in their orientation in space (Figure 3.25). In other words, the [11$\bar{1}$] direction would become the [111] direction if we made a different choice of crystallographic axes orientations. Such a set of directions, which are structurally equivalent, is called a **family of directions** and is designated by angular brackets, ⟨ ⟩. An example of body diagonals in the cubic system is

$$\langle 111 \rangle = [111], [\bar{1}11], [1\bar{1}1], [11\bar{1}], [\bar{1}\bar{1}1], [1\bar{1}\bar{1}], [\bar{1}1\bar{1}], [\bar{1}\bar{1}\bar{1}]. \qquad \textbf{(3.1)}$$

In future chapters, especially when dealing with calculations of mechanical properties, it will be useful to know the angle between directions. In general, the angles between directions can be determined by careful visualization and trigonometric calculation. In the frequently encountered cubic system, the angle can be determined from the relatively simple calculation of a dot product of two vectors. Taking directions $[uvw]$ and $[u'v'w']$ as vectors $\mathbf{D} = u\mathbf{a} + v\mathbf{b} + w\mathbf{c}$ and $\mathbf{D}' = u'\mathbf{a} + v'\mathbf{b} + w'\mathbf{c}$, you can determine the angle, δ, between these two directions by

$$\mathbf{D} \cdot \mathbf{D}' = |\mathbf{D}||\mathbf{D}'| \cos \delta \qquad \textbf{(3.2)}$$

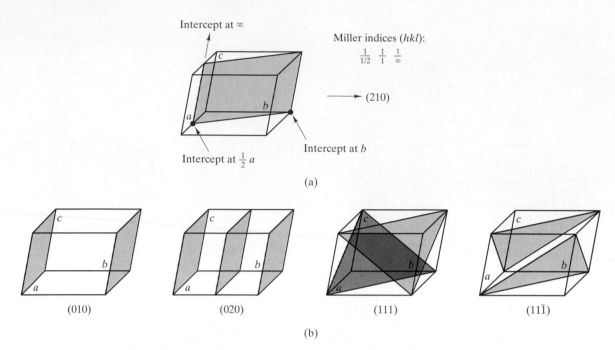

FIGURE 3.26 *Notation for lattice planes. (a) The (210) plane illustrates Miller indices (hkl). (b) Additional examples.*

or

$$\cos \delta = \frac{\mathbf{D} \cdot \mathbf{D}'}{|\mathbf{D}||\mathbf{D}'|} = \frac{uu' + vv' + ww'}{\sqrt{u^2 + v^2 + w^2}\sqrt{(u')^2 + (v')^2 + (w')^2}}. \tag{3.3}$$

It is important to remember that Equations 3.2 and 3.3 apply to the cubic system only.

Another quantity of interest in future calculations is the **linear density** of atoms along a given direction. Again, a general approach to such a calculation is careful visualization and trigonometric calculation. A streamlined approach in the case where atoms are uniformly spaced along a given direction is to determine the repeat distance, r, between adjacent atoms. The linear density is simply the inverse, r^{-1}. In making linear-density calculations for the first time, it is important to keep in mind that we are counting only those atoms whose centers lie directly on the direction line and not any that might intersect that line off-center.

Figure 3.26 illustrates the notation for describing **lattice planes**, which are planes in a crystallographic lattice. As for directions, these planes are expressed as a set of integers, known as **Miller* indices**. Obtaining these integers is a more elaborate process than was required for directions. The integers represent the inverse of axial intercepts. For example, consider the plane (210) in Figure 3.26a. As with the square brackets of direction notation, the parentheses serve as standard

---

*William Hallowes Miller (1801–1880), British crystallographer, was a major contributor along with Bravais to 19th-century crystallography. His efficient system of labeling crystallographic planes was but one of many achievements.

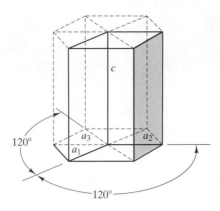

Miller–Bravais indices ($hkil$): $\frac{1}{\infty}, \frac{1}{1}, \frac{1}{-1}, \frac{1}{\infty} \rightarrow (01\bar{1}0)$
Note: $h + k = -i$

FIGURE 3.27 *Miller–Bravais indices ($hkil$) for the hexagonal system.*

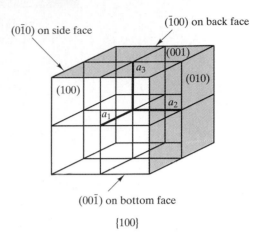

$\{100\}$

FIGURE 3.28 *Family of planes, $\{100\}$, representing all faces of unit cells in the cubic system.*

notation for planes. The (210) plane intercepts the a-axis at $\frac{1}{2}a$ and the b-axis at b, and it is parallel to the c-axis (in effect, intercepting it at ∞). The inverses of the axial intercepts are $1/\frac{1}{2}$, $1/1$, and $1/\infty$, respectively. These inverse intercepts give the 2, 1, and 0 integers leading to the (210) notation. At first, the use of these Miller indices seems like extra work. In fact, however, they provide an efficient labeling system for crystal planes and play an important role in equations dealing with diffraction measurements (Section 3.7). The general notation for Miller indices is (hkl), and it can be used for any of the seven crystal systems. Because the hexagonal system can be conveniently represented by four axes, a four-digit set of **Miller–Bravais indices** ($hkil$) can be defined as shown in Figure 3.27. Since only three axes are necessary to define the three-dimensional geometry of a crystal, one of the integers in the Miller–Bravais system is redundant. Once a plane intersects any two axes in the basal plane at the bottom of the unit cell (which contains axes a_1, a_2, and a_3 in Figure 3.27), the intersection with the third basal plane axis is determined. As a result, it can be shown that $h + k = -i$ for any plane in the hexagonal system, which also permits any such hexagonal system plane to be designated by Miller–Bravais indices ($hkil$) or by Miller indices (hkl). For the plane shown in Figure 3.27, the designation can be ($01\bar{1}0$) or (010).

As for structurally equivalent directions, we can group structurally equivalent planes as a **family of planes** with Miller or Miller–Bravais indices enclosed in braces, $\{hkl\}$ or $\{hkil\}$. Figure 3.28 illustrates that the faces of a unit cell in the cubic system are of the $\{100\}$ family with

$$\{100\} = (100), (010), (001), (\bar{1}00), (0\bar{1}0), (00\bar{1}). \qquad \textbf{(3.4)}$$

Future chapters will require calculation of **planar densities** of atoms (number per unit area) analogous to the linear densities mentioned before. As with linear densities, only those atoms centered on the plane of interest are counted.

EXAMPLE 3.8

Using Table 3.2, list the face-centered lattice-point positions for **(a)** the fcc Bravais lattice and **(b)** the *face-centered orthorhombic* (fco) lattice.

SOLUTION

(a) For the face-centered positions, $\frac{1}{2}\frac{1}{2}0$, $\frac{1}{2}0\frac{1}{2}$, $0\frac{1}{2}\frac{1}{2}$, $\frac{1}{2}\frac{1}{2}1$, $\frac{1}{2}1\frac{1}{2}$, $1\frac{1}{2}\frac{1}{2}$.

(b) Same answer as part (a). Lattice parameters do not appear in the notation for lattice positions.

EXAMPLE 3.9

Which lattice points lie on the [110] direction in the fcc and fco unit cells of Table 3.2?

SOLUTION

Sketching this case gives

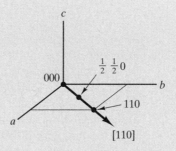

The lattice points are 000, $\frac{1}{2}\frac{1}{2}0$, and 110 for either system, fcc or fco.

EXAMPLE 3.10

List the members of the ⟨110⟩ family of directions in the cubic system.

SOLUTION

The family of directions constitutes all face diagonals of the unit cell, with two such diagonals on each face for a total of 12 members:

$$\langle 110 \rangle = [110], [1\bar{1}0], [\bar{1}10], [\bar{1}\bar{1}0], [101], [10\bar{1}], [\bar{1}01],$$
$$[\bar{1}0\bar{1}], [011], [01\bar{1}], [0\bar{1}1], [0\bar{1}\bar{1}].$$

EXAMPLE 3.11

What is the angle between the [110] and [111] directions in the cubic system?

SOLUTION

From Equation 3.3,

$$\delta = \arccos \frac{uu' + vv' + ww'}{\sqrt{u^2 + v^2 + w^2}\sqrt{(u')^2 + (v')^2 + (w')^2}}$$

$$= \arccos \frac{1 + 1 + 0}{\sqrt{2}\sqrt{3}}$$

$$= \arccos 0.816$$

$$= 35.3°.$$

EXAMPLE 3.12

Identify the axial intercepts for the $(3\bar{1}1)$ plane.

SOLUTION

For the a axis, intercept $= \frac{1}{3}a$; for the b axis, intercept $= \frac{1}{-1}b = -b$; and for the c axis, intercept $= \frac{1}{1}c = c$.

EXAMPLE 3.13

List the members of the {110} family of planes in the cubic system.

SOLUTION

$$\{110\} = (110), (1\bar{1}0), (\bar{1}10), (\bar{1}\bar{1}0), (101), 10\bar{1}), (\bar{1}01),$$

$$(\bar{1}0\bar{1}), (011), (01\bar{1}), (0\bar{1}1), (0\bar{1}\bar{1}).$$

(Compare this answer with that for Example 3.10.)

EXAMPLE 3.14

Calculate the linear density of atoms along the [111] direction in **(a)** bcc tungsten and **(b)** fcc aluminum.

SOLUTION

(a) For a bcc structure (Figure 3.4), atoms touch along the [111] direction (a body diagonal). Therefore, the repeat distance is equal to one atomic diameter. Taking data from Appendix 2, we find that the repeat distance is

$$r = d_{W\ atom} = 2r_{W\ atom}$$

$$= 2(0.137\ nm) = 0.274\ nm.$$

Therefore,

$$r^{-1} = \frac{1}{0.274 \text{ nm}} = 3.65 \text{ atoms/nm}.$$

(b) For an fcc structure, only one atom is intercepted along the body diagonal of a unit cell. To determine the length of the body diagonal, we can note that two atomic diameters equal the length of a face diagonal (see Figure 3.5). Using data from Appendix 2, we have

$$\text{face diagonal length} = 2d_{\text{Al atom}}$$
$$= 4r_{\text{Al atom}} = \sqrt{2}a$$

or the lattice parameter is

$$a = \frac{4}{\sqrt{2}} r_{\text{Al atom}} \quad \text{(see also Table 3.3)}$$
$$= \frac{4}{\sqrt{2}} (0.143 \text{ nm}) = 0.404 \text{ nm}.$$

The repeat distance is

$$r = \text{body diagonal length} = \sqrt{3}a$$
$$= \sqrt{3}(0.404 \text{ nm})$$
$$= 0.701 \text{ nm},$$

which gives a linear density of

$$r^{-1} = \frac{1}{0.701 \text{ nm}} = 1.43 \text{ atoms/nm}.$$

EXAMPLE 3.15

Calculate the planar density of atoms in the (111) plane of **(a)** bcc tungsten and **(b)** fcc aluminum.

SOLUTION
(a) For the bcc structure (Figure 3.4), the (111) plane intersects only corner atoms in the unit cell:

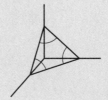

Following the calculations of Example 3.14a, we have

$$\sqrt{3}a = 4r_{W\ atom}$$

or

$$a = \frac{4}{\sqrt{3}}r_{W\ atom} = \frac{4}{\sqrt{3}}(0.137\ \text{nm}) = 0.316\ \text{nm}.$$

Face diagonal length, l, is then

$$l = \sqrt{2}a = \sqrt{2}(0.316\ \text{nm}) = 0.447\ \text{nm}.$$

The area of the (111) plane within the unit cell is

$$A = \frac{1}{2}bh = \frac{1}{2}(0.447\ \text{nm})\left(\frac{\sqrt{3}}{2} \times 0.447\ \text{nm}\right)$$

$$= 0.0867\ \text{nm}^2.$$

There is $\frac{1}{6}$ atom (i.e., $\frac{1}{6}$ of the circumference of a circle) at each corner of the equilateral triangle formed by the (111) plane in the unit cell. Therefore,

$$\text{atomic density} = \frac{3 \times \frac{1}{6}\ \text{atom}}{A}$$

$$= \frac{0.5\ \text{atom}}{0.0867\ \text{nm}^2} = 5.77\frac{\text{atoms}}{\text{nm}^2}.$$

(b) For the fcc structure (Figure 3.5), the (111) plane intersects three corner atoms plus three face-centered atoms in the unit cell:

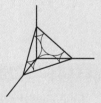

Following the calculations of Example 3.14b, we obtain the face diagonal length

$$l = \sqrt{2}a = \sqrt{2}(0.404\ \text{nm}) = 0.572\ \text{nm}.$$

The area of the (111) plane within the unit cell is

$$A = \frac{1}{2}bh = \frac{1}{2}(0.572\ \text{nm})\left(\frac{\sqrt{3}}{2}0.572\ \text{nm}\right)$$

$$= 0.142\ \text{nm}^2.$$

There are $3 \times \frac{1}{6}$ corner atoms plus $3 \times \frac{1}{2}$ face-centered atoms within this area, giving

$$\text{atomic density} = \frac{3 \times \frac{1}{6} + 3 \times \frac{1}{2} \text{ atoms}}{0.142 \text{ nm}^2} = \frac{2 \text{ atoms}}{0.142 \text{ nm}^2}$$

$$= 14.1 \text{ atoms/nm}^2.$$

EXAMPLE 3.16

Calculate the linear density of ions in the [111] direction of MgO.

SOLUTION

Figure 3.9 shows that the body diagonal of the unit cell intersects one Mg^{2+} and one O^{2-}. Following the calculations of Example 3.3, we find that the length of the body diagonal is

$$l = \sqrt{3}a = \sqrt{3}(0.420 \text{ nm}) = 0.727 \text{ nm}.$$

The ionic linear densities, then, are

$$\frac{1 \text{ Mg}^{2+}}{0.727 \text{ nm}} = 1.37 \text{ Mg}^{2+}/\text{nm}$$

and, similarly,

$$1.37 \text{ O}^{2-}/\text{nm},$$

giving

$$(1.37 \text{ Mg}^{2+} + 1.37 \text{ O}^{2-})/\text{nm}.$$

EXAMPLE 3.17

Calculate the planar density of ions in the (111) plane of MgO.

SOLUTION

There are really two separate answers to this problem. Using the unit cell of Figure 3.9, we see an arrangement comparable to an fcc metal:

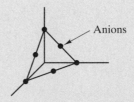

However, we could similarly define a unit cell with its origin on a cation site (rather than on an anion site as shown in Figure 3.9). In this case, the (111) plane would have a comparable arrangement of cations:

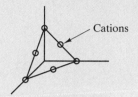

Cations

In either case, there are two ions per (111) "triangle." From Example 3.3, we know that $a = 0.420$ nm. The length of each (111) triangle side (i.e., a unit-cell face diagonal) is

$$l = \sqrt{2}a = \sqrt{2}(0.420 \text{ nm}) = 0.594 \text{ nm}.$$

The planar area, then, is

$$A = \frac{1}{2}bh = \frac{1}{2}(0.594 \text{ nm})\left(\frac{\sqrt{3}}{2}0.594 \text{ nm}\right) = 0.153 \text{ nm}^2,$$

which gives

$$\text{ionic density} = \frac{2 \text{ ions}}{0.153 \text{ nm}^2} = 13.1 \text{ nm}^{-2}$$

or

$$13.1(Mg^{2+} \text{ or } O^{2-})/\text{nm}^2.$$

EXAMPLE 3.18

Calculate the linear density of atoms along the [111] direction in silicon.

SOLUTION

We must use some caution in this problem. Inspection of Figure 3.20 indicates that atoms along the [111] direction (a body diagonal) are not uniformly spaced. Therefore, the r^{-1} calculations of Example 3.14 are not appropriate.

Referring to the comments of Example 3.6, we can see that two atoms are centered along a given body diagonal (e.g., $\frac{1}{2}$ atom at 000, 1 atom at $\frac{1}{4}\frac{1}{4}\frac{1}{4}$, and $\frac{1}{2}$ atom at 111). If we take the body diagonal length in a unit cell as l,

$$2r_{Si} = \frac{1}{4}l$$

or

$$l = 8r_{Si}.$$

From Appendix 2,

$$l = 8(0.117 \text{ nm}) = 0.936 \text{ nm}.$$

Therefore, the linear density is

$$\text{linear density} = \frac{2 \text{ atoms}}{0.936 \text{ nm}} = 2.14 \frac{\text{atoms}}{\text{nm}}.$$

EXAMPLE 3.19

Calculate the planar density of atoms in the (111) plane of silicon.

SOLUTION
Close observation of Figure 3.20 shows that the four interior atoms in the diamond cubic structure do not lie on the (111) plane. The result is that the atom arrangement in this plane is precisely that for the metallic fcc structure (see Example 3.15b). Of course, the atoms along [110]-type directions in the diamond cubic structure do not touch as in fcc metals.

As calculated in Example 3.15b, there are two atoms in the equilateral triangle bounded by sides of length $\sqrt{2}a$. From Example 3.6 and Appendix 2, we see that

$$a = \frac{8}{\sqrt{3}}(0.117 \text{ nm}) = 0.540 \text{ nm}$$

and

$$\sqrt{2}a = 0.764 \text{ nm},$$

giving a triangle area of

$$A = \frac{1}{2}bh = \frac{1}{2}(0.764 \text{ nm}) \left(\frac{\sqrt{3}}{2}0.764 \text{ nm} \right)$$

$$= 0.253 \text{ nm}^2$$

and a planar density of

$$\frac{2 \text{ atoms}}{0.253 \text{ nm}^2} = 7.91 \frac{\text{atoms}}{\text{nm}^2}.$$

PRACTICE PROBLEM 3.9

From Table 3.2, list the body-centered lattice-point positions for **(a)** the bcc Bravais lattice, **(b)** the body-centered tetragonal lattice, and **(c)** the body-centered orthorhombic lattice. (See Example 3.8.)

PRACTICE PROBLEM 3.10

Use a sketch to determine which lattice points lie along the [111] direction in the **(a)** bcc, **(b)** body-centered tetragonal, and **(c)** body-centered orthorhombic unit cells of Table 3.2. (See Example 3.9.)

PRACTICE PROBLEM 3.11

Sketch the 12 members of the ⟨110⟩ family determined in Example 3.10. (You may want to use more than one sketch.)

PRACTICE PROBLEM 3.12

(a) Determine the ⟨100⟩ family of directions in the cubic system, and **(b)** sketch the members of this family. (See Example 3.10 and Practice Problem 3.11.)

PRACTICE PROBLEM 3.13

Calculate the angles between **(a)** the [100] and [110] and **(b)** the [100] and [111] directions in the cubic system. (See Example 3.11.)

PRACTICE PROBLEM 3.14

Sketch the $(3\bar{1}1)$ plane and its intercepts. (See Example 3.12 and Figure 3.26.)

PRACTICE PROBLEM 3.15

Sketch the 12 members of the {110} family determined in Example 3.13. (To simplify matters, you will probably want to use more than one sketch.)

PRACTICE PROBLEM 3.16

Calculate the linear density of atoms along the [111] direction in **(a)** bcc iron and **(b)** fcc nickel. (See Example 3.14.)

PRACTICE PROBLEM 3.17

Calculate the planar density of atoms in the (111) plane of **(a)** bcc iron and **(b)** fcc nickel. (See Example 3.15.)

PRACTICE PROBLEM 3.18

Calculate the linear density of ions along the [111] direction for CaO. (See Example 3.16.)

PRACTICE PROBLEM 3.19

Calculate the planar density of ions in the (111) plane for CaO. (See Example 3.17.)

PRACTICE PROBLEM 3.20

Find the linear density of atoms along the [111] direction for germanium. (See Example 3.18.)

PRACTICE PROBLEM 3.21

Find the planar density of atoms in the (111) plane for germanium. (See Example 3.19.)

3.7 | X–Ray Diffraction

This chapter has introduced a large variety of crystal structures. We now end with a brief description of **x-ray diffraction**, a powerful experimental tool.

There are many ways in which x-ray diffraction is used to measure the crystal structure of engineering materials. It can be used to determine the structure of a new material, or the known structure of a common material can be used as a source of chemical identification.

Diffraction is the result of radiation's being scattered by a regular array of scattering centers whose spacing is about the same as the wavelength of the radiation. For example, parallel scratch lines spaced repeatedly about 1 μm apart cause diffraction of visible light (electromagnetic radiation with a wavelength just under 1 μm). This *diffraction grating* causes the light to be scattered with a strong intensity in a few specific directions (Figure 3.29). The precise direction of observed scattering is a function of the exact spacing between scratch lines in the diffraction grating, relative to the wavelength of the incident light. Appendix 2 shows that atoms and ions are on the order of 0.1 nm in size, so we can think of crystal structures as being diffraction gratings on a subnanometer scale. As shown in Figure 3.30, the portion of the electromagnetic spectrum with a wavelength in this range is **x-radiation** (compared to the 1,000-nm range for the wavelength of visible light). As a result, x-ray diffraction is capable of characterizing crystalline structure.

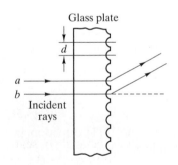

FIGURE 3.29 *Diffraction grating for visible light. Scratch lines in the glass plate serve as light-scattering centers. (From D. Halliday and R. Resnick,* Physics, *John Wiley & Sons, Inc., New York, 1962.)*

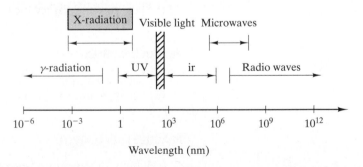

FIGURE 3.30 *Electromagnetic radiation spectrum. X-radiation represents that portion with wavelengths around 0.1 nm.*

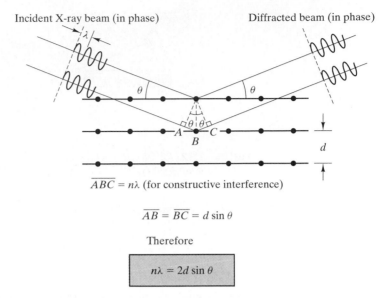

FIGURE 3.31 *Geometry for diffraction of x-radiation. The crystal structure is a three-dimensional diffraction grating. Bragg's law ($n\lambda = 2d\sin\theta$) describes the diffraction condition.*

For x-rays, atoms are the scattering centers. The specific mechanism of scattering is the interaction of a photon of electromagnetic radiation with an orbital electron in the atom. A crystal acts as a three-dimensional diffraction grating. Repeated stacking of crystal planes serves the same function as the parallel scratch lines in Figure 3.29. For a simple crystal lattice, the condition for diffraction is shown in Figure 3.31. For diffraction to occur, x-ray beams scattered off adjacent crystal planes must be in phase. Otherwise, destructive interference of waves occurs and essentially no scattering intensity is observed. At the precise geometry for constructive interference (scattered waves in phase), the difference in path length between the adjacent x-ray beams is some integral number (*n*) of radiation wavelengths (λ). The relationship that demonstrates this condition is the **Bragg** [*] **equation**

$$n\lambda = 2d\sin\theta, \tag{3.5}$$

where *d* is the spacing between adjacent crystal planes and θ is the angle of scattering as defined in Figure 3.31. The angle θ is usually referred to as the **Bragg angle**, and the angle 2θ is referred to as the **diffraction angle** because it is the angle measured experimentally (Figure 3.32).

The magnitude of **interplanar spacing** (*d* in Equation 3.5) is a direct function of the Miller indices for the plane. For a cubic system, the relationship is fairly

[*]William Henry Bragg (1862–1942) and William Lawrence Bragg (1890–1971), English physicists, were a gifted father-and-son team. They were the first to demonstrate the power of Equation 3.5 by using x-ray diffraction to determine the crystal structures of several alkali halides, such as NaCl. Since this achievement in 1912, the crystal structures of more than 200,000 materials have been catalogued (see the footnote on p. 96).

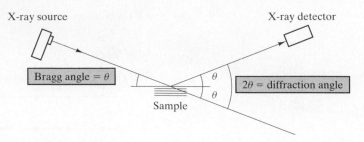

FIGURE 3.32 *Relationship of the Bragg angle (θ) and the experimentally measured diffraction angle (2θ).*

simple. The spacing between adjacent *hkl* planes is

$$d_{hkl} = \frac{a}{\sqrt{h^2 + k^2 + l^2}},$$ (3.6)

where *a* is the lattice parameter (edge length of the unit cell). For more complex unit-cell shapes, the relationship is more complex. For a *hexagonal system,*

$$d_{hkl} = \frac{a}{\sqrt{\frac{4}{3}(h^2 + hk + k^2) + l^2(a^2/c^2)}},$$ (3.7)

where *a* and *c* are the lattice parameters.

Bragg's law (Equation 3.5) is a necessary but insufficient condition for diffraction. It defines the diffraction condition for **primitive unit cells**; that is, those Bravais lattices with lattice points only at unit-cell corners, such as simple cubic and simple tetragonal. Crystal structures with **nonprimitive unit cells** have atoms at additional lattice sites located along a unit-cell edge, within a unit-cell face, or in the interior of the unit cell. The extra scattering centers can cause out-of-phase scattering to occur at certain Bragg angles. The result is that some of the diffraction predicted by Equation 3.5 does not occur. An example of this effect is given in Table 3.4, which gives the **reflection rules** for the common metal structures. These rules show which sets of Miller indices do not produce diffraction as predicted by Bragg's law. Keep in mind that *reflection* here is a casual term in that diffraction rather than true reflection is being described.

A diffraction pattern for a specimen of aluminum powder is shown in Figure 3.33. Each peak represents a solution to Bragg's law. Because the powder consists of many small crystal grains oriented randomly, a single wavelength of radiation is used to keep the number of diffraction peaks in the pattern to a small, workable number. The experiment is done in a **diffractometer** (Figure 3.34), an

TABLE 3.4

Reflection Rules of X-Ray Diffraction for the Common Metal Structures		
Crystal structure	**Diffraction does not occur when**	**Diffraction occurs when**
Body-centered cubic (bcc)	$h + k + l$ = odd number	$h + k + l$ = even number
Face-centered cubic (fcc)	h, k, l mixed (i.e., both even and odd numbers)	h, k, l unmixed (i.e., are all even numbers or all are odd numbers)
Hexagonal close packed (hcp)	$(h + 2k) = 3n, l$ odd (n is an integer)	All other cases

Note how small differences in the atomic scale dimensions of crystal structures (interplanar spacings) lead to large angular differences in the x-ray diffraction pattern.

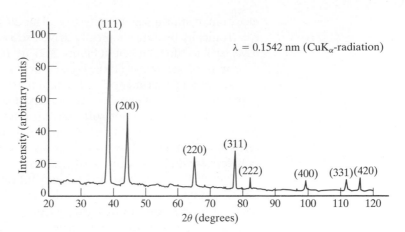

FIGURE 3.33 *Diffraction pattern of aluminum powder. Each peak (in the plot of x-ray intensity versus diffraction angle, 2θ) represents diffraction of the x-ray beam by a set of parallel crystal planes (hkl) in various powder particles.*

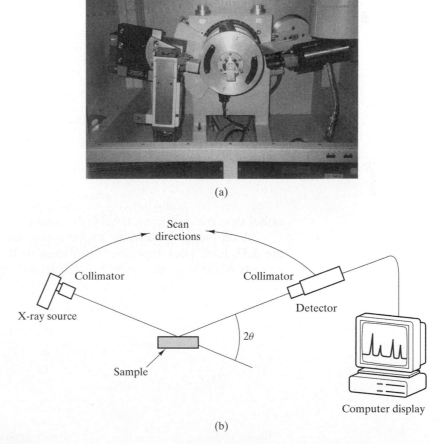

(a)

(b)

FIGURE 3.34 *(a) An x-ray diffractometer. (Courtesy of Scintag, Inc.) (b) A schematic of the experiment.*

electromechanical scanning system. The diffracted beam intensity is monitored electronically by a mechanically driven scanning radiation detector. *Powder patterns* such as those shown in Figure 3.33 are routinely used by materials engineers for comparison against a large collection of known diffraction patterns.* The comparison of an experimental diffraction pattern such as that shown in Figure 3.33 with the database of known diffraction patterns can be done in a few seconds with "search/match" computer software, an integral part of modern diffractometers such as that shown in Figure 3.34. The unique relationship between such patterns and crystal structures provides a powerful tool for chemical identification of powders and polycrystalline materials.

Standard procedure for analyzing the diffraction patterns of powder samples or polycrystalline solids involves the use of $n = 1$ in Equation 3.5. This use is justified in that the nth-order diffraction of any (hkl) plane occurs at an angle identical to the first-order diffraction of the ($nh\ nk\ nl$) plane [which is, by the way, parallel to (hkl)]. As a result, we can use an even simpler version of Bragg's law for powder diffraction:

$$\lambda = 2d \sin \theta. \tag{3.8}$$

EXAMPLE 3.20

Using Bragg's law, calculate the diffraction angles (2θ) for the first three peaks in the aluminum powder pattern of Figure 3.33.

SOLUTION

Figure 3.33 indicates that the first three (i.e., lowest-angle) peaks are for (111), (200), and (220). From Example 3.14b, we note that $a = 0.404$ nm. Therefore, Equation 3.6 yields

$$d_{111} = \frac{0.404\ \text{nm}}{\sqrt{1+1+1}} = \frac{0.404\ \text{nm}}{\sqrt{3}} = 0.234\ \text{nm},$$

$$d_{200} = \frac{0.404\ \text{nm}}{\sqrt{2^2+0+0}} = \frac{0.404\ \text{nm}}{2} = 0.202\ \text{nm, and}$$

$$d_{220} = \frac{0.404\ \text{nm}}{\sqrt{2^2+2^2+0}} = \frac{0.404\ \text{nm}}{\sqrt{8}} = 0.143\ \text{nm}.$$

Noting that $\lambda = 0.1542$ nm in Figure 3.33, Equation 3.8 gives

$$\theta = \arcsin \frac{\lambda}{2d}$$

or

$$\theta_{111} = \arcsin \frac{0.1542\ \text{nm}}{2 \times 0.234\ \text{nm}} = 19.2°$$

$$\text{or} \quad (2\theta)_{111} = 38.5°,$$

Powder Diffraction File, more than 200,000 powder diffraction patterns catalogued by the International Centre for Diffraction Data (ICDD), Newtown Square, PA.

$$\theta_{200} = \arcsin \frac{0.1542 \text{ nm}}{2 \times 0.202 \text{ nm}} = 22.4°$$

$$\text{or} \quad (2\theta)_{200} = 44.9°, \text{ and}$$

$$\theta_{220} = \arcsin \frac{0.1542 \text{ nm}}{2 \times 0.143 \text{ nm}} = 32.6°$$

$$\text{or} \quad (2\theta)_{220} = 65.3°.$$

PRACTICE PROBLEM 3.22

The diffraction angles for the first three peaks in Figure 3.33 are calculated in Example 3.20. Calculate the diffraction angles for the remainder of the peaks in Figure 3.33.

Summary

Most materials used by engineers are crystalline in nature; that is, their atomic-scale structure is regular and repeating. This regularity allows the structure to be defined in terms of a fundamental structural unit, the unit cell. There are seven crystal systems, which correspond to the possible unit-cell shapes. Based on these crystal systems, there are 14 Bravais lattices that represent the possible arrangements of points through three-dimensional space. These lattices are the "skeletons" on which the large number of crystalline atomic structures is based.

There are three primary crystal structures observed for common metals: the body-centered cubic (bcc), the face-centered cubic (fcc), and the hexagonal close packed (hcp). These are relatively simple structures, with the fcc and hcp forms representing optimum efficiency in packing equal-sized spheres (i.e., metal atoms). The fcc and hcp structures differ only in the pattern of stacking of close-packed atomic planes.

Chemically more complex than metals, ceramic compounds exhibit a wide variety of crystalline structures. Some, such as the NaCl structure, are similar to the simpler metal structures sharing a common Bravais lattice, but with more than one ion associated with each lattice point. Silica, SiO_2, and the silicates exhibit a wide array of relatively complex arrangements of silica tetrahedra (SiO_4^{4-}). In this chapter, several representative ceramic structures are displayed.

Polymers are characterized by long-chain polymeric structures. The elaborate way in which these chains must be folded to form a repetitive pattern produces two effects: (1) The resulting crystal structures are relatively complex, and (2) most commercial polymers are only partially crystalline. The unit-cell structure of polyethylene is illustrated in this chapter.

High-quality single crystals are an important part of *semiconductor* technology, which is possible largely because most semiconductors can be produced in a few relatively simple crystal structures. Elemental semiconductors, such as silicon, have the diamond cubic structure, a modification of the fcc Bravais lattice with two atoms associated with each lattice point. Many compound semiconductors are found in the closely related zinc blende (ZnS) structure, in which the diamond cubic atom positions are retained but with Zn^{2+} and S^{2-} ions alternating on those sites.

There are standard methods for describing the geometry of crystalline structures. These methods give an efficient and systematic notation for lattice positions, directions, and planes.

X-ray diffraction is the standard experimental tool for analyzing crystal structures. The regular atomic arrangement of crystals serves as a subnanometer diffraction grating for *x-radiation* (with a subnanometer wavelength). The use of Bragg's law in conjunction with the reflection rules permits a precise measurement of interplanar spacings in the crystal structure. Polycrystalline (or powdered) materials are routinely analyzed in this way.

Key Terms

General
atomic packing factor (APF) (62)
Bravais lattice (59)
crystal system (59)
family of directions (81)
family of planes (83)
ionic packing factor (66)
lattice constant (59)
lattice direction (80)
lattice parameter (59)
lattice plane (82)
lattice point (59)
lattice position (80)
lattice translation (80)
linear density (82)
Miller–Bravais indices (83)
Miller indices (82)
planar density (83)
point lattice (59)

III–V compounds (77)
II–VI compounds (77)
unit cell (59)
Structures
body-centered cubic (62)
buckminsterfullerene (72)
buckyball (72)
buckytube (72)
cesium chloride (66)
corundum (70)
cristobalite (68)
cubic close packed (62)
diamond cubic (76)
face-centered cubic (62)
fluorite (68)
fullerene (72)
hexagonal close packed (63)
kaolinite (70)
polyethylene (74)

silica (68)
silicon (76)
sodium chloride (66)
zinc blende (78)
Diffraction
Bragg angle (93)
Bragg equation (93)
Bragg's law (94)
diffraction (92)
diffraction angle (93)
diffractometer (94)
interplanar spacing (93)
nonprimitive unit cells (94)
primitive unit cells (94)
reflection rules (94)
x-radiation (92)
x-ray diffraction (92)

References

Accelrys, Inc., San Diego, CA. Computer-generated crystal structures of a wide range of materials available on CD-ROM for display on graphics workstations. Updated annually.

Barrett, C. S., and **T. B. Massalski**, *Structure of Metals*, 3rd revised ed., Pergamon Press, New York, 1980. This text includes substantial coverage of x-ray diffraction techniques.

Chiang, Y., **D. P. Birnie III**, and **W. D. Kingery**, *Physical Ceramics*, John Wiley & Sons, Inc., New York, 1997.

Cullity, B. D., and **S. R. Stock**, *Elements of X-Ray Diffraction*, 3rd ed., Prentice Hall, Upper Saddle River, NJ,

2001. A recent revision of a classic text and an especially clear discussion of the principles and applications of x-ray diffraction.

Williams, D. J., *Polymer Science and Engineering*, Prentice Hall, Inc., Englewood Cliffs, NJ, 1971.

Wyckoff, R. W. G., ed., *Crystal Structures*, 2nd ed., Vols. 1–5 and Vol. 6, Parts 1 and 2, John Wiley & Sons, Inc., New York, 1963–1971. An encyclopedic collection of crystal structure data.

Problems

Section 3.1 • Seven Systems and Fourteen Lattices

3.1. Why is the simple hexagon not a two-dimensional point lattice?

3.2. What would be an equivalent two-dimensional point lattice for the area-centered hexagon?

3.3. Why is there no base-centered cubic lattice in Table 3.2? (Use a sketch to answer.)

*3.4. (a) Which two-dimensional point lattice corresponds to the crystalline ceramic illustrated in Figure 1.8a? **(b)** Sketch the unit cell.

3.5. Under what conditions does the triclinic system reduce to the hexagonal system?

3.6. Under what conditions does the monoclinic system reduce to the orthorhombic system?

Section 3.2 • Metal Structures

3.7. Calculate the density of Mg, an hcp metal. (Note Problem 3.11 for the ideal c/a ratio.)

3.8. Calculate the APF of 0.68 for the bcc metal structure.

3.9. Calculate the APF of 0.74 for fcc metals.

3.10. Calculate the APF of 0.74 for hcp metals.

3.11. **(a)** Show that the c/a ratio (height of the unit cell divided by its edge length) is 1.633 for the ideal hcp structure. **(b)** Comment on the fact that real hcp metals display c/a ratios varying from 1.58 (for Be) to 1.89 (for Cd).

Section 3.3 • Ceramic Structures

3.12. Calculate the IPF for UO_2, which has the CaF_2 structure (Figure 3.10).

3.13. In Section 3.3, the open nature of the CaF_2 structure was given credit for the ability of UO_2 to absorb He gas atoms and thereby resist swelling. Confirm that an He atom (diameter ≈ 0.2 nm) can fit in the center of the UO_2 unit cell (see Figure 3.10 for the CaF_2 structure).

3.14. Show that the unit cell in Figure 3.14 gives the chemical formula $2(OH)_4Al_2Si_2O_5$.

3.15. Calculate the density of UO_2.

*3.16. **(a)** Derive a general relationship between the IPF of the NaCl-type structure and the radius ratio (r/R). **(b)** Over what r/R range is this relationship reasonable?

*3.17. Calculate the IPF for cristobalite (Figure 3.11).

*3.18. Calculate the IPF for corundum (Figure 3.13).

Section 3.4 • Polymeric Structures

3.19. Calculate the reaction energy involved in forming a single unit cell of polyethylene.

3.20. How many unit cells are contained in the thickness of a 10-nm-thick polyethylene platelet (Figure 3.19)?

3.21. Calculate the APF for polyethylene.

Section 3.5 • Semiconductor Structures

3.22. Calculate the IPF for the zinc blende structure (Figure 3.21).

3.23. Calculate the density of zinc blende using data from Appendixes 1 and 2.

*3.24. **(a)** Derive a general relationship between the IPF of the zinc blende structure and the radius ratio (r/R). **(b)** What is a primary limitation of such IPF calculations for these compound semiconductors?

Section 3.6 • Lattice Positions, Directions, and Planes

3.25. **(a)** Sketch, in a cubic unit cell, a [111] and a [112] lattice direction. **(b)** Use a trigonometric calculation to determine the angle between these two directions. **(c)** Use Equation 3.3 to determine the angle between these two directions.

3.26. List the lattice-point positions for the corners of the unit cell in **(a)** the base-centered orthorhombic lattice and **(b)** the triclinic lattice.

3.27. **(a)** Sketch, in a cubic unit cell, [100] and [210] directions. **(b)** Use a trigonometric calculation to determine the angle between these directions. **(c)** Use Equation 3.3 to determine this angle.

3.28. List the body-centered and base-centered lattice-point positions for **(a)** the body-centered orthorhombic lattice and **(b)** the base-centered monoclinic lattice, respectively.

*3.29. What polyhedron is formed by "connecting the dots" between a corner atom in the fcc lattice and the three adjacent face-centered positions? Illustrate your answer with a sketch.

*3.30. Repeat Problem 3.29 for the six face-centered positions on the surface of the fcc unit cell.

3.31. What [hkl] direction connects the adjacent face-centered positions $\frac{1}{2}\frac{1}{2}0$ and $\frac{1}{2}0\frac{1}{2}$? Illustrate your answer with a sketch.

3.32. A useful rule of thumb for the cubic system is that a given [hkl] direction is the normal to the (hkl) plane. Using this rule and Equation 3.3, determine which members of the ⟨110⟩ family of directions lie within the (111) plane. (*Hint:* The dot product of two perpendicular vectors is zero.)

3.33. Which members of the ⟨111⟩ family of directions lie within the (110) plane? (See the comments in Problem 3.32.)

3.34. Repeat Problem 3.32 for the ($\bar{1}$11) plane.

3.35. Repeat Problem 3.32 for the $(11\bar{1})$ plane.

3.36. Repeat Problem 3.33 for the (101) plane.

3.37. Repeat Problem 3.33 for the $(10\bar{1})$ plane.

3.38. Repeat Problem 3.33 for the $(\bar{1}01)$ plane.

3.39. Sketch the basal plane for the hexagonal unit cell that has the Miller–Bravais indices (0001) (see Figure 3.27).

3.40. List the members of the family of prismatic planes for the hexagonal unit cell $\{01\bar{1}0\}$ (see Figure 3.27).

3.41. The four-digit notation system (Miller–Bravais indices) introduced for planes in the hexagonal system can also be used for describing crystal directions. In a hexagonal unit cell, sketch **(a)** the $[0001]$ direction, and **(b)** the $[11\bar{2}0]$ direction.

3.42. The family of directions described in Practice Problem 3.12 contains six members. The size of this family will be diminished for noncubic unit cells. List the members of the $\langle 100 \rangle$ family for **(a)** the tetragonal system and **(b)** the orthorhombic system.

3.43. The comment in Problem 3.42 about families of directions also applies to families of planes. Figure 3.28 illustrates the six members of the $\{100\}$ family of planes for the cubic system. List the members of the $\{100\}$ family for **(a)** the tetragonal system, and **(b)** the orthorhombic system.

3.44. **(a)** List the first three lattice points (including the 000 point) lying on the $[112]$ direction in the fcc lattice. **(b)** Illustrate your answer to part (a) with a sketch.

3.45. Repeat Problem 3.44 for the bcc lattice.

3.46. Repeat Problem 3.44 for the bct lattice.

3.47. Repeat Problem 3.44 for the base-centered orthorhombic lattice.

3.48. In the cubic system, which of the $\langle 110 \rangle$ family of directions represents the line of intersection between the (111) and $(11\bar{1})$ planes? (Note the comment in Problem 3.32.)

3.49. Sketch the directions and planar intersection described in Problem 3.48.

3.50. Sketch the members of the $\{100\}$ family of planes in the triclinic system.

*3.51. The first eight planes that give x-ray diffraction peaks for aluminum are indicated in Figure 3.33. Sketch each plane and its intercepts relative to a cubic unit cell. (To avoid confusion, use a separate sketch for each plane.)

*3.52. **(a)** List the $\langle 112 \rangle$ family of directions in the cubic system. **(b)** Sketch this family. (You will want to use more than one sketch.)

3.53. In Figures 3.4b and 3.5b, we show atoms and fractional atoms making up a unit cell. An alternative convention is to describe the unit cell in terms of "equivalent points." For example, the two atoms in the bcc unit cell can be considered to be one corner atom at 000 and one body-centered atom at $\frac{1}{2}\frac{1}{2}\frac{1}{2}$. The one corner atom is equivalent to the eight $\frac{1}{8}$ atoms shown in Figure 3.4b. In a similar way, identify the four atoms associated with equivalent points in the fcc structure.

3.54. Identify the atoms associated with equivalent points in the hcp structure. (See Problem 3.53.)

3.55. Repeat Problem 3.54 for the body-centered orthorhombic lattice.

3.56. Repeat Problem 3.54 for the base-centered orthorhombic lattice.

3.57. Sketch the $[1\bar{1}0]$ direction within the (111) plane relative to an fcc unit cell. Include all atom-center positions within the plane of interest.

3.58. Sketch the $[1\bar{1}1]$ direction within the (110) plane relative to a bcc unit cell. Include all atom-center positions within the plane of interest.

3.59. Sketch the $[11\bar{2}0]$ direction within the (0001) plane relative to an hcp unit cell. Include all atom-center positions within the plane of interest.

*3.60. The $\frac{1}{4}\frac{1}{4}\frac{1}{4}$ position in the fcc structure is a "tetrahedral site," an interstice with fourfold atomic coordination. The $\frac{1}{2}\frac{1}{2}\frac{1}{2}$ position is an "octahedral site," an interstice with sixfold atomic coordination. How many tetrahedral and octahedral sites are there per fcc unit cell? Use a sketch to illustrate your answer.

*3.61. The first eight planes that give x-ray diffraction peaks for aluminum are indicated in Figure 3.33. Sketch each plane relative to the fcc unit cell (Figure 3.5a) and emphasize atom positions within the planes. (Note Problem 3.51 and use a separate sketch for each plane.)

3.62. Calculate the linear density of ions along the [111] direction in UO_2, which has the CaF_2 structure (Figure 3.10).

3.63. Identify the ions associated with equivalent points in the NaCl structure. (Note Problem 3.53.)

*__3.64.__ Sketch the ion positions in a (111) plane through the cristobalite unit cell (Figure 3.11).

*__3.65.__ Sketch the ion positions in a (101) plane through the cristobalite unit cell (Figure 3.11).

3.66. Calculate the linear density of ions along the [111] direction in zinc blende (Figure 3.21).

3.67. Calculate the planar density of ions along the (111) plane in zinc blende (Figure 3.21).

3.68. Identify the ions associated with equivalent points in the diamond cubic structure. (Note Problem 3.53.)

3.69. Identify the ions associated with equivalent points in the zinc blende structure. (Note Problem 3.53.)

Section 3.7 • X-Ray Diffraction

3.70. The diffraction peaks labeled in Figure 3.33 correspond to the reflection rules for an fcc metal (h, k, l unmixed, as shown in Table 3.4). What would be the (hkl) indices for the three lowest diffraction-angle peaks for a bcc metal?

3.71. Using the result of Problem 3.70, calculate the diffraction angles (2θ) for the first three peaks in the diffraction pattern of α-Fe powder using CuK_α-radiation ($\lambda = 0.1542$ nm).

3.72. Repeat Problem 3.71 using CrK_α-radiation ($\lambda = 0.2291$ nm).

3.73. Repeat Problem 3.70 for the next three lowest diffraction-angle peaks for a bcc metal.

3.74. Repeat Problem 3.71 for the next three lowest diffraction-angle peaks for α-Fe powder using CuK_α-radiation.

3.75. Assuming the relative peak heights would be the same for given (hkl) planes, sketch a diffraction pattern similar to that shown in Figure 3.33 for copper powder using CuK_α-radiation. Cover the range of $20° < 2\theta < 90°$.

3.76. Repeat Problem 3.75 for lead powder.

*__3.77.__ What would be the (hkl) indices for the three lowest diffraction-angle peaks for an hcp metal?

*__3.78.__ Using the result of Problem 3.77, calculate the diffraction angles (2θ) for the first three peaks in the diffraction pattern of magnesium powder using CuK_α-radiation ($\lambda = 0.1542$ nm). Note that the c/a ratio for Mg is 1.62.

*__3.79.__ Repeat Problem 3.78 for CrK_α-radiation ($\lambda = 0.2291$ nm).

3.80. Calculate the first six diffraction-peak positions for MgO powder using CuK_α-radiation. (This ceramic structure based on the fcc lattice shares the reflection rules of the fcc metals.)

3.81. Repeat Problem 3.80 for CrK_α-radiation ($\lambda = 0.2291$ nm).

*__3.82.__ The first three diffraction peaks of a metal powder are $2\theta = 44.4°, 64.6°$, and $81.7°$ using CuK_α-radiation. Is this a bcc or an fcc metal?

*__3.83.__ More specifically, is the metal powder in Problem 3.82 Cr, Ni, Ag, or W?

3.84. What would the positions of the first three diffraction peaks in Problem 3.82 have been using CrK_α-radiation ($\lambda = 0.2291$ nm)?

3.85. The wavelength given for CuK_α-radiation ($\lambda = 0.1542$ nm) is, in fact, an average of two closely spaced peaks ($CuK_{\alpha1}$ and $CuK_{\alpha2}$). By carefully filtering the radiation from a copper target x-ray tube, one can perform diffraction with a more precise wavelength ($CuK_{\alpha1} = 0.15406$ nm). Repeat Example 3.20 using this more precise radiation.

3.86. Calculate the percentage change in diffraction angle (2θ) for each peak in Problem 3.85 that results from using the more precise radiation, $CuK_{\alpha1}$.

3.87. As with copper radiation in Problem 3.85, chromium radiation, CrK_α ($\lambda = 0.2291$ nm), is an average of two closely spaced peaks ($CrK_{\alpha1}$ and $CrK_{\alpha2}$). Repeat Problem 3.72 using $CrK_{\alpha1}$ (= 0.22897 nm).

3.88. Calculate the percentage change in diffraction angle (2θ) for each peak in Problem 3.87 that results from using the more precise radiation, $CrK_{\alpha1}$.

CHAPTER 4
Crystal Defects and Noncrystalline Structure—Imperfection

An effective 2-dimensional physical model of defects in crystal structures can be produced with a "bubble raft," in which uniform sized soap bubbles stack together in a regular fashion and occasional fluctuations produce vacancies and grain boundaries.

Students interested in exploring the intriguing subjects of high-angle grain boundaries, quasi-crystals, and fractals can read the chapter on "advanced structural topics" on the related Web site.

Similarly, there is a chapter on materials characterization that describes the tools available for imaging the structural defects introduced here in Chapter 4.

In Chapter 3, we looked at a wide variety of atomic-scale structures characteristic of important engineering materials. The primary limitation of Chapter 3 was that it dealt only with the perfectly repetitive crystalline structures. As you have learned long before this first course in engineering materials, nothing in our world is quite perfect. No crystalline material exists that does not have at least a few structural flaws. In this chapter, we will systematically survey these imperfections.

Our first consideration is that no material can be prepared without some degree of chemical impurity. The impurity atoms or ions in the resulting *solid solution* serve to alter the structural regularity of the ideally pure material.

Independent of impurities, there are numerous structural flaws that represent a loss of crystalline perfection. The simplest type of flaw is the *point defect*, such as a missing atom (vacancy). This type of flaw is the inevitable result of the normal thermal vibration of atoms in any solid at a temperature above absolute zero. Linear defects, or *dislocations,* follow an extended and sometimes complex path through the crystal structure. *Planar defects* represent the boundary between a nearly perfect crystalline region and its surroundings. Some materials are completely lacking in crystalline order. Common window glass is such a *noncrystalline solid.*

4.1 | The Solid Solution—Chemical Imperfection

It is not possible to avoid some contamination of practical materials. Even high-purity semiconductor products have some measurable level of impurity atoms. Many engineering materials contain significant amounts of several different

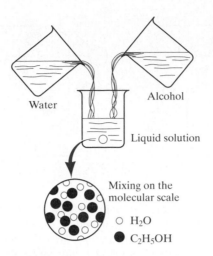

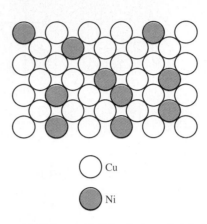

FIGURE 4.1 *Forming a liquid solution of water and alcohol. Mixing occurs on the molecular scale.*

FIGURE 4.2 *Solid solution of nickel in copper shown along a (100) plane. This is a* substitutional *solid solution with nickel atoms substituting for copper atoms on fcc atom sites.*

components. Commercial metal alloys are examples. As a result, all materials that the engineer deals with on a daily basis are actually **solid solutions**. At first, the concept of a solid solution may be difficult to grasp. In fact, it is essentially equivalent to the more familiar liquid solution, such as the water–alcohol system shown in Figure 4.1. The complete solubility of alcohol in water is the result of complete molecular mixing. A similar result is seen in Figure 4.2, which shows a solid solution of copper and nickel atoms sharing the fcc crystal structure. Nickel acts as a **solute** dissolving in the copper **solvent**. This particular configuration is referred to as a **substitutional solid solution** because the nickel atoms are substituting for copper atoms on the fcc atom sites. This configuration will tend to occur when the atoms do not differ greatly in size. The water–alcohol system shown in Figure 4.1 represents two liquids completely soluble in each other in all proportions. For this complete miscibility to occur in metallic solid solutions, the two metals must be quite similar, as defined by the **Hume–Rothery*** **rules**:

1. Less than about 15% difference in atomic radii

2. The same crystal structure

3. Similar electronegativities (the ability of the atom to attract an electron)

4. The same valence

If one or more of the Hume-Rothery rules are violated, only partial solubility is possible. For example, less than 2 at % (atomic percent) silicon is soluble in aluminum. Inspection of Appendixes 1 and 2 shows that Al and Si violate rules 1,

---

*William Hume-Rothery (1899–1968), British metallurgist, made major contributions to theoretical and experimental metallurgy as well as metallurgical education. His empirical rules of solid-solution formation have been a practical guide to alloy design for more than half a century.

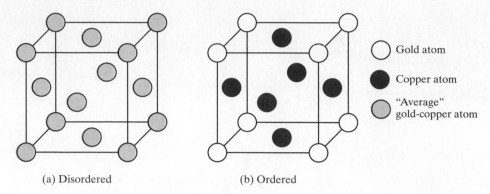

(a) Disordered   (b) Ordered

FIGURE 4.3 *Ordering of the solid solution in the AuCu₃ alloy system. (a) Above ~390°C, there is a random distribution of the Au and Cu atoms among the fcc sites. (b) Below ~390°C, the Au atoms preferentially occupy the corner positions in the unit cell, giving a simple cubic Bravais lattice. (From B. D. Cullity and S. R. Stock,* Elements of X-Ray Diffraction, *3rd ed., Prentice Hall, Upper Saddle River, NJ, 2001.)*

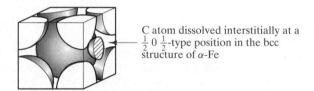

C atom dissolved interstitially at a
$\frac{1}{2} 0 \frac{1}{2}$-type position in the bcc
structure of α-Fe

FIGURE 4.4 *Interstitial solid solution of carbon in α-iron. The carbon atom is small enough to fit with some strain in the interstice (or opening) among adjacent Fe atoms in this structure of importance to the steel industry. [This unit-cell structure can be compared with that shown in Figure 3.4b.]*

2, and 4. Regarding rule 3, Figure 2.21 shows that the electronegativities of Al and Si are quite different, despite their adjacent positions on the periodic table.

Figure 4.2 shows a **random solid solution**. By contrast, some systems form **ordered solid solutions**. A good example is the alloy AuCu₃, shown in Figure 4.3. At high temperatures (above 390°C), thermal agitation keeps a random distribution of the Au and Cu atoms among the fcc sites. Below approximately 390°C, the Cu atoms preferentially occupy the face-centered positions, and the Au atoms preferentially occupy corner positions in the unit cell. Ordering may produce a new crystal structure similar to some of the ceramic compound structures. For AuCu₃ at low temperatures, the compound-like structure is based on a simple cubic Bravais lattice.

When atom sizes differ greatly, substitution of the smaller atom on a crystal structure site may be energetically unstable. In this case, it is more stable for the smaller atom simply to fit into one of the spaces, or interstices, among adjacent atoms in the crystal structure. Such an **interstitial solid solution** is displayed in Figure 4.4, which shows carbon dissolved interstitially in α-Fe. This interstitial solution is a dominant phase in steels. Although more stable than a substitutional configuration of C atoms on Fe lattice sites, the interstitial structure of Figure 4.4 produces considerable strain locally to the α-Fe crystal structure, and less than 0.1 at % C is soluble in α-Fe.

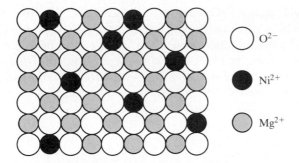

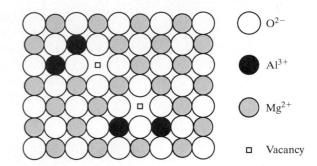

FIGURE 4.5 *Random, substitutional solid solution of NiO in MgO. The O²⁻ arrangement is unaffected. The substitution occurs among Ni²⁺ and Mg²⁺ ions.*

FIGURE 4.6 *A substitutional solid solution of Al₂O₃ in MgO is not as simple as the case of NiO in MgO (Figure 4.5). The requirement of charge neutrality in the overall compound permits only two Al³⁺ ions to fill every three Mg²⁺ vacant sites, leaving one Mg²⁺ vacancy.*

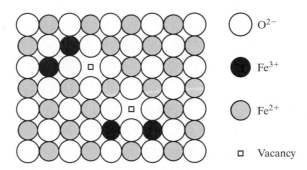

FIGURE 4.7 *Iron oxide, Fe₁₋ₓO with x ≈ 0.05, is an example of a nonstoichiometric compound. Similar to the case of Figure 4.6, both Fe²⁺ and Fe³⁺ ions occupy the cation sites, with one Fe²⁺ vacancy occurring for every two Fe³⁺ ions present.*

To this point, we have looked at solid-solution formation in which a pure metal or semiconductor solvent dissolves some solute atoms either substitutionally or interstitially. The principles of substitutional solid-solution formation in these elemental systems also apply to compounds. For example, Figure 4.5 shows a random, substitutional solid solution of NiO in MgO. Here, the O^{2-} arrangement is unaffected. The substitution occurs between Ni^{2+} and Mg^{2+}. The example of Figure 4.5 is a relatively simple one. In general, the charged state for ions in a compound affects the nature of the substitution. In other words, one could not indiscriminately replace all of the Ni^{2+} ions in Figure 4.5 with Al^{3+} ions. This replacement would be equivalent to forming a solid solution of Al_2O_3 in MgO, each having distinctly different formulas and crystal structures. The higher valence of Al^{3+} would give a net positive charge to the oxide compound, creating a highly unstable condition. As a result, an additional ground rule in forming compound solid solutions is the maintenance of charge neutrality.

Figure 4.6 shows how charge neutrality is maintained in a dilute solution of Al^{3+} in MgO by having only two Al^{3+} ions fill every three Mg^{2+} sites, which leaves one Mg^{2+} site vacancy for each two Al^{3+} substitutions. This type of vacancy and several other point defects will be discussed further in Section 4.2. This example of a defect compound suggests the possibility of an even more subtle type of solid solution. Figure 4.7 shows a **nonstoichiometric compound,** $Fe_{1-x}O$, in which x

is ~0.05. An ideally stoichiometric FeO would be identical to MgO with a NaCl-type crystal structure consisting of equal numbers of Fe^{2+} and O^{2-} ions. However, ideal FeO is never found in nature due to the multivalent nature of iron. Some Fe^{3+} ions are always present. As a result, these Fe^{3+} ions play the same role in the $Fe_{1-x}O$ structure as Al^{3+} plays in the Al_2O_3 in MgO solid solution of Figure 4.6. One Fe^{2+} site vacancy is required to compensate for the presence of every two Fe^{3+} ions in order to maintain charge neutrality.

EXAMPLE 4.1

Do Cu and Ni satisfy Hume-Rothery's first rule for complete solid solubility?

SOLUTION
From Appendix 2,

$$r_{Cu} = 0.128 \text{ nm},$$
$$r_{Ni} = 0.125 \text{ nm},$$

and

$$\% \text{ difference} = \frac{(0.128 - 0.125) \text{ nm}}{0.128 \text{ nm}} \times 100$$
$$= 2.3\% (< 15\%).$$

Therefore, yes.

In fact, all four rules are satisfied by these two neighbors from the periodic table (in agreement with the observation that they are completely soluble in all proportions).

EXAMPLE 4.2

How much "oversize" is the C atom in α-Fe? (See Figure 4.4.)

SOLUTION
By inspection of Figure 4.4, it is apparent that an ideal interstitial atom centered at $\frac{1}{2}0\frac{1}{2}$ would just touch the surface of the iron atom in the center of the unit-cell cube. The radius of such an ideal interstitial would be

$$r_{\text{interstitial}} = \frac{1}{2}a - R,$$

where a is the length of the unit-cell edge, and R is the radius of an iron atom.

Remembering Figure 3.4, we note that

$$\text{length of unit-cell body diagonal} = 4R$$
$$= \sqrt{3}a,$$

or

$$a = \frac{4}{\sqrt{3}} R,$$

as given in Table 3.3. Then,

$$r_{\text{interstitial}} = \frac{1}{2}\left(\frac{4}{\sqrt{3}}R\right) - R = 0.1547\,R.$$

From Appendix 2, $R = 0.124$ nm, giving

$$r_{\text{interstitial}} = 0.1547(0.124\ \text{nm}) = 0.0192\ \text{nm}.$$

However, Appendix 2 gives $r_{\text{carbon}} = 0.077$ nm, or

$$\frac{r_{\text{carbon}}}{r_{\text{interstitial}}} = \frac{0.077\ \text{nm}}{0.0192\ \text{nm}} = 4.01.$$

Therefore, the carbon atom is roughly four times too large to fit next to the adjacent iron atoms without strain. The severe local distortion required for this accommodation leads to the low solubility of C in α-Fe (< 0.1 at %).

PRACTICE PROBLEM 4.1

Copper and nickel (which are completely soluble in each other) satisfy the first Hume-Rothery rule of solid solubility, as shown in Example 4.1. Aluminum and silicon are soluble in each other to only a limited degree. Do they satisfy the first Hume-Rothery rule? **(Note that the solutions to all practice problems are provided on the related Web site.)**

PRACTICE PROBLEM 4.2

The interstitial site for dissolving a carbon atom in α-Fe was shown in Figure 4.4. Example 4.2 shows that a carbon atom is more than four times too large for the site and, consequently, carbon solubility in α-Fe is quite low. Consider now the case for interstitial solution of carbon in the high-temperature (fcc) structure of γ-Fe. The largest interstitial site for a carbon atom is a $\frac{1}{2}01$ type. **(a)** Sketch this interstitial solution in a manner similar to the structure shown in Figure 4.4. **(b)** Determine by how much the C atom in γ-Fe is oversize. (Note that the atomic radius for fcc iron is 0.127 nm.)

4.2 | Point Defects—Zero-Dimensional Imperfections

Structural defects exist in real materials independently of chemical impurities. Imperfections associated with the crystalline point lattice are called **point defects**. Figure 4.8 illustrates the two common types of point defects associated with elemental solids: (1) The **vacancy** is simply an unoccupied atom site in the crystal structure, and (2) the interstitial, or **interstitialcy**, is an atom occupying an interstitial site not normally occupied by an atom in the perfect crystal structure or an extra atom inserted into the perfect crystal structure such that two atoms occupy positions close to a singly occupied atomic site in the perfect structure. In the preceding section, we saw how vacancies can be produced in compounds as a response to chemical impurities and nonstoichiometric compositions. Such vacancies can also occur independently of these chemical factors (e.g., by the thermal vibration of atoms in a solid above a temperature of absolute zero).

Figure 4.9 illustrates the two analogs of the vacancy and interstitialcy for compounds. The **Schottky* defect** is a pair of oppositely charged ion vacancies. This pairing is required in order to maintain local charge neutrality in the compound's crystal structure. The **Frenkel defect**[†] is a vacancy–interstitialcy combination. Most of the compound crystal structures described in Chapter 3 were too "tight" to allow Frenkel defect formation. However, the relatively open CaF_2-type structure can accommodate cation interstitials without excessive lattice strain. Defect structures in compounds can be further complicated by charging due to "electron trapping" or "electron hole trapping" at these lattice imperfections. We

These atomic scale defects play a central role in the diffusion mechanisms of Chapter 5.

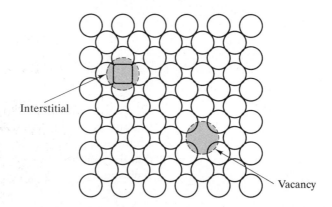

Interstitial

Vacancy

FIGURE 4.8 *Two common point defects in metal or elemental semiconductor structures are the vacancy and the interstitial.*

---

*Walter Hans Schottky (1886–1976), German physicist, was the son of a prominent mathematician. Besides identifying the *Schottky defect,* he invented the screen-grid tube (in 1915) and discovered the *Schottky effect* of thermionic emission (i.e., the current of electrons leaving a heated metal surface increases when an external electrical field is applied).

[†]Yakov Ilyich Frenkel (1894–1954), Russian physicist, made significant contributions to a wide range of areas, including solid-state physics, electrodynamics, and geophysics. Although his name is best remembered in conjunction with defect structure, he was an especially strong contributor to the understanding of ferromagnetism (a topic that is discussed in Chapter 13D on the related Web site).

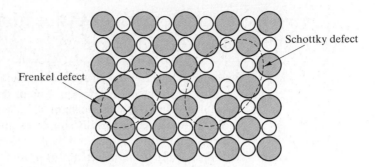

FIGURE 4.9 *Two common point defect structures in compound structures are the Schottky defect and the Frenkel defect. Note their similarity to the structures shown in Figure 4.8.*

shall not dwell on these more complex systems now, but they can have important implications for optical properties (as discussed in the supplementary Chapter 13B available on the related Web site).

EXAMPLE 4.3

The fraction of vacant lattice sites in a crystal is typically small. For example, the fraction of aluminum sites vacant at 400°C is 2.29×10^{-5}. Calculate the density of these sites (in units of m^{-3}).

SOLUTION
From Appendix 1, we find the density of aluminum to be 2.70 Mg/m^3 and its atomic mass to be 26.98 amu. The corresponding density of aluminum atoms is then

$$\text{at. density} = \frac{\rho}{\text{at. mass}} = \frac{2.70 \times 10^6 \ g/m^3}{26.98 \ g/(0.602 \times 10^{24} \ \text{atoms})}$$

$$= 6.02 \times 10^{28} \ \text{atoms·}m^{-3}.$$

Then, the density of vacant sites will be

$$\text{vac. density} = 2.29 \times 10^{-5} \ \text{atom}^{-1} \times 6.02 \times 10^{28} \ \text{atoms·}m^{-3}$$

$$= 1.38 \times 10^{24} \ m^{-3}.$$

PRACTICE PROBLEM 4.3

Calculate the density of vacant sites (in m^{-3}) for aluminum at 660°C (just below its melting point) where the fraction of vacant lattice sites is 8.82×10^{-4}. (See Example 4.3.)

4.3 | Linear Defects, or Dislocations—One-Dimensional Imperfections

We have seen that point (zero-dimensional) defects are structural imperfections resulting from thermal agitation. **Linear defects**, which are one-dimensional, are associated primarily with mechanical deformation. Linear defects are also known as **dislocations**. An especially simple example is shown in Figure 4.10. The linear defect is commonly designated by the "inverted T" symbol (⊥), which represents the edge of an *extra half-plane of atoms.* Such a configuration lends itself to a simple quantitative designation, the **Burgers*** **vector**, **b**. This parameter is simply the displacement vector necessary to close a stepwise loop around the defect. In the perfect crystal (Figure 4.11a), an $m \times n$ atomic step loop closes at the starting point. In the region of a dislocation (Figure 4.11b), the same loop fails to close. The closure vector **(b)** represents the magnitude of the structural defect. In Chapter 6, we shall see that the magnitude of **b** for the common metal structures (bcc, fcc, and hcp) is simply the repeat distance along the highest atomic density direction (the direction in which atoms are touching).

Figure 4.10 represents a specific type of linear defect, the **edge dislocation**, so named because the defect, or *dislocation line,* runs along the edge of the extra row of atoms. For the edge dislocation, the Burgers vector is perpendicular to the dislocation line. Figure 4.12 shows a fundamentally different type of linear defect, the **screw dislocation**, which derives its name from the spiral stacking of crystal planes around the dislocation line. *For the screw dislocation, the Burgers vector is parallel to the dislocation line.* The edge and screw dislocations can be considered the pure extremes of linear defect structure. Most linear defects in actual materials will be mixed, as shown in Figure 4.13. In this general case, the **mixed dislocation** has both edge and screw character. The Burgers vector for the mixed dislocation is neither perpendicular nor parallel to the dislocation line, but instead retains a fixed orientation in space consistent with the previous definitions

Nanometer scale defects such as this dislocation play a central role in the plastic deformation mechanisms of Chapter 6.

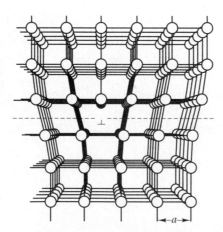

FIGURE 4.10 *Edge dislocation. The linear defect is represented by the edge of an extra half-plane of atoms. (From A. G. Guy,* Elements of Physical Metallurgy, *Addison-Wesley Publishing Co., Inc., Reading, MA, 1959.)*

*Johannes Martinus Burgers (1895–1981), Dutch-American fluid mechanician. Although his highly productive career centered on aerodynamics and hydrodynamics, a brief investigation of dislocation structure around 1940 has made Burgers's name one of the best known in materials science. He was the first to identify the convenience and utility of the closure vector for characterizing a dislocation.

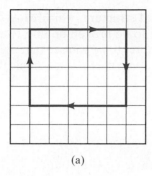

(a)

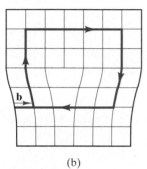

(b)

FIGURE 4.11 *Definition of the Burgers vector,* **b**, *relative to an edge dislocation. (a) In the perfect crystal, an m × n atomic step loop closes at the starting point. (b) In the region of a dislocation, the same loop does not close, and the closure vector* **(b)** *represents the magnitude of the structural defect. For the edge dislocation, the Burgers vector is perpendicular to the dislocation line.*

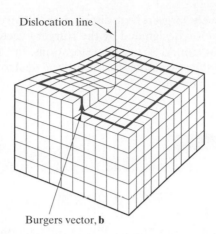

FIGURE 4.12 *Screw dislocation. The spiral stacking of crystal planes leads to the Burgers vector being* parallel *to the dislocation line.*

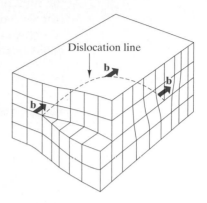

FIGURE 4.13 *Mixed dislocation. This dislocation has both edge and screw character with a single Burgers vector consistent with the pure edge and pure screw regions.*

for the pure edge and pure screw regions. The local atomic structure around a mixed dislocation is difficult to visualize, but the Burgers vector provides a convenient and simple description. In compound structures, even the basic Burgers vector designation can be relatively complicated. Figure 4.14 shows the Burgers vector for the aluminum oxide structure (Section 3.3). The complication arises

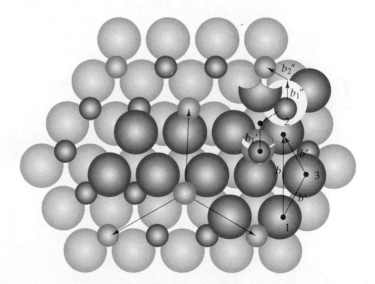

FIGURE 4.14 *Burgers vector for the aluminum oxide structure. The large repeat distance in this relatively complex structure causes the Burgers vector to be broken up into two (for O^{2-}) or four (for Al^{3+}) partial dislocations, each representing a smaller slip step. This complexity is associated with the brittleness of ceramics compared with metals. (From W. D. Kingery, H. K. Bowen, and D. R. Uhlmann,* Introduction to Ceramics, *2nd ed., John Wiley & Sons, Inc., New York, 1976.)*

from the relatively large repeat distance in this crystal structure, which causes the total dislocation designated by the Burgers vector to be broken up into two (for O^{2-}) or four (for Al^{3+}) partial dislocations. In Chapter 6, we will see that the complexity of dislocation structures has a good deal to do with the basic mechanical behavior of the material.

EXAMPLE 4.4

Calculate the magnitude of the Burgers vector for **(a)** α-Fe, **(b)** Al, and **(c)** Al_2O_3.

SOLUTION

(a) As noted in the opening of this section, $|\mathbf{b}|$ is merely the repeat distance between adjacent atoms along the highest atomic density direction. For α-Fe, a bcc metal, this distance tends to be along the body diagonal of a unit cell. We saw in Figure 3.4 that Fe atoms are in contact along the body diagonal. As a result, the atomic repeat distance is

$$r = 2R_{Fe}.$$

Using Appendix 2, we can then calculate, in a simple way,

$$|\mathbf{b}| = r = 2(0.124 \text{ nm}) = 0.248 \text{ nm}.$$

(b) Similarly, the highest atomic density direction in fcc metals such as Al tends to be along the face diagonal of a unit cell. As shown in Figure 3.5, this direction is also a line of contact for atoms in an fcc structure. Again,

$$|\mathbf{b}| = r = 2R_{Al} = 2(0.143 \text{ nm})$$
$$= 0.286 \text{ nm}.$$

(c) Figure 4.14 shows how the situation is more complex for ceramics. The total slip vector connects two O^{2-} ions (labeled 1 and 2):

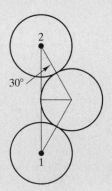

Thus,

$$|\mathbf{b}| = (2)(2R_{O^{2-}})(\cos 30°).$$

Using Appendix 2, we can calculate

$$|\mathbf{b}| = (2)(2 \times 0.132 \text{ nm})(\cos 30°)$$

$$= 0.457 \text{ nm}.$$

PRACTICE PROBLEM 4.4

Calculate the magnitude of the Burgers vector for an hcp metal, Mg. (See Example 4.4.)

4.4 | Planar Defects—Two-Dimensional Imperfections

Point defects and linear defects are acknowledgments that crystalline materials cannot be made flaw-free. These imperfections exist in the interior of each of these materials. But we must also consider that we are limited to a finite amount of any material and that material is contained within some boundary surface. This surface is, in itself, a disruption of the atomic-stacking arrangement of the crystal. There are various forms of **planar defects**. We shall briefly list them beginning with the one that is the simplest geometrically.

Figure 4.15 illustrates a **twin boundary**, which separates two crystalline regions that are, structurally, mirror images of each other. This highly symmetrical discontinuity in structure can be produced by deformation (e.g., in bcc and hcp metals) and by annealing (e.g., in fcc metals).

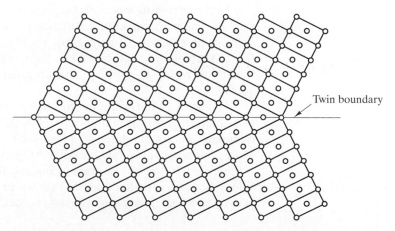

Twin boundary

FIGURE 4.15 *A twin boundary separates two crystalline regions that are, structurally, mirror images of each other.*

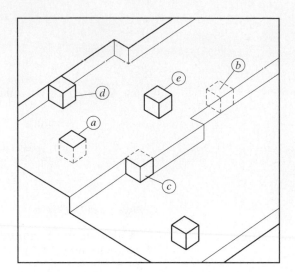

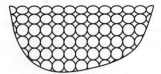

FIGURE 4.16 *Simple view of the surface of a crystalline material.*

FIGURE 4.17 *A more detailed model of the elaborate ledgelike structure of the surface of a crystalline material. Each cube represents a single atom. [From J. P. Hirth and G. M. Pound, J. Chem. Phys. 26, 1216 (1957).]*

All crystalline materials do not exhibit twin boundaries, but all must have a *surface*. A simple view of the crystalline surface is given in Figure 4.16. This surface is little more than an abrupt end to the regular atomic stacking arrangement. One should note that this schematic illustration indicates that the surface atoms are somehow different from interior (or "bulk") atoms. This is the result of different coordination numbers for the surface atoms leading to different bonding strengths and some asymmetry. A more detailed picture of atomic-scale surface geometry is shown in Figure 4.17. This **Hirth–Pound*** **model** of a crystal surface has elaborate ledge systems rather than atomically smooth planes.

The most important planar defect for our consideration in this introductory course occurs at the **grain boundary**, the region between two adjacent single crystals, or **grains**. In the most common planar defect, the grains meeting at the boundary have different orientations. Aside from the electronics industry, most practical engineering materials are polycrystalline rather than in the form of single crystals. The predominant microstructural feature of many engineering materials is the grain structure (Figure 4.18). Many materials' properties are highly sensitive to such grain structures. What, then, is the structure of a grain boundary on the atomic scale? The answer depends greatly on the relative orientations of the adjacent grains.

Figure 4.19 illustrates an unusually simple grain boundary produced when two adjacent grains are tilted only a few degrees relative to each other. This **tilt boundary** is accommodated by a few isolated edge dislocations (see Section 4.3). Most grain boundaries involve adjacent grains at some arbitrary and rather large misorientation angle. The grain-boundary structure in this general case is considerably more complex than that shown in Figure 4.19. However, considerable progress has been made in the past two decades in understanding the nature of

---

*John Price Hirth (1930–) and Guy Marshall Pound (1920–1988), American metallurgists, formulated their model of crystal surfaces in the late 1950s after careful analysis of the kinetics of vaporization.

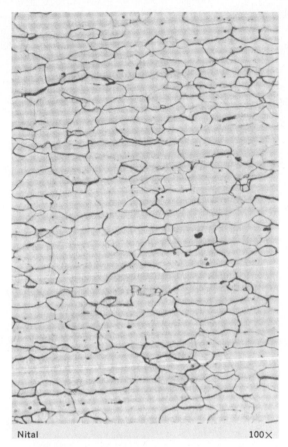

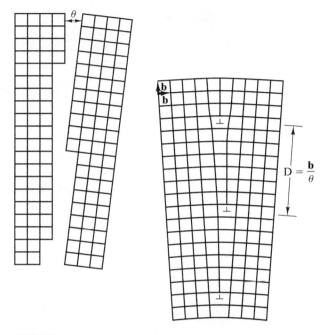

FIGURE 4.18 *Typical optical micrograph of a grain structure, 100×. The material is a low-carbon steel. The grain boundaries have been lightly etched with a chemical solution so that they reflect light differently from the polished grains, thereby giving a distinctive contrast. (From* Metals Handbook, *8th ed., Vol. 7:* Atlas of Microstructures of Industrial Alloys, *American Society for Metals, Metals Park, OH, 1972.)*

FIGURE 4.19 *Simple grain-boundary structure. This is termed a tilt boundary because it is formed when two adjacent crystalline grains are tilted relative to each other by a few degrees (θ). The resulting structure is equivalent to isolated edge dislocations separated by the distance b/θ, where b is the length of the Burgers vector,* **b***. (From W. T. Read,* Dislocations in Crystals, *McGraw-Hill Book Company, New York, 1953. Reprinted with permission of the McGraw-Hill Book Company.)*

the structure of the general, high-angle grain boundary. Advances in both electron microscopy and computer modeling techniques have played primary roles in this improved understanding.

These theoretical and experimental studies of high-angle boundaries have indicated that the simple, low-angle model of Figure 4.19 serves as a useful analogy for the high-angle case. Specifically, a grain boundary between two grains at some arbitrary, high angle will tend to consist of regions of good correspondence separated by **grain-boundary dislocations** (GBD), linear defects within the boundary plane. The GBD associated with high-angle boundaries tend to be *secondary* in that they have Burgers vectors different from those found in the bulk material (*primary* dislocations).

With atomic-scale structure in mind, we can return to the microstructural view of grain structures (e.g., Figure 4.18). In describing microstructures, it is

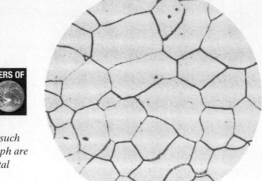

Micrometer-scale grain sizes such as those seen in this micrograph are typical of many common metal alloys.

FIGURE 4.20 *Specimen for the calculation of the grain-size number, G,* 100×. *The material is a low-carbon steel similar to that shown in Figure 4.18. (From* Metals Handbook, *8th ed., Vol. 7:* Atlas of Microstructures of Industrial Alloys, *American Society for Metals, Metals Park, OH, 1972.)*

Nital   100×

useful to have a simple index of *grain size.* A frequently used parameter standardized by the American Society for Testing and Materials (ASTM) is the **grain-size number**, *G*, defined by

$$N = 2^{G-1}, \tag{4.1}$$

Software for grain-size calculations is available on the related Web site.

where N is the number of grains observed in an area of 1 in.^{2} (= 645 mm^2) on a photomicrograph taken at a magnification of 100 times (100×), as shown in Figure 4.20. The calculation of G follows.

There are 21 grains within the field of view and 20 grains cut by the circumference, giving

$$21 + \frac{22}{2} = 32 \text{ grains}$$

in a circular area with diameter = 2.25 in. The area density of grains is

$$N = \frac{32 \text{ grains}}{\pi (2.25/2)^2 \text{ in.}^2} = 8.04 \frac{\text{grains}}{\text{in.}^2}.$$

From Equation 4.1,

$$N = 2^{(G-1)}$$

or

$$G = \frac{\ln N}{\ln 2} + 1$$

$$= \frac{\ln (8.04)}{\ln 2} + 1$$

$$= 4.01.$$

Although the grain-size number is a useful indicator of average grain size, it has the disadvantage of being somewhat indirect. It would be useful to obtain an average value of *grain diameter* from a microstructural section. A simple indicator is to count the number of grains intersected per unit length, n_L, of a random line drawn across a micrograph. The average grain size is roughly indicated by the inverse of n_L, corrected for the magnification, M, of the micrograph. Of course, one must consider that the random line cutting across the micrograph (in itself, a random plane cutting through the microstructure) will not tend, on average, to go along the maximum diameter of a given grain. Even for a microstructure of uniform size grains, a given planar slice (micrograph) will show various size grain sections (e.g., Figure 4.20), and a random line would indicate a range of segment lengths defined by grain-boundary intersections. In general, then, the true average grain diameter, d, is given by

$$d = \frac{C}{n_L M},$$ (4.2)

where C is some constant greater than 1. Extensive analysis of the statistics of grain structures has led to various theoretical values for the constant, C. For typical microstructures, a value of $C = 1.5$ is adequate.

EXAMPLE 4.5

Calculate the separation distance of dislocations in a low-angle ($\theta = 2°$) tilt boundary in aluminum.

SOLUTION
As calculated in Example 4.4b,

$$|\mathbf{b}| = 0.286 \text{ nm}.$$

From Figure 4.19, we see that

$$D = \frac{|\mathbf{b}|}{\theta}$$

$$= \frac{0.286 \text{ nm}}{2° \times (1 \text{ rad}/57.3°)} = 8.19 \text{ nm}.$$

EXAMPLE 4.6

Find the grain-size number, G, for the microstructure in Figure 4.20 if the micrograph represents a magnification of 300× rather than 100×.

SOLUTION
There would still be $21 + 11 = 32$ grains in the 3.98-in.^{2} region. But to scale this grain density to 100×, we must note that the 3.98-in.^{2} area at 300× would be comparable to an area at 100× of

$$A_{100\times} = 3.98 \text{ in.}^2 \times \left(\frac{100}{300}\right)^2 = 0.442 \text{ in.}^2.$$

Then, the grain density becomes

$$N = \frac{32 \text{ grains}}{0.442 \text{ in.}^2} = 72.4 \text{ grains/in.}^2.$$

Applying Equation 4.1 gives us

$$N = 2^{(G-1)}$$

or

$$\ln N = (G - 1)\ln 2,$$

giving

$$G - 1 = \frac{\ln N}{\ln 2}$$

and, finally,

$$G = \frac{\ln N}{\ln 2} + 1$$

$$= \frac{\ln (72.4)}{\ln 2} + 1 = 7.18$$

or

$$G = 7 + .$$

PRACTICE PROBLEM 4.5

In Example 4.5, we find the separation distance between dislocations for a 2° tilt boundary in aluminum. Repeat this calculation for **(a)** $\theta = 1°$ and **(b)** $\theta = 5°$. **(c)** Plot the overall trend of D versus θ over the range $\theta = 0$ to 5°.

PRACTICE PROBLEM 4.6

Figure 4.20 gives a sample calculation of grain-size number, G. Example 4.6 recalculates G assuming a magnification of 300× rather than 100×. Repeat this process, assuming that the micrograph in Figure 4.20 is at 50× rather than 100×.

4.5 | Noncrystalline Solids—Three-Dimensional Imperfections

Some engineering materials lack the repetitive, crystalline structure. These **noncrystalline**, or amorphous, **solids** are imperfect in three dimensions. The two-dimensional schematic of Figure 4.21a shows the repetitive structure of a hypothetical crystalline oxide. Figure 4.21b shows a noncrystalline version of this

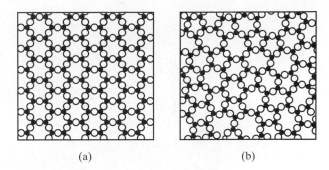

(a)   (b)

FIGURE 4.21 *Two-dimensional schematics give a comparison of (a) a crystalline oxide and (b) a noncrystalline oxide. The noncrystalline material retains short-range order (the triangularly coordinated building block), but loses long-range order (crystallinity). This illustration was also used to define glass in Chapter 1 (Figure 1.8).*

material. The latter structure is referred to as the **Zachariasen*** **model** and, in a simple way, it illustrates the important features of **oxide glass** structures. (Remember from Chapter 1 that glass generally refers to a noncrystalline material with a chemical composition comparable to that of a ceramic.) The building block of the crystal (the AO_3^{3-} "triangle") is retained in the glass; that is, **short-range order** (SRO) is retained. But **long-range order** (LRO)—that is, crystallinity—is lost in the glass. The Zachariasen model is the visual definition of the **random network theory** of glass structure, which is the analog of the point lattice associated with crystal structure.

Our first example of a noncrystalline solid was the traditional oxide glass because many oxides (especially the silicates) are easy to form in a noncrystalline state, which is the direct result of the complexity of the oxide crystal structures. Rapidly cooling a liquid silicate or allowing a silicate vapor to condense on a cool substrate effectively "freezes in" the random stacking of silicate building blocks (SiO_4^{4-} tetrahedra). Since many silicate glasses are made by rapidly cooling liquids, the term *supercooled liquid* is often used synonymously with *glass*. In fact, however, there is a distinction. The supercooled liquid is the material cooled just below the melting point, where it still behaves like a liquid (e.g., deforming by a viscous flow mechanism). The glass is the same material cooled to a sufficiently low temperature so that it has become a truly rigid solid (e.g., deforming by an elastic mechanism). The relationship of these various terms is illustrated in Figure 6.41. The atomic mobility of the material at these low temperatures is insufficient for the theoretically more stable crystalline structures to form. Those semiconductors with structures similar to some ceramics can be made in amorphous forms also. There is an economic advantage to **amorphous semiconductors** compared with preparing high-quality single crystals. A disadvantage is the greater complexity of the electronic properties. As discussed in Section 3.4, the complex polymeric structure of plastics causes a substantial fraction of their volume to be noncrystalline.

---

*William Houlder Zachariasen (1906–1980), Norwegian-American physicist, spent most of his career working in x-ray crystallography. However, his description of glass structure in the early 1930s became a standard definition for the structure of this noncrystalline material.

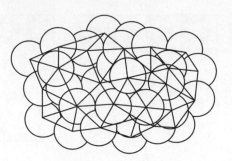

FIGURE 4.22 *Bernal model of an amorphous metal structure. The irregular stacking of atoms is represented as a connected set of polyhedra. Each polyhedron is produced by drawing lines between the centers of adjacent atoms. Such polyhedra are irregular in shape and the stacking is not repetitive.*

Perhaps the most intriguing noncrystalline solids are the newest members of the class, **amorphous metals**, also known as *metallic glasses*. Because metallic crystal structures are typically simple in nature, they can be formed quite easily. It is necessary for liquid metals to be cooled very rapidly to prevent crystallization. Cooling rates of $1°C$ per microsecond are required in typical cases. This is an expensive process, but potentially worthwhile due to the unique properties of these materials. For example, the uniformity of the noncrystalline structure eliminates the grain boundary structures associated with typical polycrystalline metals, which results in unusually high strengths and excellent corrosion resistance. Figure 4.22 illustrates a useful method for visualizing an amorphous metal structure: the **Bernal* model**, which is produced by drawing lines between the centers of adjacent atoms. The resulting polyhedra are irregular in shape, and lack any repetitive stacking arrangement.

At this point, it may be unfair to continue to use the term *imperfect* as a general description of noncrystalline solids. The Zachariasen structure (Figure 4.21b) is uniformly and "perfectly" random. Imperfections such as chemical impurities, however, can be defined relative to the uniformly noncrystalline structure, as shown in Figure 4.23. Addition of Na^+ ions to silicate glass substantially increases formability of the material in the supercooled liquid state (i.e., viscosity is reduced).

Finally, the state of the art in our understanding of the structure of noncrystalline solids is represented by Figure 4.24, which shows the nonrandom arrangement of Ca^{2+} modifier ions in a $CaO–SiO_2$ glass. What we see in Figure 4.24 is, in fact, adjacent octahedra rather than Ca^{2+} ions. Each Ca^{2+} ion is coordinated by six O^{2-} ions in a perfect octahedral pattern. In turn, the octahedra tend to be arranged in a regular, edge-sharing fashion, which is in sharp contrast to the random distribution of Na^+ ions implied in Figure 4.23. The evidence for **medium-range order** in the study represented by Figure 4.24 confirms long-standing theories of a tendency for some structural order to occur in the medium range of a few nanometers, between the well-known short-range order of the silica tetrahedra and the long-range randomness of the irregular linkage of those tetrahedra. As a practical matter, the random network model of Figure 4.21b is an adequate description of vitreous SiO_2. Medium-range order such as that shown in Figure 4.24 is, however,

---

*John Desmond Bernal (1901–1971), British physicist, was one of the pioneers in x-ray crystallography but is perhaps best remembered for his systematic descriptions of the irregular structure of liquids.

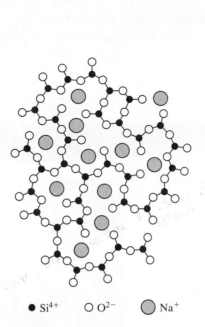

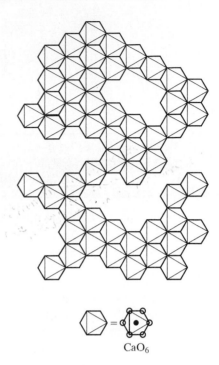

● Si^{4+} ○ O^{2-} ◉ Na^+

FIGURE 4.23 *A chemical impurity such as Na^+ is a glass modifier, breaking up the random network and leaving nonbridging oxygen ions. [From B. E. Warren, J. Am. Ceram. Soc. 24, 256 (1941).]*

FIGURE 4.24 *Schematic illustration of medium-range ordering in a CaO–SiO_2 glass. Edge-sharing CaO_6 octahedra have been identified by neutron-diffraction experiments. [From P. H. Gaskell et al., Nature 350, 675 (1991).]*

likely to be present in common glasses containing significant amounts of modifiers, such as Na_2O and CaO.

EXAMPLE 4.7

Randomization of atomic packing in amorphous metals (e.g., Figure 4.22) generally causes no more than a 1% drop in density compared with the crystalline structure of the same composition. Calculate the APF of an amorphous, thin film of nickel whose density is 8.84 g/cm³.

SOLUTION

Appendix 1 indicates that the normal density for nickel (which would be in the crystalline state) is 8.91 g/cm³. The APF for the fcc metal structure is 0.74 (see Section 3.2). Therefore, the APF for this amorphous nickel would be

$$APF = (0.74) \times \frac{8.84}{8.91} = 0.734.$$

PRACTICE PROBLEM 4.7

Estimate the APF of amorphous silicon if its density is reduced by 1% relative to the crystalline state. (Review Example 3.6.)

THE MATERIAL WORLD

Nanotechnology

When President Clinton extolled the potential of *nanotechnology* in his State of the Union Speech in 2000, he introduced a new and exciting scientific concept to many members of the general public. In response, materials scientists and engineers welcomed the public's awareness of a field that many of them had focused on for years. Many of the structural features in Chapter 4 fall within the range of 1 to 100 nm that has now come to define the *nanoscale*. As a practical matter, nanotechnology generally means more than simply observing features in this size range; instead, it means the control of those features in a functional engineering design.

The concept of performing science and engineering at the nanoscale originated with a famous lecture by physicist Richard Feynman in 1959 entitled "There's Plenty of Room at the Bottom." Subsequent pioneering efforts involving atomic imaging and nanofabrication justified the presidential boost in 2000 that was manifested in the U.S. National Nanotechnology Initiative and similar efforts in Europe and Japan. A good example of nanotechnology is shown as follows [i.e., nanoscale cantilevers similar to those used in atomic-force microscopes (AFM)]. The sharp AFM probe is mounted on a thin cantilever much like a small-scale phonograph needle on the stylus of an old-fashioned record player. These cantilevers have become increasingly fine-scaled, with the researchers at the IBM Zurich Research Laboratory, birthplace of atomic-scale microscopy, generally leading the way. In the case shown here, the cantilevers are 500 nanometers long by 100 nanometers wide, and the application is not microscopy but to measure the bending of cantilevers coated with DNA chains when exposed

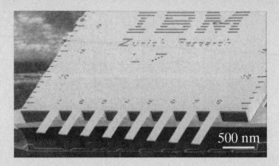

(Courtesy of International Business Machines Corporation. Unauthorized use not permitted.)

to an environment of other DNA molecules. In this way, the coated cantilevers can serve as sensitive probes for specific DNA sequences, an important application for the biotechnology field.

Materials scientists and engineers are generally interested in more than just creating spectacularly small-scale devices. As we found in Chapter 1, the relationship of structure and properties is a cornerstone of the field of materials science and engineering. The importance of the nanoscale to this relationship is becoming increasingly evident to researchers. Mechanical properties, especially strength, can be significantly improved when grain sizes can be maintained at 100 nm and below. In Chapter 13, we will find that nanoscale quantum wells, wires, and dots can provide exceptionally high operating speeds in semiconductor devices. Other researchers are finding that the quantum dots also provide unique optical properties. In general, the steady progress in the miniaturization of integrated circuits, as chronicled in Chapter 13, is requiring the move from the micrometer scale to the nanometer scale.

Summary

No real material used in engineering is as perfect as the structural descriptions of Chapter 3 would imply. There is always some contamination in the form of solid solution. When the impurity, or solute, atoms are similar to the solvent atoms, substitutional solution takes place in which impurity atoms rest on crystal-lattice sites. Interstitial solution takes place when a solute atom is small enough to occupy open spaces among adjacent atoms in the crystal structure. Solid solution in ionic compounds must account for charge neutrality of the material as a whole.

Point defects can be missing atoms or ions (vacancies) or extra atoms or ions (interstitialcies). Charge neutrality must be maintained locally for point defect structures in ionic compounds.

Linear defects, or dislocations, correspond to an extra half-plane of atoms in an otherwise perfect crystal. Although dislocation structures can be complex, they can also be characterized with a simple parameter, the Burgers vector.

Planar defects include any boundary surface surrounding a crystalline structure. Twin boundaries divide two mirror-image regions. The exterior surface has a characteristic structure involving an elaborate ledge system. The predominant microstructural feature for many engineering materials is grain structure, where each grain is a region with a characteristic crystal-structure orientation. A grain-size number (G) is used to quantify this microstructure. The structure of the region of mismatch between adjacent grains (i.e., the grain boundary) depends on the relative orientation of the grains.

Noncrystalline solids, on the atomic scale, are lacking in any long-range order (LRO), but may exhibit short-range order (SRO) associated with structural building blocks such as SiO_4^{4-} tetrahedra. Relative to a perfectly random structure, one can define solid solution, just as was done relative to perfectly crystalline structures. Recently, medium-range order has been found for the distribution of modifier ions such as Na^+ and Ca^{2+} in silicate glasses.

Key Terms

amorphous metal (120)
amorphous semiconductor (119)
Bernal model (120)
Burgers vector (110)
dislocation (110)
edge dislocation (110)
Frenkel defect (108)
grain (114)
grain boundary (114)
grain-boundary dislocation (GBD) (115)
grain-size number (116)
Hirth–Pound model (114)

Hume–Rothery rules (103)
interstitialcy (108)
interstitial solid solution (104)
linear defect (110)
long-range order (119)
medium-range order (120)
mixed dislocation (110)
noncrystalline solid (118)
nonstoichiometric compound (105)
ordered solid solution (104)
oxide glass (119)
planar defect (113)
point defect (108)

random network theory (119)
random solid solution (104)
Schottky defect (108)
screw dislocation (110)
short-range order (119)
solid solution (103)
solute (103)
solvent (103)
substitutional solid solution (103)
tilt boundary (114)
twin boundary (113)
vacancy (108)
Zachariasen model (119)

References

Chiang, **Y.**, **D. P. Birnie III**, and **W. D. Kingery**, *Physical Ceramics*, John Wiley & Sons, Inc., New York, 1997.

Hull, **D.**, and **D. J. Bacon**, *Introduction to Dislocations*, 4th ed., Butterworth-Heinemann, Boston, 2001.

Williams, **D. B.**, **A. R. Pelton**, and **R. Gronsky**, Eds., *Images of Materials*, Oxford University Press, New York, 1991. A comprehensive as well as beautiful example of the microscopic tools available for imaging the various structures introduced in this chapter.

Problems

Section 4.1 • The Solid Solution—Chemical Imperfection

4.1. In Chapter 9, we shall find a phase diagram for the Al–Cu system that indicates that these two metals do not form a complete solid solution. Which of the Hume-Rothery rules can you identify for Al–Cu that are violated? (For electronegativity data relative to rule 3, consult Figure 2.21.)

4.2. The Al–Mg system shows incomplete solid solution. Which of the Hume–Rothery rules are violated? (See Problem 4.1.)

4.3. For the Cu–Zn system with a phase diagram in Chapter 9 showing incomplete solid solution, which of the Hume–Rothery rules are violated? (See Problem 4.1.)

4.4. For the Pb–Sn system with a phase diagram in Chapter 9 showing incomplete solid solution, which of the Hume–Rothery rules are violated? (See Problem 4.1.)

4.5. Sketch the pattern of atoms in the (111) plane of the ordered $AuCu_3$ alloy shown in Figure 4.3. (Show an area at least five atoms wide by five atoms high.)

4.6. Sketch the pattern of atoms in the (110) plane of the ordered $AuCu_3$ alloy shown in Figure 4.3. (Show an area at least five atoms wide by five atoms high.)

4.7. Sketch the pattern of atoms in the (200) plane of the ordered $AuCu_3$ alloy shown in Figure 4.3. (Show an area at least five atoms wide by five atoms high.)

4.8. What are the equivalent points for ordered $AuCu_3$ (Figure 4.3)? (Note Problem 3.53.)

4.9. Although the Hume–Rothery rules apply strictly only to metals, the concept of similarity of cations corresponds to the complete solubility of NiO in MgO (Figure 4.5). Calculate the percent difference between cation sizes in this case.

4.10. Calculate the percent difference between cation sizes for Al_2O_3 in MgO (Figure 4.6), a system that does not exhibit complete solid solubility.

4.11. Calculate the number of Mg^{2+} vacancies produced by the solubility of 1 mol of Al_2O_3 in 99 mol of MgO (see Figure 4.6).

4.12. Calculate the number of Fe^{2+} vacancies in 1 mol of $Fe_{0.95}O$ (see Figure 4.7).

4.13. In Chapter 13, we shall be especially interested in "doped" semiconductors, in which small levels of impurities are added to an essentially pure semiconductor in order to produce desirable electrical properties. For silicon with 5×10^{21} aluminum atoms per cubic meter in solid solution, calculate **(a)** the atomic percent of aluminum atoms and **(b)** the weight percent of aluminum atoms.

4.14. For 5×10^{21} aluminum atoms/m^3 in solid solution in germanium, calculate **(a)** the atomic percent of aluminum atoms and **(b)** the weight percent of aluminum atoms.

4.15. For 5×10^{21} phosphorus atoms/m^3 in solid solution in silicon, calculate **(a)** the atomic percent of phosphorous atoms and **(b)** the weight percent of phosphorous atoms.

4.16. One way to determine a structural defect model (such as that shown in Figure 4.6 for a solid solution of Al_2O_3 in MgO) is to make careful density measurements. What would be the percent change in density for a 5 at % solution of Al_2O_3 in MgO (compared with pure, defect-free MgO)?

Section 4.2 • Point Defects—Zero-Dimensional Imperfections

4.17. Calculate the density of vacant sites (in m^{-3}) in a single crystal of silicon if the fraction of vacant lattice sites is 1×10^{-7}.

4.18. Calculate the density of vacant sites (in m^{-3}) in a single crystal of germanium if the fraction of vacant lattice sites is 1×10^{-7}.

4.19. Calculate the density of Schottky pairs (in m^{-3}) in MgO if the fraction of vacant lattice sites is 5×10^{-6}. (The density of MgO is 3.60 Mg/m^3.)

4.20. Calculate the density of Schottky pairs (in m^{-3}) in CaO if the fraction of vacant lattice sites is 5×10^{-6}. (The density of CaO is 3.45 Mg/m^3.)

Section 4.3 • Linear Defects, or Dislocations—One-Dimensional Imperfections

4.21. The energy necessary to generate a dislocation is proportional to the square of the length of the Burgers vector, $|\mathbf{b}|^2$. This relationship means that the most stable (lowest energy) dislocations have the minimum length, $|\mathbf{b}|$. For the bcc metal structure, calculate (relative to $E_{\mathbf{b}=[111]}$) the dislocation energies for **(a)** $E_{\mathbf{b}=[110]}$ and **(b)** $E_{\mathbf{b}=[100]}$.

4.22. The comments in Problem 4.21 also apply for the fcc metal structure. Calculate (relative to $E_{\mathbf{b}=[110]}$) the dislocation energies for **(a)** $E_{\mathbf{b}=[111]}$ and **(b)** $E_{\mathbf{b}=[100]}$.

4.23. The comments in Problem 4.21 also apply for the hcp metal structure. Calculate (relative to $E_{\mathbf{b}=[11\bar{2}0]}$) the dislocation energies for **(a)** $E_{\mathbf{b}=[1\bar{1}00]}$ and **(b)** $E_{\mathbf{b}=[0001]}$.

*__4.24.__ Figure 4.14 illustrates how a Burgers vector can be broken up into partials. The Burgers vector for an fcc metal can be broken up into two partials. **(a)** Sketch the partials relative to the full dislocation, and **(b)** identify the magnitude and crystallographic orientation of each partial.

Section 4.4 • Planar Defects—Two-Dimensional Imperfections

4.25. Determine the grain-size number, G, for the microstructure shown in Figure 4.18. (Keep in mind that the precise answer will depend on your choice of an area of sampling.)

4.26. Calculate the grain-size number for each of the two microstructures in the feature box in Chapter 1 (page 8).

4.27. Using Equation 4.2, estimate the average grain diameter of the two microstructures in the feature box in Chapter 1 (page 8) using "random lines" cutting across the diagonal of each figure from its lower-left corner to its upper-right corner. (See Problem 4.26.)

Section 4.5 • Noncrystalline Solids—Three-Dimensional Imperfections

4.28. Figure 4.21b is a useful schematic for simple B_2O_3 glass, composed of rings of BO_3^{3-} triangles. To appreciate the openness of this glass structure, calculate the size of the interstice (i.e., largest inscribed circle) of a regular six-membered ring of BO_3^{3-} triangles.

4.29. In amorphous silicates, a useful indication of the lack of crystallinity is the "ring statistics." For the schematic illustration in Figure 4.21b, plot a histogram of the n-membered rings of O^{2-} ions, where n = number of O^{2-} ions in a loop surrounding an open interstice in the network structure. [*Note:* In Figure 4.21a, all rings are six-membered ($n = 6$).] (**Hint**: Ignore incomplete rings at the edge of the illustration.)

*__4.30.__ As Figure 4.22 shows, noncrystalline metal has a range of space-filling polyhedra comprising its structure. Similarly, the fcc structure can be represented by a repetitive polyhedra structure, which is an alternative to our usual unit-cell configuration. In other words, the fcc structure can be equally represented by a space-filling stacking of regular polyhedra (tetrahedra and octahedra in a ratio of 2:1). **(a)** Sketch a typical tetrahedron (four-sided figure) on a perspective sketch such as that shown in Figure 3.5a. **(b)** Similarly, show a typical octahedron (eight-sided figure). (Note also Problem 3.60.)

*__4.31.__ In Problem 4.30, a tetrahedron and octahedron were identified as the appropriate polyhedra to define an fcc structure. For the hcp structure, the tetrahedron and octahedron are also the appropriate polyhedra. **(a)** Sketch a typical tetrahedron on a perspective sketch such as that shown in Figure 3.6a. **(b)** Similarly, show a typical octahedron. (Again, we are dealing with a crystalline solid in this example, but, as Figure 4.22 shows, the noncrystalline, amorphous metal has a range of such polyhedra that fill space.)

*__4.32.__ There are several polyhedra that can occur in noncrystalline solids, as discussed relative to Figure 4.22. The tetrahedron and octahedron treated in Problems 4.30 and 4.31 are the simplest. The next simplest is the pentagonal bipyramid, which consists of 10 equilateral triangle faces. Sketch this polyhedron as accurately as you can.

4.33. Sketch a few adjacent CaO_6 octahedra in the pattern shown in Figure 4.24. Indicate both the nearest neighbor $Ca^{2+}-Ca^{2+}$ distance, R_1, and the next-nearest neighbor $Ca^{2+}-Ca^{2+}$ distance, R_2.

4.34. Diffraction measurements on the CaO-SiO$_2$ glass represented by Figure 4.24 show that the nearest neighbor $Ca^{2+}-Ca^{2+}$ distance, R_1, is 0.375 nm. What would be the next-nearest neighbor $Ca^{2+}-Ca^{2+}$ distance, R_2? (Note the results of Problem 4.33.)

CHAPTER 5
Diffusion

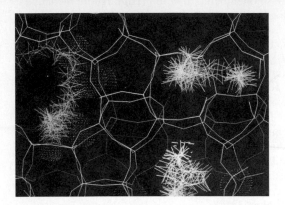

The principles introduced in this chapter can be applied to complex systems of relevance to modern technology. This computer simulation illustrates methane diffusing through a zeolite catalyst, an aluminosilicate ceramic with large channels that permit the transport of hydrocarbon molecules.

A link to diffusion modules is available on the related Web site.

During production and application, the chemical composition of engineering materials is often changed as a result of the movement of atoms, or *solid-state diffusion*. In some cases, atoms are redistributed within the microstructure of the material. In other cases, atoms are added from the material's environment, or atoms from the material may be discharged into the environment. Understanding the nature of the movement of atoms within the material can be critically important both in producing the material and in applying it successfully within an engineering design.

In Chapter 4, we were introduced to a variety of point defects, such as the vacancy. These defects were said to result typically from the thermal vibration of the atoms in the material. In this chapter, we will see the detailed relationship between temperature and the number of these defects. Specifically, the concentration of such defects rises exponentially with increasing temperature. The flow of atoms in engineering materials occurs by the movement of point defects, and, as a result, the rate of this solid-state diffusion increases exponentially with temperature. The mathematics of diffusion allows a precise description of the variation of the chemical composition within materials as a result of various diffusional processes. An important example is the *carburization* of steels, in which the surface is hardened by the diffusion of carbon atoms from a carbon-rich environment.

After some time, the chemical concentration profile within a material may become linear, and the corresponding mathematics for this *steady-state diffusion* is relatively simple.

Although we generally consider diffusion within the entire volume of a material, there are some cases in which the atomic transport occurs primarily along grain boundaries (by *grain boundary diffusion*) or along the surface of the material (by *surface diffusion*).

5.1 | Thermally Activated Processes

A large number of processes in materials science and engineering share a common feature—the process rate rises exponentially with temperature. The diffusivity of elements in metal alloys, the rate of creep deformation in structural materials,

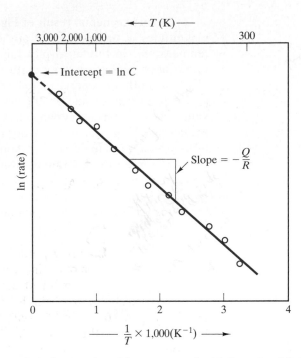

FIGURE 5.1 *Typical Arrhenius plot of data compared with Equation 5.2. The slope equals* $-Q/R$, *and the intercept (at* $1/T = 0$) *is* ln C.

and the electrical conductivity of semiconductors are a few examples that will be covered in this book. The general equation that describes these various processes is of the form

$$\text{rate} = Ce^{-Q/RT}, \tag{5.1}$$

where C is a **preexponential constant** (independent of temperature), Q is the **activation energy**, R is the universal gas constant, and T is the absolute temperature. It should be noted that the universal gas constant is as important for the solid state as it is for the gaseous state. The term *gas constant* derives from its role in the perfect gas law ($pV = nRT$) and related gas-phase equations. In fact, R is a fundamental constant that appears frequently in this book devoted to the solid state.

Equation 5.1 is generally referred to as the **Arrhenius* equation**. Taking the logarithm of each side of Equation 5.1 gives

$$\ln(\text{rate}) = \ln C - \frac{Q}{R}\frac{1}{T}. \tag{5.2}$$

By making a semilog plot of ln (rate) versus the reciprocal of absolute temperature ($1/T$), one obtains a straight-line plot of rate data (Figure 5.1). The slope of the resulting **Arrhenius plot** is $-Q/R$. Extrapolation of the Arrhenius plot to $1/T = 0$ (or $T = \infty$) gives an intercept equal to ln C.

---

*Svante August Arrhenius (1859–1927), Swedish chemist, made numerous contributions to physical chemistry, including the experimental demonstration of Equation 5.1 for chemical reaction rates.

The experimental result of Figure 5.1 is a very powerful one. Knowing the magnitudes of process rate at any two temperatures allows the rate at a third temperature (in the linear-plot range) to be determined. Similarly, knowledge of a process rate at any temperature and of the activation energy, Q, allows the rate at any other temperature to be determined. A common use of the Arrhenius plot is to obtain a value of Q from measurement of the slope of the plot. This value of activation energy can indicate the mechanism of the process. In summary, Equation 5.2 contains two constants. Therefore, only two experimental observations are required to determine them.

To appreciate why rate data show the characteristic behavior of Figure 5.1, we must explore the concept of the activation energy, Q. As used in Equation 5.1, Q has units of energy per mole. It is possible to rewrite this equation by dividing both Q and R by Avogadro's number (N_{AV}), giving

$$\text{rate} = Ce^{-q/kT}, \tag{5.3}$$

where q ($= Q/N_{AV}$) is the activation energy per atomic scale unit (e.g., atom, electron, and ion) and k ($= R/N_{AV}$) is Boltzmann's* constant ($13.8 \times 10^{-24}\ J/K$). Equation 5.3 provides for an interesting comparison with the high-energy end of the **Maxwell–Boltzmann**[†] **distribution** of molecular energies in gases,

$$P \propto e^{-\Delta E/kT}, \tag{5.4}$$

where P is the probability of finding a molecule at an energy ΔE greater than the average energy characteristic of a particular temperature, T. Herein lies the clue to the nature of the activation energy. It is the energy barrier that must be overcome by **thermal activation**. Although Equation 5.4 was originally developed for gases, it applies to solids as well. As temperature increases, a larger number of atoms (or any other species involved in a given process, such as electrons or ions) is available to overcome a given energy barrier, q. Figure 5.2 shows a *process path* in which a single atom overcomes an energy barrier, q. Figure 5.3 shows a simple mechanical model of activation energy in which a box is moved from one position to another by going through an increase in potential energy, ΔE, analogous to the q in Figure 5.2.

In the many processes described in the text where an Arrhenius equation applies, particular values of activation energy will be found to be characteristic of process mechanisms. In each case, it is useful to remember that various possible mechanisms may be occurring simultaneously within the material, and each mechanism has a characteristic activation energy. The fact that one activation energy is representative of the experimental data means simply that a single mechanism is dominant. If the process involves several sequential steps, the slowest step will be

---

*Ludwig Edward Boltzmann (1844–1906), Austrian physicist, is associated with many major scientific achievements of the 19th century (prior to the development of modern physics). The constant that bears his name plays a central role in the statistical statement of the second law of thermodynamics. Some ideas are difficult to put aside. His second law equation is carved on his tombstone.

[†]James Clerk Maxwell (1831–1879), Scottish mathematician and physicist, was an unusually brilliant and productive individual. His equations of electromagnetism are among the most elegant in all of science. He developed the kinetic theory of gases (including Equation 5.4) independently of his contemporary, Ludwig Edward Boltzmann.

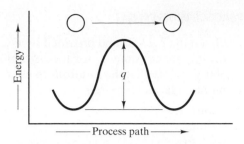

FIGURE 5.2 *Process path showing how an atom must overcome an activation energy, q, to move from one stable position to a similar adjacent position.*

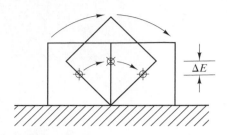

FIGURE 5.3 *Simple mechanical analog of the process path of Figure 5.2. The box must overcome an increase in potential energy, ΔE, in order to move from one stable position to another.*

the **rate-limiting step**. The activation energy of the rate-limiting step, then, will be the activation energy for the overall process.

EXAMPLE 5.1

The rate at which a metal alloy oxidizes in an oxygen-containing atmosphere is a typical example of the practical utility of the Arrhenius equation (Equation 5.1). For example, the rate of oxidation of a magnesium alloy is represented by a rate constant, k. The value of k at 300°C is 1.05×10^{-8} kg/(m$^4 \cdot$ s). At 400°C, the value of k rises to 2.95×10^{-4} kg/(m$^{-4} \cdot$ s). Calculate the activation energy, Q, for this oxidation process (in units of kJ/mol).

SOLUTION
For this specific case, Equation 5.1 has the form

$$k = Ce^{-Q/RT}.$$

Taking the ratio of rate constants at 300°C (= 573 K) and 400°C (= 673 K), we conveniently cancel out the unknown preexponential constant, C, and obtain

$$\frac{2.95 \times 10^{-4} \text{ kg/[m}^4 \cdot \text{s]}}{1.05 \times 10^{-8} \text{ kg/[m}^4 \cdot \text{s]}} = \frac{e^{-Q/(8.314 \text{ J/[mol·K]})(673 \text{ K})}}{e^{-Q/(8.314 \text{ J/[mol·K]})(573 \text{ K})}}$$

or

$$2.81 \times 10^4 = e^{\{-Q/(8.314 \text{ J/[mol·K]})\}\{1/(673 \text{ K})-1/(573 \text{ K})\}},$$

giving

$$Q = 328 \times 10^3 \text{ J/mol} = 328 \text{ kJ/mol}.$$

PRACTICE PROBLEM 5.1

Given the background provided by Example 5.1, calculate the value of the rate constant, k, for the oxidation of the magnesium alloy at $500°C$. **(Note that the solutions to all practice problems are provided on the related Web site.)**

5.2 | Thermal Production of Point Defects

Point defects occur as a direct result of the periodic oscillation, or **thermal vibration**, of atoms in the crystal structure. As temperature increases, the intensity of this vibration increases and, thereby, the likelihood of structural disruption and the development of point defects increase. At a given temperature, the thermal energy of a given material is fixed, but this is an average value. The thermal energy of individual atoms varies over a wide range, as indicated by the Maxwell–Boltzmann distribution. At a given temperature, a certain fraction of the atoms in the solid have sufficient thermal energy to produce point defects. An important consequence of the Maxwell–Boltzmann distribution is that this fraction increases exponentially with absolute temperature. As a result, the concentration of point defects increases exponentially with temperature; that is,

$$\frac{n_{\text{defects}}}{n_{\text{sites}}} = Ce^{-(E_{\text{defect}})/kT}, \tag{5.5}$$

where $n_{\text{defects}}/n_{\text{sites}}$ is the ratio of point defects to ideal crystal-lattice sites, C is a preexponential constant, E_{defect} is the energy needed to create a single-point defect in the crystal structure, k is Boltzmann's constant, and T is the absolute temperature.

The temperature sensitivity of point-defect production depends on the type of defect being considered; that is, E_{defect} for producing a vacancy in a given crystal structure is different from E_{defect} for producing an interstitialcy.

Figure 5.4 illustrates the thermal production of vacancies in aluminum. The slight difference between the thermal expansion measured by overall sample dimensions ($\Delta L/L$) and by x-ray diffraction ($\Delta a/a$) is the result of vacancies. The x-ray value is based on unit-cell dimensions measured by x-ray diffraction (Section 3.7). The increasing concentration of empty lattice sites (vacancies) in the material at temperatures approaching the melting point produces a measurably greater thermal expansion as measured by overall dimensions. The concentration of vacancies (n_v/n_{sites}) follows the Arrhenius expression of Equation 5.5,

$$\frac{n_v}{n_{\text{sites}}} = Ce^{-Ev/kT}, \tag{5.6}$$

where C is a preexponential constant and E_V is the energy of formation of a single vacancy. As discussed previously, this expression leads to a convenient semilog plot of data. Taking the logarithm of each side of Equation 5.6 gives

$$\ln\frac{n_v}{n_{\text{sites}}} = \ln C - \frac{E_V}{k}\frac{1}{T}. \tag{5.7}$$

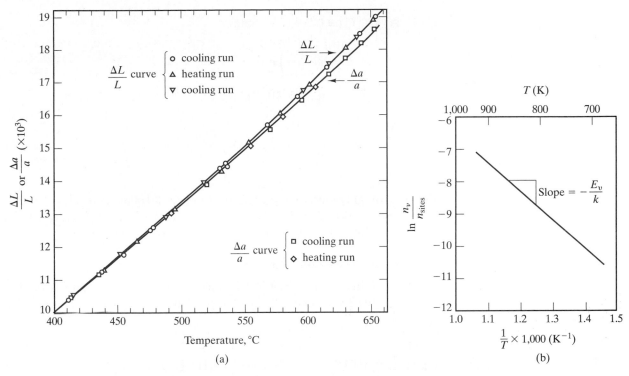

FIGURE 5.4 *(a) The overall thermal expansion $(\Delta L/L)$ of aluminum is measurably greater than the lattice parameter expansion $(\Delta a/a)$ at high temperatures because vacancies are produced by thermal agitation. (b) A semilog (Arrhenius-type) plot of ln (vacancy concentration) versus $1/T$ based on the data of part (a). The slope of the plot $(-E_v/k)$ indicates that 0.76 eV of energy is required to create a single vacancy in the aluminum crystal structure. (From P. G. Shewmon,* Diffusion in Solids, *McGraw-Hill Book Company, New York, 1963.)*

Figure 5.4 shows the linear plot of ln (n_v/n_{sites}) versus $1/T$. The slope of this Arrhenius plot is $-E_V/k$. These experimental data indicate that the energy required to create one vacancy in the aluminum crystal structure is 0.76 eV.

EXAMPLE 5.2

At 400°C, the fraction of aluminum lattice sites vacant is 2.29×10^{-5}. Calculate the fraction at 660°C (just below its melting point).

SOLUTION

From the text discussion relative to Figure 5.4, we have $E_V = 0.76$ eV. Using Equation 5.5, we have

$$\frac{n_v}{n_{\text{sites}}} = Ce^{-E_V/kT}.$$

At 400°C (= 673 K), we obtain

$$C = \left(\frac{n_v}{n_{sites}} \right) e^{+E_V/kT}$$

$$= (2.29 \times 10^{-5}) e^{+0.76 \text{ eV}/(86.2 \times 10^{-6} \text{ eV/K})(673 \text{ K})} = 11.2.$$

At 660°C (= 933 K),

$$\frac{n_v}{n_{sites}} = (11.2) e^{-0.76 \text{ eV}/(86.2 \times 10^{-6} \text{ eV/K})(933 \text{ K})} = 8.82 \times 10^{-4},$$

or roughly nine vacancies occur for every 10,000 lattice sites.

PRACTICE PROBLEM 5.2

Calculate the fraction of aluminum lattice sites vacant at **(a)** 500°C, **(b)** 200°C, and **(c)** room temperature (25°C). (See Example 5.2.)

5.3 | Point Defects and Solid–State Diffusion

At sufficient temperatures, atoms and molecules can be quite mobile in both liquids and solids. Watching a drop of ink fall into a beaker of water and spread out until all the water is evenly colored gives a simple demonstration of **diffusion**, the movement of molecules from an area of higher concentration to an area of lower concentration. But diffusion is not restricted to different materials. At room temperature, H_2O molecules in pure water are in continuous motion and are migrating through the liquid as an example of **self-diffusion**. This atomic-scale motion is relatively rapid in liquids and relatively easy to visualize. It is more difficult to visualize diffusion in rigid solids. Nonetheless, diffusion does occur in the solid state. A primary difference between solid-state and liquid-state diffusion is the low rate of diffusion in solids. Looking back at the crystal structures of Chapter 3, we can appreciate that diffusion of atoms or ions through those generally tight structures is difficult. In fact, the energy requirements to squeeze most atoms or ions through perfect crystal structures are so high as to make diffusion nearly impossible. To make solid-state diffusion practical, point defects are generally required. Figure 5.5 illustrates how atomic migration becomes possible without major crystal-structure distortion by means of a **vacancy migration** mechanism. It is important to note that the overall direction of material flow is opposite to the direction of vacancy flow.

Figure 5.6 shows diffusion by an interstitialcy mechanism and illustrates effectively the **random-walk** nature of atomic migration. This randomness does not preclude the net flow of material when there is an overall variation in chemical composition. This frequently occurring case is illustrated in Figures 5.7 and 5.8. Although each atom of solid A has an equal probability of randomly "walking" in any direction, the higher initial concentration of A on the left side of the system will cause such random motion to produce *interdiffusion*, a net flow of A atoms

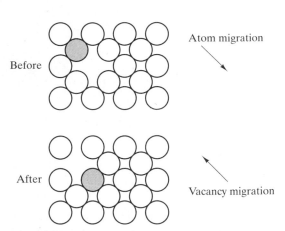

Before | Atom migration

After | Vacancy migration

FIGURE 5.5 *Atomic migration occurs by a mechanism of vacancy migration. Note that the overall direction of material flow (the atom) is opposite to the direction of vacancy flow.*

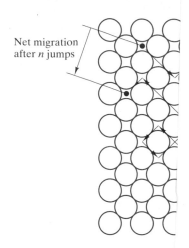

Net migration after *n* jumps

FIGURE 5.6 *Diffusion by an in〈 illustrating the random-walk na〈*

**POWERS OF
1**

*The schematic ill〈
show how the ato〈
4.8 facilitate diffu〈
these mechanism〈
will be seen in C〈*

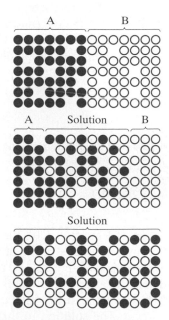

A   B

A   Solution   B

Solution

FIGURE 5.7 *The interdiffusion of materials A and B. Although any given A or B atom is equally likely to "walk" in any random direction (see Figure 5.6), the concentration gradients of the two materials can result in a net flow of A atoms into the B material, and vice versa. (From W. D. Kingery, H. K. Bowen, and D. R. Uhlmann,* Introduction to Ceramics, *2nd ed., John Wiley & Sons, Inc., New York, 1976.)*

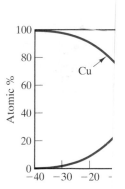

Original Cu–Ni interfa〈

FIGURE 5.8 *The int〈 an atomic scale was 〈 This interdiffusion o〈 comparable example〈*

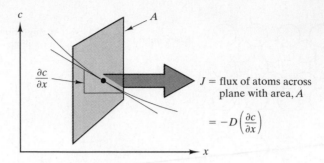

FIGURE 5.9 *Geometry of Fick's first law (Equation 5.8).*

into solid B. Similarly, solid B diffuses into solid A. The formal mathematical treatment of such diffusional flow begins with an expression known as **Fick's**[*] **first law**,

$$J_x = -D\frac{\partial c}{\partial x}, \tag{5.8}$$

where J_x is the *flux*, or flow rate, of the diffusing species in the x-direction due to a **concentration gradient** $(\partial c/\partial x)$. The proportionality coefficient, D, is called the **diffusion coefficient** or, simply, the **diffusivity**. The geometry of Equation 5.8 is illustrated in Figure 5.9. Figure 5.7 reminds us that the concentration gradient at a specific point along the diffusion path changes with time, t. This transient condition is represented by a second-order differential equation also known as **Fick's second law**,

$$\frac{\partial c_x}{\partial t} = \frac{\partial}{\partial x}\left(D\frac{\partial c_x}{\partial x}\right). \tag{5.9}$$

For many practical problems, one can assume that D is independent of c, leading to a simplified version of Equation 5.9:

$$\frac{\partial c_x}{\partial t} = D\frac{\partial^2 c_x}{\partial x^2}. \tag{5.10}$$

Figure 5.10 illustrates a common application of Equation 5.10, the diffusion of material into a semi-infinite solid while the surface concentration of the diffusing species, c_s, remains constant. Two examples of this system would be the plating of metals and the saturation of materials with reactive atmospheric gases. Specifically, steel surfaces are often hardened by **carburization**, the diffusion of carbon atoms into the steel from a carbon-rich environment. The solution to this differential equation with the given boundary conditions is

$$\frac{c_x - c_0}{c_s - c_0} = 1 - \text{erf}\left(\frac{x}{2\sqrt{Dt}}\right), \tag{5.11}$$

----

[*]Adolf Eugen Fick (1829–1901), German physiologist. The medical sciences frequently apply principles previously developed in the fields of mathematics, physics, and chemistry. However, Fick's work in the "mechanistic" school of physiology was so excellent that it served as a guide for the physical sciences. He developed the diffusion laws as part of a study of blood flow.

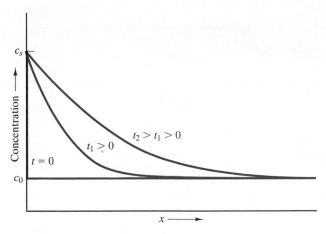

FIGURE 5.10 *Solution to Fick's second law (Equation 5.10) for the case of a semi-infinite solid, constant surface concentration of the diffusing species c_s, initial bulk concentration c_0, and a constant diffusion coefficient, D.*

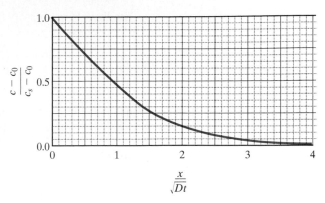

FIGURE 5.11 *Master plot summarizing all of the diffusion results of Figure 5.10 on a single curve.*

where c_0 is the initial bulk concentration of the diffusing species and erf refers to the **Gaussian* error function**, based on the integration of the "bell-shaped" curve with values readily available in mathematical tables. Representative values are given in Table 5.1. A great power of this analysis is that the result (Equation 5.11) allows all of the concentration profiles of Figure 5.10 to be redrawn on a single master plot (Figure 5.11). Such a plot permits rapid calculation of the time necessary for relative saturation of the solid as a function of x, D, and t. Figure 5.12 shows similar *saturation curves* for various geometries. It is important to keep in mind that these results are but a few of the large number of solutions that have been obtained by materials scientists for diffusion geometries in various practical processes.

The preceding mathematical analysis of diffusion implicitly assumed a fixed temperature. Our previous discussion of the dependence of diffusion on point defects causes us to expect a strong temperature dependence for diffusivity by analogy to Equation 5.5—and this is precisely the case. Diffusivity data are perhaps the best known examples of an Arrhenius equation

$$D = D_0 e^{-q/kT}, \tag{5.12}$$

where D_0 is the preexponential constant and q is the activation energy for defect motion. In general, q is not equal to the E_{defect} of Equation 5.5. E_{defect} represents the energy required for defect formation, while q represents the energy required for movement of that defect through the crystal structure ($E_{defect\ motion}$) for *interstitial diffusion*. For the vacancy mechanism, vacancy formation is an integral part of the diffusional process (see Figure 5.5), and $q = E_{defect} + E_{defect\ motion}$.

---

*Johann Karl Friedrich Gauss (1777–1855), German mathematician, was one of the great geniuses in the history of mathematics. In his teens, he developed the method of least squares for curve-fitting data. Much of his work in mathematics was similarly applied to physical problems, such as astronomy and geomagnetism. His contribution to the study of magnetism led to the unit of magnetic-flux density being named in his honor.

TABLE 5.1

The Error Function

z	$erf(z)$	z	$erf(z)$
0.00	0.0000	0.70	0.6778
0.01	0.0113	0.75	0.7112
0.02	0.0226	0.80	0.7421
0.03	0.0338	0.85	0.7707
0.04	0.0451	0.90	0.7969
0.05	0.0564	0.95	0.8209
0.10	0.1125	1.00	0.8427
0.15	0.1680	1.10	0.8802
0.20	0.2227	1.20	0.9103
0.25	0.2763	1.30	0.9340
0.30	0.3286	1.40	0.9523
0.35	0.3794	1.50	0.9661
0.40	0.4284	1.60	0.9763
0.45	0.4755	1.70	0.9838
0.50	0.5205	1.80	0.9891
0.55	0.5633	1.90	0.9928
0.60	0.6039	2.00	0.9953
0.65	0.6420		

Source: Handbook of Mathematical Functions, M. Abramowitz and I. A. Stegun, Eds., National Bureau of Standards, Applied Mathematics Series 55, Washington, DC, 1972.

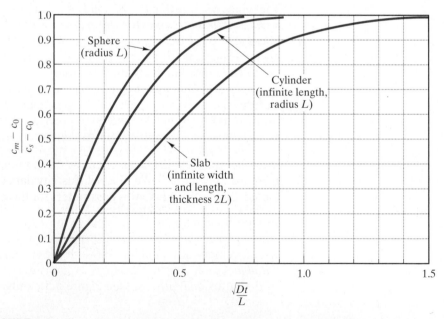

FIGURE 5.12 Saturation curves similar to that shown in Figure 5.11 for various geometries. The parameter c_m is the average concentration of diffusing species within the sample. Again, the surface concentration, c_s, and diffusion coefficient, D, are assumed to be constant. (From W. D. Kingery, H. K. Bowen, and D. R. Uhlmann, Introduction to Ceramics, 2nd ed., John Wiley & Sons, Inc., New York, 1976.)

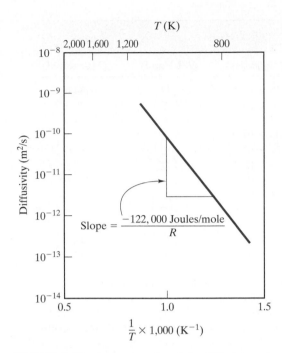

FIGURE 5.13 *Arrhenius plot of the diffusivity of carbon in α-iron over a range of temperatures. Note also related Figures 4.4 and 5.6 and other metallic diffusion data in Figure 5.14.*

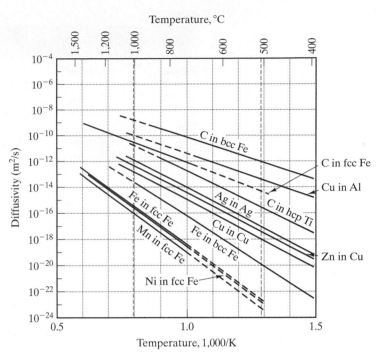

FIGURE 5.14 *Arrhenius plot of diffusivity data for a number of metallic systems. (From L. H. Van Vlack,* Elements of Materials Science and Engineering, *4th ed., Addison-Wesley Publishing Co., Inc., Reading, MA, 1980.)*

It is more common to tabulate diffusivity data in terms of molar quantities (i.e., with an activation energy, Q, per mole of diffusing species),

$$D = D_0 e^{-Q/RT}, \tag{5.13}$$

where R is the universal gas constant ($= N_{AV}k$, as discussed previously). Figure 5.13 shows an Arrhenius plot of the diffusivity of carbon in α-Fe over a range of temperatures, which is an example of an interstitialcy mechanism as sketched in Figure 5.6. Figure 5.14 collects diffusivity data for a number of metallic systems. Table 5.2 gives the Arrhenius parameters for these data. It is useful to compare different data sets. For instance, C can diffuse by an interstitialcy mechanism through bcc Fe more readily than through fcc Fe ($Q_{bcc} < Q_{fcc}$ in Table 5.2). The greater openness of the bcc structure (Section 3.2) makes this difference understandable. Similarly, the self-diffusion of Fe by a vacancy mechanism is greater in bcc Fe than in fcc Fe. Figure 5.15 and Table 5.3 give comparable diffusivity data for several nonmetallic systems. In many compounds, such as Al_2O_3, the smaller ionic species (e.g., Al^{3+}) diffuse much more readily through the system. The Arrhenius behavior of ionic diffusion in ceramic compounds is especially analogous to the temperature dependence of semiconductors to be discussed in Chapter 13. It is this ionic transport mechanism that is responsible for the semiconducting behavior of certain ceramics such as ZnO; that is, charged ions rather than electrons produce the measured electrical conductivity. Polymer data are not included with

TABLE 5.2

Diffusivity Data for a Number of Metallic Systems[a]				
Solute	**Solvent**	$D_0\,(m^2/s)$	Q **(kJ/mol)**	Q **(kcal/mol)**
Carbon	Fcc iron	20×10^{-6}	142	34.0
Carbon	Bcc iron	220×10^{-6}	122	29.3
Iron	Fcc iron	22×10^{-6}	268	64.0
Iron	Bcc iron	200×10^{-6}	240	57.5
Nickel	Fcc iron	77×10^{-6}	280	67.0
Manganese	Fcc iron	35×10^{-6}	282	67.5
Zinc	Copper	34×10^{-6}	191	45.6
Copper	Aluminum	15×10^{-6}	126	30.2
Copper	Copper	20×10^{-6}	197	47.1
Silver	Silver	40×10^{-6}	184	44.1
Carbon	Hcp titanium	511×10^{-6}	182	43.5

[a]See Equation 5.13.

Source: Data from L. H. Van Vlack, *Elements of Materials Science and Engineering*, 4th Ed., Addison-Wesley Publishing Co., Inc., Reading, MA, 1980.

the other nonmetallic systems of Figure 5.15 and Table 5.3 because most commercially important diffusion mechanisms in polymers involve the liquid state or the amorphous solid state, where the point-defect mechanisms of this section do not apply.

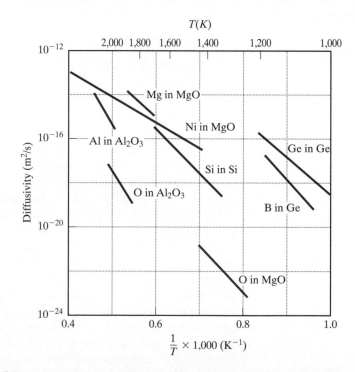

FIGURE 5.15 *Arrhenius plot of diffusivity data for a number of nonmetallic systems. (From P. Kofstad,* Nonstoichiometry, Diffusion, and Electrical Conductivity in Binary Metal Oxides, *John Wiley & Sons, Inc., NY, 1972; and S. M. Hu in* Atomic Diffusion in Semiconductors, *D. Shaw, Ed., Plenum Press, New York, 1973.)*

TABLE 5.3

Diffusivity Data for a Number of Nonmetallic Systems[a]				
Solute	**Solvent**	$D_0 (\text{m}^2/\text{s})$	Q (kJ/mol)	Q (kcal/mol)
Al	Al_2O_3	2.8×10^{-3}	477	114.0
O	Al_2O_3	0.19	636	152.0
Mg	MgO	24.9×10^{-6}	330	79.0
O	MgO	4.3×10^{-9}	344	82.1
Ni	MgO	1.8×10^{-9}	202	48.3
Si	Si	0.18	460	110.0
Ge	Ge	1.08×10^{-3}	291	69.6
B	Ge	1.1×10^3	439	105.0

[a]See Equation 5.13.

Source: Data from P. Kofstad, *Nonstoichiometry, Diffusion, and Electrical Conductivity in Binary Metal Oxides,* John Wiley & Sons, Inc., New York, 1972; and S. M. Hu, in *Atomic Diffusion in Semiconductors*, D. Shaw, Ed., Plenum Press, New York, 1973.

EXAMPLE 5.3

Steel surfaces can be hardened by *carburization,* as discussed relative to Figure 5.10. During one such treatment at 1,000°C, there is a drop in carbon concentration from 5 to 4 at % carbon between 1 and 2 mm from the surface of the steel. Estimate the flux of carbon atoms into the steel in this near-surface region. (The density of γ-Fe at 1,000°C is 7.63 g/cm^3.)

SOLUTION
First, we approximate

$$\frac{\partial c}{\partial x} \simeq \frac{\Delta c}{\Delta x} = \frac{5 \text{ at } \% - 4 \text{ at } \%}{1 \text{ mm} - 2 \text{ mm}}$$
$$= -1 \text{ at } \%/\text{mm}.$$

To obtain an absolute value for carbon-atom concentration, we must first know the concentration of iron atoms. From the given data and Appendix 1,

$$\rho = 7.63 \frac{g}{\text{cm}^3} \times \frac{0.6023 \times 10^{24} \text{ atoms}}{55.85 \text{ g}} = 8.23 \times 10^{22} \frac{\text{atoms}}{\text{cm}^3}.$$

Therefore,

$$\frac{\Delta c}{\Delta x} = -\frac{0.01(8.23 \times 10^{22} \text{ atoms/cm}^3)}{1 \text{ mm}} \times \frac{10^6 \text{ cm}^3}{\text{m}^3} \times \frac{10^3 \text{ mm}}{\text{m}}$$
$$= -8.23 \times 10^{29} \text{ atoms/m}^4.$$

From Table 5.2,

$$D_{c \text{ in } \gamma-\text{Fe}, 1000°\text{C}} = D_0 e^{-Q/RT}$$
$$= (20 \times 10^{-6} \text{ m}^2/\text{s}) e^{-(142,000 \text{ J/mol})/(8.314 \text{ J/mol/K})(1273 \text{ K})}$$
$$= 2.98 \times 10^{-11} \text{ m}^2/\text{s}.$$

Using Equation 5.8 gives us

$$J_x = -D\frac{\partial c}{\partial x}$$

$$\simeq -D\frac{\Delta c}{\Delta x}$$

$$= -(2.98 \times 10^{-11}\text{ m}^2/\text{s})(-8.23 \times 10^{29}\text{ atoms/m}^4)$$

$$= 2.45 \times 10^{19}\text{ atoms}/(\text{m}^2 \cdot \text{s}).$$

EXAMPLE 5.4

The diffusion result described by Equation 5.11 can apply to the carburization process (Example 5.3). The carbon environment (a hydrocarbon gas) is used to set the surface-carbon content (c_s) at 1.0 wt %. The initial carbon content of the steel (c_0) is 0.2 wt %. Using the error-function table, calculate how long it would take at $1,000°$C to reach a carbon content of 0.6 wt % [i.e., $(c - c_0)/(c_s - c_0) = 0.5$] at a distance of 1 mm from the surface.

SOLUTION
Using Equation 5.11, we get

$$\frac{c_x - c_0}{c_s - c_0} = 0.5 = 1 - \text{erf}\left(\frac{x}{2\sqrt{Dt}}\right)$$

or

$$\text{erf}\left(\frac{x}{2\sqrt{Dt}}\right) = 1 - 0.5 = 0.5.$$

Interpolating from Table 5.1 gives

$$\frac{0.5 - 0.4755}{0.5205 - 0.4755} = \frac{z - 0.45}{0.50 - 0.45}$$

or

$$z = \frac{x}{2\sqrt{Dt}} = 0.4772$$

or

$$t = \frac{x^2}{4(0.4772)^2 D}.$$

Using the diffusivity calculation from Example 5.3, we obtain

$$t = \frac{(1 \times 10^{-3}\text{ m})^2}{4(0.4772)^2(2.98 \times 10^{-11}\text{ m}^2/\text{s})}$$

$$= 3.68 \times 10^4\text{ s} \times \frac{1\text{ h}}{3.6 \times 10^3\text{ s}}$$

$$= 10.2\text{ h}.$$

EXAMPLE 5.5

Recalculate the carburization time for the conditions of Example 5.4 using the master plot of Figure 5.11 rather than the error-function table.

SOLUTION
From Figure 5.11, we see that the condition for $(c-c_0)/(c_s-c_0) = 0.5$ is

$$\frac{x}{\sqrt{Dt}} \simeq 0.95$$

or

$$t = \frac{x^2}{(0.95)^2 D}.$$

Using the diffusivity calculation from Example 5.3, we obtain

$$t = \frac{(1 \times 10^{-3} \text{ m})^2}{(0.95)^2 (2.98 \times 10^{-11} \text{ m}^2/\text{s})}$$

$$= 3.72 \times 10^4 \text{ s} \times \frac{1 \text{ h}}{3.6 \times 10^3 \text{ s}}$$

$$= 10.3 \text{ h}.$$

Note. There is appropriately close agreement with the calculation of Example 5.4. Exact agreement is hindered by the need for graphical interpretation (in this problem) and tabular interpolation (in the previous problem).

EXAMPLE 5.6

For a carburization process similar to that in Example 5.5, a carbon content of 0.6 wt % is reached at 0.75 mm from the surface after 10 h. What is the carburization temperature? (Assume, as before, that $c_s = 1.0$ wt % and $c_0 = 0.2$ wt %.)

SOLUTION
As in Example 5.5,

$$\frac{x}{\sqrt{Dt}} \simeq 0.95$$

or

$$D = \frac{x^2}{(0.95)^2 t}$$

with the following data given:

$$D = \frac{(0.75 \times 10^{-3} \text{ m})^2}{(0.95)^2 (3.6 \times 10^4 \text{ s})} = 1.73 \times 10^{-11} \text{ m}^2/\text{s}.$$

From Table 5.2, for C in γ-Fe,

$$D = (20 \times 10^{-6} \text{ m}^2/\text{s})e^{-(142,000 \text{ J/mol})/[8.314 \text{ J/(mol·K)}](T)}.$$

Equating the two values for D gives

$$1.73 \times 10^{-11}\frac{\text{m}^2}{\text{s}} = 20 \times 10^{-6}\frac{\text{m}^2}{\text{s}}\, e^{-1.71 \times 10^4/T}$$

or

$$T^{-1} = \frac{-\ln{(1.73 \times 10^{-11}/20 \times 10^{-6})}}{1.71 \times 10^4}$$

or

$$T = 1225 \text{ K} = 952°\text{C}.$$

PRACTICE PROBLEM 5.3

Suppose that the carbon-concentration gradient described in Example 5.3 occurred at 1,100°C rather than 1,000°C. Calculate the carbon-atom flux for this case.

PRACTICE PROBLEM 5.4

In Example 5.4, the time to generate a given carbon concentration profile is calculated using the error-function table. The carbon content at the surface was 1.0 wt % and at 1 mm from the surface was 0.6 wt %. For this diffusion time, what is the carbon content at a distance **(a)** 0.5 mm from the surface and **(b)** 2 mm from the surface?

PRACTICE PROBLEM 5.5

Repeat Practice Problem 5.4 using the graphical method of Example 5.5.

PRACTICE PROBLEM 5.6

In Example 5.6, a carburization temperature is calculated for a given carbon-concentration profile. Calculate the carburization temperature if the given profile were obtained in 8 hours rather than 10 hours, as originally stated.

5.4 | Steady-State Diffusion

The change in the concentration profile with time for processes such as carburization was shown in Figure 5.10. A similar observation for a process with slightly different boundary conditions is shown in Figure 5.16. In this case, the relatively high surface concentration of the diffusing species, c_h, is held constant with time, just as c_s was held constant in Figure 5.10, but the relatively low concentration,

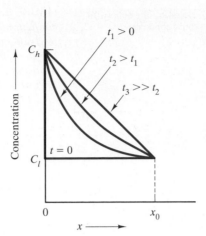

FIGURE 5.16 *Solution to Fick's second law (Equation 5.10) for the case of a solid of thickness x_0, constant surface concentrations of the diffusing species c_h and c_l, and a constant diffusion coefficient D. For long times (e.g., t_3), the linear concentration profile is an example of steady-state diffusion.*

c_l, at x_0 is also held constant with time. As a result, the nonlinear concentration profiles at times greater than zero (e.g., at t_1 and t_2 in Figure 5.16) approach a straight line after a relatively long time (e.g., at t_3 in Figure 5.16). This *linear* concentration profile is unchanging with additional time as long as c_h and c_l remain fixed. This limiting case is an example of **steady-state diffusion** (i.e., mass transport that is unchanging with time). The concentration gradient defined by Equation 5.8 takes an especially simple form in this case:

$$\frac{\partial c}{\partial x} = \frac{\Delta c}{\Delta x} = \frac{c_h - c_l}{0 - x_0} = -\frac{c_h - c_l}{x_0}. \tag{5.14}$$

In the case of carburization represented by Figure 5.10, the surface concentration, c_s, was held fixed by maintaining a fixed carbon-source atmospheric pressure at the $x = 0$ surface. In this same way, both c_h and c_l are maintained fixed in the case represented by Figure 5.16. A plate of material with a thickness of x_0 is held between two gas atmospheres: a high-pressure atmosphere on the $x = 0$ surface, which produces the fixed concentration c_h, and a low-pressure atmosphere on the $x = x_0$ surface, which produces the concentration c_l (Figure 5.17).

A common application of steady-state diffusion is the use of materials as gas-purification membranes. For example, a thin sheet of palladium metal is permeable to hydrogen gas, but not to common atmospheric gases such as oxygen, nitrogen, and water vapor. By introducing the "impure" gas mixture on the high-pressure side of Figure 5.17b and maintaining a constant, reduced hydrogen pressure on the low-pressure side, a steady flow of purified hydrogen passes through the palladium sheet.

EXAMPLE 5.7

A 5-mm-thick sheet of palladium with a cross-sectional area of 0.2 m^2 is used as a steady-state diffusional membrane for purifying hydrogen. If the hydrogen concentration on the high-pressure (impure gas) side of the sheet is 0.3 kg/m^3 and the diffusion coefficient for hydrogen in Pd is 1.0×10^{-8} m^2/s, calculate the mass of hydrogen being purified per hour.

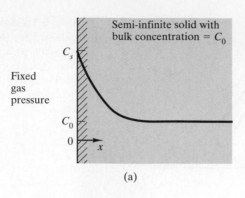

Semi-infinite solid with bulk concentration = C_0

Fixed gas pressure

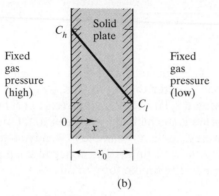

(a)

Fixed gas pressure (high)

Solid plate

Fixed gas pressure (low)

(b)

FIGURE 5.17 *Schematic of sample configurations in gas environments that lead, after a long time, to the diffusion profiles representative of (a) nonsteady state diffusion (Figure 5.10) and (b) steady-state diffusion (Figure 5.16).*

SOLUTION

Fick's first law (Equation 5.8) is simplified by the steady-state concentration gradient of Equation 5.14, giving

$$J_x = -D(\partial c/\partial x) = [-(c_h - c_l)/x_0]$$

$$= -(1.0 \times 10^{-8} \text{ m}^2/\text{s}) \left[-(0.3 \text{ kg/m}^3 - 0 \text{ kg/m}^3)/(5 \times 10^{-3} \text{ m}) \right]$$

$$= 0.6 \times 10^{-6} \text{ kg/m}^2 \cdot \text{s} \times 3.6 \times 10^3 \text{ s/h} = 2.16 \times 10^{-3} \text{ kg/m}^2 \cdot \text{h}.$$

The total mass of hydrogen being purified will then be this flux times the membrane area:

$$m = J_x \times A = 2.16 \times 10^{-3} \text{ kg/m}^2 \cdot \text{h} \times 0.2 \text{ m}^2 = 0.432 \times 10^{-3} \text{ kg/h}.$$

PRACTICE PROBLEM 5.7

For the purification membrane in Example 5.7, how much hydrogen would be purified per hour if the membrane used is 3 mm thick, with all other conditions unchanged?

THE MATERIAL WORLD

Diffusion in Fuel Cells

William Grove, an English jurist and amateur physicist, discovered the principle of the fuel cell in 1839. Grove used four large cells, each containing hydrogen and oxygen, to produce electric power that was then used to split water in another cell into hydrogen and oxygen gases. In 1842, he wrote to the great English chemist and physicist Michael Faraday: "I cannot but regard the experiment as an important one." Grove's confidence was appropriate, but somewhat premature.

More than a century passed until, in 1959, NASA demonstrated the potential of fuel cells to provide power during space flight. This demonstration stimulated industry to explore applications in the 1960s, but technical barriers and high investment costs prevented commercialization. In 1984, the Office of Transportation Technologies at the U.S. Department of Energy began supporting research and development in fuel-cell technology. Today, hundreds of companies worldwide, including several major automakers, are vigorously pursuing fuel-cell technology.

Diffusion plays a key role in modern fuel-cell designs. As the accompanying figure shows, a fuel cell is an electrochemical energy conversion device. (A major attraction is that it is between two and three times more efficient than an internal-combustion engine in converting fuel to power.) As hydrogen flows into the fuel cell on the anode side, the hydrogen gas is separated into hydrogen ions (protons) and electrons at the surface of a platinum catalyst. The hydrogen ions diffuse across the membrane and combine with oxygen and electrons at the surface of another platinum catalyst on the cathode side (producing water, the only emission from a hydrogen fuel cell!).

The electrons cannot pass through the nonconductive membrane and, instead, produce a current from the anode to the cathode via an external circuit that provides about enough power for a single light-bulb. By stacking enough cells in series, sufficient power to run automobiles and other large-scale systems is possible.

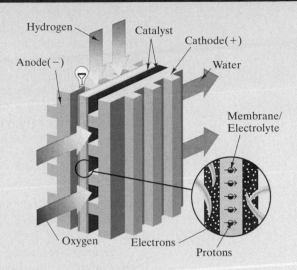

(From *Fuel Cells–Green Power,* Los Alamos National Laboratory Report LA-UR-99-3231.)

Substantial research is under way on optimizing the membrane material. Both polymeric and ceramic materials are under investigation. Related research involves fuels other than pure hydrogen and the infrastructure required to deliver these fuels on a wide scale. The significant potential of fuel-cell technology is quite promising, more than 160 years after William Grove's "important" experiment, and governments and private industry are together investing billions of dollars in these efforts.

Prototype fuel-cell-powered Toyota sport utility vehicle under evaluation at the University of California, Davis. (Courtesy of the University of California, Davis.)

5.5 | Alternate Diffusion Paths

A final word of caution is in order about using specific diffusivity data to analyze a particular material process. Figure 5.18 shows that the self-diffusion coefficients for silver vary by several orders of magnitude, depending on the route for diffusional transport. To this point, we have considered **volume diffusion**, or **bulk diffusion**, through a material's crystal structure by means of some defect mechanism. However, there can be "short circuits" associated with easier diffusion paths. As seen in Figure 5.18, diffusion is much faster (with a lower Q) along a grain boundary. As we saw in Section 4.4, this region of mismatch between adjacent crystal grains in the material's microstructure is a more open structure, allowing enhanced grain-boundary diffusion. The crystal surface is an even more open region, and **surface diffusion** allows easier atom transport along the free surface less hindered by adjacent atoms. The overall result is that

$$Q_{\text{volume}} > Q_{\text{grain boundary}} > Q_{\text{surface}} \text{ and } D_{\text{volume}} < D_{\text{grain boundary}} < D_{\text{surface}}.$$

This result does not mean that surface diffusion is always the important process just because D_{surface} is greatest. More important is the amount of diffusing region available. In most cases, volume diffusion dominates. For a material with a small average grain size (see Section 4.4) and therefore a large grain-boundary area, **grain-boundary diffusion** can dominate. Similarly, in a fine-grained powder with a large surface area, surface diffusion can dominate.

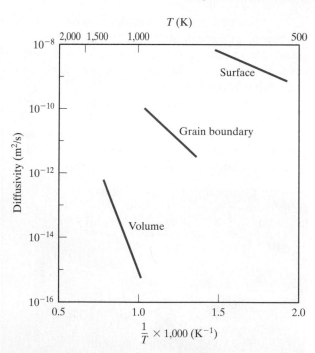

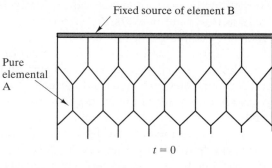

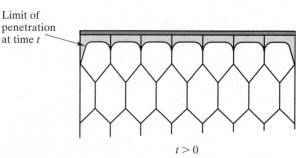

FIGURE 5.18 *Self-diffusion coefficients for silver depend on the diffusion path. In general, diffusivity is greater through less-restrictive structural regions. (From J. H. Brophy, R. M. Rose, and J. Wulff,* The Structure and Properties of Materials, *Vol. 2:* Thermodynamics of Structure, *John Wiley & Sons, Inc., New York, 1964.)*

FIGURE 5.19 *Schematic illustration of how a coating of impurity B can penetrate more deeply into grain boundaries and even further along a free surface of polycrystalline A, consistent with the relative values of diffusion coefficients (*$D_{\text{volume}} < D_{\text{grain boundary}} < D_{\text{surface}}$*).*

For a given polycrystalline microstructure, the penetration of a diffusing species will tend to be greater along grain boundaries and even greater along the free surface of the sample (Figure 5.19).

EXAMPLE 5.8

We can approximate the extent of grain-boundary penetration in Figure 5.19 by using the semi-infinite diffusion expression of Equation 5.11. **(a)** Taking $D_{\text{grain boundary}} = 1.0 \times 10^{-10}$ m²/s, calculate the penetration of B into A along the grain boundary after 1 hour, defined as the distance, x, at which $c_x = 0.01c_s$ (with $c_0 = 0$ for initially pure A). **(b)** For comparison, calculate the penetration defined in the same way within the bulk grain for which $D_{\text{volume}} = 1.0 \times 10^{-14}$ m²/s.

SOLUTION

(a) We can simplify Equation 5.11 as

$$\frac{c_x - c_0}{c_s - c_0} = 1 - \text{erf}\left(\frac{x}{2\sqrt{Dt}}\right) = \frac{c_x - 0}{c_s - 0} = \frac{0.01c_s}{c_s} = 0.01,$$

or

$$\text{erf}\left(\frac{x}{2\sqrt{Dt}}\right) = 1 - 0.01 = 0.99.$$

Interpolating from Table 5.1 gives

$$\frac{0.9928 - 0.99}{0.9928 - 0.9891} = \frac{1.90 - z}{1.90 - 1.80},$$

or

$$z = \frac{x}{2\sqrt{Dt}} = 1.824,$$

and then

$$x = 2(1.824)\sqrt{Dt}$$
$$= 2(1.824)\sqrt{(1.0 \times 10^{-10}\ \text{m}^2/\text{s})(1\ \text{h})(3.6 \times 10^3\ \text{s/h})}$$
$$= 2.19 \times 10^{-3}\ \text{m} = 2.19\ \text{mm}.$$

(b) For comparison,

$$x = 2(1.824)\sqrt{(1.0 \times 10^{-14}\ \text{m}^2/\text{s})(1\ \text{h})(3.6 \times 10^3\ \text{s/h})}$$
$$= 21.9 \times 10^{-6}\ \text{m} = 21.9\ \mu\text{m}.$$

Note. The amount of grain-boundary penetration calculated in **(a)** is exaggerated, as the excess buildup of impurity, B, in the boundary leads to some of that material being "drained away" by side diffusion into the A grains, as indicated by Figure 5.19.

In Example 5.8, we calculated the extent of impurity penetration for volume and grain-boundary paths. For further comparison, calculate the penetration for surface diffusion for which $D_{surface} = 1.0 \times 10^{-8}$ m^2/s.

Summary

The point defects introduced in Chapter 4 are seen to play a central role in the movement of atoms by solid-state diffusion. Several practical problems in the production and application of engineering materials involve these diffusional processes.

Specifically, we find that point-defect concentrations increase exponentially with absolute temperature following an Arrhenius expression. As solid-state diffusion in crystalline materials occurs via a point-defect mechanism, the diffusivity, as defined by Fick's laws, also increases exponentially with absolute temperature in another Arrhenius expression.

The mathematics of diffusion allows a relatively precise description of the chemical concentration pro-

files of diffusing species. For some sample geometries, the concentration profile approaches a simple, linear form after a relatively long time. This steady-state diffusion is well illustrated by gas transport across thin membranes.

In the case of fine-grained polycrystalline materials or powders, material transport may be dominated by grain-boundary diffusion or surface diffusion, respectively, because, in general, $D_{volume} < D_{grain\ boundary} < D_{surface}$. Another result is that, for a given polycrystalline solid, impurity penetration will be greater along grain boundaries and even greater along the free surface.

Key Terms

activation energy (127)
Arrhenius equation (127)
Arrhenius plot (127)
bulk diffusion (146)
carburization (134)
concentration gradient (134)
diffusion (132)
diffusion coefficient (134)
diffusivity (134)

Fick's first law (134)
Fick's second law (134)
Gaussian error function (135)
grain-boundary diffusion (146)
Maxwell–Boltzmann distribution (128)
preexponential constant (127)
random walk (132)
rate-limiting step (129)

self-diffusion (132)
steady-state diffusion (143)
surface diffusion (146)
thermal activation (128)
thermal vibration (130)
vacancy migration (132)
volume diffusion (146)

References

Chiang, Y., D. P. Birnie III, and **W. D. Kingery**, *Physical Ceramics*, John Wiley & Sons, Inc., New York, 1997.

Crank, J., *The Mathematics of Diffusion*, 2nd ed., Clarendon Press, Oxford, 1999.

Shewmon, P. G., *Diffusion in Solids*, 2nd ed., Minerals, Metals, and Materials Society, Warrendale, PA, 1989.

Problems

Section 5.1 • Thermally Activated Processes

5.1. In steel-making furnaces, temperature-resistant ceramic bricks (refractories) are used as linings to contain the molten metal. A common byproduct of the steel-making process is a calcium aluminosilicate (slag) that is chemically corrosive to the refractories. For alumina refractories, the corrosion rate is 2.0×10^{-8} m/s at $1{,}425°$C and 8.95×10^{-8} m/s at $1{,}500°$C. Calculate the activation energy for the corrosion of these alumina refractories.

5.2. For a steel furnace similar to that described in Problem 5.1, silica refractories have corrosion rates of 2.0×10^{-7} m/s at $1{,}345°$C and 9.0×10^{-7} m/s at $1{,}510°$C. Calculate the activation energy for the corrosion of these silica refractories.

5.3. Manufacturing traditional clayware ceramics typically involves driving off the water of hydration in the clay minerals. The rate constant for the dehydration of kaolinite, a common clay mineral, is 1.0×10^{-4} s^{-1} at $485°$C and 1.0×10^{-3} s^{-1} at $525°$C. Calculate the activation energy for the dehydration of kaolinite.

5.4. For the thermally activated process described in Problem 5.3, calculate the rate constant at $600°$C, assuming that to be the upper temperature limit specified for the dehydration process.

Section 5.2 • Thermal Production of Point Defects

5.5. Verify that the data represented by Figure 5.4b correspond to an energy of formation of 0.76 eV for a defect in aluminum.

5.6. What type of crystal direction corresponds to the movement of interstitial carbon in α-Fe between equivalent ($\frac{1}{2}0\frac{1}{2}$-type) interstitial positions? Illustrate your answer with a sketch.

5.7. Repeat Problem 5.6 for the movement between equivalent interstices in γ-Fe. (Note Practice Problem 4.2.)

5.8. What crystallographic positions and directions are indicated by the migration shown in Figure 5.5? [Assume the atoms are in a (100) plane of an fcc metal.]

Section 5.3 • Point Defects and Solid-State Diffusion

5.9. Verify that the data represented by Figure 5.13 correspond to an activation energy of 122,000 J/mol for the diffusion of carbon in α-iron.

5.10. Carburization was described in Example 5.3. The decarburization of a steel can also be described by using the error function. Starting with Equation 5.11 and taking $c_s = 0$, derive an expression to describe the concentration profile of carbon as it diffuses out of a steel with initial concentration, c_0. (This situation can be produced by placing the steel in a vacuum at elevated temperature.)

5.11. Using the decarburization expression derived in Problem 5.10, plot the concentration profile of carbon within 1 mm of the carbon-free surface after 1 hour in a vacuum at $1{,}000°$C. Take the initial carbon content of the steel to be 0.3 wt %.

5.12. A *diffusion couple* is formed when two different materials are allowed to interdiffuse at an elevated temperature (see Figure 5.8). For a block of pure metal A adjacent to a block of pure metal B, the concentration profile of A (in at %) after interdiffusion is given by

$$c_x = 50 \left[1 - \operatorname{erf} \left(\frac{x}{2\sqrt{Dt}} \right) \right],$$

where x is measured from the original interface. For a diffusion couple with $D = 10^{-14}$ m^2/s, plot the concentration profile of metal A over a range of 20 μm on either side of the original interface ($x = 0$) after a time of 1 hour. [Note that erf $(-z) = -$erf(z).]

5.13. Use the information from Problem 5.12 to plot the progress of interdiffusion of two metals, X and Y, with $D = 10^{-12}$ m^2/s. Plot the concentration profile of metal X over a range of 300 μm on either side of the original interface after times of 1, 2, and 3 h.

5.14. Using the results of Problem 5.12 and assuming that profile occurred at a temperature of $1{,}000°$C, superimpose the concentration profile of metal A for the same diffusion couple for 1 hour but heated at $1{,}200°$C at which $D = 10^{-13}$ m^2/s.

5.15. Given the information in Problems 5.12 and 5.14, calculate the activation energy for the interdiffusion of metals A and B.

5.16. Use the result of Problem 5.15 to calculate the diffusion coefficient for the interdiffusion of metals A and B at 1,400°C.

5.17. Using data in Table 5.2, calculate the self-diffusivity for iron in bcc iron at 900°C.

5.18. Using data in Table 5.2, calculate the self-diffusivity for iron in fcc iron at 1,000°C.

5.19. Using data in Table 5.2, calculate the self-diffusivity for copper in copper at 1,000°C.

5.20. The diffusivity of copper in a commercial brass alloy is 10^{-20} m^2/s at 400°C. The activation energy for diffusion of copper in this system is 195 kJ/mol. Calculate the diffusivity at 600°C.

5.21. The diffusion coefficient of nickel in an austenitic (fcc structure) stainless steel is 10^{-22} m^2/s at 500°C and 10^{-15} m^2/s at 1,000°C. Calculate the activation energy for the diffusion of nickel in this alloy over this temperature range.

*5.22. Show that the relationship between vacancy concentration and fractional dimension changes for the case shown in Figure 5.4 is approximately

$$\frac{n_v}{n_{\text{sites}}} = 3\left(\frac{\Delta L}{L} - \frac{\Delta a}{a}\right).$$

[Note that $(1+x)^3 \simeq 1 + 3x$ for small x.]

*5.23. A popular use of diffusion data in materials science is to identify mechanisms for certain phenomena. This identification is done by comparison of activation energies. For example, consider the oxidation of an aluminum alloy. The rate-controlling mechanism is the diffusion of ions through an Al$_2$O$_3$ surface layer, which means that the rate of growth of the oxide layer thickness is directly proportional to a diffusion coefficient. We can specify whether oxidation is controlled by Al^{3+} diffusion or O^{2-} diffusion by comparing the activation energy for oxidation with the activation energies of the two species as given in Table 5.3. Given that the rate constant for oxide growth is 4.00×10^{-8} kg/(m$^4 \cdot$ s) at 500°C and 1.97×10^{-4} kg/(m$^4 \cdot$ s) at 600°C, determine whether the oxidation process is controlled by Al^{3+} diffusion or O^{2-} diffusion.

5.24. *Diffusion length*, λ, is a popular term in characterizing the production of semiconductors by the controlled diffusion of impurities into a high-purity material. The value of λ is taken as $2\sqrt{Dt}$, where λ represents the extent of diffusion for an impurity with a diffusion coefficient, D, over a period of time, t. Calculate the diffusion length for B in Ge for a total diffusion time of 30 minutes at a temperature of **(a)** 800°C and **(b)** 900°C.

Section 5.4 • Steady-State Diffusion

5.25. A differential nitrogen pressure exists across a 2-mm-thick steel furnace wall. After some time, steady-state diffusion of the nitrogen is established across the wall. Given that the nitrogen concentration on the high-pressure surface of the wall is 2 kg/m^3 and on the low-pressure surface is 0.2 kg/m^3, calculate the flow of nitrogen through the wall (in kg/m$^2\cdot$h) if the diffusion coefficient for nitrogen in this steel is 1.0×10^{-10} m^2/s at the furnace operating temperature.

5.26. For the furnace described in Problem 5.25, design changes are made that include a thicker wall (3 mm) and a lower operating temperature that reduces the nitrogen diffusivity to 5.0×10^{-11} m^2/s. What would be the steady-state nitrogen flow across the wall in this case?

5.27. Many laboratory furnaces have small glass windows that provide visual access to samples. The leakage of furnace atmospheres through the windows can be a problem. Consider a 3-mm-thick window of vitreous silica in a furnace containing an inert, helium atmosphere. For a window with a cross-sectional area of 600 mm^2, calculate the steady-state flow of helium gas (in atoms/s) across the window if the concentration of helium on the high-pressure (furnace) surface of the window is 6.0×10^{23} atoms/m^3 and on the low-pressure (outside) surface is essentially zero. The diffusion coefficient for helium in vitreous silica at this wall temperature is 1.0×10^{-10} m^2/s.

5.28. Figure 4.23 shows that Na$_2$O additions to vitreous silica tends to "tighten" the structure, as Na$^+$ions fill in open voids in the silica structure. This structural feature can have a significant effect on the gas diffusion described in Problem 5.27. Consider the replacement of the vitreous silica window with a sodium silicate window (containing 30 mol. % Na$_2$O) of the same dimensions. For the "tighter" sodium silicate glass, the concentration of helium on the high-pressure surface is reduced to 3.0×10^{22} atoms/m^3. Similarly, the diffusion coefficient for helium in the sodium silicate glass at this wall temperature is reduced to 2.5×10^{-12} m^2/s. Calculate the steady-state flow of helium gas (in atoms/s) across this replacement window.

Section 5.5 • Alternate Diffusion Paths

5.29. The endpoints of the Arrhenius plot of $D_{\text{grain boundary}}$ in Figure 5.18 are $D_{\text{grain boundary}} = 3.2 \times 10^{-12}$ m^2/s at a temperature of 457°C and $D_{\text{grain boundary}} = 1.0 \times 10^{-10}$ m^2/s at a temperature of 689°C. Using these data, calculate the activation energy for grain-boundary diffusion in silver.

5.30. The endpoints of the Arrhenius plot of D_{surface} in Figure 5.18 are $D_{\text{surface}} = 7.9 \times 10^{-10}$ m^2/s at a temperature of 245°C and $D_{\text{surface}} = 6.3 \times 10^{-9}$ m^2/s at a temperature of 398°C. Using these data, calculate the activation energy for surface diffusion in silver.

5.31. The contribution of grain-boundary diffusion can sometimes be seen from diffusivity measurements made on polycrystalline samples of increasingly small grain size. As an example, plot the following data (as ln D versus ln[grain size]) for the diffusion coefficient of Ni^{2+} in NiO at 480°C, measured as a function of sample grain size.

Grain size (μm)	D (m^2/s)
1	1.0×10^{-19}
10	1.0×10^{-20}
100	1.0×10^{-21}

5.32. Using the plot from Problem 5.31, estimate the diffusion coefficient of Ni^{2+} in NiO at 480°C for a 20-μm grain-size material.

CHAPTER 6
Mechanical Behavior

6.1 Stress Versus Strain

6.2 Elastic Deformation

6.3 Plastic Deformation

6.4 Hardness

6.5 Creep and Stress Relaxation

6.6 Viscoelastic Deformation

Hardness testing of materials is a relatively simple but effective method for predicting their mechanical behavior in practical applications. These contemporary testing machines provide various types of hardness testing described in this chapter. (Courtesy of Instron Corporation.)

Mechanical properties data are available on the related Web site along with related experiments in a laboratory manual.

As discussed in Chapter 1, probably no materials are more closely associated with the engineering profession than metals such as structural steel. In this chapter, we shall explore some of the major mechanical properties of metals: stress versus strain, hardness, and creep. Although this chapter will provide an introduction to these properties, an appreciation of the versatility of metals will be presented in Chapters 9 and 10. Microstructural development as related to phase diagrams will be dealt with in Chapter 9. Heat treatment based on the kinetics of solid-state reactions will be covered in Chapter 10. Each of these topics deals with methods to "fine tune" the properties of given alloys within a broad range of values.

Many of the important mechanical properties of metals apply to ceramics as well, although the values of those properties may be very different for the ceramics. For example, brittle fracture and creep play important roles in the structural applications of ceramics. The liquidlike structure of glass leads to high-temperature deformation by a viscous flow mechanism. The production of fracture-resistant tempered glass depends on precise control of that viscosity.

We follow our discussions of the mechanical properties of the inorganic materials, metals and ceramics, with a similar discussion of the mechanical properties of the organic polymers. An important trend in engineering design is the increased concentration on so-called *engineering polymers,* which have sufficiently high strength and stiffness to be able to substitute for traditional structural metals. Often, we find that polymers exhibit behavior associated with their long-chain or network molecular structure. Viscoelastic deformation is an important example.

6.1 | Stress Versus Strain

Metals are used in engineering designs for many reasons, but they generally serve as structural elements. Thus, we shall initially concentrate on the mechanical properties of metals in this chapter.

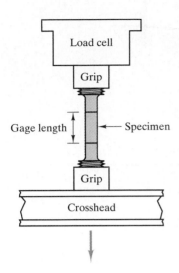

FIGURE 6.1 *Tensile test.*

METALS

The mechanical properties discussed next are not exhaustive but instead are intended to cover the major factors in selecting a durable material for structural applications under a variety of service conditions. As we go through these various properties, an attempt will be made to use a consistent and comprehensive set of sample metals and alloys to demonstrate typical data and, especially, important data trends.

Perhaps the simplest questions that a design engineer can ask about a structural material are (1) "How strong is it?" and (2) "How much deformation must I expect given a certain load?" This basic description of the material is obtained by the *tensile test*. Figure 6.1 illustrates this simple pull test. The load necessary to produce a given elongation is monitored as the specimen is pulled in tension at a constant rate. A load-versus-elongation curve (Figure 6.2) is the immediate result of such a test. A more general statement about material characteristics is obtained by normalizing the data of Figure 6.2 for geometry. The resulting *stress-versus-strain curve* is given in Figure 6.3. Here, the **engineering stress**, σ, is defined as

$$\sigma = \frac{P}{A_0},\qquad (6.1)$$

where P is the load on the sample with an original (zero-stress) cross-sectional area, A_0. *Sample cross section* refers to the region near the center of the specimen's length. Specimens are machined such that the cross-sectional area in this region is uniform and smaller than at the ends gripped by the testing machine. This smallest area region, referred to as the **gage length**, experiences the largest stress concentration so that any significant deformation at higher stresses is localized there. The **engineering strain**, ϵ, is defined as

$$\epsilon = \frac{l - l_0}{l_0} = \frac{\Delta l}{l_0},\qquad (6.2)$$

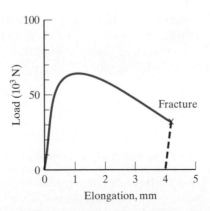

FIGURE 6.2 *Load-versus-elongation curve obtained in a tensile test. The specimen was aluminum 2024-T81.*

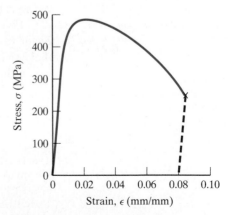

FIGURE 6.3 *Stress-versus-strain curve obtained by normalizing the data of Figure 6.2 for specimen geometry.*

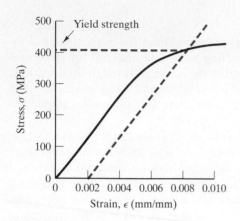

FIGURE 6.4 *The yield strength is defined relative to the intersection of the stress–strain curve with a "0.2% offset." Yield strength is a convenient indication of the onset of plastic deformation.*

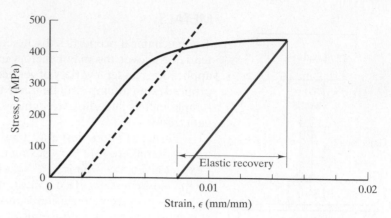

FIGURE 6.5 *Elastic recovery occurs when stress is removed from a specimen that has already undergone plastic deformation.*

The related Web site contains the chapter "Uniaxial Tension Testing" from the ASM Handbook, Vol. 8 (Mechanical Testing and Evaluation), ASM International, Materials Park, OH, 2000. Used by permission.

where l is the gage length at a given load, and l_0 is the original (zero-stress) length. Figure 6.3 is divided into two distinct regions: (1) elastic deformation and (2) plastic deformation. **Elastic deformation** is temporary deformation. It is fully recovered when the load is removed. The elastic region of the stress–strain curve is the initial linear portion. **Plastic deformation** is permanent deformation. It is not recovered when the load is removed, although a small elastic component is recovered. The plastic region is the nonlinear portion generated once the total strain exceeds its elastic limit. It is often difficult to specify precisely the point at which the stress–strain curve deviates from linearity and enters the plastic region. The usual convention is to define as **yield strength** (Y.S.) the intersection of the deformation curve with a straight line parallel to the elastic portion and offset 0.2% on the strain axis (Figure 6.4). The yield strength represents the stress necessary to generate this small amount (0.2%) of permanent deformation. Figure 6.5 indicates the small amount of elastic recovery that occurs when a load well into the plastic region is released.

Figure 6.6 summarizes the key mechanical properties obtained from the tensile test. The slope of the stress–strain curve in the elastic region is the **modulus of elasticity**, E, also known as **Young's*** **modulus**. The linearity of the stress–strain plot in the elastic region is a graphical statement of **Hooke's†** **law:**

$$\sigma = E\epsilon. \tag{6.3}$$

---

*Thomas Young (1773–1829), English physicist and physician, was the first to define the modulus of elasticity. Although this contribution made his name a famous one in solid mechanics, his most brilliant achievements were in optics. He was largely responsible for the acceptance of the wave theory of light.

†Robert Hooke (1635–1703), English physicist, was one of the most brilliant scientists of the 17th century as well as one of its most cantankerous personalities. His quarrels with fellow scientists such as Sir Isaac Newton did not diminish his accomplishments, which included the elastic behavior law (Equation 6.3) and the coining of the word "cell" to describe the structural building blocks of biological systems that he discovered in optical-microscopy studies.

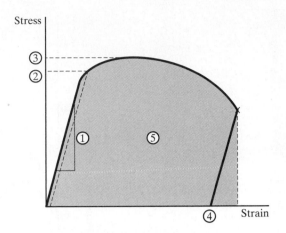

FIGURE 6.6 *The key mechanical properties obtained from a tensile test: 1, modulus of elasticity, E; 2, yield strength, Y.S.; 3, tensile strength, T.S.; 4, ductility, $100 \times \epsilon_{\text{failure}}$ (note that elastic recovery occurs after fracture); and 5, toughness $= \int \sigma \, d\epsilon$ (measured under load; hence, the dashed line is vertical).*

The modulus, E, is a highly practical piece of information. It represents the stiffness of the material (i.e., its resistance to elastic strain), which manifests itself as the amount of deformation in normal use below the yield strength and the springiness of the material during forming. As with E, the yield strength has major practical significance. It shows the resistance of the metal to permanent deformation and indicates the ease with which the metal can be formed by rolling and drawing operations.

Although we are concentrating on the behavior of metals under tensile loads, the testing apparatus illustrated in Figure 6.1 is routinely used in a reversed mode to produce a compressive test. The elastic modulus, in fact, tends to be the same for metal alloys tested in either tensile or compressive modes. Later, we will see that the elastic behavior under shear loads is also related to the tensile modulus.

One should note that many design engineers, especially in the aerospace field, are more interested in strength-per-unit density than strength or density individually. (If two alloys each have adequate strength, the one with lower density is preferred for potential fuel savings.) The strength-per-unit density is generally termed **specific strength**, or **strength-to-weight ratio**, and is discussed relative to composite properties in Chapter 12. Another term of practical engineering importance is **residual stress**, defined as the stress remaining within a structural material after all applied loads are removed. This stress commonly occurs following various thermomechanical treatments such as welding and machining.

As the plastic deformation represented in Figure 6.6 continues at stresses above the yield strength, the engineering stress continues to rise toward a maximum. This maximum stress is termed the *ultimate tensile strength*, or simply the **tensile strength** (T.S.). Within the region of the stress–strain curve between Y.S. and T.S., the phenomenon of increasing strength with increasing deformation is referred to as **strain hardening**. Strain hardening is an important factor in shaping metals by **cold working** (i.e., plastic deformation occurring well below one-half times the absolute melting point). It might appear from Figure 6.6 that plastic deformation beyond T.S. softens the material because the engineering stress falls.

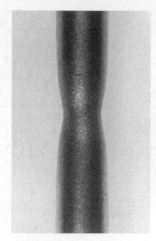

FIGURE 6.7 *Neck down of a tensile test specimen within its gage length after extension beyond the tensile strength. (Courtesy of R. S. Wortman.)*

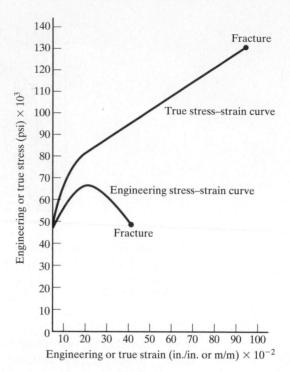

FIGURE 6.8 *True stress (load divided by actual area in the necked-down region) continues to rise to the point of fracture, in contrast to the behavior of engineering stress. (From R. A. Flinn and P. K. Trojan,* Engineering Materials and Their Applications, *2nd ed., Houghton Mifflin Company, 1981, used by permission.)*

Instead, this drop in stress is simply the result of the fact that engineering stress and strain are defined relative to original sample dimensions. At the ultimate tensile strength, the sample begins to *neck down* within the gage length (Figure 6.7). The true stress ($\sigma_{tr} = P/A_{actual}$) continues to rise to the point of fracture (see Figure 6.8).

For many metals and alloys, the region of the true stress (σ_T) versus true strain (ϵ_T) curve between the onset of plastic deformation (corresponding to the yield stress in the engineering stress versus engineering strain curve) and the onset of necking (corresponding to the tensile stress in the engineering stress versus engineering strain curve) can be approximated by

$$\sigma_T = K\epsilon_T^n \qquad (6.4)$$

where K and n are constants with values for a given metal or alloy depending on its thermomechanical history (e.g., degree of mechanical working or heat treatment). In other words, the true stress versus true strain curve in this region is nearly straight when plotted on logarithmic coordinates. The slope of the log–log plot is the parameter n, which is termed the **strain-hardening exponent**. For low-carbon steels used to form complex shapes, the value of n will normally be approximately 0.22. Higher values, up to 0.26, indicate an improved ability to be deformed during the shaping process without excess thinning or fracture of the piece.

The engineering stress at failure in Figure 6.6 is lower than T.S. and occasionally even lower than Y.S. Unfortunately, the complexity of the final stages of neck

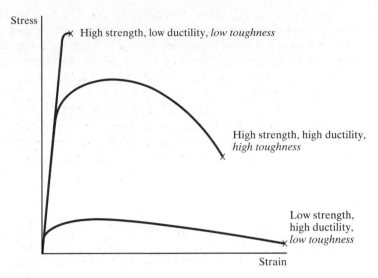

FIGURE 6.9 *The toughness of an alloy depends on a combination of strength and ductility.*

down causes the value of the failure stress to vary substantially from specimen to specimen. More useful is the strain at failure. **Ductility** is frequently quantified as the percent elongation at failure ($= 100 \times \epsilon_{\text{failure}}$). A less-used definition is the percent reduction in area $[= (A_0 - A_{\text{final}})/A_0]$. The values for ductility from the two different definitions are not, in general, equal. It should also be noted that the value of percent elongation at failure is a function of the gage length used. Tabulated values are frequently specified for a gage length of 2 in. Ductility indicates the general ability of the metal to be plastically deformed. Practical implications of this ability include formability during fabrication and relief of locally high stresses at crack tips during structural loading (see the discussion of fracture toughness in Chapter 8).

It is also useful to know whether an alloy is both strong and ductile. A high-strength alloy that is also highly brittle may be as unusable as a deformable alloy with unacceptably low strength. Figure 6.9 compares these two extremes with an alloy with both high strength and substantial ductility. The term **toughness** is used to describe this combination of properties. Figure 6.6 shows that this is conveniently defined as the total area under the stress–strain curve. Since integrated $\sigma - \epsilon$ data are not routinely available, we shall be required to monitor the relative magnitudes of strength (Y.S. and T.S.) and ductility (percent elongation at fracture).

Values of four of the five basic tensile test parameters (defined in Figure 6.6) for various alloys are given in Table 6.1. Values of the strain-hardening parameters of Equation 6.4, K and n, are given in Table 6.2.

The general appearance of the stress-versus-strain curve in Figure 6.3 is typical of a wide range of metal alloys. For certain alloys (especially low-carbon steels), the curve of Figure 6.10 is obtained. The obvious distinction for this latter case is a distinct break from the elastic region at a **yield point**, also termed an **upper yield point**. The distinctive ripple pattern following the yield point is associated with nonhomogeneous deformation that begins at a point of stress concentration (often near the specimen grips). A **lower yield point** is defined at the end of the ripple pattern and at the onset of general plastic deformation.

TABLE 6.1

Alloy*	E [GPa (psi)]	Y.S. [MPa (ksi)]	T.S. [MPa (ksi)]	Percent elongation at failure
Tensile Test Data for Some Typical Metal Alloys				
1. 1040 carbon steel	200 (29 × 10^6)	600 (87)	750 (109)	17
2. 8630 low-alloy steel		680 (99)	800 (116)	22
3. a. 304 stainless steel	193 (28 × 10^6)	205 (30)	515 (75)	40
b. 410 stainless steel	200 (29 × 10^6)	700 (102)	800 (116)	22
4. L2 tool steel		1,380 (200)	1,550 (225)	12
5. Ferrous superalloy (410)	200 (29 × 10^6)	700 (102)	800 (116)	22
6. a. Ductile iron, quench	165 (24 × 10^6)	580 (84)	750 (108)	9.4
b. Ductile iron, 60–40–18	169 (24.5 × 10^6)	329 (48)	461 (67)	15
7. a. 3003-H14 aluminum	70 (10.2 × 10^6)	145 (21)	150 (22)	8–16
b. 2048, plate aluminum	70.3 (10.2 × 10^6)	416 (60)	457 (66)	8
8. a. AZ31B magnesium	45 (6.5 × 10^6)	220 (32)	290 (42)	15
b. AM100A casting magnesium	45 (6.5 × 10^6)	83 (12)	150 (22)	2
9. a. Ti–5Al–2.5Sn	107–110 (15.5–16 × 10^6)	827 (120)	862 (125)	15
b. Ti–6Al–4V	110 (16 × 10^6)	825 (120)	895 (130)	10
10. Aluminum bronze, 9% (copper alloy)	110 (16.1 × 10^6)	320 (46.4)	652 (94.5)	34
11. Monel 400 (nickel alloy)	179 (26 × 10^6)	283 (41)	579 (84)	39.5
12. AC41A zinc			328 (47.6)	7
13. 50:50 solder (lead alloy)		33 (4.8)	42 (6.0)	60
14. Nb–1 Zr (refractory metal)	68.9 (10 × 10^6)	138 (20)	241 (35)	20
15. Dental gold alloy (precious metal)			310–380 (45–55)	20–35

*Alloy designations and associated properties cited in this table and Tables 6.3 and 6.10 are from *Metals Handbook*, 8th ed., Vol. 1, and 9th ed., Vols. 1–3, American Society for Metals, Metals Park, OH, 1961, 1978, 1979, and 1980.

TABLE 6.2

Alloy	K [MPa (ksi)]	n
Typical Values of Strain Hardening Parameters[a] for Various Metals and Alloys		
Low-carbon steel (annealed)	530 (77)	0.26
4340 low-alloy steel (annealed)	640 (93)	0.15
304 stainless steel (annealed)	1,275 (185)	0.45
Al (annealed)	180 (26)	0.20
2024 aluminum alloy (heat treated)	690 (100)	0.16
Cu (annealed)	315 (46)	0.54
Brass, 70Cu–30Zn (annealed)	895 (130)	0.49

[a]Defined by Equation 6.4.

Source: Data from S. Kalpakjian, *Manufacturing Processes for Engineering Materials*, Addison-Wesley Publishing Company, Reading, MA, 1984.

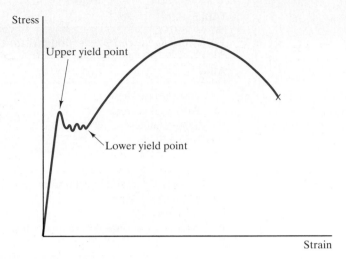

FIGURE 6.10 *For a low-carbon steel, the stress-versus-strain curve includes both an upper and lower yield point.*

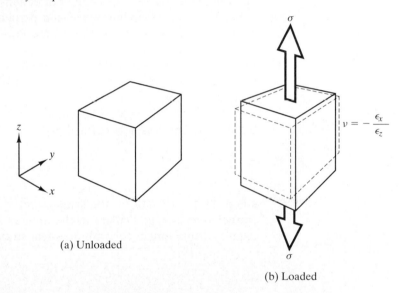

(a) Unloaded

(b) Loaded

FIGURE 6.11 *The Poisson's ratio (ν) characterizes the contraction perpendicular to the extension caused by a tensile stress.*

Figure 6.11 illustrates another important feature of elastic deformation, namely, a contraction perpendicular to the extension caused by a tensile stress. This effect is characterized by the **Poisson's* ratio**, ν, where

$$\nu = -\frac{\epsilon_x}{\epsilon_z} \tag{6.5}$$

---

*Simeon-Denis Poisson (1781–1840), French mathematician, succeeded Fourier in a faculty position at the Ecole Polytechnique. Although he did not generate original results in the way Fourier had, Poisson was a master of applying a diligent mathematical treatment to the unresolved questions raised by others. He is best known for the Poisson distribution dealing with probability for large number systems.

TABLE 6.3

Poisson's Ratio and Shear Modulus for the Alloys of Table 6.1			
Alloy	ν	G (GPa)	G/E
1. 1040 carbon steel	0.30		
2. 8630 carbon steel	0.30		
3. a. 304 stainless steel	0.29		
6. b. Ductile iron, 60–40–18	0.29		
7. a. 3003-H14 aluminum	0.33	25	0.36
8. a. AZ31B magnesium	0.35	17	0.38
b. AM100A casting magnesium	0.35		
9. a. Ti–5Al–2.5Sn	0.35	48	0.44
b. Ti–6Al–4V	0.33	41	0.38
10. Aluminum bronze, 9% (copper alloy)	0.33	44	0.40
11. Monel 400 (nickel alloy)	0.32		

and where the strains in the x and z directions are defined as shown in Figure 6.11. (There is a corresponding expansion perpendicular to the compression caused by a compressive stress.) Although the Poisson's ratio does not appear directly on the stress-versus-strain curve, it is, along with the elastic modulus, the most fundamental description of the elastic behavior of engineering materials. Table 6.3 summarizes values of ν for several common alloys. Note that values fall within the relatively narrow band of 0.26 to 0.35.

Figure 6.12 illustrates the nature of elastic deformation in a pure shear loading. The **shear stress**, τ, is defined as

$$\tau = \frac{P_s}{A_s},\qquad (6.6)$$

where P_S is the load on the sample and A_s is the area of the sample parallel (rather than perpendicular) to the applied load. The shear stress produces an angular displacement (α) with the **shear strain**, γ, being defined as

$$\gamma = \tan\alpha,\qquad (6.7)$$

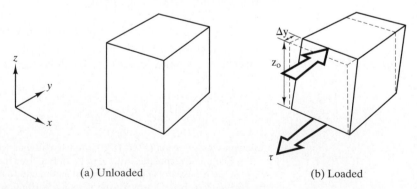

(a) Unloaded   (b) Loaded

FIGURE 6.12 *Elastic deformation under a shear load.*

which is equal to $\Delta y/z_0$ in Figure 6.12. The **shear modulus**, or **modulus of rigidity**, G, is defined (in a manner comparable to Equation 6.3) as

$$G = \frac{\tau}{\gamma}. \tag{6.8}$$

The shear modulus, G, and the elastic modulus, E, are related, for small strains, by Poisson's ratio; namely,

$$E = 2G(1 + \nu). \tag{6.9}$$

Typical values of G are given in Table 6.3. As the two moduli are related by ν (Equation 6.9) and ν falls within a narrow band, the ratio of G/E is relatively fixed for most alloys at about 0.4 (see Table 6.3).

EXAMPLE 6.1

From Figure 6.3, calculate E, Y.S., T.S., and percent elongation at failure for the aluminum 2024-T81 specimen.

SOLUTION
To obtain the modulus of elasticity, E, note that the strain at $\sigma = 300$ MPa is 0.0043 (as shown in the following figure). Then,

$$E = \frac{\sigma}{\epsilon} = \frac{300 \times 10^6 \text{ Pa}}{0.0043} = 70 \text{ GPa.}$$

The 0.2% offset construction gives

$$\text{Y.S.} = 410 \text{ MPa.}$$

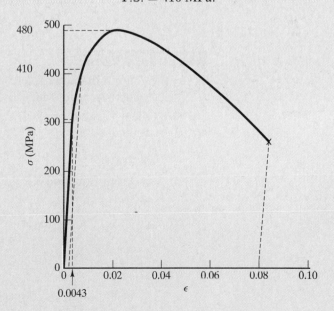

The maximum for the stress–strain curve gives

$$T.S. = 480 \text{ MPa}.$$

Finally, the strain at fracture is $\epsilon_f = 0.08$, giving

$$\% \text{ elongation at failure} = 100 \times \epsilon_f = 8\%.$$

EXAMPLE 6.2

A 10-mm-diameter bar of 1040 carbon steel (see Table 6.1) is subjected to a tensile load of 50,000 N, taking it beyond its yield point. Calculate the elastic recovery that would occur upon removal of the tensile load.

SOLUTION
Using Equation 6.1 to calculate engineering stress gives

$$\sigma = \frac{P}{A_0} = \frac{50,000 \text{ N}}{\pi (5 \times 10^{-3} \text{ m})^2} = 637 \times 10^6 \frac{N}{m^2}$$
$$= 637 \text{ MPa},$$

which is between the Y.S. (600 MPa) and the T.S. (750 MPa) for this alloy (Table 6.1).

The elastic recovery can be calculated from Hooke's law (Equation 6.3) using the elastic modulus of Table 6.1:

$$\epsilon = \frac{\sigma}{E}$$
$$= \frac{637 \times 10^6 \text{ Pa}}{200 \times 10^9 \text{ Pa}}$$
$$= 3.18 \times 10^{-3}.$$

EXAMPLE 6.3

(a) A 10-mm-diameter rod of 3003-H14 aluminum alloy is subjected to a 6-kN tensile load. Calculate the resulting rod diameter.

(b) Calculate the diameter if this rod is subjected to a 6-kN compressive load.

SOLUTION
(a) From Equation 6.1, the engineering stress is

$$\sigma = \frac{P}{A_0}$$
$$= \frac{6 \times 10^3 \text{ N}}{\pi (\frac{10}{2} \times 10^{-3} \text{ m})^2} = 76.4 \times 10^6 \frac{N}{m^2} = 76.4 \text{ MPa}.$$

From Table 6.1, we see that this stress is well below the yield strength (145 MPa) and, as a result, the deformation is elastic.

From Equation 6.3, we can calculate the tensile strain using the elastic modulus from Table 6.1:

$$\epsilon = \frac{\sigma}{E} = \frac{76.4 \text{ MPa}}{70 \times 10^3 \text{ MPa}} = 1.09 \times 10^{-3}.$$

If we use Equation 6.5 and the value for ν from Table 6.3, the strain for the diameter can be calculated as

$$\epsilon_{\text{diameter}} = -\nu\epsilon_z = -(0.33)(1.09 \times 10^{-3})$$
$$= -3.60 \times 10^{-4}.$$

The resulting diameter can then be determined (analogous to Equation 6.2) from

$$\epsilon_{\text{diameter}} = \frac{d_f - d_0}{d_0}$$

or

$$d_f = d_0(\epsilon_{\text{diameter}} + 1) = 10 \text{ mm}(-3.60 \times 10^{-4} + 1)$$
$$= 9.9964 \text{ mm}.$$

(b) For a compressive stress, the diameter strain will be of equal magnitude but of opposite sign; that is,

$$\epsilon_{\text{diameter}} = +3.60 \times 10^{-4}.$$

As a result, the final diameter will be

$$d_f = d_0(\epsilon_{\text{diameter}} + 1) = 10 \text{ mm}(+3.60 \times 10^{-4} + 1)$$
$$= 10.0036 \text{ mm}.$$

PRACTICE PROBLEM 6.1

In Example 6.1, the basic mechanical properties of a 2024-T81 aluminum are calculated based on its stress–strain curve (Figure 6.3). Given at the top of page 164 are load-elongation data for a type 304 stainless steel similar to that presented in Figure 6.2. This steel is similar to alloy 3(a) in Table 6.1 except that it has a different thermomechanical history, giving it slightly higher strength with lower ductility. **(a)** Plot these data in a manner comparable to the plot shown in Figure 6.2. **(b)** Replot these data as a stress–strain curve similar to that shown in Figure 6.3. **(c)** Replot the initial strain data on an expanded scale, similar to that used for Figure 6.4. Using the results of parts **(a)**–**(c)**, calculate **(d)** E, **(e)** Y.S., **(f)** T.S., and **(g)** percent elongation at failure for this 304 stainless steel. For parts **(d)**–**(f)**, express answers in both Pa and psi units. **(Note that the solutions to all practice problems are provided on the related Web site.)**

For the 304 stainless steel introduced in Practice Problem 6.1, calculate the elastic recovery for the specimen upon removal of the load of **(a)** 35,720 N and **(b)** 69,420 N. (See Example 6.2.)

Load (N)	Gage length (mm)	Load (N)	Gage length (mm)
0	50.8000	35,220	50.9778
4,890	50.8102	35,720	51.0032
9,779	50.8203	40,540	51.8160
14,670	50.8305	48,390	53.3400
19,560	50.8406	59,030	55.8800
24,450	50.8508	65,870	58.4200
27,620	50.8610	69,420	60.9600
29,390	50.8711	69,670 (maximum)	61.4680
32,680	50.9016	68,150	63.5000
33,950	50.9270	60,810 (fracture)	66.0400 (after fracture)
34,580	50.9524		

Original, specimen diameter: 12.7 mm.

For the alloy in Example 6.3, calculate the rod diameter at the (tensile) yield stress indicated in Table 6.1.

CERAMICS AND GLASSES

Many of the mechanical properties discussed for metals are equally important to ceramics or glasses used in structural applications. In addition, the different nature of these nonmetals leads to some unique mechanical behavior.

Metal alloys generally demonstrate a significant amount of plastic deformation in a typical tensile test. In contrast, ceramics and glasses generally do not. Figure 6.13 shows characteristic results for uniaxial loading of dense, polycrystalline Al_2O_3. In Figure 6.13a, failure of the sample occurred in the elastic region. This **brittle fracture** is characteristic of ceramics and glasses. An equally important characteristic is illustrated by the difference between the parts of Figure 6.13. Figure 6.13a illustrates the breaking strength in a tensile test (280 MPa), while Figure 6.13b is the same for a compressive test (2,100 MPa). This is an especially dramatic example of the fact that ceramics are relatively weak in tension, but relatively strong in compression. This behavior is shared by some cast irons (Chapter 11) and concrete (Chapter 12B on the related Web site). Table 6.4 summarizes moduli of elasticity and strengths for several ceramics and glasses. The strength parameter is the modulus of rupture, a value calculated from data in a bending test. The **modulus of rupture** (**MOR**) is given by

$$MOR = \frac{3FL}{2bh^2},$$ (6.10)

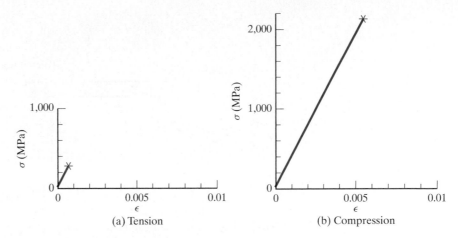

FIGURE 6.13 *The brittle nature of fracture in ceramics is illustrated by these stress–strain curves, which show only linear, elastic behavior. In (a), fracture occurs at a* tensile *stress of 280 MPa. In (b) a* compressive *strength of 2,100 MPa is observed. The sample in both tests is a dense, polycrystalline Al_2O_3.*

where F is the applied force and b, h, and L are dimensions defined in Figure 6.14. The MOR is sometimes referred to as the flexural strength (F.S.) and is similar in magnitude to the tensile strength, as the failure mode in bending is tensile (along the outermost edge of the sample). The bending test, illustrated in

TABLE 6.4

Modulus of Elasticity and Strength (Modulus of Rupture) for Some Ceramics and Glasses

		E (MPa)	MOR (MPa)
1.	Mullite (aluminosilicate) porcelain	69×10^3	69
2.	Steatite (magnesia aluminosilicate) porcelain	69×10^3	140
3.	Superduty fireclay (aluminosilicate) brick	97×10^3	5.2
4.	Alumina (Al_2O_3) crystals	380×10^3	340–1,000
5.	Sintered[a] alumina (~5% porosity)	370×10^3	210–340
6.	Alumina porcelain (90–95% alumina)	370×10^3	340
7.	Sintered[a] magnesia (~5% porosity)	210×10^3	100
8.	Magnesite (magnesia) brick	170×10^3	28
9.	Sintered[a] spinel (magnesia aluminate) (~5% porosity)	238×10^3	90
10.	Sintered[a] stabilized zirconia (~5% porosity)	150×10^3	83
11.	Sintered[a] beryllia (~5% porosity)	310×10^3	140–280
12.	Dense silicon carbide (~5% porosity)	470×10^3	170
13.	Bonded silicon carbide (~20% porosity)	340×10^3	14
14.	Hot-pressed[b] boron carbide (~5% porosity)	290×10^3	340
15.	Hot-pressed[b] boron nitride (~5% porosity)	83×10^3	48–100
16.	Silica glass	72.4×10^3	107
17.	Borosilicate glass	69×10^3	69

[a] *Sintering* refers to fabrication of the product by the bonding of powder particles by solid-state diffusion at high temperature (> one-half the absolute melting point). See Section 10.6 for a more detailed description.

[b] *Hot pressing* is sintering accompanied by high-pressure application.

Source: W. D. Kingery, H. K. Bowen, and D. R. Uhlmann, *Introduction to Ceramics*, 2nd ed., John Wiley & Sons, Inc., New York, 1976.

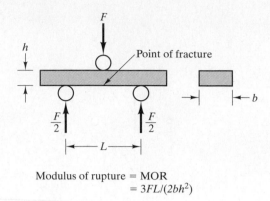

Modulus of rupture = MOR
$$= 3FL/(2bh^2)$$

FIGURE 6.14 *The bending test that generates a modulus of rupture. This strength parameter is similar in magnitude to a tensile strength. Fracture occurs along the outermost sample edge, which is under a tensile load.*

TABLE 6.5

Poisson's Ratio for Some Ceramics and Glasses	
	ν
1. Al$_2$O$_3$	0.26
2. BeO	0.26
3. CeO$_2$	0.27–0.31
4. Cordierite (2MgO $\cdot$ 2Al$_2$O$_3$ $\cdot$ 5SiO$_2$)	0.31
5. Mullite (3Al$_2$O$_3$ $\cdot$ 2SiO$_2$)	0.25
6. SiC	0.19
7. Si$_3$N$_4$	0.24
8. TaC	0.24
9. TiC	0.19
10. TiO$_2$	0.28
11. Partially stabilized ZrO$_2$	0.23
12. Fully stabilized ZrO$_2$	0.23–0.32
13. Glass-ceramic (MgO–Al$_2$O$_3$–SiO$_2$)	0.24
14. Borosilicate glass	0.2
15. Glass from cordierite	0.26

Source: Data from *Ceramic Source '86* and *Ceramic Source '87*, American Ceramic Society, Columbus, OH, 1985 and 1986.

Figure 6.14, is frequently easier to conduct on brittle materials than the traditional tensile test. Values of Poisson's ratio are given in Table 6.5. One can note from comparing Tables 6.3 and 6.5 that ν for metals is typically $\approx \frac{1}{3}$ and for ceramics $\approx \frac{1}{4}$.

To appreciate the reason for the mechanical behavior of structural ceramics, we must consider the stress concentration at crack tips. For purely brittle

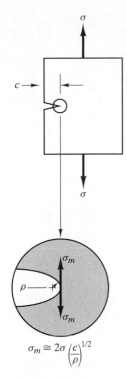

$$\sigma_m \cong 2\sigma \left(\frac{c}{\rho}\right)^{1/2}$$

FIGURE 6.15 *Stress (σ_m) at the tip of a Griffith crack.*

materials, the simple **Griffith* crack model** is applicable. Griffith assumed that in any real material there will be numerous elliptical cracks at the surface and/or in the interior. It can be shown that the highest stress (σ_m) at the tip of such a crack is

$$\sigma_m \simeq 2\sigma \left(\frac{c}{\rho}\right)^{1/2}, \qquad (6.11)$$

where σ is the applied stress, c is the crack length as defined in Figure 6.15, and ρ is the radius of the crack tip. Since the crack-tip radius can be as small as an interatomic spacing, the stress intensification can be quite large. Routine production and handling of ceramics and glasses make Griffith flaws inevitable. Hence, these materials are relatively weak in tension. A compressive load tends to close, not open, the Griffith flaws and consequently does not diminish the inherent strength of the ionically and covalently bonded material.

The drawing of small-diameter glass fibers in a controlled atmosphere is one way to avoid Griffith flaws. The resulting fibers can demonstrate tensile strengths approaching the theoretical atomic bond strength of the material, which helps to make them excellent reinforcing fibers for composite systems.

EXAMPLE 6.4

A glass plate contains an atomic-scale surface crack. (Take the crack-tip radius $\simeq$ diameter of an O^{2-} ion.) Given that the crack is 1-μm long and the theoretical strength of the defect-free glass is 7.0 GPa, calculate the breaking strength of the plate.

SOLUTION
This is an application of Equation 6.11:

$$\sigma_m = 2\sigma \left(\frac{c}{\rho}\right)^{1/2}.$$

Rearranging gives us

$$\sigma = \frac{1}{2}\sigma_m \left(\frac{\rho}{c}\right)^{1/2}.$$

Using Appendix 2, we have

$$\rho = 2r_{O^{2-}} = 2(0.132 \text{ nm})$$
$$= 0.264 \text{ nm}.$$

---

*Alan Arnold Griffith (1893–1963), British engineer. Griffith's career was spent primarily in aeronautical engineering. He was one of the first to suggest that the gas turbine would be a feasible propulsion system for aircraft. In 1920, he published his research on the strength of glass fibers that was to make his name one of the best known in the field of materials engineering.

Then,

$$\sigma = \frac{1}{2}(7.0 \times 10^9 \text{ Pa})\left(\frac{0.264 \times 10^{-9} \text{ m}}{1 \times 10^{-6} \text{ m}}\right)^{1/2}$$

$$= 57 \text{ MPa}.$$

PRACTICE PROBLEM 6.4

Calculate the breaking strength of a given glass plate containing **(a)** a 0.5-μm-long surface crack and **(b)** a 5-μm-long surface crack. Except for the length of the crack, use the conditions described in Example 6.4.

POLYMERS

As with ceramics, the mechanical properties of polymers can be described with much of the vocabulary introduced for metals. Tensile strength and modulus of elasticity are important design parameters for polymers as well as inorganic structural materials.

With the increased availability of engineering polymers for metals substitution, a greater emphasis has been placed on presenting the mechanical behavior of polymers in a format similar to that used for metals. Primary emphasis is on stress-versus-strain data. Although strength and modulus values are important parameters for these materials, design applications frequently involve a bending, rather than tensile, mode. As a result, flexural strength and flexural modulus are frequently quoted.

As noted earlier, the **flexural strength** (F.S.) is equivalent to the modulus of rupture defined for ceramics in Equation 6.10 and Figure 6.14. For the same test-specimen geometry, the **flexural modulus**, or **modulus of elasticity in bending** (E_{flex}), is

$$E_{\text{flex}} = \frac{L^3 m}{4bh^3}, \tag{6.12}$$

where m is the slope of the tangent to the initial straight-line portion of the load-deflection curve and all other terms are defined relative to Equation 6.10 and Figure 6.14. An important advantage of the flexural modulus for polymers is that it describes the combined effects of compressive deformation (adjacent to the point of applied load in Figure 6.14) and tensile deformation (on the opposite side of the specimen). For metals, as noted before, tensile and compressive moduli are generally the same. For many polymers, the tensile and compressive moduli differ significantly.

Some polymers, especially the elastomers, are used in structures for the purpose of isolation and absorption of shock and vibration. For such applications, a "dynamic" elastic modulus is more useful to characterize the performance of the polymer under an oscillating mechanical load. For elastomers in general, the dynamic modulus is greater than the static modulus. For some compounds, the

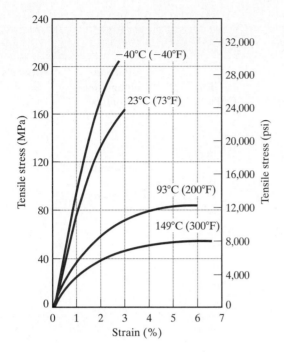

FIGURE 6.16 *Stress-versus-strain curves for a polyester engineering polymer. (From* Design Handbook for Du Pont Engineering Plastics, *used by permission.)*

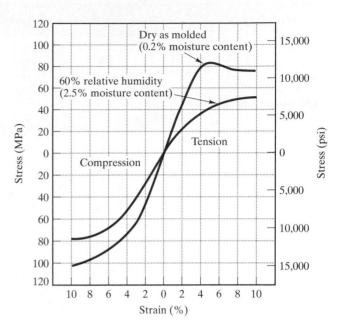

FIGURE 6.17 *Stress-versus-strain curves for a nylon 66 at 23°C showing the effect of relative humidity. (From* Design Handbook for Du Pont Engineering Plastics, *used by permission.)*

two moduli may differ by a factor of two. The **dynamic modulus of elasticity**, E_{dyn} (in MPa), is

$$E_{\text{dyn}} = CIf^2, \tag{6.13}$$

where C is a constant dependent upon specific test geometry, I is the moment of inertia (in kg · m²) of the beam and weights used in the dynamic test, and f is the frequency of vibration (in cycles/s) for the test. Equation 6.13 holds for both compressive and shear measurements, with the constant C having a different value in each case.

Figure 6.16 shows typical stress-versus-strain curves for an engineering polymer, polyester. Although these plots look similar to common stress-versus-strain plots for metals, there is a strong effect of temperature. Nonetheless, this mechanical behavior is relatively independent of atmospheric moisture. Both polyester and acetal engineering polymers have this advantage. However, relative humidity is a design consideration for the use of nylons, as shown in Figure 6.17. Also demonstrated in Figure 6.17 is the difference in elastic modulus (slope of the plots near the origin) for tensile and compressive loads. (Recall that this point was raised in the introduction of flexural modulus.) Table 6.6 gives mechanical properties of the thermoplastic polymers (those that become soft and deformable upon heating). Table 6.7 gives similar properties for the thermosetting polymers (those that become hard and rigid upon heating). Note that the dynamic modulus values in Table 6.7 are not, in general, greater than the *tensile* modulus values. The statement that the dynamic modulus of an elastomer is generally greater than the

TABLE 6.6

Mechanical Properties Data for Some Thermoplastic Polymers

Polymer	E^a [MPa (ksi)]	E_{flex}^b [MPa (ksi)]	T.S. [MPa (ksi)]	Percent elongation at failure	Poisson's ratio ν
General-use polymers					
Polyethylene					
High-density	830 (120)		28 (4)	15–100	
Low-density	170 (25)		14 (2)	90–800	
Polyvinylchloride	2,800 (400)		41 (6)	2–30	
Polypropylene	1,400 (200)		34 (5)	10–700	
Polystyrene	3,100 (450)		48 (7)	1–2	
Polyesters	—(—)	8,960 (1,230)	158 (22.9)	2.7	
Acrylics (Lucite)	2,900 (420)		55 (8)	5	
Polyamides (nylon 66)	2,800 (410)	2,830 (410)	82.7 (12.0)	60	0.41
Cellulosics	3,400–28,000 (500–4,000)		14–55 (2–8)	5–40	
Engineering polymers					
Acrylonitrile-butadiene-styrene	2,100 (300)		28–48 (4–7)	20–80	
Polycarbonates	2,400 (350)		62 (9)	110	
Acetals	3,100 (450)	2,830 (410)	69 (10)	50	0.35
Polytetrafluoroethylene (Teflon)	410 (60)		17 (2.5)	100–350	
Thermoplastic elastomers					
Polyester-type		585 (85)	46 (6.7)	400	

[a]Low strain data (in tension).

[b]In shear.

Source: From data collections in R. A. Flinn and P. K. Trojan, *Engineering Materials and Their Applications*, 2nd ed., Houghton Mifflin Company, Boston, 1981; M. F. Ashby and D. R. H. Jones, *Engineering Materials*, Pergamon Press, Inc., Elmsford, NY, 1980; and *Design Handbook for Du Pont Engineering Plastics*.

static modulus is valid for a given mode of stress application. The dynamic *shear* modulus values are, in general, greater than static *shear* modulus values.

EXAMPLE 6.5

The following data are collected in a flexural test of a nylon to be used in the fabrication of lightweight gears:

Test piece geometry: 7 mm × 13 mm × 100 mm,

Distance between supports = L = 50 mm,

and

Initial slope of load-deflection curve = 404×10^3 N/m.

Calculate the flexural modulus for this engineering polymer.

TABLE 6.7

Mechanical Properties Data for Some Thermosetting Polymers

Polymer	E^a [MPa (ksi)]	E_{Dyn}^b [MPa (ksi)]	T.S. [MPa (ksi)]	Percent elongation at failure
Thermosets				
Phenolics (phenolformaldehyde)	6,900 (1,000)	—	52 (7.5)	0
Urethanes	—	—	34 (5)	—
Urea-melamine	10,000 (1,500)	—	48 (7)	0
Polyesters	6,900 (1,000)	—	28 (4)	0
Epoxies	6,900 (1,000)	—	69 (10)	0
Elastomers				
Polybutadiene/polystyrene copolymer				
Vulcanized	1.6 (0.23)	0.8 (0.12)	1.4–3.0 (0.20–0.44)	440–600
Vulcanized with 33% carbon black	3–6 (0.4–0.9)	8.7 (1.3)	17–28 (2.5–4.1)	400–600
Polyisoprene				
Vulcanized	1.3 (0.19)	0.4 (0.06)	17–25 (2.5–3.6)	750–850
Vulcanized with 33% carbon black	3.0–8.0 (0.44–1.2)	6.2 (0.90)	25–35 (3.6–5.1)	550–650
Polychloroprene				
Vulcanized	1.6 (0.23)	0.7 (0.10)	25–38 (3.6–5.5)	800–1,000
Vulcanized with 33% carbon black	3–5 (0.4–0.7)	2.8 (0.41)	21–30 (3.0–4.4)	500–600
Polyisobutene/polyisoprene copolymer				
Vulcanized	1.0 (0.15)	0.4 (0.06)	18–21 (2.6–3.0)	750–950
Vulcanized with 33% carbon black	3–4 (0.4–0.6)	3.6 (0.52)	18–21 (2.6–3.0)	650–850
Silicones	—	—	7 (1)	4,000
Vinylidene fluoride/ hexafluoropropylene	—	—	12.4 (1.8)	—

[a]Low strain data (in tension).

[b]In shear.

Source: From data collections in R. A. Flinn and P. K. Trojan, *Engineering Materials and Their Applications*, 2nd ed., Houghton Mifflin Company, Boston, 1981; M. F. Ashby and D. R. H. Jones, *Engineering Materials*, Pergamon Press, Inc., Elmsford, NY, 1980; and J. Brandrup and E. H. Immergut, Eds., *Polymers Handbook*, 2nd ed., John Wiley & Sons, Inc., New York, 1975.

SOLUTION

Referring to Figure 6.14 and Equation 6.12, we find that

$$E_{flex} = \frac{L^3 m}{4bh^3}$$

$$= \frac{(50 \times 10^{-3} \text{ m})^3 (404 \times 10^3 \text{ N/m})}{4(13 \times 10^{-3} \text{ m})(7 \times 10^{-3} \text{ m})^3}$$

$$= 2.83 \times 10^9 \text{ N/m}^2 = 2,830 \text{ MPa}.$$

EXAMPLE 6.6

A small, uniaxial stress of 1 MPa (145 psi) is applied to a rod of high-density polyethylene.

(a) What is the resulting strain?

(b) Repeat for a rod of vulcanized isoprene.

(c) Repeat for a rod of 1040 steel.

SOLUTION

(a) For this modest stress level, we can assume Hooke's law behavior:

$$\epsilon = \frac{\sigma}{E}.$$

Table 6.6 gives $E = 830$ MPa. So,

$$\epsilon = \frac{1 \text{ MPa}}{830 \text{ MPa}} = 1.2 \times 10^{-3}.$$

(b) Table 6.7 gives $E = 1.3$ MPa, or

$$\epsilon = \frac{1 \text{ MPa}}{1.3 \text{ MPa}} = 0.77.$$

(c) Table 6.1 gives $E = 200$ GPa $= 2 \times 10^5$ MPa, or

$$\epsilon = \frac{1 \text{ MPa}}{2 \times 10^5 \text{ MPa}}$$
$$= 5.0 \times 10^{-6}.$$

Note. The dramatic difference between elastic moduli of polymers and inorganic solids is used to advantage in composite materials (Chapter 12).

PRACTICE PROBLEM 6.5

The data in Example 6.5 permit the flexural modulus to be calculated. For the configuration described, an applied force of 680 N causes fracture of the nylon sample. Calculate the corresponding flexural strength.

PRACTICE PROBLEM 6.6

In Example 6.6, strain is calculated for various materials under a stress of 1 MPa. While the strain is relatively large for polymers, there are some high-modulus polymers with substantially lower results. Calculate the strain in a cellulosic fiber with a modulus of elasticity of 28,000 MPa (under a uniaxial stress of 1 MPa).

6.2 | Elastic Deformation

Before leaving the discussion of stress-versus-strain behavior for materials, it is appropriate to look at the atomic-scale mechanisms involved. Figure 6.18 shows that the fundamental mechanism of elastic deformation is the stretching of atomic bonds. The fractional deformation of the material in the initial elastic region is small so that, on the atomic scale, we are dealing only with the portion of the force–atom separation curve in the immediate vicinity of the equilibrium atom separation distance (a_0 corresponding to $F = 0$). The nearly straight-line plot of F versus a across the a axis implies that similar elastic behavior will be observed in a compressive, or push, test as well as in tension. This similarity is often the case, especially for metals.

EXAMPLE 6.7

In the absence of stress, the center-to-center atomic separation distance of two Fe atoms is 0.2480 nm (along a $\langle 111 \rangle$ direction). Under a tensile stress of 1,000 MPa along this direction, the atomic separation distance increases to 0.2489 nm. Calculate the modulus of elasticity along the $\langle 111 \rangle$ directions.

Here we see how the bonding force and energy curves of Figures 2.7 and 2.8 play a central role in understanding the nature of elastic deformation.

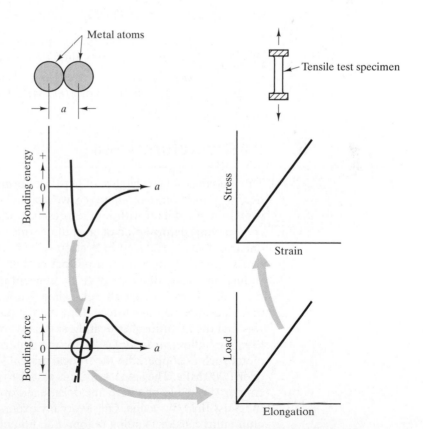

FIGURE 6.18 *Relationship of elastic deformation to the stretching of atomic bonds.*

SOLUTION
From Hooke's law (Equation 6.3),

$$E = \frac{\sigma}{\epsilon},$$

with

$$\epsilon = \frac{(0.2489 - 0.2480)\ \text{nm}}{0.2480\ \text{nm}} = 0.00363,$$

giving

$$E = \frac{1{,}000\ \text{MPa}}{0.00363} = 280\ \text{GPa}.$$

Note. This modulus represents the maximum value in the iron crystal structure. The minimum value of E is 125 GPa in the $\langle 100 \rangle$ direction. In polycrystalline iron with random grain orientations, an average modulus of 205 GPa occurs, which is close to the value for most steels (Table 6.1).

PRACTICE PROBLEM 6.7

(a) Calculate the center-to-center separation distance of two Fe atoms along the $\langle 100 \rangle$ direction in unstressed α-iron.

(b) Calculate the separation distance along that direction under a tensile stress of 1,000 MPa. (See Example 6.7.)

6.3 | Plastic Deformation

The fundamental mechanism of plastic deformation is the distortion and reformation of atomic bonds. In Chapter 5, we saw that atomic diffusion in crystalline solids is extremely difficult without the presence of point defects. Similarly, the plastic (permanent) deformation of crystalline solids is difficult without dislocations, the linear defects introduced in Section 4.3. Frenkel first calculated the mechanical stress necessary to deform a perfect crystal. This deformation would occur by sliding one plane of atoms over an adjacent plane, as shown in Figure 6.19. The shear stress associated with this sliding action can be calculated with knowledge of the periodic bonding forces along the slip plane. The result obtained by Frenkel was that the theoretical **critical shear stress** is roughly one order of magnitude less than the bulk *shear modulus, G,* for the material (see Equation 6.8). For a typical metal such as copper, the theoretical critical shear stress represents a value well over 1,000 MPa. The actual stress necessary to plastically deform a sample of pure copper (i.e., slide atomic planes past each other) is at least an order of magnitude less than this value. Our everyday experience with metallic alloys (opening aluminum cans or bending automobile fenders) represents deformations generally requiring stress levels of only a few hundred megapascals. What, then, is the

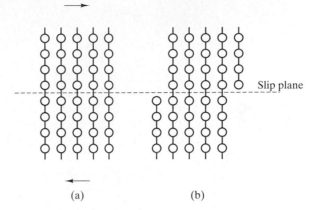

FIGURE 6.19 *Sliding of one plane of atoms past an adjacent one. This high-stress process is necessary to plastically (permanently) deform a perfect crystal.*

basis of the mechanical deformation of metals, which requires only a fraction of the theoretical strength? The answer, to which we have already alluded, is the dislocation. Figure 6.20 illustrates the role a dislocation can play in the shear of a crystal along a slip plane. The key point to observe is that only a relatively small shearing force needs to operate in the immediate vicinity of the dislocation in order to produce a step-by-step shear that eventually yields the same overall deformation as the high-stress mechanism of Figure 6.19. A perspective view of a shearing mechanism involving a more general, mixed dislocation (see Figure 4.13) is given in Figure 6.21.

We can appreciate this defect mechanism of slip by considering a simple analogy. Figure 6.22 introduces Goldie the caterpillar. It is impractical to force Goldie to slide along the ground in a perfect straight line (Figure 6.22a). However, Goldie "slips along" nicely by passing a "dislocation" along the length of her body (Figure 6.22b).

Reflecting on Figure 6.20, we can appreciate that the stepwise slip mechanism would tend to become more difficult as the individual atomic step distances are increased. As a result, slip is more difficult on a low-atomic-density plane than on a high-atomic-density plane. Figure 6.23 shows this difference schematically. In general, the micromechanical mechanism of slip—dislocation motion—will occur in high-atomic-density planes and in high-atomic-density directions. A combination of families of crystallographic planes and directions corresponding to dislocation motion is referred to as a **slip system**. In Figure 6.24, we can label the slip systems in (a) fcc aluminum and (b) hcp magnesium. Aluminum and its alloys are characteristically ductile (deformable) due to the large number (12) of high-density plane–direction combinations. Magnesium and its alloys are typically brittle (fracturing with little deformation) due to the smaller number (3) of such combinations. Table 6.8 summarizes the major slip systems in typical metal structures.

Several basic concepts of the mechanical behavior of crystalline materials relate directly to simple models of dislocation motion. The *cold working* of metals involves deliberate deformation of the metal at relatively low temperatures (see Section 6.1 and Chapter 10). While forming a stronger product, an important feature of cold working is that the metal becomes more difficult to deform as the extent of deformation increases. The basic micromechanical reason for this is that

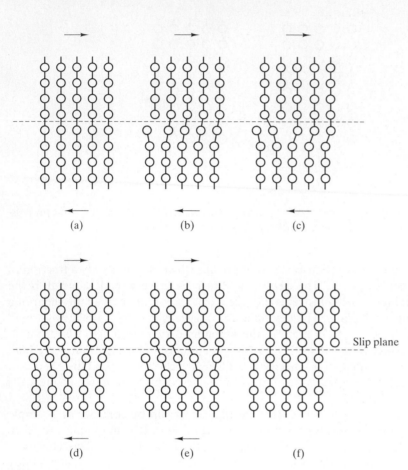

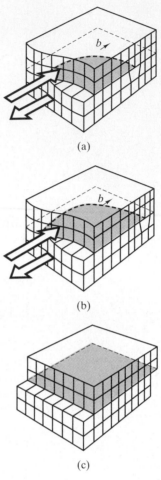

FIGURE 6.20 *A low-stress alternative for plastically deforming a crystal involves the motion of a dislocation along a slip plane.*

FIGURE 6.21 *Schematic illustration of the motion of a dislocation under the influence of a shear stress. The net effect is an increment of plastic (permanent) deformation. (Compare Figure 6.21a with Figure 4.13.)*

The schematic illustrations on this page show how the dislocation structures introduced in Section 4.3 play a central role in understanding the nature of plastic deformation.

a dislocation hinders the motion of another dislocation. The slip mechanism of Figure 6.20 proceeds most smoothly when the slip plane is free of obstructions. Cold working generates dislocations that serve as such obstacles. In fact, cold working generates so many dislocations that the configuration is referred to as a "forest of dislocations" (Figure 6.25). Foreign atoms can also serve as obstacles to dislocation motion. Figure 6.26 illustrates this micromechanical basis of **solution hardening** of alloys (i.e., restricting plastic deformation by forming solid solutions). Hardening, or increasing strength, occurs because the elastic region is extended, producing a higher yield strength. These concepts will be discussed further in Section 6.4. Obstacles to dislocation motion harden metals, but high temperatures can help to overcome these obstacles and thereby soften the metals. An example of this concept is the *annealing* process, a stress-relieving heat treatment to be described in Chapter 10. The micromechanical mechanism here is rather straightforward.

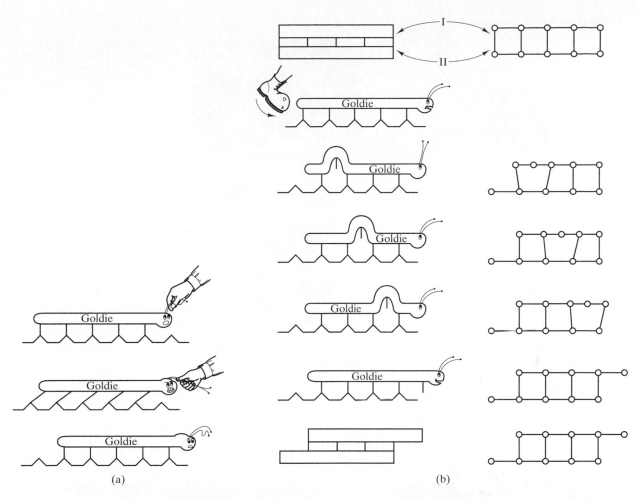

FIGURE 6.22 *Goldie the caterpillar illustrates (a) how difficult it is to move along the ground without (b) a "dislocation" mechanism. (From W. C. Moss, Ph.D. thesis, University of California, Davis, CA, 1979.)*

At sufficiently high temperatures, atomic diffusion is sufficiently great to allow highly stressed crystal grains produced by cold working to be restructured into more nearly perfect crystalline structures. The dislocation density is dramatically reduced with increasing temperature, which permits the relatively simple deformation mechanism of Figure 6.20 to occur free of the forest of dislocations. At this point, we have seen an important blending of the concepts of solid-state diffusion (from Chapter 5) and mechanical deformation. There will be many other examples in later chapters. In each case, a useful rule of thumb will apply: The temperature at which atomic mobility is sufficient to affect mechanical properties is approximately one-third to one-half times the absolute melting point, T_m.

One additional basic concept of mechanical behavior is that the more complex crystal structures correspond to relatively brittle materials. Common examples are intermetallic compounds (e.g., Ag_3Al) and ceramics (e.g., Al_2O_3). Relatively large Burgers vectors combined with the difficulty in creating obstacle-free slip planes create a limited opportunity for dislocation motion. The

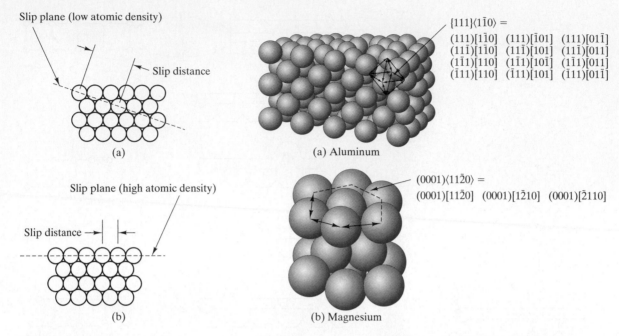

Slip plane (low atomic density)

Slip distance

(a)

Slip plane (high atomic density)

Slip distance →

(b)

FIGURE 6.23 *Dislocation slip is more difficult along (a) a low-atomic-density plane than along (b) a high-atomic-density plane.*

{111}⟨1̄10⟩ =
(111)[1̄10] (111)[1̄01] (111)[01̄1]
(11̄1)[1̄10] (11̄1)[101] (11̄1)[011]
(1̄1̄1)[110] (1̄1̄1)[101̄] (1̄1̄1)[011]
(1̄11)[110] (1̄11)[101] (1̄11)[01̄1]

(0001)⟨112̄0⟩ =
(0001)[112̄0] (0001)[1̄21̄0] (0001)[2̄110]

(a) Aluminum

(b) Magnesium

FIGURE 6.24 *Slip systems for (a) fcc aluminum and (b) hcp magnesium.*

TABLE 6.8

Major Slip Systems in the Common Metal Structures					
Crystal structure	Slip plane	Slip direction	Number of slip systems	Unit-cell geometry	Examples
bcc	{110}	⟨1̄11⟩	6 × 2 = 12	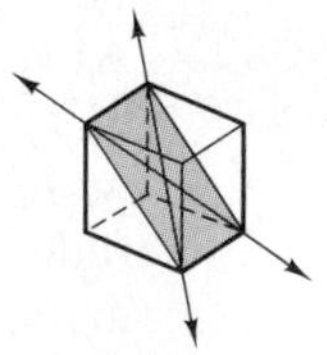	α-Fe, Mo, W
fcc	{111}	⟨1̄10⟩	4 × 3 = 12	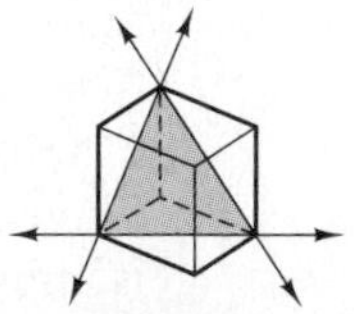	Al, Cu, γ-Fe, Ni
hcp	(0001)	⟨112̄0⟩	1 × 3 = 3	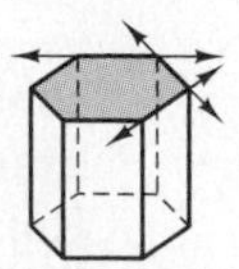	Cd, Mg, α-Ti, Zn

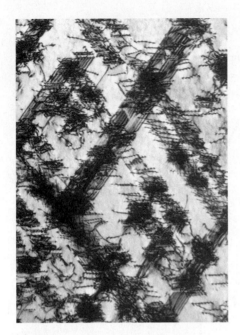

FIGURE 6.25 *Forest of dislocations in a stainless steel as seen by a transmission electron microscope [Courtesy of Chuck Echer, Lawrence Berkeley National Laboratory, National Center for Electron Microscopy.]*

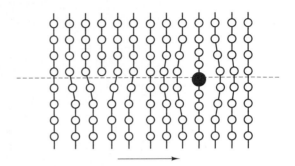

Direction of "attempted" dislocation motion

FIGURE 6.26 *How an impurity atom generates a strain field in a crystal lattice, thereby causing an obstacle to dislocation motion.*

formation of *brittle intermetallics* is a common concern in high-temperature designs involving interfaces between dissimilar metals. Ceramics, as we saw in Section 6.1, are characteristically brittle materials. Inspection of Figure 4.14 confirms the statement about large Burgers vectors. An additional consideration adding to the brittleness of ceramics is that many slip systems are not possible, owing to the charged state of the ions. The sliding of ions with like charges past one another can result in high coulombic repulsive forces. As a result, even ceramic compounds with relatively simple crystal structures exhibit significant dislocation mobility only at a relatively high temperature.

We close this section with a macroscopic calculation of the deformation stress for a crystalline material relative to the microscopic mechanism of a slip system. Figure 6.27 defines the **resolved shear stress**, τ, which is the actual stress operating on the slip system (in the slip plane and in the slip direction) resulting from the application of a simple tensile stress, $\sigma (= F/A)$, where F is the externally applied force perpendicular to the cross-sectional area (A) of the single-crystal sample. The important concept here is that the fundamental deformation mechanism is a shearing action based on the projection of the applied force onto the slip system. The component of the applied force (F) operating in the slip direction is ($F \cos \lambda$). The projection of the sample's cross-sectional area (A) onto the slip plane gives an area of ($A/\cos \varphi$). As a result, the resolved shear stress, τ, is

$$\tau = \frac{F \cos \lambda}{A/\cos \varphi} = \frac{F}{A} \cos \lambda \cos \varphi = \sigma \cos \lambda \cos \varphi, \qquad (6.14)$$

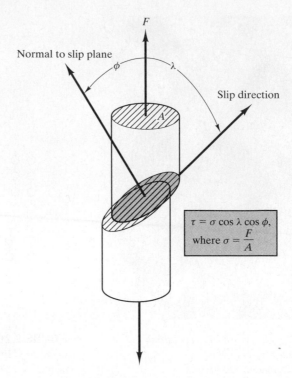

FIGURE 6.27 *Definition of the resolved shear stress, τ, which directly produces plastic deformation (by a shearing action) as a result of the external application of a simple tensile stress, σ.*

where σ is the applied tensile stress ($= F/A$) and λ and φ are defined in Figure 6.27. Equation 6.14 identifies the resolved shear stress, τ, resulting from a given applied stress. A value of τ great enough to produce slip by dislocation motion is called the **critical resolved shear stress** and is given by

$$\tau_c = \sigma_c \cos \lambda \cos \varphi, \qquad (6.15)$$

where σ_c is, of course, the applied stress necessary to produce this deformation. In considering plastic deformation, we should always keep in mind this connection between macroscopic stress values and the micromechanical mechanism of dislocation slip.

EXAMPLE 6.8

A zinc single crystal is being pulled in tension, with the normal to its basal plane (0001) at 60° to the tensile axis and with the slip direction [11$\bar{2}$0] at 40° to the tensile axis.

(a) What is the resolved shear stress, τ, acting in the slip direction when a tensile stress of 0.690 MPa (100 psi) is applied?

(b) What tensile stress is necessary to reach the critical resolved shear stress, τ_c, of 0.94 MPa (136 psi)?

SOLUTION

(a) From Equation 6.14,

$$\tau = \sigma \cos \lambda \cos \varphi$$

$$= (0.690 \text{ MPa}) \cos 40° \cos 60°$$

$$= 0.264 \text{ MPa (38.3 psi)}.$$

(b) From Equation 6.15,

$$\tau_c = \sigma_c \cos \lambda \cos \varphi,$$

or

$$\sigma_c = \frac{\tau_c}{\cos \lambda \cos \varphi} = \frac{0.94 \text{ MPa}}{\cos 40° \cos 60°}$$

$$= 2.45 \text{ MPa (356 psi)}.$$

PRACTICE PROBLEM 6.8

Repeat Example 6.8, assuming that the two directions are 45° rather than 60° and 40°.

6.4 | Hardness

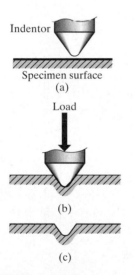

Indentor

Specimen surface
(a)

Load

(b)

(c)

FIGURE 6.28 *Hardness test. The analysis of indentation geometry is summarized in Table 6.9.*

The *hardness test* (Figure 6.28) is available as a relatively simple alternative to the tensile test of Figure 6.1. The resistance of the material to indentation is a qualitative indication of its strength. The indenter can be either rounded or pointed and is made of a material much harder than the test piece, for example, hardened steel, tungsten carbide, or diamond. Table 6.9 summarizes the common types of hardness tests with their characteristic indenter geometries. Empirical *hardness numbers* are calculated from appropriate formulas using indentation geometry measurements. Microhardness measurements are made using a high-power microscope. **Rockwell* hardness** is widely used with many scales (e.g., Rockwell A and Rockwell B) available for different hardness ranges. Correlating hardness with depth of penetration allows the hardness number to be conveniently shown on a dial or digital display. In this chapter, we shall often quote **Brinell[†] hardness numbers** (BHN) because a single scale covers a wide range of material hardness and a fairly linear correlation with strength can be found. Table 6.10 gives BHN values for the alloys of Table 6.1. Figure 6.29a shows a clear trend of BHN with tensile strength for these alloys. Figure 6.29b shows that the correlation is more precise for given families of alloys. The tensile strength is generally used for this correlation rather than yield strength because the hardness test includes a substantial component of plastic deformation. There will be further discussions of hardness in relation to heat treatments in Chapter 10. Typical hardness values for a variety of polymers are given in Table 6.11.

*The Rockwell hardness tester was invented in 1919 by Stanley P. Rockwell, an American metallurgist. The word *Rockwell*, as applied to the tester and reference standards, is a registered trademark in several countries, including the United States.

[†]Johan August Brinell (1849–1925), Swedish metallurgist, was an important contributor to the metallurgy of steels. His apparatus for hardness testing was first displayed in 1900 at the Paris Exposition. Current "Brinell testers" are essentially unchanged in design.

TABLE 6.9

Common Types of Hardness–Test Geometries

Test	Indenter	Shape of indentation Side view	Top view	Load	Formula for hardness number
Brinell	10-mm sphere of steel or tungsten carbide	D, d	d	P	$\text{BHN} = \dfrac{2P}{\pi D\left[D - \sqrt{D^2 - d^2}\right]}$
Vickers	Diamond pyramid	136°	d_1, d_1	P	$\text{VHN} = 1.854 P/d_1^2$
Knoop microhardness	Diamond pyramid	t, $l/b = 7.11$, $b/t = 4.00$	b, l	P	$\text{KHN} = 14.2 P/l^2$
Rockwell					
A	Diamond cone 120°			60 kg	$R_A =$ ⎫
C				150 kg	$R_C =$ ⎬ 100 − 500t
D				100 kg	$R_D =$ ⎭
B	$\frac{1}{16}$ in.-diameter steel sphere			100 kg	$R_B =$ ⎫
F				60 kg	$R_F =$ ⎬ 130 − 500t
G				150 kg	$R_G =$ ⎭
E	$\frac{1}{8}$ in.-diameter steel sphere			100 kg	$R_E =$ ⎫
H				60 kg	$R_H =$ ⎬

Source: H. W. Hayden, W. G. Moffatt, and J. Wulff, *The Structure and Properties of Materials*, Vol. 3: *Mechanical Behavior*, John Wiley & Sons, Inc., New York, 1965.

TABLE 6.10

Comparison of Brinell Hardness Numbers (BHN) with Tensile Strength (T.S.) for the Alloys of Table 6.1

Alloy		BHN	T.S. (MPa)
1.	1040 carbon steel	235	750
2.	8630 low-alloy steel	220	800
3.	b. 410 stainless steel	250	800
5.	Ferrous superalloy (410)	250	800
6.	b. Ductile iron, 60–40–18	167	461
7.	a. 3003-H14 aluminum	40	150
8.	a. AZ31B magnesium	73	290
	b. AM100A casting magnesium	53	150
9.	a. Ti–5Al–2.5Sn	335	862
10.	Aluminum bronze, 9% (copper alloy)	165	652
11.	Monel 400 (nickel alloy)	110–150	579
12.	AC41A zinc	91	328
13.	50:50 solder (lead alloy)	14.5	42
15.	Dental gold alloy (precious metal)	80–90	310–380

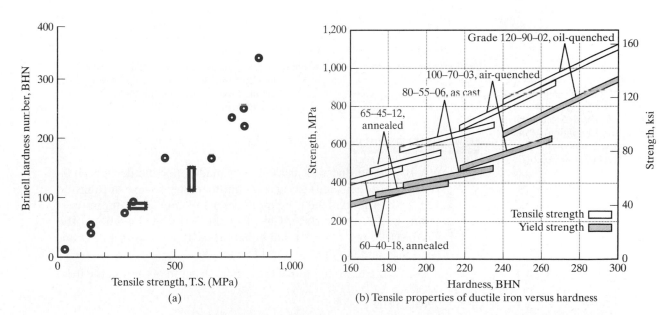

FIGURE 6.29 *(a) Plot of data from Table 6.10. A general trend of BHN with T.S. is shown. (b) A more precise correlation of BHN with T.S. (or Y.S.) is obtained for given families of alloys. [Part (b) from* Metals Handbook, *9th ed., Vol. 1, American Society for Metals, Metals Park, OH, 1978.]*

TABLE 6.11

Hardness Data for Various Polymers

Polymer	Rockwell hardness R scale[a]
Thermoplastic polymers	
General-use polymers	
Polyethylene	
High-density	40
Low-density	10
Polyvinylchloride	110
Polypropylene	90
Polystyrene	75
Polyesters	120
Acrylics (Lucite)	130
Polyamides (nylon 66)	121
Cellulosics	50 to 115
Engineering polymers	
ABS	95
Polycarbonates	118
Acetals	120
Polytetrafluoroethylene (Teflon)	70
Thermosetting Polymers	
Phenolics (phenolformaldehyde)	125
Urea-melamine	115
Polyesters	100
Epoxies	90

[a]For relatively soft materials: indenter radius $\frac{1}{2}$ in. and load of 60 kg.

Source: From data collections in R. A. Flinn and P. K. Trojan, *Engineering Materials and Their Applications,* 2nd ed., Houghton Mifflin Company, Boston, 1981; M. F. Ashby and D. R. H. Jones, *Engineering Materials,* Pergamon Press, Inc., Elmsford, NY, 1980; and *Design Handbook for Du Pont Engineering Plastics.*

The related Web site contains the chapter "Macroindentation Hardness Testing" from the ASM Handbook, *Vol. 8* (Mechanical Testing and Evaluation), *ASM International, Materials Park, OH, 2000. Used by permission.*

EXAMPLE 6.9

(a) A Brinell hardness measurement is made on a ductile iron (100–70–03, air-quenched) using a 10-mm-diameter sphere of tungsten carbide. A load of 3,000 kg produces a 3.91-mm-diameter impression in the iron surface. Calculate the BHN of this alloy. (The correct units for the Brinell equation of Table 6.9 are kilograms for load and millimeters for diameters.)

(b) Use Figure 6.29b to predict the tensile strength of this ductile iron.

SOLUTION

(a) From Table 6.9,

$$\text{BHN} = \frac{2P}{\pi D\left(D - \sqrt{D^2 - d^2}\,\right)}$$

$$= \frac{2(3{,}000)}{\pi(10)\left(10 - \sqrt{10^2 - 3.91^2}\,\right)} = 240.$$

(b) From Figure 6.29b,

$$(\text{T.S.})_{\text{BHN}=240} = 800 \text{ MPa}.$$

PRACTICE PROBLEM 6.9

Suppose that a ductile iron (100–70–03, air-quenched) has a tensile strength of 700 MPa. What diameter impression would you expect the 3,000-kg load to produce with the 10-mm-diameter ball? (See Example 6.9.)

6.5 | Creep and Stress Relaxation

The tensile test alone cannot predict the behavior of a structural material used at elevated temperatures. The strain induced in a typical metal bar loaded below its yield strength at room temperature can be calculated from Hooke's law (Equation 6.3). This strain will not generally change with time under a fixed load (Figure 6.30). Repeating this experiment at a "high" temperature (T greater than one-third to one-half times the melting point on an absolute temperature scale) produces dramatically different results. Figure 6.31 shows a typical test design, and Figure 6.32 shows a typical **creep curve** in which the strain, ϵ, gradually increases with time after the initial elastic loading. **Creep** can be defined as plastic (permanent) deformation occurring at high temperature under constant load over a long time period.

After the initial elastic deformation at $t \simeq 0$, Figure 6.32 shows three stages of creep deformation. The *primary stage* is characterized by a decreasing strain rate (slope of the ϵ vs. t curve). The relatively rapid increase in length induced during this early time period is the direct result of enhanced deformation mechanisms. A common example for metal alloys is **dislocation climb**, as illustrated in Figure 6.33. As discussed in Section 6.3, this enhanced deformation comes from thermally activated atom mobility, giving dislocations additional slip planes in which to move. The *secondary stage* of creep deformation is characterized by straight-line, constant-strain-rate data (Figure 6.32). In this region, the increased ease of slip due to high-temperature mobility is balanced by increasing resistance to slip due to the buildup of dislocations and other microstructural barriers. In the final *tertiary stage*, strain rate increases due to an increase in true stress. This

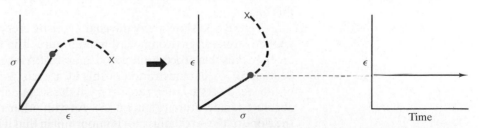

FIGURE 6.30 *Elastic strain induced in an alloy at room temperature is independent of time.*

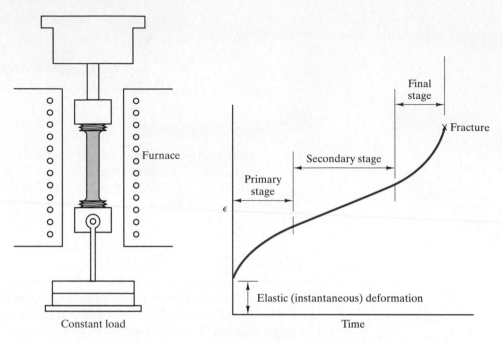

FIGURE 6.31 *Typical creep test.*

FIGURE 6.32 *Creep curve. In contrast to Figure 6.30, plastic strain occurs over time for a material stressed at high temperatures (above about one-half the absolute melting point).*

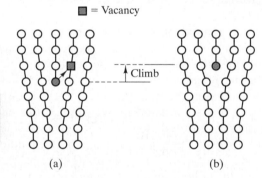

Here we see how both the atomic scale point defects introduced in Section 4.2 and the nanoscale dislocations introduced in Section 4.3 help us understand an important mechanical behavior, viz. creep deformation.

FIGURE 6.33 *Mechanism of dislocation climb. Obviously, many adjacent atom movements are required to produce climb of an entire dislocation line.*

increase results from cross-sectional area reduction due to necking or internal cracking. In some cases, fracture occurs in the secondary stage, thus eliminating this final stage.

Figure 6.34 shows how the characteristic creep curve varies with changes in applied stress or environmental temperature. The thermally activated nature of creep makes this process another example of Arrhenius behavior, as discussed in Section 5.1. A demonstration of this idea is an Arrhenius plot of the logarithm of the steady-state creep rate ($\dot{\epsilon}$) from the secondary stage against the inverse of absolute temperature (Figure 6.35). As with other thermally activated processes, the slope of the Arrhenius plot is important in that it provides an activation energy,

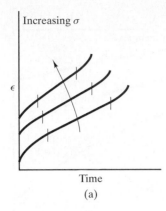

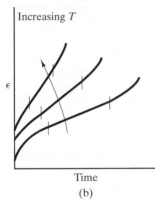

FIGURE 6.34 *Variation of the creep curve with (a) stress or (b) temperature. Note how the steady-state creep rate ($\dot{\epsilon}$) in the secondary stage rises sharply with temperature (see also Figure 6.35).*

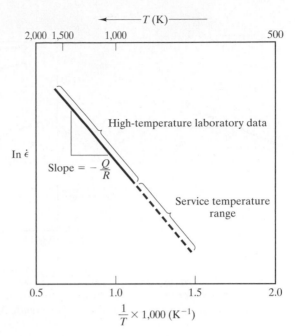

FIGURE 6.35 *Arrhenius plot of* $\ln \dot{\epsilon}$ *versus* $1/T$, *where* $\dot{\epsilon}$ *is the secondary-stage creep rate and* T *is the absolute temperature. The slope gives the activation energy for the creep mechanism. Extension of high-temperature, short-term data permits prediction of long-term creep behavior at lower service temperatures.*

Q, for the creep mechanism from the Arrhenius expression

$$\dot{\epsilon} = Ce^{-Q/RT}, \tag{6.16}$$

where C is the preexponential constant, R is the universal gas constant, and T is the absolute temperature. Another powerful aspect of the Arrhenius behavior is its predictive ability. The dashed line in Figure 6.35 shows how high-temperature strain-rate data, which can be gathered in short-time laboratory experiments, can be extrapolated to predict long-term creep behavior at lower service temperatures. This extrapolation is valid as long as the same creep mechanism operates over the entire temperature range. Many elaborate semiempirical plots have been developed based on this principle to guide design engineers in material selection.

A shorthand characterization of creep behavior is given by the secondary-stage strain rate ($\dot{\epsilon}$) and the time-to-creep rupture (t), as shown in Figure 6.36. Plots of these parameters, together with applied stress (σ) and temperature (T), provide another convenient data set for design engineers responsible for selecting materials for high-temperature service (e.g., Figure 6.37).

Creep is probably more important in ceramics than in metals because high-temperature applications are so widespread. The role of diffusion mechanisms in the creep of ceramics is more complex than in the case of metals because diffusion, in general, is more complex in ceramics. The requirement of charge neutrality and different diffusivities for cations and anions contribute to this complexity. As a result, grain boundaries frequently play a dominant role in the creep

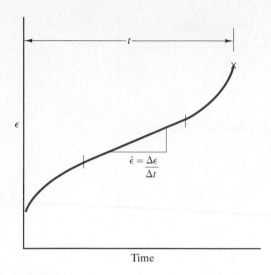

FIGURE 6.36 *Simple characterization of creep behavior is obtained from the secondary-stage strain rate (ε̇) and the time to creep rupture (t).*

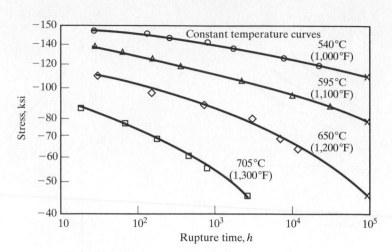

FIGURE 6.37 *Creep rupture data for the nickel-based superalloy Inconel 718. (From* Metals Handbook, *9th ed., Vol. 3, American Society for Metals, Metals Park, OH, 1980.)*

of ceramics. Sliding of adjacent grains along these boundaries provides for microstructural rearrangement during creep deformation. In some relatively impure refractory ceramics, a substantial layer of glassy phase may be present at the grain boundaries. In that case, creep can again occur by the mechanism of grain-boundary sliding due to the viscous deformation of the glassy phase. This "easy" sliding mechanism is generally undesirable due to the resulting weakness at high temperatures. In fact, the term *creep* is not applied to bulk glasses themselves. The subject of viscous deformation of glasses is discussed separately in Section 6.6.

Creep-rate data for some common ceramics at a fixed temperature are given in Table 6.12. An Arrhenius-type plot of creep-rate data at various temperatures (load fixed) is given in Figure 6.38.

TABLE 6.12

Creep-Rate Data for Various Polycrystalline Ceramics	
Material	$\dot{\epsilon}$ **at 1,300°C, 1,800 psi (12.4 MPa)** **[mm/(mm · h) $\times 10^6$]**
Al_2O_3	1.3
BeO	300.0
MgO (slip cast)	330.0
MgO (hydrostatic pressed)	33.0
$MgAl_2O_4$ (2–5 μm)	263.0
$MgAl_2O_4$ (1–3 mm)	1.0
ThO_2	1,000.0
ZrO_2 (stabilized)	30.0

Source: W. D. Kingery, H. K. Bowen, and D. R. Uhlmann, *Introduction to Ceramics,* 2nd ed., John Wiley & Sons, Inc., New York, 1976.

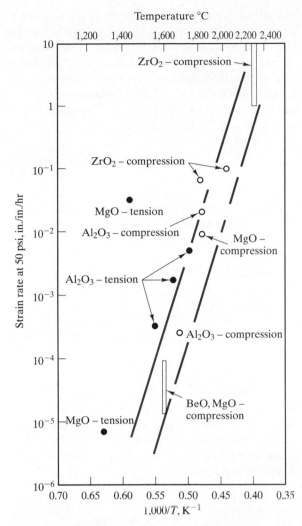

FIGURE 6.38 *Arrhenius-type plot of creep-rate data for several polycrystalline oxides under an applied stress of 50 psi (345 × 10³ Pa). Note that the inverse temperature scale is reversed (i.e., temperature increases to the right). (From W. D. Kingery, H. K. Bowen, and D. R. Uhlmann,* Introduction to Ceramics, *2nd ed., John Wiley & Sons, Inc., New York, 1976.)*

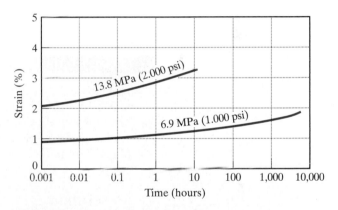

FIGURE 6.39 *Creep data for a nylon 66 at 60°C and 50% relative humidity. (From* Design Handbook for Du Pont Engineering Plastics, *used by permission.)*

For metals and ceramics, we have found creep deformation to be an important phenomenon at high temperatures (greater than one-half the absolute melting point). Creep is a significant design factor for polymers given their relatively low melting points. Figure 6.39 shows creep data for nylon 66 at moderate temperature and load. A related phenomenon, termed **stress relaxation**, is also an important design consideration for polymers. A familiar example is the rubber band, under stress for a long period of time, which does not snap back to its original size upon stress removal.

Creep deformation involves increasing strain with time for materials under constant stresses. By contrast, stress relaxation involves decreasing stress with time for polymers under constant strains. The mechanism of stress relaxation is viscous flow (i.e., molecules gradually sliding past each other over an extended period of time). Viscous flow converts some of the fixed elastic strain into nonrecoverable plastic deformation. Stress relaxation is characterized by a **relaxation time**, τ, defined as the time necessary for the stress (σ) to fall to 0.37 ($= 1/e$) of the initial stress (σ_0). The exponential decay of stress with time (t) is given by

$$\sigma = \sigma_0 e^{-t/\tau}. \tag{6.17}$$

In general, stress relaxation is an Arrhenius phenomenon, as was creep for metals and ceramics. The form of the Arrhenius equation for stress relaxation is

$$\frac{1}{\tau} = Ce^{-Q/RT}, \tag{6.18}$$

where C is a preexponential constant, Q is the activation energy (per mole) for viscous flow, R is the universal gas constant, and T is the absolute temperature.

EXAMPLE 6.10

In a laboratory creep experiment at 1,000°C, a steady-state creep rate of 5×10^{-1} % per hour is obtained in a metal alloy. The creep mechanism for this alloy is known to be dislocation climb with an activation energy of 200 kJ/mol. Predict the creep rate at a service temperature of 600°C. (Assume that the laboratory experiment duplicated the service stress.)

SOLUTION
Using the laboratory experiment to determine the preexponential constant in Equation 6.16, we obtain

$$C = \dot{\epsilon}e^{+Q/RT}$$

$$= (5 \times 10^{-1} \text{ % per hour})e^{+(2\times10^5 \text{ J/mol})/[8.314 \text{ J/mol·K}](1,273 \text{ K})}$$

$$= 80.5 \times 10^6 \text{ % per hour.}$$

Applying this amount to the service temperature yields

$$\dot{\epsilon} = (80.5 \times 10^6 \text{ % per hour})e^{-(2\times10^5)/(8.314)(873)}$$

$$= 8.68 \times 10^{-5} \text{ % per hour.}$$

Note. We have assumed that the creep mechanism remains the same between 1,000 and 600°C.

D   **EXAMPLE 6.11**

Starting at this point and continuing throughout the remainder of the book, problems that deal with materials in the engineering design process will be identified with a design icon, **D**.

In designing a pressure vessel for the petrochemical industry, an engineer must estimate the temperature to which Inconel 718 could be subjected and still provide a service life of 10,000 h under a service stress of 690 MPa (100,000 psi) before failing by creep rupture. What is the service temperature?

SOLUTION

Using Figure 6.37, we must replot the data, noting that the failure stress for a rupture time of 10^4 h varies with temperature as follows:

σ (ksi)	T (°C)
125	540
95	595
65	650

Plotting gives

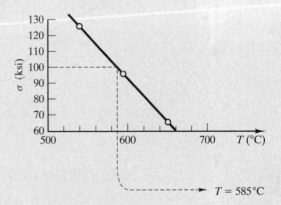

$T = 585$°C

EXAMPLE 6.12

The relaxation time for a rubber band at 25°C is 60 days.

(a) If it is stressed to 2 MPa initially, how many days will be required before the stress relaxes to 1 MPa?

(b) If the activation energy for the relaxation process is 30 kJ/mol, what is the relaxation time at 35°C?

SOLUTION

(a) From Equation 6.17,

$$\sigma = \sigma_0 e^{-t/\tau}$$

and

$$1 \text{ MPa} = 2 \text{ MPa}e^{-t/(60 \text{ d})}.$$

Rearranging the terms yields

$$t = -(60 \text{ days})\left(\ln \tfrac{1}{2}\right) = 41.5 \text{ days}.$$

(b) From Equation 6.18,

$$\frac{1}{\tau} = Ce^{-Q/RT},$$

or

$$\frac{1/\tau_{25°C}}{1/\tau_{35°C}} = \frac{e^{-Q/R(298 \text{ K})}}{e^{-Q/R(308 \text{ K})}},$$

or

$$\tau_{35°C} = \tau_{25°C}\exp\left[\frac{Q}{R}\left(\frac{1}{308 \text{ K}} - \frac{1}{298 \text{ K}}\right)\right],$$

giving, finally,

$$\tau_{35°C} = (60 \text{ days})\exp\left[\frac{30 \times 10^3 \text{ J/mol}}{8.314 \text{ J/(mol} \cdot \text{K)}}\left(\frac{1}{308 \text{ K}} - \frac{1}{298 \text{ K}}\right)\right]$$
$$= 40.5 \text{ days}.$$

PRACTICE PROBLEM 6.10

Using an Arrhenius equation, we are able to predict the creep rate for a given alloy at 600°C in Example 6.10. For the same system, calculate the creep rate at **(a)** 700°C, **(b)** 800°C, and **(c)** 900°C. **(d)** Plot the results on an Arrhenius plot similar to that shown in Figure 6.35.

PRACTICE PROBLEM 6.11

In Example 6.11, we are able to estimate a maximum service temperature for Inconel 718 in order to survive a stress of 690 MPa (100,000 psi) for 10,000 h. What is the maximum service temperature for this pressure-vessel design that will allow this alloy to survive **(a)** 100,000 h and **(b)** 1,000 h at the same stress?

PRACTICE PROBLEM 6.12

In Example 6.12a, the time for relaxation of stress to 1 MPa at 25°C is calculated. **(a)** Calculate the time for stress to relax to 0.5 MPa at 25°C. **(b)** Repeat part (a) for 35°C using the result of Example 6.12b.

6.6 | Viscoelastic Deformation

We shall see in the next chapter on thermal behavior that materials generally expand upon heating. This thermal expansion is monitored as an incremental increase in length, ΔL, divided by its initial length, L_0. Two unique mechanical responses are found in measuring the thermal expansion of an inorganic glass or an organic polymer (Figure 6.40). First, there is a distinct break in the expansion curve at the temperature T_g. There are two different thermal-expansion coefficients (slopes) above and below T_g. The thermal-expansion coefficient below T_g is comparable to that of a crystalline solid of the same composition. The thermal-expansion coefficient above T_g is comparable to that for a liquid. As a result, T_g is referred to as the **glass transition temperature**. Below T_g, the material is a true glass (a rigid solid), and above T_g it is a supercooled liquid (see Section 4.5). In terms of mechanical behavior, elastic deformation occurs below T_g, while **viscous** (liquidlike) **deformation** occurs above T_g. Continuing to measure thermal expansion above T_g leads to a precipitous drop in the data curve at the temperature T_s. This **softening temperature** marks the point where the material has become so fluid that it can no longer support the weight of the length-monitoring probe (a small refractory rod). A plot of specific volume versus temperature is given in Figure 6.41. This plot is closely related to the thermal-expansion curve of Figure 6.40. The addition of data for the crystalline material (of the same composition as the glass) gives a pictorial definition of a glass in comparison to a supercooled liquid and a crystal.

The viscous behavior of glasses (organic or inorganic) can be described by the **viscosity**, η, which is defined as the proportionality constant between a shearing force per unit area (F/A) and velocity gradient (dv/dx),

$$\frac{F}{A} = \eta \frac{dv}{dx}, \tag{6.19}$$

with the terms illustrated in Figure 6.42. The units for viscosity are traditionally the poise [$= 1$ g/(cm·s)], which is equal to 0.1 Pa·s.

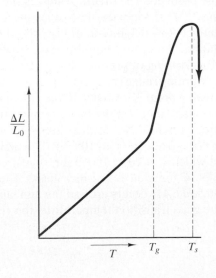

FIGURE 6.40 *Typical thermal-expansion measurement of an inorganic glass or an organic polymer indicates a glass transition temperature, T_g, and a softening temperature, T_s.*

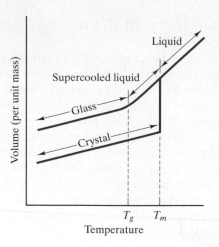

FIGURE 6.41 *Upon heating, a crystal undergoes modest thermal expansion up to its melting point (T_m), at which a sharp increase in specific volume occurs. Upon further heating, the liquid undergoes a greater thermal expansion. Slow cooling of the liquid would allow crystallization abruptly at T_m and a retracing of the melting plot. Rapid cooling of the liquid can suppress crystallization producing a supercooled liquid. In the vicinity of the glass transition temperature (T_g), gradual solidification occurs. A true glass is a rigid solid with thermal expansion similar to the crystal, but an atomic-scale structure similar to the liquid (see Figure 4.21).*

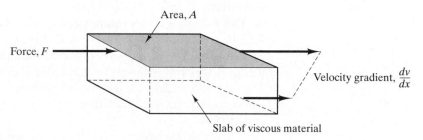

FIGURE 6.42 *Illustration of terms used to define viscosity, η, in Equation 6.19.*

INORGANIC GLASSES

The viscosity of a typical soda–lime–silica glass from room temperature to 1,500°C is summarized in Figure 6.43, which serves as an example of **viscoelastic deformation** in that the plot goes from room temperature where the glass is elastic to above the glass transition temperature where it is viscous in nature. A good deal of useful processing information is contained in Figure 6.43 relative to the manufacture of glass products. The **melting range** is the temperature range (between about 1,200 and 1,500°C for soda–lime–silica glass), where η is between 50 and 500 P. This relatively low magnitude of viscosity represents a very fluid material for a silicate liquid. Water and liquid metals, however, have viscosities of only about 0.01 P. The forming of product shapes is practical in the viscosity range of 10^4 to 10^8 P, the **working range** (between about 700 and 900°C for soda–lime–silica glass). The **softening point** is formally defined at an η value of $10^{7.6}$ P ($\sim$700°C for soda–lime–silica glass) and is at the lower temperature end of the working range. After a glass product is formed, residual stresses can be relieved by holding in the *annealing range* of η from $10^{12.5}$ to $10^{13.5}$ P. The **annealing point** is defined as the temperature at which $\eta = 10^{13.4}$ P and internal stresses can be relieved in about 15 minutes ($\sim$450°C for soda–lime–silica glass). The glass transition temperature (of Figures 6.40 and 6.41) occurs around the annealing point.

Above the glass transition temperature, the viscosity data follow an Arrhenius form with

$$\eta = \eta_0 e^{+Q/RT}, \tag{6.20}$$

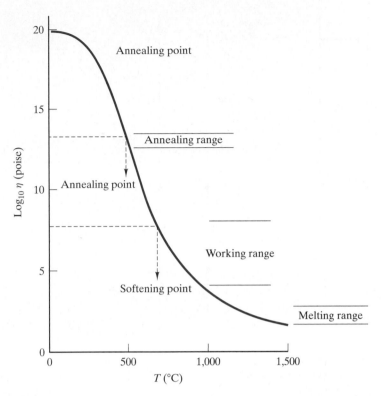

FIGURE 6.43 *Viscosity of a typical soda–lime–silica glass from room temperature to 1,500°C. Above the glass transition temperature (~450°C in this case), the viscosity decreases in the Arrhenius fashion (see Equation 6.20).*

where η_0 is the preexponential constant, Q is the activation energy for viscous deformation, R is the universal gas constant, and T is the absolute temperature. Note that the exponential term has a positive sign rather than the usual negative sign associated with diffusivity data. The reason for this difference is simply the nature of the definition of viscosity, which decreases rather than increases with temperature. Fluidity, which could be defined as $1/\eta$, would, by definition, have a negative exponential sign comparable to the case for diffusivity.

A creative application of viscous deformation is **tempered glass**. Figure 6.44 shows how the glass is first equilibrated above the glass transition temperature, T_g, followed by a surface quench that forms a rigid surface "skin" at a temperature below T_g. Because the interior is still above T_g, interior compressive stresses are largely relaxed, although a modest tensile stress is present in the surface "skin." Slow cooling to room temperature allows the interior to contract considerably more than the surface, causing a net compressive residual stress on the surface balanced by a smaller tensile residual stress in the interior. This situation is ideal for a brittle ceramic. Susceptible to surface Griffith flaws, the material must be subjected to a significant tensile load before the residual compressive load can be neutralized. An additional tensile load is necessary to fracture the material. The breaking strength becomes the normal (untempered) breaking strength plus the magnitude of the surface residual stress. A chemical rather than thermal technique to achieve the same result is to chemically exchange larger radius K^+ ions

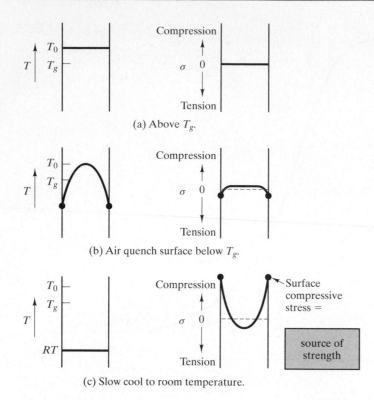

(a) Above T_g.

(b) Air quench surface below T_g.

(c) Slow cool to room temperature.

FIGURE 6.44 *Thermal and stress profiles occurring during the production of tempered glass. The high breaking strength of this product is due to the residual compressive stress at the material surfaces.*

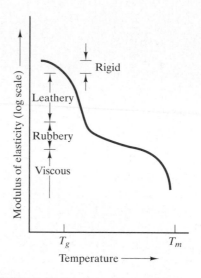

FIGURE 6.45 *Modulus of elasticity as a function of temperature for a typical thermoplastic polymer with 50% crystallinity. There are four distinct regions of viscoelastic behavior: (1) rigid, (2) leathery, (3) rubbery, and (4) viscous.*

for the Na^+ ions in the surface of a sodium-containing silicate glass. The compressive stressing of the silicate network produces a product known as **chemically strengthened glass**.

ORGANIC POLYMERS

For inorganic glasses, the variation in viscosity was plotted against temperature (Figure 6.43). For organic polymers, the *modulus of elasticity* is usually plotted instead of viscosity. Figure 6.45 illustrates the drastic and complicated drop in modulus with temperature for a typical commercial thermoplastic with approximately 50% crystallinity. The magnitude of the drop is illustrated by the use of a logarithmic scale for modulus, which was also necessary for viscosity in Figure 6.43.

Figure 6.45 shows four distinct regions. At low temperatures (well below T_g), a rigid modulus occurs corresponding to mechanical behavior reminiscent of metals and ceramics. However, the substantial component of secondary bonding in the polymers causes the modulus for these materials to be substantially lower than the ones found for metals and ceramics, which were fully bonded by primary

THE MATERIAL WORLD

The Mechanical Behavior of Safety Glass

Even the most routine materials engineered for our surroundings can be the basis of health and safety concerns. Common examples include the window glass in buildings and in automobiles. Window glass is available in three basic configurations: annealed, laminated, and tempered. As discussed in this chapter with regard to the viscoelastic behavior of glass, annealing is a thermal treatment that largely removes the residual stresses of the manufacturing process. Modern windows are made largely by the float-glass method of sheet-glass manufacturing introduced by Pilkington Brothers, Ltd. in England in the 1950s. Annealing effectively removes processing stresses and allows the glass plate to be cut, ground, drilled, and beveled as required. Unfortunately, annealed glass has only moderate strength and is brittle. As a result, thermal gradients, wind loads, or impact can produce characteristic dagger-shaped shards radiating out from the failure origin, as illustrated.

The obvious danger of injury from the breakage of annealed glass has led to widespread legislation requiring *safety glass* in buildings and vehicles. Laminated and tempered glasses serve this purpose. Laminated glass consists of two pieces of ordinary annealed glass with a central layer of polymer [polyvinyl butyral (PVB)] sandwiched between them. As shown, the annealed glass sheets break in the same fashion as ordinary annealed glass, but the shards adhere to the PVB layer, reducing the danger of injury.

The tempering of glass is introduced in this chapter as a relatively sophisticated application of the viscoelastic nature of the material. A direct benefit of tempering is that the bending strength of tempered glass is up to five times greater than that of annealed glass. More important to its safe application, tempered glass breaks into small chunks with relatively harmless blunt shapes and dull edges. This highly desirable break pattern, as shown, is the result of the nearly instantaneous running and bifurcation of cracks initiated at the point of fracture. The energy required to propagate the break pattern comes from the strain energy associated with the residual tensile stresses in the interior of the plate. As the midplane stress rises above some threshold value, the characteristic break pattern occurs with an increasingly fine-scale fracture particle size with further increase in tensile stress.

(Courtesy of Tamglass, Ltd.)

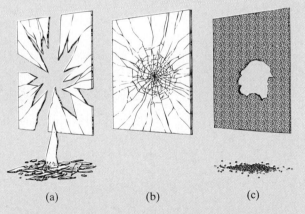

(a) (b) (c)

Break pattern of three states of glass used in commercial and consumer applications: (a) annealed, (b) laminated, and (c) tempered. (From R. A. McMaster, D. M. Shetterly, and A. G. Bueno, "Annealed and Tempered Glass," in Engineered Materials Handbook, *Vol. 4,* Ceramics and Glasses, *ASM International, Materials Park, OH, 1991.)*

chemical bonds (metallic, ionic, and covalent). In the glass transition temperature (T_g) range, the modulus drops precipitously, and the mechanical behavior is *leathery*. The polymer can be extensively deformed and slowly returns to its original shape upon stress removal. Just above T_g, a *rubbery* plateau is observed. In this region, extensive deformation is possible with rapid spring back to the original shape when stress is removed. These last two regions (leathery and rubbery) extend our understanding of elastic deformation. For metals and ceramics, elastic deformation meant a relatively small strain directly proportional to applied stress. For polymers, extensive nonlinear deformation can be fully recovered and is, by definition, elastic. This concept will be explored further when we discuss elastomers, those polymers with a predominant rubbery region. Returning to Figure 6.45, we see that, as the melting point (T_m) is approached, the modulus again drops precipitously as we enter the liquidlike viscous region. (It should be noted that, in many cases, it is more precise to define a "decomposition point" rather than a true melting point. The term *melting point* is nonetheless generally used.)

Figure 6.45 represents a linear, thermoplastic polymer with approximately 50% crystallinity. Figure 6.46 shows how that behavior lies midway between that for a fully amorphous material and a fully crystalline one. The curve for the fully amorphous polymer displays the general shape shown in Figure 6.45. The fully crystalline polymer, on the other hand, is relatively rigid up to its melting point, which is consistent with the behavior of crystalline metals and ceramics. Another structural feature that can affect mechanical behavior in polymers is the **cross-linking** of adjacent linear molecules to produce a more rigid, network structure (Figure 6.47). Figure 6.48 shows how increased cross-linking produces an effect comparable with increased crystallinity. The similarity is due to the increased rigidity of the cross-linked structure, in that cross-linked structures are generally noncrystalline.

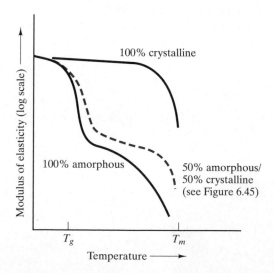

FIGURE 6.46 *In comparison with the plot of Figure 6.45, the behavior of the completely amorphous and completely crystalline thermoplastics falls below and above that for the 50% crystalline material. The completely crystalline material is similar to a metal or ceramic in remaining rigid up to its melting point.*

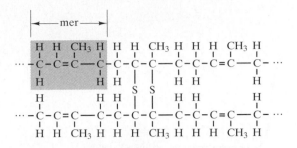

The atomic scale mechanism shown here (cross linking) results in a profound effect on the mechanical behavior of polymers, as seen in the figure below.

FIGURE 6.47 *Cross-linking produces a network structure by the formation of primary bonds between adjacent linear molecules. The classic example shown here is the vulcanization of rubber. Sulfur atoms form primary bonds with adjacent polyisoprene mers, which is possible because the polyisoprene chain molecule still contains double bonds after polymerization. [It should be noted that sulfur atoms can themselves bond together to form a molecule chain. Sometimes, cross-linking occurs by an* $(S)_n$ *chain, where* $n > 1$.]

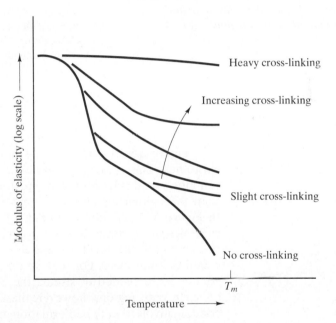

FIGURE 6.48 *Increased cross-linking of a thermoplastic polymer produces increased rigidity of the material.*

ELASTOMERS

Figure 6.46 showed that a typical linear polymer exhibits a rubbery deformation region. For the polymers known as **elastomers**, the rubbery plateau is pronounced and establishes the normal, room-temperature behavior of these materials. (For them, the glass transition temperature is below room temperature.) Figure 6.49 shows the plot of log (modulus) versus temperature for an elastomer. This subgroup of thermoplastic polymers includes the natural and synthetic rubbers, such as polyisoprene. These materials provide a dramatic example of the uncoiling of a linear polymer (Figure 6.50). As a practical matter, the complete uncoiling of the molecule is not achieved, but huge elastic strains do occur.

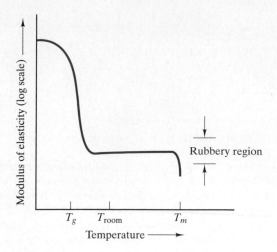

FIGURE 6.49 *The modulus of elasticity versus temperature plot of an elastomer has a pronounced rubbery region.*

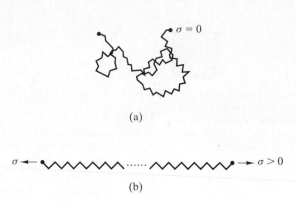

FIGURE 6.50 *Schematic illustration of the uncoiling of (a) an initially coiled linear molecule under (b) the effect of an external stress. This illustration indicates the molecular-scale mechanism for the stress versus strain behavior of an elastomer, as shown in Figure 6.51.*

Figure 6.51 shows a stress–strain curve for the elastic deformation of an elastomer. This curve is in dramatic contrast to the stress–strain curve for a common metal (Figures 6.3 and 6.4). In that case, the elastic modulus was constant throughout the elastic region. (Stress was directly proportional to strain.) In Figure 6.51, the elastic modulus (slope of the stress-strain curve) increases with increasing strain. For low strains (up to $\approx 15\%$), the modulus is low, corresponding to the small forces needed to overcome secondary bonding and to uncoil the molecules. For high strains, the modulus rises sharply, indicating the greater force needed to stretch the primary bonds along the molecular backbone. In both regions, however, there is a significant component of secondary bonding involved in the deformation mechanism, and the moduli are much lower than those for common metals and ceramics. Tabulated values of moduli for elastomers are generally for the low-strain region in which the materials are primarily used. Finally, it is important to emphasize that we are talking about elastic or temporary deformation. The uncoiled polymer molecules of an elastomer recoil to their original lengths upon removal of the stress. However, as the dashed line in Figure 6.51 indicates, the recoiling of the molecules (during unloading) has a slightly different path in the stress-versus-strain plot than does the uncoiling (during loading). The different plots for loading and unloading define **hysteresis**.

Figure 6.52 shows modulus of elasticity versus temperature for several commercial polymers. These data can be compared with the general curves in Figures 6.46 and 6.48. The deflection temperature under load (DTUL), illustrated by Figure 6.52, corresponds to the glass transition temperature.

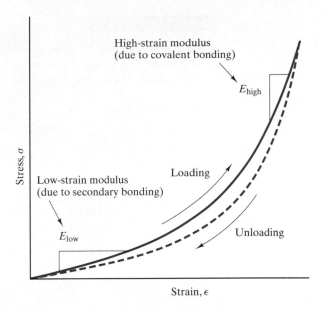

FIGURE 6.51 *The stress–strain curve for an elastomer is an example of nonlinear elasticity. The initial low-modulus (i.e., low-slope) region corresponds to the uncoiling of molecules (overcoming weak, secondary bonds), as illustrated by Figure 6.50. The high-modulus region corresponds to elongation of extended molecules (stretching primary, covalent bonds), as shown by Figure 6.50b. Elastomeric deformation exhibits hysteresis; that is, the plots during loading and unloading do not coincide.*

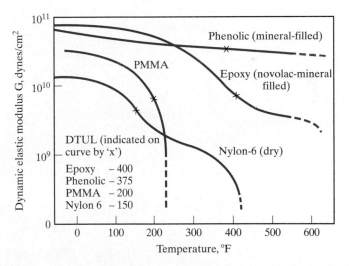

FIGURE 6.52 *Modulus of elasticity versus temperature for a variety of common polymers. The dynamic elastic modulus in this case was measured in a torsional pendulum (a shear mode). The DTUL is the deflection temperature under load, the load being 264 psi. This parameter is frequently associated with the glass transition temperature. (From* Modern Plastics Encyclopedia, 1981–82, *Vol. 58, No. 10A, McGraw-Hill Book Company, New York, October 1981.)*

EXAMPLE 6.13

A soda–lime–silica glass used to make lamp bulbs has an annealing point of 514°C and a softening point of 696°C. Calculate the working range and the melting range for this glass.

SOLUTION

The following is an application of Equation 6.20:

$$\eta = \eta_0 e^{+Q/RT}.$$

Given

$$\text{annealing point} = 514 + 273 = 787\text{K} \quad \text{for } \eta = 10^{13.4}\text{P}$$

and

$$\text{softening point} = 696 + 273 = 969\text{K} \quad \text{for } \eta = 10^{7.6}\text{P},$$

$$10^{13.4}\text{P} = \eta_0 e^{+Q/[8.314 \text{ J/(mol·K)}](787 \text{ K})},$$

$$10^{7.6}\text{P} = \eta_0 e^{+Q/[8.314 \text{ J/(mol·K)}](969 \text{ K})},$$

and

$$\frac{10^{13.4}}{10^{7.6}} = e^{+Q/[8.314 \text{ J/(mol·K)}](1/787 - 1/969)K^{-1}},$$

or

$$Q = 465 \text{ kJ/mol}$$

and

$$\eta_0 = (10^{13.4}\text{ P})e^{-(465 \times 10^3 \text{ J/mol})/[8.314 \text{ J/(mol·K)}](787 \text{ K})}$$

$$= 3.31 \times 10^{-18} \text{ P}.$$

The working range is bounded by $\eta = 10^4$ P and $\eta = 10^8$ P. In general,

$$T = \frac{Q}{R \ln(\eta/\eta_0)}.$$

For $\eta = 10^4$ P,

$$T = \frac{465 \times 10^3 \text{ J/mol}}{[8.314 \text{ J/(mol · K)}] \ln (10^4/[3.31 \times 10^{-18}])}$$

$$= 1{,}130 \text{ K} = 858°\text{C}.$$

For $\eta = 10^8$ P,

$$T = \frac{465 \times 10^3 \text{ J/mol}}{[8.314 \text{ J/(mol · K)}] \ln (10^8/[3.31 \times 10^{-18}])}$$

$$= 953 \text{ K} = 680°\text{C}.$$

Therefore,

$$\text{working range} = 680 \text{ to } 858°C.$$

For the melting range, $\eta = 50$ to 500 P. For $\eta = 50$ P,

$$T = \frac{465 \times 10^3 \text{ J/mol}}{[8.314 \text{ J/(mol} \cdot \text{K)}] \ln (50/[3.31 \times 10^{-18}])}$$
$$= 1{,}266 \text{ K} = 993°C.$$

For $\eta = 500$ P,

$$T = \frac{465 \times 10^3 \text{ J/mol}}{[8.314 \text{ J/(mol} \cdot \text{K)}] \ln (500/[3.31 \times 10^{-18}])}$$
$$= 1{,}204 \text{ K} = 931°C.$$

Therefore,

$$\text{melting range} = 931 \text{ to } 993°C.$$

PRACTICE PROBLEM 6.13

In Example 6.13, various viscosity ranges are characterized for a soda–lime–silica glass. For this material, calculate the annealing range (see Figure 6.43).

Summary

The wide use of metals as structural elements leads us to concentrate on their mechanical properties. The tensile test gives the most basic design data, including modulus of elasticity, yield strength, tensile strength, ductility, and toughness. Closely related elastic properties are Poisson's ratio and the shear modulus. The fundamental mechanism of elastic deformation is the stretching of atomic bonds. Dislocations play a critical role in the plastic deformation of crystalline metals. They facilitate atom displacement by slipping in high-density atomic planes along high-density atomic directions. Without dislocation slip, exceptionally high stresses are required to deform these materials permanently. Many of the mechanical properties discussed in this chapter are explained in terms of the micromechanical mechanism of dislocation slip. The hardness test is a simple alternative to the tensile test that provides an indication of alloy strength. The creep test indicates that above a temperature of about one-half

times the absolute melting point, an alloy has sufficient atomic mobility to deform plastically at stresses below the room-temperature yield stress.

Several mechanical properties play important roles in the structural applications and processing of ceramics and glasses. Both ceramics and glasses are characterized by brittle fracture, although they typically have compressive strengths significantly higher than their tensile strengths. Creep plays an important role in the application of ceramics in high-temperature service. Diffusional mechanisms combine with grain-boundary sliding to provide the possibility of extensive deformation. Below the glass transition temperature (T_g), glasses deform by an elastic mechanism. Above T_g, they deform by a viscous flow mechanism. The exponential change of viscosity with temperature provides a guideline for routine processing of glass products as well as the development of fracture-resistant tempered glass.

The major mechanical properties of polymers include many of those of importance to metals and ceramics. The wide use of polymers in design applications involving bending and shock absorption requires emphasis on the flexural modulus and the dynamic modulus, respectively. An analog of creep is stress relaxation. Due to the low melting points of polymers, these phenomena can be observed at room temperature and below. Like creep, stress relaxation is an Arrhenius process. As with glasses, viscoelastic defor-

mation is important to polymers. There are four distinct regions of viscoelastic deformation for polymers: (1) rigid (below the glass transition temperature T_g), (2) leathery (near T_g), (3) rubbery (above T_g), and (4) viscous (near the melting temperature, T_m). For typical thermosetting polymers, rigid behavior holds nearly to the melting (or decomposition) point. Polymers with a pronounced rubbery region are termed elastomers. Natural and synthetic rubbers are examples. They exhibit substantial, nonlinear elasticity.

Key Terms

annealing point (194)
Brinell hardness number (181)
brittle fracture (164)
chemically strengthened glass (196)
cold working (155)
creep (185)
creep curve (185)
critical resolved shear stress (180)
critical shear stress (174)
cross-linking (198)
dislocation climb (185)
ductility (157)
dynamic modulus of elasticity (169)
elastic deformation (154)
elastomer (199)
engineering strain (153)
engineering stress (153)
flexural modulus (168)
flexural strength (168)
gage length (153)

glass transition temperature (193)
Griffith crack model (167)
Hooke's law (154)
hysteresis (200)
lower yield point (157)
melting range (194)
modulus of elasticity (154)
modulus of elasticity in bending (168)
modulus of rigidity (161)
modulus of rupture (164)
plastic deformation (154)
Poisson's ratio (159)
relaxation time (190)
residual stress (155)
resolved shear stress (179)
Rockwell hardness (181)
shear modulus (161)
shear strain (160)
shear stress (160)
slip system (175)

softening point (194)
softening temperature (193)
solution hardening (176)
specific strength (155)
strain hardening (155)
strain-hardening exponent (156)
strength-to-weight ratio (155)
stress relaxation (189)
tempered glass (195)
tensile strength (155)
toughness (157)
upper yield point (157)
viscoelastic deformation (194)
viscosity (193)
viscous deformation (193)
working range (194)
yield point (157)
yield strength (154)
Young's modulus (154)

References

Ashby, M. F., and **D. R. H. Jones**, *Engineering Materials— 1 An Introduction to Properties, Applications, and Design*, 3rd ed., Butterworth-Heinemann, Boston, 2005.

ASM Handbook, Vols. 1 (*Properties and Selection: Irons, Steels, and High-Performance Alloys*), and 2 (*Properties and Selection: Nonferrous Alloys and Special-Purpose Metals*), ASM International, Materials Park, OH, 1990 and 1991.

Chiang, Y., **D. P. Birnie III**, and **W. D. Kingery**, *Physical Ceramics*, John Wiley & Sons, Inc., New York, 1997.

Courtney, T. H., *Mechanical Behavior of Materials*, 2nd ed., McGraw-Hill Book Company, New York, 2000.

Davis, J. R., Ed., *Metals Handbook*, Desk Ed., 2nd ed., ASM International, Materials Park, OH, 1998. A one-volume summary of the extensive *Metals Handbook* series.

Engineered Materials Handbook, Desk Edition, ASM International, Materials Park, OH, 1995.

Hull, D., and **D. J. Bacon**, *Introduction to Dislocations*, 4th ed., Butterworth-Heinemann, Boston, 2001.

Problems

As noted on page 191, problems that deal with materials in the engineering design process will be identified with a design icon, **D**.

Section 6.1 • Stress versus Strain

6.1. The following three σ–ϵ data points are provided for a titanium alloy for aerospace applications: $\epsilon = 0.002778$ (at $\sigma = 300$ MPa), 0.005556 (600 MPa), and 0.009897 (900 MPa). Calculate E for this alloy.

6.2. If the Poisson's ratio for the alloy in Problem 6.1 is 0.35, calculate **(a)** the shear modulus G, and **(b)** the shear stress τ necessary to produce an angular displacement α of 0.2865°.

6.3. In Section 6.1, the point was made that the theoretical strength (i.e., critical shear strength) of a material is roughly 0.1 G. **(a)** Use the result of Problem 6.2a to estimate the theoretical critical shear strength of the titanium alloy. **(b)** Comment on the relative value of the result in (a) compared with the apparent yield strength implied by the data given in Problem 6.1.

D **6.4.** Consider the 1040 carbon steel listed in Table 6.1. **(a)** A 20-mm-diameter bar of this alloy is used as a structural member in an engineering design. The unstressed length of the bar is precisely 1 m. The structural load on the bar is 9×10^4 N in tension. What will be the length of the bar under this structural load? **(b)** A design engineer is considering a structural change that will increase the tensile load on this member. What is the maximum tensile load that can be permitted without producing extensive plastic deformation of the bar? Give your answers in both newtons (N) and pounds force (lb_f).

D **6.5.** Heat treatment of the alloy in the design application of Problem 6.4 does not significantly affect the modulus of elasticity, but does change strength and ductility. For a particular heat treatment, the corresponding mechanical property data are

$$Y.S. = 1,100 \text{ MPa (159 ksi)},$$
$$T.S. = 1,380 \text{ MPa (200 ksi)},$$

and

$$\% \text{ elongation at failure} = 12.$$

Again, considering a 20-mm-diameter by 1-m-long bar of this alloy, what is the maximum tensile load that can be permitted without producing extensive plastic deformation of the bar?

D **6.6.** Repeat Problem 6.4 for a structural design using the 2024-T81 aluminum illustrated in Figure 6.3 and Example 6.1.

6.7. In normal motion, the load exerted on the hip joint is 2.5 times body weight. **(a)** Calculate the corresponding stress (in MPa) on an artificial hip implant with a cross-sectional area of 5.64 cm^2 in a patient weighing 150 lb_f. **(b)** Calculate the corresponding strain if the implant is made of Ti–6Al–4V, which has an elastic modulus of 124 GPa.

6.8. Repeat Problem 6.7 for the case of an athlete who undergoes a hip implant. The same alloy is used, but because the athlete weighs 200 lb_f, a larger implant is required (with a cross-sectional area of 6.90 cm^2). Also, consider the situation in which the athlete expends his maximum effort exerting a load of five times his body weight.

D **6.9.** Suppose that you were asked to select a material for a spherical pressure vessel to be used in an aerospace application. The stress in the vessel wall is

$$\sigma = \frac{pr}{2t},$$

where p is the internal pressure, r is the outer radius of the sphere, and t is the wall thickness. The mass of the vessel is

$$m = 4\pi r^2 t \rho,$$

where ρ is the material density. The operating stress of the vessel will always be

$$\sigma \leq \frac{Y.S.}{S},$$

where S is a safety factor. **(a)** Show that the minimum mass of the pressure vessel will be

$$m = 2S\pi pr^3 \frac{\rho}{Y.S.}.$$

(b) Given Table 6.1 and the following data, select the alloy that will produce the lightest vessel.

Alloy	ρ (Mg/m^3)	Cost^a ($/kg)
1040 carbon steel	7.80	0.63
304 stainless steel	7.80	3.70
3003-H14 aluminum	2.73	3.00
Ti–5Al–2.5Sn	4.46	15.00

^{a}Approximate in U.S. dollars.

(c) Given Table 6.1 and the data in the preceding table, select the alloy that will produce the minimum cost vessel.

6.10. Prepare a table comparing the tensile strength-per-unit density of the aluminum alloys of Table 6.1 with the 1040 steel in the same table. Take the densities of 1040 steel and the 2048 and 3003 alloys to be 7.85, 2.91, and 2.75 Mg/m^3, respectively.

6.11. Expand on Problem 6.10 by including the magnesium alloys and the titanium alloys of Table 6.1 in the comparison of strength-per-unit density. (Take the densities of the AM100A and AZ31B alloys and the Ti–6Al–4V and Ti–5Al–2.5Sn alloys to be 1.84, 1.83, 4.43, and 4.49 Mg/m^3, respectively.)

D **6.12.** **(a)** Select the alloy in the pressure vessel design of Problem 6.9 with the maximum tensile strength-per-unit density. (Note Problem 6.10 for a discussion of this quantity.) **(b)** Select the alloy in Problem 6.9 with the maximum (tensile strength-per-unit density)/unit cost.

*6.13.** In analyzing residual stress by x-ray diffraction, the stress constant, K_1, is used, where

$$K_1 = \frac{E \cot \theta}{2(1 + \nu) \sin^2 \psi}$$

and where E and ν are the elastic constants defined in this chapter, θ is a Bragg angle (see Section 3.7), and ψ is an angle of rotation of the sample during the x-ray diffraction experiment (generally $\psi = 45°$). To maximize experimental accuracy, one prefers to use the largest possible Bragg angle, θ. However, hardware configuration (Figure 3.34) prevents θ from being greater than 80°. **(a)** Calculate the maximum θ for a 1040 carbon steel using CrK_α radiation ($\lambda = 0.2291$ nm). (Note that 1040 steel is nearly pure iron, which is a bcc metal, and that the reflection rules for a bcc metal are given in Table 3.4.) **(b)** Calculate the value of the stress constant for 1040 steel.

*6.14.** Repeat Problem 6.13 for 2048 aluminum, which for purposes of the diffraction calculations can be approximated by pure aluminum. (Note that aluminum is an fcc metal and that the reflection rules for such materials are given in Table 3.4.)

6.15. **(a)** The following data are collected for a modulus of rupture test on an MgO refractory brick (refer to Equation 6.10 and Figure 6.14):

$$F = 7.0 \times 10^4 N,$$

$$L = 178 \text{ mm},$$

$$b = 114 \text{ mm},$$

and

$$h = 76 \text{ mm}.$$

Calculate the modulus of rupture. **(b)** Suppose that you are given a similar MgO refractory with the same strength and same dimensions except that its height, h, is only 64 mm. What would be the load (F) necessary to break this thinner refractory?

6.16. A single crystal Al_2O_3 rod (precisely 6 mm diameter × 50 mm long) is used to apply loads to small samples in a high-precision dilatometer (a length-measuring device). Calculate the resulting rod dimensions if the crystal is subjected to a 25-kN axial compression load.

6.17. A freshly drawn glass fiber (100 μm diameter) breaks under a tensile load of 40 N. After subsequent handling, a similar fiber breaks under a tensile load of 0.15 N. Assuming the first fiber was free of defects and that the second fiber broke due to an atomically sharp surface crack, calculate the length of that crack.

6.18. A nondestructive testing program can ensure that a given 80-μm-diameter glass fiber will have no surface cracks longer than 5 μm. Given that the theoretical strength of the fiber is 5 GPa, what can you say about the expected breaking strength of this fiber?

6.19. The following data are collected in a flexural test of a polyester to be used in the exterior trim of an automobile:

Test piece geometry: 6 mm × 15 mm × 50 mm,

Distance between supports = $L = 60$ mm,

and

Initial slope of load-deflection curve = 538×10^3 N/m.

Calculate the flexural modulus of this engineering polymer.

6.20. The following data are collected in a flexural test of a polyester to be used in the fabrication of molded office furniture:

Test piece geometry: 10 mm × 30 mm × 100 mm,

Distance between supports = $L = 50$ mm,

and

Load at fracture = 6,000 N.

Calculate the flexural strength of this engineering polymer.

6.21. Figure 6.17 illustrates the effect of humidity on stress-versus-strain behavior for a nylon 66. In addition, the distinction between tensile and compressive behavior is shown. Approximating the data between 0 and 20 MPa as a straight line, calculate **(a)** the initial elastic modulus in tension and **(b)** the initial elastic modulus in compression for the nylon at 60% relative humidity.

6.22. An acetal disk precisely 5 mm thick by 25 mm diameter is used as a cover plate in a mechanical loading device. If a 20-kN load is applied to the disk, calculate the resulting dimensions.

Section 6.2 • Elastic Deformation

6.23. The maximum modulus of elasticity for a copper crystal is 195 GPa. What tensile stress is required along the corresponding crystallographic direction in order to increase the interatomic separation distance by 0.05%?

6.24. Repeat Problem 6.23 for the crystallographic direction corresponding to the minimum modulus of elasticity for copper, which is 70 GPa.

6.25. An expression for the van der Waals bonding energy as a function of interatomic distance is given in Example 2.13. Derive an expression for the slope of the force curve at the equilibrium bond length, a_0. (As shown in Figure 6.18, that slope is directly related to the elastic modulus of solid argon, which exists at cryogenic temperatures.)

6.26. Using the result of Problem 6.25 and data in Example 2.13, calculate the value of the slope of the force curve at the equilibrium bond length, a_0, for solid argon. (Note that the units will be N/m rather than MPa in that we are dealing with the slope of the force-versus-elongation curve rather than the stress-versus-strain curve.)

Section 6.3 • Plastic Deformation

6.27. A crystalline grain of aluminum in a metal plate is situated so that a tensile load is oriented along the [111] crystal direction. If the applied stress is 0.5 MPa (72.5 psi), what will be the resolved shear stress, τ, along the [101] direction within the $(11\bar{1})$ plane? (Review the comments in Problem 3.32.)

6.28. In Problem 6.27, what tensile stress is required to produce a critical resolved shear stress, τ_c, of 0.242 MPa?

6.29. A crystalline grain of iron in a metal plate is situated so that a tensile load is oriented along the [110] crystal direction. If the applied stress is 50 MPa (7.25 $\times 10^3$ psi), what will be the resolved shear stress, τ, along the $[11\bar{1}]$ direction within the (101) plane? (Review the comments in Problem 3.32.)

6.30. In Problem 6.29, what tensile stress is required to produce a critical resolved shear stress, τ_c, of 31.1 MPa?

****6.31.** Consider the slip systems for aluminum shown in Figure 6.24. For an applied tensile stress in the [111] direction, which slip system(s) would be most likely to operate?

****6.32.** Figure 6.24 lists the slip systems for an fcc and an hcp metal. For each case, this list represents all unique combinations of close-packed planes and close-packed directions (contained within the close-packed planes). Make a similar list for the 12 slip systems in the bcc structure (see Table 6.8). (A few important hints: It will help to first verify the list for the fcc metal. Note that each slip system involves a plane $(h_1 k_1 l_1)$ and a direction $[h_2 k_2 l_2]$ whose indices give a dot product of zero (i.e., $h_1 h_2 + k_1 k_2 + l_1 l_2 = 0$). Further, all members of the $\{hkl\}$ family of planes are not listed. Because a stress involves simultaneous force application in two antiparallel directions, only nonparallel planes need to be listed. Similarly, antiparallel crystal directions are redundant. You may want to review Problems 3.32 to 3.34.)

6.33. Sketch the atomic arrangement and Burgers vector orientations in the slip plane of a bcc metal. (Note the shaded area of Table 6.8.)

6.34. Sketch the atomic arrangement and Burgers vector orientation in the slip plane of an fcc metal. (Note the shaded area of Table 6.8.)

6.35. Sketch the atomic arrangement and Burgers vector orientation in the slip plane of an hcp metal. (Note the shaded area of Table 6.8.)

****6.36.** In some bcc metals, an alternate slip system operates, namely, the $\{211\}\langle \bar{1}11 \rangle$. This system has the same Burgers vector but a lower-density slip plane, as compared with the slip system in Table 6.8. Sketch the unit-cell geometry for this alternate slip system in the manner used in Table 6.8.

****6.37.** Identify the 12 individual slip systems for the alternate system given for bcc metals in Problem 6.36. (Recall the comments in Problem 6.32.)

*6.38. Sketch the atomic arrangement and Burgers vector orientation in a (211) slip plane of a bcc metal. (Note Problems 6.36 and 6.37.)

Section 6.4 • Hardness

6.39. You are provided with an unknown alloy with a measured BHN of 100. Having no other information than the data of Figure 6.29a, estimate the tensile strength of the alloy. (Express your answer in the form $x \pm y$.)

6.40. Show that the data of Figure 6.29b are consistent with the plot of Figure 6.29a.

Ⓓ 6.41. A ductile iron (65–45–12, annealed) is to be used in a spherical pressure vessel. The specific alloy obtained for the vessel has a BHN of 200. The design specifications for the vessel include a spherical outer radius of 0.30 m, wall thickness of 20 mm, and a safety factor of 2. Using the information in Figure 6.29 and Problem 6.9, calculate the maximum operating pressure, p, for this vessel design.

Ⓓ 6.42. Repeat Problem 6.41 for another ductile iron (grade 120–90–02, oil-quenched) with a BHN of 280.

6.43. The simple expressions for Rockwell hardness numbers in Table 6.9 involve indentation, t, expressed in millimeters. A given steel with a BHN of 235 is also measured by a Rockwell hardness tester. Using a $\frac{1}{16}$-in.-diameter steel sphere and a load of 100 kg, the indentation t is found to be 0.062 mm. What is the Rockwell hardness number?

6.44. An additional Rockwell hardness test is made on the steel considered in Problem 6.43. Using a diamond cone under a load of 150 kg, an indentation t of 0.157 mm is found. What is the resulting alternative Rockwell hardness value?

6.45. You are asked to measure nondestructively the yield strength and tensile strength of an annealed 65–45–12 cast iron structural member. Fortunately, a small hardness indentation in this structural design will not impair its future usefulness, which is a working definition of *nondestructive*. A 10-mm-diameter tungsten carbide sphere creates a 4.26-mm-diameter impression under a 3,000-kg load. What are the yield and tensile strengths?

6.46. As in Problem 6.45, calculate the yield and tensile strengths for the case of a 4.48-mm-diameter impression under identical conditions.

6.47. The Ti–6Al–4V orthopedic implant material introduced in Problem 6.7 gives a 3.27-mm-diameter impression when a 10-mm-diameter tungsten carbide sphere is applied to the surface with a 3,000-kg load. What is the BHN of this alloy?

6.48. In Section 6.4, a useful correlation between hardness and tensile strength was demonstrated for metallic alloys. Plot hardness versus tensile strength for the data given in Table 6.11, and comment on whether a similar trend is shown for these common thermoplastic polymers. (You can compare this plot with the plot shown in Figure 6.29a.)

Section 6.5 • Creep and Stress Relaxation

6.49. An alloy is evaluated for potential creep deformation in a short-term laboratory experiment. The creep rate ($\dot{\epsilon}$) is found to be 1% per hour at 800°C and 5.5×10^{-2} % per hour at 700°C. **(a)** Calculate the activation energy for creep in this temperature range. **(b)** Estimate the creep rate to be expected at a service temperature of 500°C. **(c)** What important assumption underlies the validity of your answer in part (b)?

6.50. The inverse of time to reaction (t_R^{-1}) can be used to approximate a rate and, consequently, can be estimated using the Arrhenius expression (Equation 6.16). The same is true for time-to-creep rupture, as defined in Figure 6.36. If the time to rupture for a given superalloy is 2,000 h at 650°C and 50 h at 700°C, calculate the activation energy for the creep mechanism.

6.51. Estimate the time to rupture at 750°C for the superalloy of Problem 6.50.

*6.52. Figure 6.34 indicates the dependence of creep on both stress (σ) and temperature (T). For many alloys, such dependence can be expressed in a modified form of the Arrhenius equation,

$$\dot{\epsilon} = C_1 \sigma^n e^{-Q/RT},$$

where $\dot{\epsilon}$ is the steady-state creep rate, C_1 is a constant, and n is a constant that usually lies within the range of 3 to 8. The exponential term ($e^{-Q/RT}$) is the same as in other Arrhenius expressions (see Equation 6.16). The product of $C_1\sigma^n$ is a temperature-independent term equal to the preexponential constant, C, in Equation 6.16. The presence of the σ^n term gives the name *power-law creep* to this expression. Given the power-law creep relationship with $Q = 250$ kJ/mol and $n = 4$, calculate what percentage increase in stress will be necessary to produce the same increase in $\dot{\epsilon}$ as a 10°C increase in temperature from 1,000 to 1,010°C.

6.53. Using Table 6.12, calculate the lifetime of **(a)** a slip-cast MgO refractory at 1,300°C and 12.4 MPa if 1% total strain is permissible. **(b)** Repeat the calculation for a hydrostatically pressed MgO refractory. **(c)** Comment on the effect of processing on the relative performance of these two refractories.

6.54. Assume the activation energy for the creep of Al_2O_3 is 425 kJ/mol. **(a)** Predict the creep rate, $\dot{\epsilon}$, for Al_2O_3 at 1,000°C and 1,800 psi applied stress. (See Table 6.12 for data at 1,300°C and 1,800 psi.) **(b)** Calculate the lifetime of an Al_2O_3 furnace tube at 1,000°C and 1,800 psi if 1% total strain is permissible.

6.55. In Problem 6.52 power-law creep was introduced, in which

$$\dot{\epsilon} = c_1 \sigma^n e^{-Q/RT}.$$

(a) For a value of $n = 4$, calculate the creep rate, $\dot{\epsilon}$, for Al_2O_3 at 1,300°C and 900 psi. **(b)** Calculate the lifetime of an Al_2O_3 furnace tube at 1,300°C and 900 psi if 1% total strain is permissible.

***6.56.** **(a)** The creep plot in Figure 6.38 indicates a general "band" of data roughly falling between the two parallel lines. Calculate a general activation energy for the creep of oxide ceramics using the slope indicated by those parallel lines. **(b)** Estimate the uncertainty in the answer to part (a) by considering the maximum and minimum slopes within the band between temperatures of 1,400 and 2,200°C.

6.57. The stress on a rubber disk is seen to relax from 0.75 to 0.5 MPa in 100 days. **(a)** What is the relaxation time, τ, for this material? **(b)** What will be the stress on the disk after (i) 50 days, (ii) 200 days, or (iii) 365 days? (Consider time = 0 to be at the stress level of 0.75 MPa.)

6.58. Increasing temperature from 20 to 30°C decreases the relaxation time for a polymeric fiber from 3 to 2 days. Determine the activation energy for relaxation.

6.59. Given the data in Problem 6.58, calculate the expected relaxation time at 40°C.

6.60. A spherical pressure vessel is fabricated from nylon 66 and will be used at 60°C and 50% relative humidity. The vessel dimensions are 50-mm outer radius and 2-mm wall thickness. **(a)** What internal pressure is required to produce a stress in the vessel wall of 6.9 MPa (1,000 psi)? (The stress in the vessel wall is

$$\sigma = \frac{pr}{2t},$$

where p is the internal pressure, r is the outer radius of the sphere, and t is the wall thickness.) **(b)** Calculate the circumference of the sphere after 10,000 h at this pressure. (Note Figure 6.39.)

Section 6.6 • Viscoelastic Deformation

6.61. A borosilicate glass used for sealed-beam headlights has an annealing point of 544°C and a softening point of 780°C. Calculate **(a)** the activation energy for viscous deformation in this glass, **(b)** its working range, and **(c)** its melting range.

6.62. The following viscosity data are available on a borosilicate glass used for vacuum-tight seals:

T (°C)	η (poise)
700	4.0×10^7
1,080	1.0×10^4

Determine the temperatures at which this glass should be **(a)** melted and **(b)** annealed.

D 6.63. For the vacuum-sealing glass described in Problem 6.62, assume you have traditionally annealed the product at the viscosity of 10^{13} poise. After a cost–benefit analysis, you realize that it is more economical to anneal for a longer time at a lower temperature. If you decide to anneal at a viscosity of $10^{13.4}$ poise, how many degrees (°C) should your annealing furnace operator lower the furnace temperature?

D 6.64. You are asked to help design a manufacturing furnace for a new optical glass. Given that it has an annealing point of 460°C and a softening point of 647°C, calculate the temperature range in which the product shape would be formed (i.e., the working range).

CHAPTER 7
Thermal Behavior

Metal casting operations require the use of refractories, high-temperature resistant ceramics with low values of thermal expansion and thermal conductivity. (Alamy Images)

7.1 Heat Capacity

7.2 Thermal Expansion

7.3 Thermal Conductivity

7.4 Thermal Shock

In the previous chapter, we surveyed a range of properties that define the mechanical behavior of materials. In a similar way, we will now survey various properties that define the thermal behavior of materials, indicating how materials respond to the application of heat.

Both *heat capacity* and *specific heat* indicate a material's ability to absorb heat from its surroundings. The energy imparted to the material from the external heat source produces an increase in the thermal vibration of the atoms in the material. Most materials increase slightly in size as they are heated. This *thermal expansion* is a direct result of the greater separation distance between the centers of adjacent atoms as the thermal vibration of the individual atoms increases with increasing temperature.

In describing the flow of heat through a material, the *thermal conductivity* is the proportionality constant between the heat flow rate and the temperature gradient, exactly analogous to the diffusivity, defined in Chapter 5 as the proportionality constant between the mass flow rate and the concentration gradient.

There can be mechanical consequences from the flow of heat in materials. *Thermal shock* refers to the fracture of a material due to a temperature change, usually a sudden cooling.

7.1 | Heat Capacity

As a material absorbs heat from its environment, its temperature rises. This common observation can be quantified with a fundamental material property, the **heat capacity**, C, defined as the amount of heat required to raise its temperature by 1 K ($=1°C$), where

$$C = \frac{Q}{\Delta T}, \tag{7.1}$$

with Q being the amount of heat producing a temperature change ΔT. It is important to note that, for incremental temperature changes, the magnitude of ΔT is the same in either the Kelvin (K) or Celsius (°C) temperature scales.

The magnitude of C will depend on the amount of material. The heat capacity is ordinarily specified for a basis of 1 gram-atom (for elements) or 1 mole (for compounds) in units of J/g-atom·K or J/mol·K. A common alternative is the **specific heat** using a basis of unit mass, such as J/kg·K. Along with the heat and the mass, the specific heat is designated by lowercase letters:

$$c = \frac{q}{m\Delta T}.$$ **(7.2)**

There are two common ways in which heat capacity (or specific heat) is measured. One is while maintaining a constant volume, $C_v(c_v)$, and the other is while maintaining a constant pressure, $C_p(c_p)$. The magnitude of C_p is always greater than C_v, but the difference is minor for most solids at room temperature or below. Because we ordinarily use mass-based amounts of engineering materials under a fixed, atmospheric pressure, we will tend to use c_p data in this book. Such values of specific heat for a variety of engineering materials are given in Table 7.1.

Fundamental studies of the relationship between atomic vibrations and heat capacity in the early part of the 20th century led to the discovery that, at very low temperatures, C_v rises sharply from zero at 0 K as

$$C_v = AT^3,$$ **(7.3)**

where A is a temperature-independent constant. Furthermore, above a **Debye** * **temperature** (θ_D), the value of C_v was found to level off at approximately $3R$, where R is the universal gas constant. (As in Section 5.1, we see that R is a fundamental constant equally important to the solid state, even though it carries the label *gas constant* due to its presence in the perfect gas law.) Figure 7.1 summarizes how C_v rises to an asymptotic value of $3R$ above θ_D. As θ_D is below room temperature for many solids and $C_p \approx C_v$, we have a useful rule of thumb for the value of the heat capacity of many engineering materials.

Finally, it can be noted that there are other energy-absorbing mechanisms besides atomic vibrations that can contribute to the magnitude of heat capacity. An example is the energy absorption by free electrons in metals. On the whole, however, the general behavior summarized by Figure 7.1 and Table 7.1 will be adequate for most engineering materials in ordinary applications.

EXAMPLE 7.1

Show that the rule of thumb that the heat capacity of a solid is approximately $3R$ is consistent with the specific heat value for aluminum in Table 7.1.

---

*Peter Joseph Wilhelm Debye (1884–1966), Dutch-American physical chemist, developed the results displayed in Figure 7.1 as a refinement of Einstein's theory of specific heats, incorporating the then newly developed quantum theory and the elastic constants of the material. Debye made numerous contributions to the fields of physics and chemistry, including pioneering work on x-ray diffraction of powdered materials, thereby helping to lay the foundation for the data in Section 3.7, along with the Braggs.

TABLE 7.1

Values of Specific Heat for a Variety of Materials

Material	c_p [J/kg·K]
Metals[a]	
Aluminum	900
Copper	385
Gold	129
Iron (α)	444
Lead	159
Nickel	444
Silver	237
Titanium	523
Tungsten	133
Ceramics[a,b]	
Al_2O_3	160
MgO	457
SiC	344
Carbon (diamond)	519
Carbon (graphite)	711
Polymers[a]	
Nylon 66	1260–2090
Phenolic	1460–1670
Polyethylene (high density)	1920–2300
Polypropylene	1880
Polytetraflouroethylene (PTFE)	1050

Source: Data from [a]J. F. Shackelford and W. Alexander, *The CRC Materials Science and Engineering Handbook*, 3rd ed., CRC Press, Boca Raton, FL, 2001, and [b]W. D. Kingery, H. K. Bowen, and D. R. Uhlmann, *Introduction to Ceramics*, 2nd ed., John Wiley & Sons, Inc., New York, 1976.

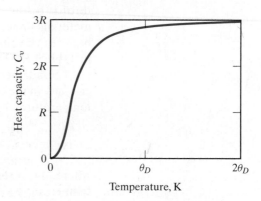

FIGURE 7.1 *The temperature dependence of the heat capacity at constant volume, C_v. The magnitude of C_v rises sharply with a temperature near 0 K and above the Debye temperature (θ_D) levels off at a value of approximately 3R.*

SOLUTION

From Appendix 3, we have

$$3R = 3(8.314 \text{ J/mol·K})$$

$$= 24.94 \text{ J/mol·K}.$$

From Appendix 1, we see that, for aluminum, there are 26.98 g per g-atom, which corresponds, for this elemental solid, to a mole. Then,

$$3R = (24.94 \text{ J/mol·K})(1 \text{ mol}/26.98 \text{ g})(1,000 \text{ g/kg})$$

$$= 924 \text{ J/kg·K},$$

which is in reasonable agreement with the value of 900 J/kg·K in Table 7.1.

PRACTICE PROBLEM 7.1

Show that a heat capacity of $3R$ is a reasonable approximation to the specific heat of copper given in Table 7.1. (See Example 7.1.). **(Note that the solutions to all practice problems are provided on the related Web site.)**

7.2 | Thermal Expansion

An increase in temperature leads to greater thermal vibration of the atoms in a material and an increase in the average separation distance of adjacent atoms (Figure 7.2). In general, the overall dimension of the material in a given direction, L, will increase with increasing temperature, T. This relationship is reflected by the **linear coefficient of thermal expansion**, α, given by

$$\alpha = \frac{dL}{LdT},\tag{7.4}$$

with α having units of mm/(mm · °C). Thermal-expansion data for various materials are given in Table 7.2.

Note that the thermal-expansion coefficients of ceramics and glasses are generally smaller than those for metals, which are, in turn, smaller than those for polymers. The differences are related to the asymmetrical shape of the energy well in Figure 7.2. The ceramics and glasses generally have deeper wells (i.e., higher bonding energies) associated with their ionic and covalent-type bonding. The

A demonstration of thermal expansion and related thermal property data are available on the related Web site.

Here we see how the bonding energy curve of Figure 2.8 plays a central role in understanding the nature of thermal expansion.

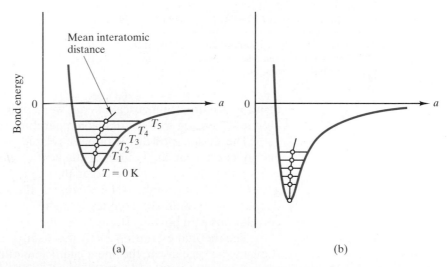

(a)   (b)

FIGURE 7.2 *Plot of atomic bonding energy versus interatomic distance for (a) a weakly bonded solid and (b) a strongly bonded solid. Thermal expansion is the result of a greater interatomic distance with increasing temperature. The effect (represented by the coefficient of thermal expansion in Equation 7.4) is greater for the more asymmetrical energy well of the weakly bonded solid. As shown in Table 7.3, melting point and elastic modulus increase with increasing bond strength.*

TABLE 7.2

Values of Linear Coefficient of Thermal Expansion for a Variety of Materials			
	α [mm/(mm·°C) × 10^6]		
Material	**Temperature = 27°C (300 K)**	**527°C (800 K)**	**0–1,000°C**
Metals[a]			
Aluminum	23.2	33.8	
Copper	16.8	20.0	
Gold	14.1	16.5	
Nickel	12.7	16.8	
Silver	19.2	23.4	
Tungsten	4.5	4.8	
Ceramics and glasses[a,b]			
Mullite (3Al$_2$O$_3$·2SiO$_2$)			5.3
Porcelain			6.0
Fireclay refractory			5.5
Al$_2$O$_3$			8.8
Spinel (MgO·Al$_2$O$_3$)			7.6
MgO			13.5
UO$_2$			10.0
ZrO$_2$ (stabilized)			10.0
SiC			4.7
Silica glass			0.5
Soda–lime–silica glass			9.0
Polymers[a]			
Nylon 66	30–31		
Phenolic	30–45		
Polyethylene (high-density)	149–301		
Polypropylene	68–104		
Polytetrafluoroethylene (PTFE)	99		

Source: Data from [a]J. F. Shackelford and W. Alexander, *The CRC Materials Science and Engineering Handbook*, 3rd ed., CRC Press, Boca Raton, FL, 2001, and [b]W. D. Kingery, H. K. Bowen, and D. R. Uhlmann, *Introduction to Ceramics*, 2nd ed., John Wiley & Sons, Inc., New York, 1976.

result is a more symmetrical energy well, with relatively less increase in interatomic separation with increasing temperature, as shown in Figure 7.2b.

The elastic modulus is directly related to the derivative of the bonding-energy curve near the bottom of the well (Figure 6.18), and it follows that the deeper the energy well, the larger the value of that derivative and hence the greater the elastic modulus. Furthermore, the stronger bonding associated with deeper energy wells corresponds to higher melting points. These various useful correlations with bonding strength are summarized in Table 7.3.

The thermal-expansion coefficient itself is a function of temperature. A plot showing the variation in the linear coefficient of thermal expansion of some common ceramic materials over a wide temperature range is shown in Figure 7.3.

Crystallites of β-eucryptite are an important part of the microstructure of some of the glass-ceramics that will be discussed in Chaper 11. The β-eucryptite (Li$_2$O·Al$_2$O$_3$·SiO$_2$) has a *negative* coefficient of thermal expansion that helps to give the overall material a low thermal-expansion coefficient and, therefore, excellent resistance to thermal shock, a problem discussed in Section 7.4. In exceptional

TABLE 7.3

Correlation of Bonding Strength with Material Properties	
Weakly bonded solids	**Strongly bonded solids**
Low melting point	High melting point
Low elastic modulus	High elastic modulus
High thermal-expansion coefficient	Low thermal-expansion coefficient

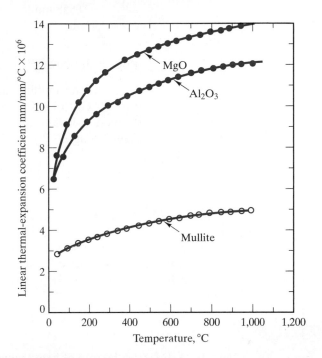

FIGURE 7.3 *Linear thermal-expansion coefficient as a function of temperature for three ceramic oxides (mullite = 3Al$_2$O$_3$ · 2SiO$_2$). (From W. D. Kingery, H. K. Bowen, and D. R. Uhlmann,* Introduction to Ceramics, *2nd ed., John Wiley & Sons, Inc., New York, 1976.)*

cases such as β-eucryptite, the overall atomic architecture "relaxes" in an accordion style as the temperature rises.

EXAMPLE 7.2

A 0.1-m-long Al$_2$O$_3$ furnace tube is heated from room temperature (25°C) to 1,000°C. Assuming the tube is not mechanically constrained, calculate the increase in length produced by this heating.

SOLUTION
Rearranging Equation 7.4,

$$dL = \alpha L \, dT.$$

We can assume linear thermal expansion using the overall thermal-expansion coefficient for this temperature range given in Table 7.2. Then,

$$\Delta L = \alpha L_0 \Delta T$$
$$= [8.8 \times 10^{-6}\ \text{mm/(mm·°C)}](0.1\ \text{m})(1{,}000 - 25)°\text{C}$$
$$= 0.858 \times 10^{-3}\ \text{m}$$
$$= 0.858\ \text{mm}.$$

PRACTICE PROBLEM 7.2

A 0.1-m-long mullite furnace tube is heated from room temperature (25°C) to 1,000°C. Assuming the tube is not mechanically constrained, calculate the increase in length produced by this heating. (See Example 7.2.)

7.3 | Thermal Conductivity

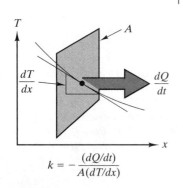

$$k = -\frac{(dQ/dt)}{A(dT/dx)}$$

FIGURE 7.4 *Heat transfer is defined by Fourier's law (Equation 7.5).*

The mathematics for the conduction of heat in solids is analogous to that for diffusion (see Section 5.3). The analog for diffusivity, D, is **thermal conductivity**, k, which is defined by **Fourier's* law**,

$$k = -\frac{dQ/dt}{A(dT/dx)},\qquad(7.5)$$

where dQ/dt is the rate of heat transfer across an area, A, due to a temperature gradient dT/dx. Figure 7.4 relates the various terms of Equation 7.5 and should be compared with the illustration of Fick's first law in Figure 5.9. The units for k are J/(s · m · K). For steady-state heat conduction through a flat slab, the differentials of Equation 7.5 become average terms:

$$k = -\frac{\Delta Q/\Delta t}{A(\Delta T/\Delta x)}.\qquad(7.6)$$

Equation 7.6 is appropriate for describing heat flow through refractory walls in high-temperature furnaces.

Thermal conductivity data are shown in Table 7.4. Like the thermal-expansion coefficient, thermal conductivity is a function of temperature. A plot of thermal conductivity for several common ceramic materials over a wide temperature range is shown in Figure 7.5.

---

*Jean Baptiste Joseph Fourier (1768–1830), French mathematician, left us with some of the most useful concepts in applied mathematics. His demonstration that complex waveforms can be described as a series of trigonometric functions brought him his first great fame (and the title "baron," bestowed by Napoleon). In 1822, his master work on heat flow, entitled *Analytical Theory of Heat,* was published.

TABLE 7.4

Values of Thermal Conductivity for a Variety of Materials

Material	k [J/(s · m · K)]			
	Temperature = 27°C (300 K)	100°C	527°C (800 K)	1,000°C
Metals[a]				
Aluminum	237		220	
Copper	398		371	
Gold	315		292	
Iron	80		43	
Nickel	91		67	
Silver	427		389	
Titanium	22		20	
Tungsten	178		128	
Ceramics and glasses[a,b]				
Mullite ($3Al_2O_3 \cdot 2SiO_2$)		5.9		3.8
Porcelain		1.7		1.9
Fireclay refractory		1.1		1.5
Al_2O_3		30.0		6.3
Spinel ($MgO \cdot Al_2O_3$)		15.0		5.9
MgO		38.0		7.1
ZrO_2 (stabilized)		2.0		2.3
TiC		25.0		5.9
Silica glass		2.0		2.5
Soda–lime–silica glass		1.7		—
Polymers[a]				
Nylon 66	2.9			
Phenolic	0.17–0.52			
Polyethylene (high-density)	0.33			
Polypropylene	2.1–2.4			
Polytetrafluoroethylene (PTFE)	0.24			

Source: Data from [a]J. F. Shackelford and W. Alexander, *The CRC Materials Science and Engineering Handbook*, 3rd ed., CRC Press, Boca Raton, FL, 2001, and [b]W. D. Kingery, H. K. Bowen, and D. R. Uhlmann, *Introduction to Ceramics*, 2nd ed., John Wiley & Sons, Inc., New York, 1976.

The conduction of heat in engineering materials involves two primary mechanisms, atomic vibrations and the conduction of free electrons. For poor electrical conductors such as ceramics and polymers, thermal energy is transported primarily by the vibration of atoms. For electrically conductive metals, the kinetic energy of the conducting (or "free") electrons can provide substantially more efficient conduction of heat than atomic vibrations.

In Chapter 13, we shall look in greater detail at the mechanism of electrical conduction. A general feature of this mechanism is that the electron can be viewed as a wave as well as a particle. For a wave, any structural disorder interferes with the movement of the waveform. The increasing vibration of the crystal lattice at increasing temperature, then, generally results in a decrease in thermal conductivity. Similarly, the structural disorder created by chemical impurities results in a similar decrease in thermal conductivity. As a result, metal alloys tend to have lower thermal conductivities than pure metals.

For ceramics and polymers, atomic vibrations are the predominant source of thermal conductivity, given the very small number of conducting electrons. These

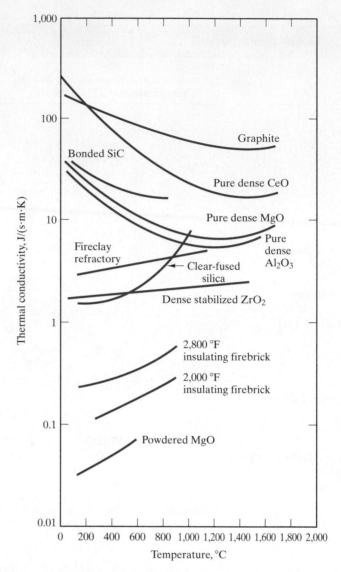

FIGURE 7.5 *Thermal conductivity of several ceramics over a range of temperatures. (From W. D. Kingery, H. K. Bowen, and D. R. Uhlmann,* Introduction to Ceramics, *2nd ed., John Wiley & Sons, Inc., New York, 1976.)*

lattice vibrations are, however, also wavelike in nature and are similarly impeded by structural disorder. As a result, glasses will tend to have a lower thermal conductivity than crystalline ceramics of the same chemical composition. In the same way, amorphous polymers will tend to have a lower thermal conductivity than crystalline polymers of comparable compositions. Also, the thermal conductivities of ceramics and polymers will drop with increasing temperature due to the increasing disorder caused by the increasing degree of atomic vibration. For some ceramics, conductivity will eventually begin to rise with further increase in temperature due to radiant heat transfer. Significant amounts of infrared radiation can be transmitted through ceramics, which tend to be optically transparent. These issues are discussed further in the supplementary Chapter 13B, available on the related Web site.

The thermal conductivity of ceramics and polymers can be further reduced by the presence of porosity. The gas in the pores has a very low thermal conductivity, giving a low net conductivity to the overall microstructure. Prominent examples include the sophisticated space-shuttle tile (discussed in the feature box in this chapter) and the common foamed polystyrene (Styrofoam) drinking cup.

THE MATERIAL WORLD

Thermal Protection System for the Space Shuttle Orbiter

The National Aeronautics and Space Administration (NASA) Space Transportation System (STS), more commonly known as the Space Shuttle Orbiter, presents exceptionally demanding thermal-insulation requirements. The Space Shuttle is a rocket-launched, reusable space vehicle that carries a wide variety of cargo, from scientific experiments to commercial satellites. At the end of an orbital mission, the space craft reenters the atmosphere and experiences enormous frictional heating. The Space Shuttle eventually lands in a manner similar to a normal aircraft.

The successful development of a fully reusable outer skin to serve as a thermal protection system (TPS) was a major part of the overall Space Shuttle design. High-performance thermal insulating materials previously available in the aerospace industry proved inadequate for the Space Shuttle's design specifications because they were either not reusable or were too dense. The system must also provide an aerodynamically smooth outer surface, resist severe thermomechanical loads, and resist moisture and other atmospheric contaminants between missions. Finally, the TPS must be attached to an aluminum-alloy airframe.

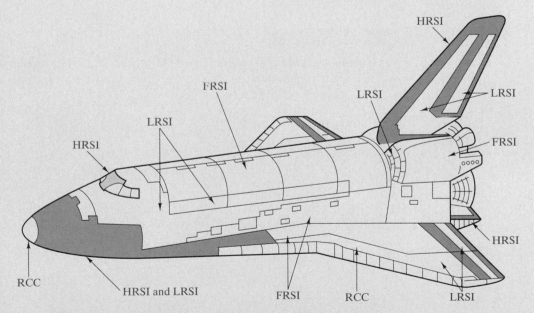

Schematic illustration of the distribution of the components of the thermal protection system for the Space Shuttle Orbiter: felt reusable surface insulation (FRSI), low-temperature reusable surface insulation (LRSI), high-temperature reusable surface insulation (HRSI), and reinforced carbon–carbon composite (RCC). (From L. J. Korb, et al., Bull. Am. Ceram. Soc. 61, *1189 [1981].)*

(Continued)

As the schematic of the Space Shuttle shows, a variety of specific materials has been used to provide the appropriate thermal insulation, depending on the local, maximum skin temperature. Approximately 70% of the Space Shuttle surface must protect against temperatures between 400°C and 1,260°C. For this major portion of the TPS, ceramic tiles are used. For example, in the range of 400°C to 650°C, a low-temperature reusable surface insulation (LRSI) is used. LRSI tiles are generally composed of high-purity vitreous silica fibers with diameters ranging between 1 and 4 μm and fiber lengths of approximately 3,000 μm. Loose packings of these fibers are sintered together to form a highly porous and lightweight material, as seen in the accompanying micrograph. Ceramic and glass materials are inherently good thermal insulators, and combining them with the extremely high porosity (approximately 93% by volume) of the following microstructure results in exceptionally low thermal-conductivity values. It is worth noting that we refer to these tiles as *ceramic* even though their

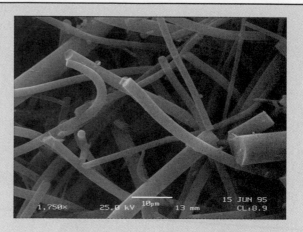

A scanning electron micrograph of sintered silica fibers in a Space Shuttle Orbiter ceramic tile. (Courtesy of Daniel Leiser, National Aeronautics and Space Administration [NASA].)

central component is generally a glass (vitreous silica). This is because glass is often taken as a subset of ceramic, and because some tiles use aluminoborosilicate fibers that can devitrify to become a true crystalline ceramic.

EXAMPLE 7.3

Calculate the steady-state heat transfer rate (in J/m^2·s) through a sheet of copper 10 mm thick if there is a 50°C temperature drop (from 50°C to 0°C) across the sheet.

SOLUTION
Rearranging Equation 7.6,

$$(\Delta Q/\Delta t)/A = -k(\Delta T/\Delta x).$$

Over this temperature range (average $T = 25°C = 298$ K), we can use the thermal conductivity for copper at 300 K given in Table 7.4, giving

$$(\Delta Q/\Delta t)/A = -(398 \text{ J/s} \cdot \text{m} \cdot \text{K})([0°C - 50°C]/[10 \times 10^{-3} \text{ m}])$$

$$= -(398 \text{ J/s} \cdot \text{m} \cdot \text{K})(-5 \times 10^{-3}°C/\text{m}).$$

The K and °C units cancel each other out given that we are dealing with an incremental change in temperature, so

$$(\Delta Q/\Delta t)/A = 1.99 \times 10^6 \text{ J/m}^2\cdot\text{s}.$$

PRACTICE PROBLEM 7.3

Calculate the steady-state heat transfer rate through a 10-mm-thick sheet of copper for a 50°C temperature drop from 550°C to 500°C. (See Example 7.3.)

7.4 | Thermal Shock

The common use of some inherently brittle materials, especially ceramics and glasses, at high temperatures leads to a special engineering problem called **thermal shock**. Thermal shock can be defined as the fracture (partial or complete) of the material as a result of a temperature change (usually a sudden cooling).

The mechanism of thermal shock can involve both thermal expansion and thermal conductivity. Thermal shock follows from these properties in one of two ways. First, a failure stress can be built up by constraint of uniform thermal expansion. Second, rapid temperature changes produce temporary temperature gradients in the material with resulting internal residual stress. Figure 7.6 shows a simple illustration of the first case. It is equivalent to allowing free expansion followed by mechanical compression of the rod back to its original length. More than one furnace design has been flawed by inadequate allowance for expansion of refractory ceramics during heating. Similar consideration must be given to expansion-coefficient matching of coating and substrate for glazes (glass coatings on ceramics) and enamels (glass coatings on metals).

Even without external constraint, thermal shock can occur due to the temperature gradients created because of a finite thermal conductivity. Figure 7.7 illustrates how rapid cooling of the surface of a high-temperature wall is accompanied by surface tensile stresses. The surface contracts more than the interior, which is still relatively hot. As a result, the surface "pulls" the interior into compression and is itself "pulled" into tension. With the inevitable presence of Griffith flaws at the surface, this surface tensile stress creates the clear potential for brittle fracture. The ability of a material to withstand a given temperature change depends on a complex combination of thermal expansion, thermal conductivity, overall geometry, and the inherent brittleness of that material. Figure 7.8 shows the kinds of thermal quenches (temperature drops) necessary to fracture various ceramics

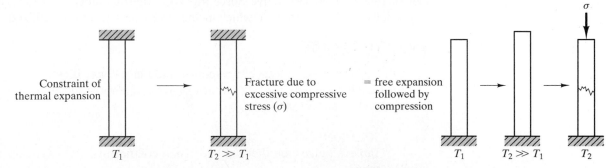

FIGURE 7.6 *Thermal shock resulting from constraint of uniform thermal expansion. This process is equivalent to free expansion followed by mechanical compression back to the original length.*

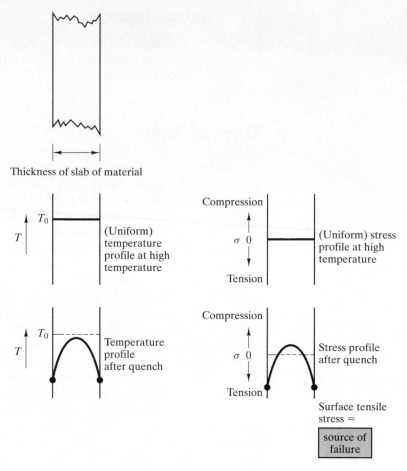

FIGURE 7.7 *Thermal shock resulting from temperature gradients created by a finite thermal conductivity. Rapid cooling produces surface tensile stresses.*

and glasses by thermal shock. Our discussion of thermal shock has been independent of the contribution of phase transformations. In Chapter 9, we shall see the effect of a phase transformation on the structural failure of unstabilized zirconia (ZrO_2). In such cases, even moderate temperature changes through the transformation range can be destructive. Susceptibility to thermal shock is also a limitation of partially stabilized zirconia, which includes small grains of unstabilized phase.

EXAMPLE 7.4

Consider an Al_2O_3 furnace tube constrained in the way illustrated in Figure 7.6. Calculate the stress that would be generated in the tube if it were heated to 1,000°C.

SOLUTION
Table 7.2 gives the thermal-expansion coefficient for Al_2O_3 over the range

$$\alpha = 8.8 \times 10^{-6} \text{ mm/(mm} \cdot {}^\circ\text{C)}.$$

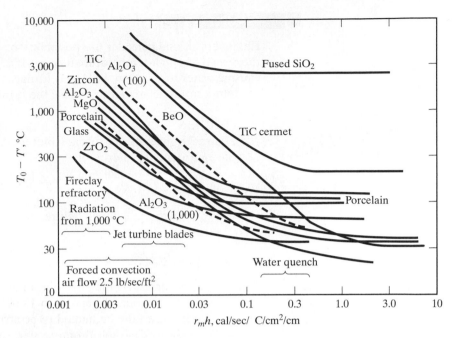

FIGURE 7.8 *Thermal quenches that produce failure by thermal shock are illustrated. The temperature drop necessary to produce fracture ($T_0 - T'$) is plotted against a heat-transfer parameter ($r_m h$). More important than the values of $r_m h$ are the regions corresponding to given types of quench (e.g.,* water quench *corresponds to an $r_m h$ around 0.2 to 0.3). (From W. D. Kingery, H. K. Bowen, and D. R. Uhlmann,* Introduction to Ceramics, *2nd ed., John Wiley & Sons, Inc., New York, 1976.)*

If we take room temperature as 25°C, the unconstrained expansion associated with heating to 1,000°C is

$$\epsilon = \alpha \, \Delta T$$
$$= [8.8 \times 10^{-6} \text{ mm/(mm} \cdot \text{° C)}](1{,}000 - 25)\text{°C}$$
$$= 8.58 \times 10^{-3}.$$

The compressive stress resulting from constraining that expansion is

$$\sigma = E\epsilon.$$

Table 6.4 gives an E for sintered Al_2O_3 as $E = 370 \times 10^3$ MPa. Then,

$$\sigma = (370 \times 10^3 \text{ MPa})(8.58 \times 10^{-3})$$
$$= 3170 \text{ MPa (compressive).}$$

This value is substantially above the failure stress for alumina ceramics (see Figure 6.13).

> **D   EXAMPLE 7.5**
>
> Engineers should consider the possibility of an accident occurring in the design of a high-temperature furnace. If a cooling water line breaks, causing water to spray on the Al_2O_3 furnace tube at 1,000°C, estimate the temperature drop that will cause the furnace tube to crack.
>
> **SOLUTION**
> Figure 7.8 gives the appropriate plot for Al_2O_3 at 1,000°C. In the range of $r_m h$ around 0.2, a drop of
>
> $$T_0 - T' \simeq 50°C$$
>
> will cause a thermal-shock failure.

PRACTICE PROBLEM 7.4

In Example 7.4, the stress in an Al_2O_3 tube is calculated as a result of constrained heating to 1,000°C. To what temperature could the furnace tube be heated to be stressed to an acceptable (but not necessarily desirable) compressive stress of 2,100 MPa?

PRACTICE PROBLEM 7.5

In Example 7.5, a temperature drop of approximately 50°C caused by a water spray is seen to be sufficient to fracture an Al_2O_3 furnace tube originally at 1,000°C. Approximately what temperature drop due to a 2.5-lb/(s · ft^2) airflow would cause a fracture?

Summary

A variety of properties describe the way in which materials respond to the application of heat. The heat capacity indicates the amount of heat necessary to raise the temperature of a given amount of material. The term *specific heat* is used when the property is determined for a unit mass of the material. Fundamental understanding of the mechanism of heat absorption by atomic vibrations leads to a useful rule of thumb for estimating heat capacity of materials at room temperature and above ($C_p \approx C_V \approx 3R$).

The increasing vibration of atoms with increasing temperature leads to increasing interatomic separations and, generally, a positive coefficient of thermal expansion. A careful inspection of the relationship of this expansion to the atomic bonding energy curve reveals that strong bonding correlates with low thermal expansion as well as high elastic modulus and high melting point.

Heat conduction in materials can be described with a thermal conductivity, k, in the same way as mass transport was described in Chapter 5 using the diffusivity, D. The mechanism of thermal conductivity in metals is largely associated with their conductive electrons, whereas the mechanism for ceramics and polymers is largely associated with atomic vibrations. Due to the wavelike nature of both mechanisms, increasing temperature and structural disorder both tend to diminish thermal conductivity. Porosity is especially effective in diminishing thermal conductivity.

The inherent brittleness of ceramics and glasses, combined with thermal-expansion mismatch or low thermal conductivities, can lead to mechanical failure by thermal shock. Sudden cooling is especially effective in creating excessive surface tensile stress and subsequent fracture.

Key Terms

Debye temperature (211)
Fourier's law (216)
heat capacity (210)

linear coefficient of thermal
  expansion (213)
specific heat (211)

thermal conductivity (216)
thermal shock (221)

References

Bird, **R. B.**, **W. E. Stewart**, and **E. N. Lightfoot**, *Transport Phenomena*, rev. 2nd ed. John Wiley & Sons, Inc., New York, 2006.

Chiang, **Y.**, **D. P. Birnie III**, and **W. D. Kingery**, *Physical Ceramics*, John Wiley & Sons, Inc., New York, 1997.

Kubaschewski, **O.**, **C. B. Alcock**, and **P. J. Spencer**, *Materials Thermochemistry*, Oxford and Pergamon Press, New York, 1993.

Problems

7.1 • Heat Capacity

7.1. Estimate the amount of heat (in J) required to raise 2 kg of **(a)** α-iron, **(b)** graphite, and **(c)** polypropylene from room temperature (25°C) to 100°C.

7.2. The specific heat of silicon is 702 J/kg·K. How many J of heat are required to raise the temperature of a silicon chip (of volume = 6.25×10^{-9} m³) from room temperature (25°C) to 35°C?

7.3. A house designed for passive solar heating has a substantial amount of brickwork in its interior to serve as a heat absorber. Each brick weighs 2.0 kg and has a specific heat of 850 J/kg·K. How many bricks are needed to absorb 5.0×10^4 kJ of heat by a temperature increase of 10°C?

7.4. How many liters of water would be required to provide the same heat storage as the bricks in Problem 7.3? The specific heat of water is 1.0 cal/g·K, and its density is 1.0 Mg/m³. (Note that 1 liter = 10^{-3} m³.)

7.2 • Thermal Expansion

7.5. A 0.01-m-long bar of nickel is placed in a laboratory furnace and heated from room temperature (25°C) to 500°C. What will be the length of the bar at 500°C? (Take the coefficient of thermal expansion over this temperature range to be the average of the two values given in Table 7.2.)

7.6. Repeat Problem 7.5 for the case of a tungsten bar of the same length heated over the same temperature range.

7.7. At room temperature (25°C), a 5.000-mm-diameter tungsten pin is too large for a 4.999-mm-diameter hole in a nickel bar. To what temperature must these two parts be heated in order for the pin to just fit?

7.8. The thermal expansion of aluminum is plotted against temperature in Figure 5.4. Measure the plot at 800 K and see how well the result compares with the data in Table 7.2.

7.3 • Thermal Conductivity

7.9. Calculate the rate of heat loss per square meter through the fireclay refractory wall of a furnace operated at 1,000°C. The external face of the furnace wall is at 100°C, and the wall is 10 cm thick.

7.10. Repeat Problem 7.9 for a 5-cm-thick refractory wall.

7.11. Repeat Problem 7.9 for a 10-cm-thick mullite refractory wall.

7.12. Calculate the rate of heat loss per cm² through a stabilized zirconia lining of a high-temperature laboratory furnace operated at 1,400°C. The external face of the lining is at 100°C, and its thickness is 1 cm. (Assume the data for stabilized zirconia in Table 7.4 are linear with temperature and can be extrapolated to 1,400°C.)

7.4 • Thermal Shock

7.13. What would be the stress developed in a mullite furnace tube constrained in the way illustrated in Figure 7.6 if it were heated to 1,000°C?

7.14. Repeat Problem 7.13 for magnesia (MgO).

7.15. Repeat Problem 7.13 for silica glass.

*__7.16.__ A textbook on the mechanics of materials gives the following expression for the stress due to thermal-expansion mismatch in a coating (of thickness a) on a substrate (of thickness b) at a temperature T:

$$\sigma = \frac{E}{1-\nu}(T_0 - T)(\alpha_c - \alpha_s)\left[1 - 3\left(\frac{a}{b}\right) + 6\left(\frac{a}{b}\right)^2\right].$$

In the preceding expression, E and ν are the elastic modulus and Poisson's ratio of the coating, respectively; T_0 is the temperature at which the coating is applied (and the coating stress is initially zero); and α_c and α_s are the thermal-expansion coefficients of the coating and the substrate, respectively. Calculate the room-temperature (25°C) stress in a thin soda–lime–silica glaze applied at 1,000°C on a porcelain ceramic. (Take $E = 65 \times 10^3$ MPa and $\nu = 0.24$, and see Table 7.2 for relevant thermal-expansion data.)

*__7.17.__ Repeat Problem 7.16 for a special high-silica glaze with an average thermal expansion coefficient of 3×10^{-6}°C^{-1}. (Take $E = 72 \times 10^3$ MPa and $\nu = 0.24$.)

D 7.18. **(a)** A processing engineer suggests that a fused SiO_2 crucible be used for a water quench from 500°C. Would you endorse this plan? Explain. **(b)** Another processing engineer suggests that a porcelain crucible be used for the water quench from 500°C. Would you endorse this plan? Again, explain.

D 7.19. In designing an automobile engine seal made of stabilized zirconia, an engineer must consider the possibility of the seal being subjected to a sudden spray of cooling oil corresponding to a heat-transfer parameter ($r_m h$) of 0.1 (see Figure 7.8). Will a temperature drop of 30°C fracture this seal?

D 7.20. For the stabilized zirconia described in Problem 7.19, will a temperature drop of 100°C fracture the seal?

7.21. As pointed out in the feature box in this chapter, the thermal protection system for NASA's Space Shuttle Orbiter is effective due to its exceptional thermal-insulation properties. Calculate the heat flow (per square meter) across a 50-mm-thick ceramic tile with an outer surface temperature of 700°C and an inner surface temperature of 30°C, given a thermal conductivity of 0.0837 J/(s · m · K).

7.22. Repeat Problem 7.21 for a more demanding surface area in which the outer surface temperature is 1,200°C and the inner surface temperature is 30°C, given a thermal conductivity of 0.113 J/(s · m · K).

7.23. To appreciate the effectiveness of the thermal insulation in Problem 7.21, determine how many times greater would be the heat flow through fully dense aluminum oxide, given the same thickness and temperature gradient. (Estimate the thermal conductivity of fully dense aluminum oxide by using the value in Figure 7.5 that corresponds to the median temperature between the outer and inner surfaces.)

7.24. To appreciate the effectiveness of the thermal insulation in Problem 7.22, determine how many times greater would be the heat flow through fully dense aluminum oxide, given the same thickness and temperature gradient. (Estimate the thermal conductivity of fully dense aluminum oxide by using the value in Figure 7.5 that corresponds to the median temperature between the outer and inner surfaces.)

CHAPTER 8
Failure Analysis and Prevention

8.1 Impact Energy
8.2 Fracture Toughness
8.3 Fatigue
8.4 Nondestructive Testing
8.5 Failure Analysis and Prevention

Impact testing of materials and components is an important part of preventing failures in engineering designs. These contemporary testing machines allow samples to be dropped under carefully controlled conditions producing a wide range of impact energies. (Courtesy of Instron Corporation.)

In Chapters 6 and 7, we saw numerous examples of the failure of engineering materials. At room temperature, metal alloys and polymers stressed beyond their elastic limit eventually fracture following a period of nonlinear plastic deformation. Brittle ceramics and glasses typically break following elastic deformation, without plastic deformation. The inherent brittleness of ceramics and glasses combined with their common use at high temperatures make thermal shock a major concern. With continuous service at relatively high temperatures, any engineering material can fracture when creep deformation reaches its limit.

In this chapter, we will look at additional ways in which materials fail. For the rapid application of stress to materials with preexisting surface flaws, the measurement of the *impact energy* corresponds to the measurement of toughness, or area under the stress-versus-strain curve. Monitoring impact energy as a function of temperature reveals that, for bcc metal alloys, there is a distinctive *ductile-to-brittle transition temperature*, below which otherwise ductile materials can fail in a catastrophic, brittle fashion.

The general analysis of the failure of structural materials with preexisting flaws is termed *fracture mechanics*. The key material property coming from fracture mechanics is *fracture toughness*, which is large for materials such as pressure-vessel steels and small for brittle materials such as typical ceramics and glasses.

Under cyclic loading conditions, otherwise ductile metal alloys and engineering polymers can eventually fail in a brittle fashion, a phenomenon appropriately termed *fatigue*. Ceramics and glasses can exhibit *static fatigue* without cyclic loading due to a chemical reaction with atmospheric moisture.

Nondestructive testing, the evaluation of engineering materials without impairing their future usefulness, is an important technology for identifying microstructural flaws in engineering systems. Since such flaws, including surface and internal cracks, play a central role in the failure of materials, nondestructive testing is a critical component of failure analysis and prevention programs. *Failure analysis* can be defined as the systematic study of the nature of the various modes of material failure. The related goal of *failure prevention* is to apply the understanding provided by failure analysis to avoid future disasters.

8.1 | Impact Energy

In Section 6.4, hardness was seen to be the analog of strength measured by the tensile test. **Impact energy**, the energy necessary to fracture a standard test piece under an impact load, is a similar analog of toughness. The most common laboratory measurement of impact energy is the **Charpy* test**, illustrated in Figure 8.1. The test principle is straightforward. The energy necessary to fracture the test piece is directly calculated from the difference in initial and final heights of the swinging pendulum. To provide control over the fracture process, a stress-concentrating notch is machined into the side of the sample subjected to maximum tensile stress. The net test result is to subject the sample to elastic deformation, plastic

Millimeter scale flaws such as the notch illustrated in this impact test can lead to the catastrophic failure of materials.

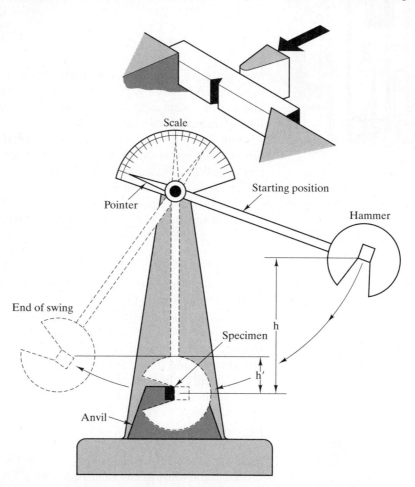

FIGURE 8.1 *Charpy test of impact energy. (From H. W. Hayden, W. G. Moffatt, and J. Wulff,* The Structure and Properties of Materials, *Vol. 3:* Mechanical Behavior, *John Wiley & Sons, Inc., New York, 1965.)*

---

*Augustin Georges Albert Charpy (1865–1945), French metallurgist. Trained as a chemist, Charpy became one of the pioneering metallurgists of France and was highly productive in this field. He developed the first platinum-resistance furnace and the silicon steel routinely used in modern electrical equipment, as well as the impact test that bears his name.

TABLE 8.1

Impact Test (Charpy) Data for Some of the Alloys of Table 6.1	
Alloy	**Impact energy [J (ft·lb)]**
1. 1040 carbon steel	180 (133)
2. 8630 low-alloy steel	55 (41)
3. b. 410 stainless steel	34 (25)
4. L2 tool steel	26 (19)
5. Ferrous superalloy (410)	34 (25)
6. a. Ductile iron, quench	9 (7)
7. b. 2048, plate aluminum	10.3 (7.6)
8. a. AZ31B magnesium	4.3 (3.2)
b. AM100A casting magnesium	0.8 (0.6)
9. a. Ti–5Al–2.5Sn	23 (17)
10. Aluminum bronze, 9% (copper alloy)	48 (35)
11. Monel 400 (nickel alloy)	298 (220)
13. 50:50 solder (lead alloy)	21.6 (15.9)
14. Nb–1 Zr (refractory metal)	174 (128)

Data from Charpy impact tests are provided on the related Web site.

deformation, and fracture in rapid succession. Although rapid, the deformation mechanisms involved are the same as those involved in tensile testing the same material. The load impulse must approach the ballistic range before fundamentally different mechanisms come into play.

In effect, a Charpy test takes the tensile test to completion very rapidly. The impact energy from the Charpy test correlates with the area under the total stress–strain curve (i.e., toughness). Table 8.1 gives Charpy impact energy data for the alloys of Table 6.1. In general, we expect alloys with large values of both strength (Y.S. and T.S.) and ductility (percent elongation at fracture) to have large-impact fracture energies. Although this is frequently so, the impact data are sensitive to test conditions. For instance, increasingly sharp notches can give lower impact-energy values due to the stress concentration effect at the notch tip. The nature of stress concentration at notch and crack tips is explored further in the next section.

Impact-energy data for a variety of polymers are given in Table 8.2. For polymers, the impact energy is typically measured with the **Izod* test** rather than the Charpy. These two standardized tests differ primarily in the configuration of the notched test specimen. Impact test temperature can also be a factor. The fcc alloys generally show ductile fracture modes in Charpy testing, and hcp alloys are generally brittle (Figure 8.2). However, bcc alloys show a dramatic variation in fracture mode with temperature. In general, they fail in a brittle mode at relatively low temperatures and in a ductile mode at relatively high temperatures. Figure 8.3 shows this behavior for two series of low-carbon steels. The ductile-to-brittle transition for bcc alloys can be considered a manifestation of the slower dislocation mechanics for these alloys compared with that for fcc and hcp alloys. (In bcc metals, slip occurs on non-close-packed planes.) Increasing yield strength combined with

---

*E. G. Izod, "Testing Brittleness of Steels," *Engr. 25* (September 1903).

TABLE 8.2

Impact Test (Izod) Data for Various Polymers	
Polymer	Impact energy [J (ft·lb)]
General-use polymers	
Polyethylene	
High-density	1.4–16 (1–12)
Low-density	22 (16)
Polyvinylchloride	1.4 (1)
Polypropylene	1.4–15 (1–11)
Polystyrene	0.4 (0.3)
Polyesters	1.4 (1)
Acrylics (Lucite)	0.7 (0.5)
Polyamides (nylon 66)	1.4 (1)
Cellulosics	3–11 (2–8)
Engineering polymers	
ABS	1.4–14 (1–10)
Polycarbonates	19 (14)
Acetals	3 (2)
Polytetrafluoroethylene (Teflon)	5 (4)
Thermosets	
Phenolics (phenolformaldehyde)	0.4 (0.3)
Urea-melamine	0.4 (0.3)
Polyesters	0.5 (0.4)
Epoxies	1.1 (0.8)

Source: From data collections in R. A. Flinn and P. K. Trojan, *Engineering Materials and Their Applications*, 2nd ed., Houghton Mifflin Company, Boston, 1981; M. F. Ashby and D. R. H. Jones, *Engineering Materials*, Pergamon Press, Inc., Elmsford, NY, 1980; and *Design Handbook for Du Pont Engineering Plastics.*

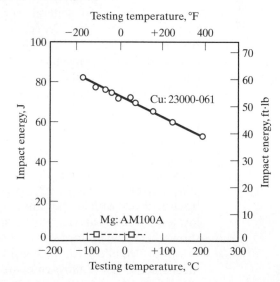

FIGURE 8.2 *Impact energy for a ductile fcc alloy (copper C23000–061, "red brass") is generally high over a wide temperature range. Conversely, the impact energy for a brittle hcp alloy (magnesium AM100A) is generally low over the same range. (From* Metals Handbook, *9th ed., Vol. 2, American Society for Metals, Metals Park, OH, 1979.)*

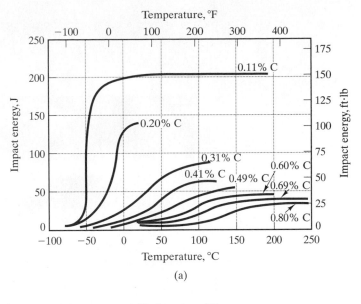

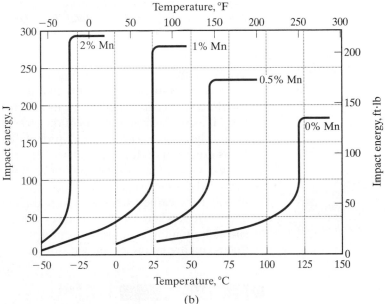

FIGURE 8.3 *Variation in ductile-to-brittle transition temperature with alloy composition. (a) Charpy V-notch impact energy with temperature for plain-carbon steels with various carbon levels (in weight percent). (b) Charpy V-notch impact energy with temperature for Fe–Mn–0.05C alloys with various manganese levels (in weight percent). (From* Metals Handbook, *9th ed., Vol. 1, American Society for Metals, Metals Park, OH, 1978.)*

decreasing dislocation velocities at decreasing temperatures eventually leads to brittle fracture. The microscopic fracture surface of the high-temperature ductile failure has a dimpled texture with many cuplike projections of deformed metal, and brittle fracture is characterized by cleavage surfaces (Figure 8.4). Near the transition temperature between brittle and ductile behavior, the fracture surface

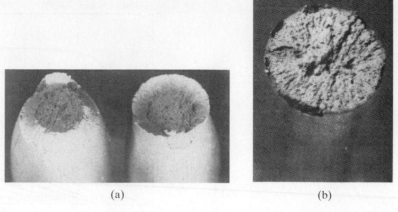

(a)   (b)

FIGURE 8.4 *(a) Typical "cup and cone" ductile fracture surface. Fracture originates near the center and spreads outward with a dimpled texture. Near the surface, the stress state changes from tension to shear, with fracture continuing at approximately 45°. (From* Metals Handbook, *9th ed., Vol. 12, ASM International,* Metals *Park, OH, 1987.) (b) Typical cleavage texture of a brittle fracture surface. (From* Metals Handbook, *8th ed., Vol. 9, American Society for Metals, Metals Park, OH, 1974.)*

exhibits a mixed texture. The **ductile-to-brittle transition temperature** is of great practical importance. The alloy that exhibits a ductile-to-brittle transition loses toughness and is susceptible to catastrophic failure below this transition temperature. Because a large fraction of the structural steels are included in the bcc alloy group, the ductile-to-brittle transition is a design criterion of great importance. The transition temperature can fall between roughly −100 and +100°C, depending on alloy composition and test conditions. Several disastrous failures of Liberty ships occurred during World War II because of this phenomenon. Some literally split in half. Low-carbon steels that were ductile in room-temperature tensile tests became brittle when exposed to lower-temperature ocean environments. Figure 8.3 shows how alloy composition can dramatically shift the transition temperature. Such data are an important guide in material selection.

D | **EXAMPLE 8.1**

You are required to use a furnace-cooled Fe–Mn–0.05 C alloy in a structural design that may see service temperatures as low as 0°C. Suggest an appropriate Mn content for the alloy.

SOLUTION
Figure 8.3 provides the specific guidance we need. A 1% Mn alloy is relatively brittle at 0°C, whereas a 2% Mn alloy is highly ductile. Therefore, a secure choice (based on notch-toughness considerations only) would be

$$\text{Mn content} = 2\%.$$

Find the necessary carbon level to ensure that a plain-carbon steel will be relatively ductile down to 0°C. (See Example 8.1.). **(Note that the solutions to all practice problems are provided on the related Web site.)**

8.2 | Fracture Toughness

A substantial effort has been made to quantify the nature of material failures such as the Liberty ship disasters just described. The term **fracture mechanics** has come to mean the general analysis of failure of structural materials with preexisting flaws. This broad field is the focus of much active research. We shall concentrate on a material property that is the most widely used single parameter from fracture mechanics. **Fracture toughness** is represented by the symbol K_{IC} (pronounced "kay-one-cee") and is the critical value of the stress intensity factor at a crack tip necessary to produce catastrophic failure under simple uniaxial loading. The subscript "I" stands for *mode I* (uniaxial) loading, and "C" stands for *critical*. A simple example of the concept of fracture toughness comes from blowing up a balloon containing a small pinhole. When the internal pressure of the balloon reaches a critical value, catastrophic failure originates at the pinhole (i.e., the balloon pops). In general, the value of fracture toughness is given by

$$K_{IC} = Y\sigma_f\sqrt{\pi a},\qquad\qquad (8.1)$$

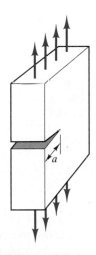

FIGURE 8.5 *Fracture-toughness test.*

where Y is a dimensionless geometry factor on the order of 1, σ_f is the overall applied stress at failure, and a is the length of a surface crack (or one-half the length of an internal crack). Fracture toughness (K_{IC}) has units of MPa $\sqrt{m}$. Figure 8.5 shows a typical measurement of K_{IC}, and Table 8.3 gives values for various materials. It must be noted that K_{IC} is associated with so-called plane strain conditions in which the specimen thickness (Figure 8.5) is relatively large compared with the notch dimension. For thin specimens (*plane stress* conditions), fracture toughness is denoted K_C and is a sensitive function of specimen thickness. Plane strain conditions generally prevail when thickness $\geq 2.5(K_{IC}/\mathrm{Y.S.})^2$.

The microscopic concept of toughness indicated by K_{IC} is consistent with that expressed by the macroscopic measurements of tensile and impact testing. Highly brittle materials, with little or no ability to deform plastically in the vicinity of a crack tip, have low K_{IC} values and are susceptible to catastrophic failures. By contrast, highly ductile alloys can undergo substantial plastic deformation on both a microscopic and a macroscopic scale prior to fracture. The major use of fracture mechanics in metallurgy is to characterize those alloys of intermediate ductility that can undergo catastrophic failure below their yield strength due to the stress-concentrating effect of structural flaws. In designing a pressure vessel, for example, it is convenient to plot operating stress (related to operating pressure) as a function of flaw size. [It is usually possible to ensure that flaws above a given size are not present by a careful inspection program involving nondestructive testing techniques (see Section 8.4).] **General yielding** (independent of a flaw) was covered in Section 6.1. **Flaw-induced fracture** is described by Equation 8.1. Taking Y in that equation as 1 gives the schematic design plot of Figure 8.6. An

TABLE 8.3

Typical Values of Fracture Toughness (K_{IC}) for Various Materials	
Material	K_{IC} (MPa $\sqrt{m}$)
Metal or alloy	
Mild steel	140
Medium-carbon steel	51
Rotor steels (A533; Discalloy)	204–214
Pressure-vessel steels (HY130)	170
High-strength steels (HSS)	50–154
Cast iron	6–20
Pure ductile metals (e.g., Cu, Ni, Ag, Al)	100–350
Be (brittle, hcp metal)	4
Aluminum alloys (high strength–low strength)	23–45
Titanium alloys (Ti–6Al–4V)	55–115
Ceramic or glass	
Partially stabilized zirconia	9
Electrical porcelain	1
Alumina (Al_2O_3)	3–5
Magnesia (MgO)	3
Cement/concrete, unreinforced	0.2
Silicon carbide (SiC)	3
Silicon nitride (Si_3N_4)	4–5
Soda glass ($Na_2O–SiO_2$)	0.7–0.8
Polymer	
Polyethylene	
High-density	2
Low-density	1
Polypropylene	3
Polystyrene	2
Polyesters	0.5
Polyamides (nylon 66)	3
ABS	4
Polycarbonates	1.0–2.6
Epoxy	0.3–0.5

Source: Data from M. F. Ashby and D. R. H. Jones, *Engineering Materials—An Introduction to Their Properties and Applications*, Pergamon Press, Inc., Elmsford, NY, 1980; GTE Laboratories, Waltham, MA; and *Design Handbook for Dupont Engineering Plastics.*

important practical point about the design plot is that failure by general yielding is preceded by observable deformation, whereas flaw-induced fracture occurs rapidly with no such warning. As a result, flaw-induced fracture is sometimes referred to as **fast fracture**.

Some progress has been made in improving the fracture toughness and, hence, the range of applications of structural ceramics. Figure 8.7 summarizes two microstructural techniques for significantly raising fracture toughness. Figure 8.7a illustrates the mechanism of **transformation toughening** in partially stabilized zirconia (PSZ). Having second-phase particles of tetragonal zirconia in a matrix of cubic zirconia is the key to improved toughness. A propagating crack creates a local stress field that induces a transformation of tetragonal zirconia particles to the monoclinic structure in that vicinity. The slightly larger specific volume of the

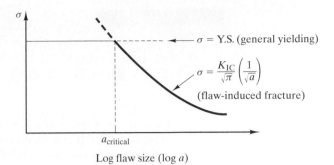

FIGURE 8.6 *A design plot of stress versus flaw size for a pressure-vessel material in which general yielding occurs for flaw sizes less than a critical size,* a_{critical}, *but catastrophic* fast fracture *occurs for flaws larger than* a_{critical}.

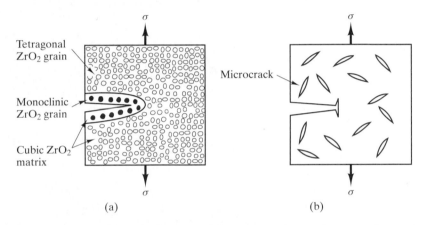

FIGURE 8.7 *Two mechanisms for improving fracture toughness of ceramics by crack arrest. (a) Transformation toughening of partially stabilized zirconia involves the stress-induced transformation of tetragonal grains to the monoclinic structure, which has a larger specific volume. The result is a local volume expansion at the crack tip, squeezing the crack shut and producing a residual compressive stress. (b) Microcracks produced during fabrication of the ceramic can blunt the advancing crack tip.*

monoclinic phase causes an effective compressive load locally and, in turn, the "squeezing" of the crack shut. Another technique of crack arrest is shown in Figure 8.7b. Microcracks purposely introduced by internal stresses during processing of the ceramic are available to blunt the tip of an advancing crack. The expression associated with Griffith cracks (Equation 6.1) indicates that the larger tip radius can dramatically reduce the local stress at the crack tip. Another technique, involving reinforcing fibers, will be discussed in Chapter 12 relative to ceramic-matrix composites.

The absence of plastic deformation in traditional ceramics and glass on the macroscopic scale (the stress–strain curve) is matched by a similar absence on the microscopic scale. This is reflected in the characteristically low fracture toughness (K_{IC}) values (≤ 5 MPa $\sqrt{\text{m}}$) for traditional ceramics and glass, as shown in Table 8.3. Most K_{IC} values are lower than those of the most brittle metals listed there. Only the recently developed transformation-toughened PSZ is competitive with some of the moderate toughness metal alloys. Further improvement in toughness will be demonstrated by some ceramic-matrix composites in Chapter 12.

EXAMPLE 8.2

A high-strength steel has a yield strength of 1,460 MPa and a K_{IC} of 98 MPa $\sqrt{m}$. Calculate the size of a surface crack that will lead to catastrophic failure at an applied stress of $\frac{1}{2}$ Y.S.

SOLUTION

We may use Equation 8.1 with the realization that we are assuming an ideal case of plane strain conditions. In lieu of specific geometrical information, we are forced to take $Y = 1$. Within these limitations, we can calculate

$$K_{IC} = Y\sigma_f\sqrt{\pi a}.$$

With $Y = 1$ and $\sigma_f = 0.5$ Y.S.,

$$K_{IC} = 0.5 \text{ Y.S. } \sqrt{\pi a},$$

or

$$a = \frac{1}{\pi}\frac{K_{IC}^2}{(0.5 \text{ Y.S.})^2}$$

$$= \frac{1}{\pi}\frac{(98 \text{ MPa}\sqrt{m})^2}{[0.5(1,460 \text{ MPa})]^2}$$

$$= 5.74 \times 10^{-3} \text{ m}$$

$$= 5.74 \text{ mm}.$$

EXAMPLE 8.3

Given that a quality-control inspection can ensure that a structural ceramic part will have no flaws greater than 25 μm in size, calculate the maximum service stress available with **(a)** SiC and **(b)** partially stabilized zirconia.

SOLUTION

In lieu of more specific information, we can treat this problem as a general fracture mechanics problem using Equation 8.1 with $Y = 1$, in which case

$$\sigma_f = \frac{K_{IC}}{\sqrt{\pi a}}.$$

This problem assumes that the maximum service stress will be the fracture stress for a part with flaw size $= a = 25$ μm. Values of K_{IC} are given in Table 8.3.

(a) For SiC,

$$\sigma_f = \frac{3\ \text{MPa}\sqrt{\text{m}}}{\sqrt{\pi \times 25 \times 10^{-6}\ \text{m}}} = 339\ \text{MPa}.$$

(b) For PSZ,

$$\sigma_f = \frac{9\ \text{MPa}\sqrt{\text{m}}}{\sqrt{\pi \times 25 \times 10^{-6}\ \text{m}}} = 1{,}020\ \text{MPa}.$$

PRACTICE PROBLEM 8.2

What crack size is needed to produce catastrophic failure in the alloy in Example 8.2 at **(a)** $\frac{1}{3}$ Y.S. and **(b)** $\frac{3}{4}$ Y.S.?

PRACTICE PROBLEM 8.3

In Example 8.3, maximum service stress for two structural ceramics is calculated based on the assurance of no flaws greater than 25 μm in size. Repeat these calculations given that a more economical inspection program can only guarantee detection of flaws greater than 100 μm in size.

8.3 | Fatigue

Up to this point, we have characterized the mechanical behavior of metals under a single load application either slowly (e.g., the tensile test) or rapidly (e.g., the impact test). Many structural applications involve cyclic rather than static loading, and a special problem arises. **Fatigue** is the general phenomenon of material failure after several cycles of loading to a stress level below the ultimate tensile stress (Figure 8.8). Figure 8.9 illustrates a common laboratory test used to cycle a

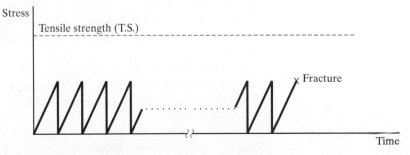

FIGURE 8.8 *Fatigue corresponds to the brittle fracture of an alloy after a total of N cycles to a stress below the tensile strength.*

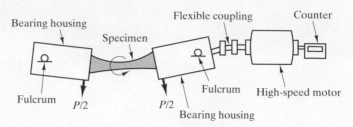

FIGURE 8.9 *Fatigue test. (From C. A. Keyser,* Materials Science in Engineering, *4th ed., Charles E. Merrill Publishing Company, Columbus, OH, 1986.)*

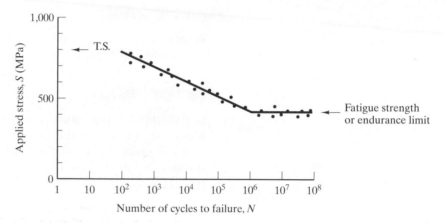

FIGURE 8.10 *Typical fatigue curve. (Note that a log scale is required for the horizontal axis.)*

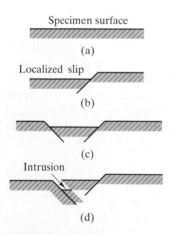

FIGURE 8.11 *An illustration of how repeated stress applications can generate localized plastic deformation at the alloy surface leading eventually to sharp discontinuities.*

test piece rapidly to a predetermined stress level. A typical **fatigue curve** is shown in Figure 8.10. This plot of stress (S) versus number of cycles (N), on a logarithmic scale, at a given stress is also called the *S–N curve*. The data indicate that while the material can withstand a stress of 800 MPa (T.S.) in a single loading ($N = 1$), it fractures after 10,000 applications ($N = 10^4$) of a stress less than 600 MPa. The reason for this decay in strength is a subtle one. Figure 8.11 shows how repeated stress applications can create localized plastic deformation at the metal surface, eventually manifesting as sharp discontinuities (extrusions and intrusions). These intrusions, once formed, continue to grow into cracks, reducing the load-carrying ability of the material and serving as stress concentrators (see the preceding section).

Fracture mechanics studies of cyclic loading provide substantial, quantitative understanding of the nature of crack growth. In particular, crack growth continues until the crack length reaches the critical value as defined by Equation 8.1 and Figure 8.6.

At low stress levels or for small crack sizes, preexisting cracks do not grow during cyclic loading. Figure 8.12 shows how, once the stress exceeds some threshold value, crack length increases, as indicated by the slope of the plot

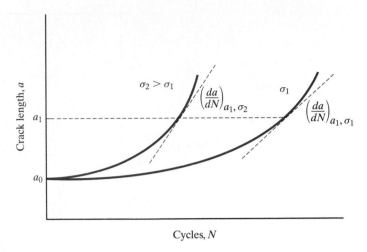

FIGURE 8.12 *Illustration of crack growth with number of stress cycles, N, at two different stress levels. Note that, at a given stress level, the crack growth rate, da/dN, increases with increasing crack length, and, for a given crack length such as a_1, the rate of crack growth is significantly increased with increasing magnitude of stress.*

(da/dN), the rate of crack growth. Figure 8.12 also shows that, at a given stress level, the crack growth rate increases with increasing crack length, and, for a given crack length, the rate of crack growth is significantly increased with increasing magnitude of stress. The overall growth of a fatigue crack as a function of the stress intensity factor, K, is illustrated in Figure 8.13. Region I in Figure 8.13 corresponds to the absence of crack growth mentioned earlier in conjunction with low stress and/or small cracks. Region II corresponds to the relationship

$$(da/dN) = A(\Delta K)^m, \tag{8.2}$$

where A and m are material parameters dependent on environment, test frequency, and the ratio of minimum and maximum stress applications, and ΔK is the stress-intensity factor range at the crack tip. Relative to Equation 8.1,

$$\Delta K = K_{\max} - K_{\min}$$
$$= Y\Delta\sigma\sqrt{\pi a} = Y(\sigma_{\max} - \sigma_{\min})\sqrt{\pi a}. \tag{8.3}$$

In Equations 8.2 and 8.3, one should note that the K is the more general **stress intensity factor** rather than the more specific fracture toughness, K_{IC}, and N is the number of cycles associated with a given crack length prior to failure rather than the total number of cycles to fatigue failure associated with an S–N curve. In Region II of Figure 8.13, Equation 8.2 implies a linear relationship between the logarithm of crack growth rate, da/dN, and the stress intensity factor range, ΔK, with the slope being m. Values of m typically range between 1 and 6. Region III corresponds to accelerated crack growth just prior to fast fracture.

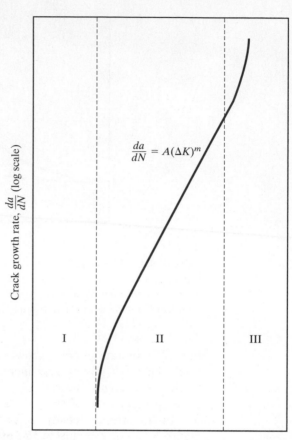

FIGURE 8.13 *Illustration of the logarithmic relationship between crack growth rate, da/dN, and the stress intensity factor range, ΔK. Region I corresponds to nonpropagating fatigue cracks. Region II corresponds to a linear relationship between log da/dN and log ΔK. Region III represents unstable crack growth prior to catastrophic failure.*

The fatigue fracture surface has a characteristic texture shown in Figure 8.14. The smoother portion of the surface is referred to as *clamshell* or *beachmark* texture. The concentric line pattern is a record of the slow, cyclic buildup of crack growth from a surface intrusion. The granular portion of the fracture surface identifies the rapid crack propagation at the time of catastrophic failure. Even for normally ductile materials, fatigue failure can occur by a characteristically brittle mechanism.

Figure 8.10 showed that the decay in strength with increasing numbers of cycles reaches a limit. This **fatigue strength**, or *endurance limit,* is characteristic of ferrous alloys. Nonferrous alloys tend not to have such a distinct limit, although the rate of decay decreases with N (Figure 8.15). As a practical matter, the fatigue strength of a nonferrous alloy is defined as the strength value after an arbitrarily large number of cycles (usually $N = 10^8$, as illustrated in Figure 8.15). Fatigue strength usually falls between one-fourth and one-half of the tensile strength, as illustrated by Table 8.4 and Figure 8.16 for the alloys of Table 6.1. For a given alloy,

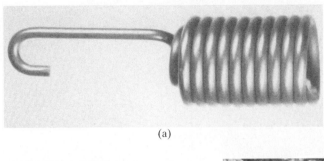

(a)

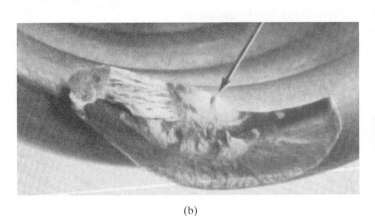

(b)

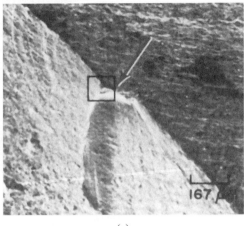

(c)

FIGURE 8.14 *Characteristic fatigue fracture surface. (a) Photograph of an aircraft throttle-control spring ($1\frac{1}{2}\times$) that broke in fatigue after 274 h of service. The alloy is 17–7PH stainless steel. (b) Optical micrograph ($10\times$) of the fracture origin (arrow) and the adjacent smooth region containing a concentric line pattern as a record of cyclic crack growth (an extension of the surface discontinuity shown in Figure 8.11). The granular region identifies the rapid crack propagation at the time of failure. (c) Scanning electron micrograph ($60\times$), showing a closeup of the fracture origin (arrow) and adjacent "clamshell" pattern. (From* Metals Handbook, *8th ed., Vol. 9:* Fractography and Atlas of Fractographs, *American Society for Metals, Metals Park, OH, 1974.)*

the resistance to fatigue will be increased by prior mechanical deformation (cold working) or reduction of structural discontinuities (Figure 8.17).

Metal fatigue has been defined as a loss of strength created by microstructural damage generated during cyclic loading. The fatigue phenomenon is also observed for ceramics and glasses, but without cyclic loading. The reason is that a chemical rather than mechanical mechanism is involved. Figure 8.18 illustrates the phenomenon of **static fatigue** for common silicate glasses. Two key observations can be made about this phenomenon: (1) It occurs in water-containing environments, and (2) it occurs around room temperature. The role of water in static fatigue is shown in Figure 8.19. By chemically reacting with the silicate network, an H_2O molecule generates two Si–OH units. The hydroxyl units are not bonded to each other, leaving a break in the silicate network. When this reaction occurs at the tip of a surface crack, the crack is lengthened by one atomic-scale step.

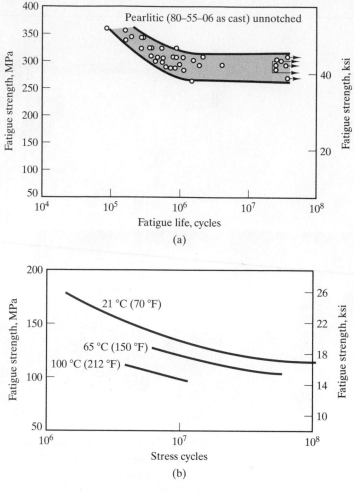

FIGURE 8.15 *Comparison of fatigue curves for (a) ferrous and (b) nonferrous alloys. The ferrous alloy is a ductile iron. The nonferrous alloy is C11000 copper wire. The nonferrous data do not show a distinct endurance limit, but the failure stress at $N = 10^8$ cycles is a comparable parameter. (After* Metals Handbook, *9th ed., Vols. 1 and 2, American Society for Metals, Metals Park, OH, 1978, 1979.)*

TABLE 8.4

Comparison of Fatigue Strength (F.S.) and Tensile Strength (T.S.) for Some of the Alloys of Table 6.1		
Alloy	**F.S. (MPa)**	**T.S. (MPa)**
1. 1040 carbon steel	280	750
2. 8630 low-alloy steel	400	800
3. a. 304 stainless steel	170	515
7. a. 3003-H14 aluminum	62	150
8. b. AM100A casting magnesium	69	150
9. a. Ti–5Al–2.5Sn	410	862
10. Aluminum bronze, 9% (copper alloy)	200	652
11. Monel 400 (nickel alloy)	290	579
12. AC41A zinc	56	328

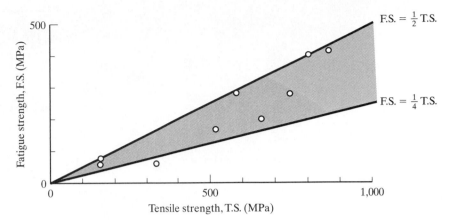

FIGURE 8.16 *Plot of data from Table 8.4 showing how fatigue strength is generally one-fourth to one-half of the tensile strength.*

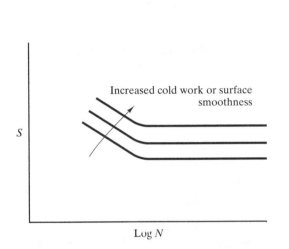

FIGURE 8.17 *Fatigue strength is increased by prior mechanical deformation or reduction of structural discontinuities.*

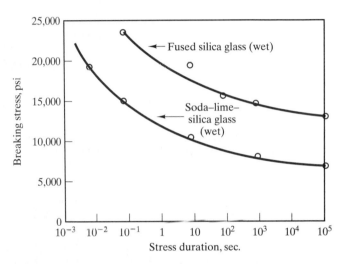

FIGURE 8.18 *The drop in strength of glasses with duration of load (and without cyclic-load applications) is termed* static fatigue. *(From W. D. Kingery,* Introduction to Ceramics, *John Wiley & Sons, Inc., New York, 1960.)*

Cyclic fatigue in metals and static fatigue in ceramics are compared in Figure 8.20. Because of the chemical nature of the mechanism in ceramics and glasses, the phenomenon is found predominantly around room temperature. At relatively high temperatures (above about 150°C), the hydroxyl reaction is so fast that the effects are difficult to monitor. At those temperatures, other factors such as viscous deformation can also contribute to static fatigue. At low temperatures (below about −100°C), the rate of hydroxyl reaction is too low to produce a significant effect in practical time periods. Analogies to static fatigue in metals would be stress-corrosion cracking and hydrogen embrittlement, involving crack-growth mechanisms under severe environments.

Fatigue in polymers is treated similarly to fatigue in metal alloys. Acetal polymers are noted for good fatigue resistance. Figure 8.21 summarizes S–N curves

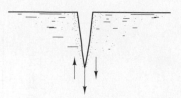

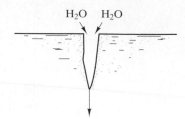

Crack growth by local shearing mechanism

(a)

Crack growth by chemical breaking of oxide network

(b)

FIGURE 8.20 *Comparison of (a) cyclic fatigue in metals and (b) static fatigue in ceramics.*

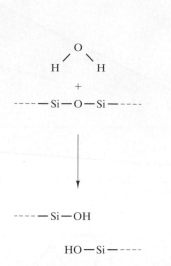

FIGURE 8.19 *The role of H_2O in static fatigue depends on its reaction with the silicate network. One H_2O molecule and one –Si–O–Si– segment generate two Si–OH units, which is equivalent to a break in the network.*

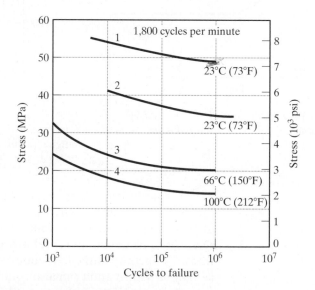

1. Tensile stress only
2. Completely reversed tensile and compressive stress
3. Completely reversed tensile and compressive stress
4. Completely reversed tensile and compressive stress

FIGURE 8.21 *Fatigue behavior for an acetal polymer at various temperatures. (From Design Handbook for Du Pont Engineering Plastics, used by permission.)*

for such a material at various temperatures. The fatigue limit for polymers is generally reported at 10^6 cycles rather than 10^8 cycles, as commonly used for nonferrous alloys (e.g., Figure 8.15).

EXAMPLE 8.4

Given only that the alloy for a structural member has a tensile strength of 800 MPa, estimate a maximum permissible service stress knowing that the loading will be cyclic in nature and that a safety factor of 2 is required.

SOLUTION

If we use Figure 8.16 as a guide, a conservative estimate of the fatigue strength will be

$$\text{F.S.} = \tfrac{1}{4}\text{T.S.} = \tfrac{1}{4}(800 \text{ MPa}) = 200 \text{ MPa}.$$

Using a safety factor of 2 will give a permissible service stress of

$$\text{service stress} = \frac{\text{F.S.}}{2} = \frac{200 \text{ MPa}}{2} = 100 \text{ MPa}.$$

Note. The safety factor helps to account for, among other things, the approximate nature of the relationship between F.S. and T.S.

EXAMPLE 8.5

Static fatigue depends on a chemical reaction (Figure 8.19) and, as a result, is another example of Arrhenius behavior. Specifically, at a given load, the inverse of time to fracture has been shown to increase exponentially with temperature. The activation energy associated with the mechanism of Figure 8.19 is 78.6 kJ/mol. If the time to fracture for a soda–lime–silica glass is 1 s at $+50°C$ at a given load, what is the time to fracture at $-50°C$ at the same load?

SOLUTION

As stated, we can apply the Arrhenius equation (Equation 5.1). In this case,

$$t^{-1} = Ce^{-Q/RT},$$

where t is the time to fracture.
At 50°C (323 K),

$$t^{-1} = 1 \text{ s}^{-1} = Ce^{-(78.6 \times 10^3 \text{ J/mol})/[8.314 \text{ J/mol·K}](323 \text{ K})},$$

giving

$$C = 5.15 \times 10^{12} \text{ s}^{-1}.$$

Then,

$$t^{-1}_{-50°C} = (5.15 \times 10^{12} \text{ s}^{-1})e^{-(78.6\times10^3 \text{ J/mol})/[8.314 \text{ J/(mol·K)}](223 \text{ K})}$$

$$= 1.99 \times 10^{-6} \text{ s}^{-1},$$

or

$$t = 5.0 \times 10^5 \text{ s}$$

$$= 5.0 \times 10^5 \text{ s} \times \frac{1 \text{ h}}{3.6 \times 10^3 \text{ s}}$$

$$= 140 \text{ h}$$

$$= 5 \text{ days, } 20 \text{ h.}$$

PRACTICE PROBLEM 8.4

In Example 8.4, a service stress is calculated with consideration for fatigue loading. Using the same considerations, estimate a maximum permissible service stress for an 80–55–06 as-cast ductile iron with a BHN of 200 (see Figure 6.29).

PRACTICE PROBLEM 8.5

For the system discussed in Example 8.5, what would be the time to fracture **(a)** at 0°C and **(b)** at room temperature, 25°C?

8.4 | Nondestructive Testing

Nondestructive testing is the evaluation of engineering materials without impairing their usefulness. A central focus of many of the nondestructive testing techniques is the identification of potentially critical flaws, such as surface and internal cracks. As with fracture mechanics, nondestructive testing can serve to analyze an existing failure or it can be used to prevent future failures. The dominant techniques of this field are x-radiography and ultrasonics.

Although diffraction allows dimensions on the order of the x-ray wavelength (typically < 1 nm) to be measured, **x-radiography** produces a *shadowgraph* of the internal structure of a part with a much coarser resolution, typically on the order of 1 mm (Figure 8.22). The medical chest x-ray is a common example. Industrial x-radiography is widely used for inspecting castings and weldments. For a given material being inspected by a given energy x-ray beam, the intensity of the beam, I, transmitted through a thickness of material, x, is given by Beer's* law,

$$I = I_0e^{-\mu x}, \tag{8.4}$$

---

*August Beer (1825–1863), German physicist. Beer graduated from the University of Bonn, where he was to remain as a teacher for the remainder of his relatively short life. He is primarily remembered for the law first stated relative to his observations on the absorption of visible light.

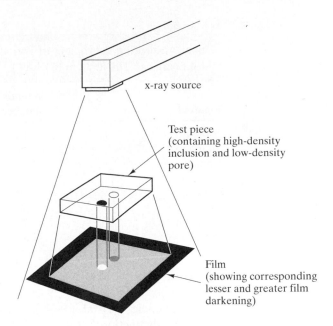

FIGURE 8.22 *A schematic of x-radiography.*

A gallery of x-radiographic images is available on the related Web site.

where I_0 is the incident beam intensity and μ is the linear absorption coefficient for the material. The intensity is proportional to the number of photons in the beam and is distinct from the energy of photons in the beam. The absorption coefficient is a function of the beam energy and of the elemental composition of the material. Experimental values for the μ of iron as a function of energy are given in Table 8.5. There is a general drop in the magnitude of μ with increasing beam energy primarily due to mechanisms of photon absorption and scattering. The dependence of the linear absorption coefficient on elemental composition is illustrated by the data of Table 8.6. Note that μ for a given beam energy generally increases with atomic number, causing low-atomic-number metals such as aluminum to be relatively transparent and high-atomic-number metals such as lead to be relatively opaque.

While x-radiography is based on a portion of the electromagnetic spectrum with relatively short wavelengths in comparison with the visible region, **ultrasonic testing** is based on a portion of the acoustic spectrum (typically 1 to 25 MHz)

TABLE 8.5

Linear Absorption Coefficient of Iron as a Function of X–Ray Beam Energy	
Energy (MeV)	μ **(mm^{-1})**
0.05	1.52
0.10	0.293
0.50	0.0662
1.00	0.0417
2.00	0.0334
4.00	0.0260

Source: Selected data from D. E. Bray and R. K. Stanley, *Nondestructive Evaluation*, McGraw-Hill Book Co., New York, 1989.

TABLE 8.6

Linear Absorption Coefficient of Various Elements for an X-Ray Beam with Energy = 100 keV (= 0.1 MeV)

Element	Atomic number	μ (mm^{-1})
Aluminum	13	0.0459
Titanium	22	0.124
Iron	26	0.293
Nickel	28	0.396
Copper	29	0.410
Zinc	30	0.356
Tungsten	74	8.15
Lead	82	6.20

Source: Selected data from D. E. Bray and R. K. Stanley, *Nondestructive Evaluation*, McGraw-Hill Book Co., New York, 1989.

with frequencies well above those of the audible range (20 to 20,000 Hz). An important distinction between x-radiography and ultrasonic testing is that the ultrasonic waves are mechanical in nature, requiring a transmitting medium, while electromagnetic waves can be transmitted in a vacuum. A typical ultrasonic source involving a piezoelectric transducer is shown in Section 4 of Chapter 13A on the related Web site.

X-ray attenuation is a dominant factor in x-radiography, but typical engineering materials are relatively transparent to ultrasonic waves. The key factor in ultrasonic testing is the reflection of the ultrasonic waves at interfaces of dissimilar materials. The high degree of reflectivity by a typical flaw, such as an internal crack, is the basis for defect inspection. Figure 8.23 illustrates a typical *pulse echo* ultrasonic inspection. This technique is not suitable for use on complex-shaped parts, and there is a tendency for ultrasonic waves to scatter due to microstructural features such as porosity and precipitates.

EXAMPLE 8.6

Calculate the fraction of x-ray beam intensity transmitted through a 10-mm-thick plate of low-carbon steel. Take the beam energy to be 100 keV. Because of the small amount of carbon and its inherently low absorption of x-rays, the steel can be approximated as elemental iron.

SOLUTION

Using Equation 8.4 and the attenuation coefficient from Table 8.6,

$$I = I_0 e^{-\mu x},$$

or

$$I/I_0 = e^{-\mu x}$$

$$= e^{-(0.293 \text{ mm}^{-1})(10 \text{ mm})}$$

$$= e^{-2.93} = 0.0534.$$

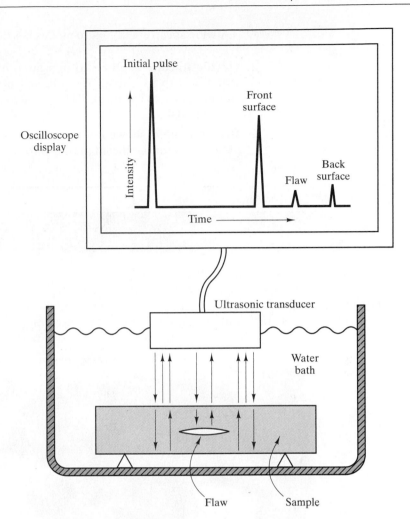

Note that a common goal of nondestructive tests is to find millimeter scale flaws such as the one shown here. Even careful visual inspections can sometimes find these flaws when present on the surface of a structure.

FIGURE 8.23 *A schematic of a* pulse echo *ultrasonic test.*

PRACTICE PROBLEM 8.6

For a 100-keV x-ray beam, calculate the fraction of beam intensity transmitted through a 10-mm-thick plate of **(a)** titanium and **(b)** lead. (See Example 8.6.)

8.5 | Failure Analysis and Prevention

Failure analysis and prevention are important components of the application of materials in engineering design. There is now a well-established, systematic methodology for **failure analysis** of engineering materials. The related issue of **failure prevention** is equally important for avoiding future disasters. Ethical and legal issues are moving the field of materials science and engineering into a central role in the broader topic of engineering design.

A wide spectrum of failure modes has been identified.

1. **Ductile fracture** is observed in a large number of the failures occurring in metals due to "overload" (i.e., taking a material beyond the elastic limit and, subsequently, to fracture). The microscopic result of ductile fracture is shown in Figure 8.4a.
2. **Brittle fracture**, shown in Figure 8.4b, is characterized by rapid crack propagation without significant plastic deformation on a macroscopic scale.

THE MATERIAL WORLD

Analysis of the Titanic Failure

We see in this chapter both the utility and the necessity of analyzing engineering failures. Such information can help prevent the recurrence of such disasters. In other cases, the analysis may serve a largely historical purpose, providing insight about the nature of a famous catastrophe of the past. An example of the latter is available for one of the most dramatic events of the 20th century, the

(Courtesy of Paramount Pictures and Twentieth Century Fox.)

(Continued)

sinking of the Royal Mail Ship *Titanic* on the night of April 12, 1912, during its maiden voyage across the North Atlantic from England to New York. The wreckage was not discovered until September 1, 1985, when Robert Ballard observed it on the ocean floor at a depth of 3,700 m. During an expedition to the wreckage on August 15, 1996, researchers retrieved some steel from the ship's hull, allowing metallurgical analysis at the University of Missouri–Rolla.

The general failure analysis is straightforward. The *Titanic* sank as a result of striking an iceberg three to six times larger than the ship itself. The six forward compartments of the ship were ruptured, leading to flooding that caused the ship to sink in less than 3 hours with a loss of more than 1,500 lives. There is some benefit, however, from closer inspection of the material from which the ship's hull was constructed. Chemical analysis of the steel showed that it was similar to that of contemporary 1020 steel, defined in Chapter 11 as a common, low-carbon steel consisting primarily of iron with 0.20 wt % carbon. The following table, from the UM–R analysis, shows that the mechanical behavior of the *Titanic* hull alloy is also similar to that of 1020 steel, although the *Titanic* material has a slightly lower yield strength and higher elongation at failure associated with a larger average grain size (approximately 50 μm versus approximately 25 μm).

More important than the basic tensile properties, the ductile-to-brittle transition temperature measured in Charpy impact tests of the *Titanic* steel is significantly higher than that of contemporary alloys. Modern steels in this composition range tend to have higher manganese contents and lower sulfur contents. The significantly higher Mn:S ratio reduces the ductile-to-brittle transition temperature

substantially. The ductile-to-brittle transition temperature measured at an impact energy of 20 J is −27°C for a comparable contemporary alloy and is 32°C and 56°C for the *Titanic* plate specimens cut in directions longitudinal and transverse, respectively, to the hull configuration. Given that the sea-water temperature at the time of the collision was −2°C, the materials selection clearly contributed to the failure. On the other hand, the choice was not inappropriate in the early part of the 20th century. The *Titanic* steel was likely the best plain carbon ship plate available at the time.

In a related study, metallurgists at Johns Hopkins University and the National Institute of Standards and Technology have suggested that the ship's builder, under construction deadlines, used inferior iron rivets in the area of the hull that impacted the iceberg. (J. H. McCarty and T. Foecke, *What Really Sank the Titanic*, Citadel Press, 2008.)

In addition to significantly better alloy selection, contemporary ship passengers have the benefit of superior navigational aides that decrease the probability of such collisions. Finally, it is worth noting that the *Titanic*'s sister ship, the *Olympic*, had a successful career of more than 20 years with similar hull steel and iron rivets, but it never encountered a large iceberg.

Comparison of Tensile Properties of *Titanic* Steel and SAE 1020

Property	Titanic	SAE 1020
Yield strength	193 MPa	207 MPa
Tensile strength	417 MPa	379 MPa
Elongation	29%	26%
Reduction in area	57%	50%

Source: K. Felkins, H. P. Leigh, Jr., and A. Jankovic, *Journal of Materials*, January 1998, pp. 12–18.

3. **Fatigue failure** by a mechanism of slow crack growth gives the distinctive "clamshell" fatigue-fracture surface shown in Figure 8.14.

4. **Corrosion-fatigue failure** is due to the combined actions of a cyclic stress and a corrosive environment. In general, the fatigue strength of the metal will be decreased in the presence of an aggressive, chemical environment.

5. **Stress-corrosion cracking** (SCC) is another combined mechanical and chemical failure mechanism in which a noncyclic tensile stress (below the yield strength) leads to the initiation and propagation of fracture in a relatively

mild chemical environment. Stress-corrosion cracks may be intergranular, transgranular, or a combination of the two.

6. **Wear failure** is a term encompassing a broad range of relatively complex, surface-related damage phenomena. Both surface damage and wear debris can constitute failure of materials intended for sliding-contact applications.

7. **Liquid-erosion failure** is a special form of wear damage in which a liquid is responsible for the removal of material. Liquid-erosion damage typically results in a pitted or honeycomb-like surface region.

8. **Liquid-metal embrittlement** involves the material losing some degree of ductility or fracturing below its yield stress in conjunction with its surface being wetted by a lower-melting-point liquid metal.

9. **Hydrogen embrittlement** is perhaps the most notorious form of catastrophic failure in high-strength steels. A few parts per million of hydrogen dissolved in these materials can produce substantial internal pressure, leading to fine, hairline cracks and a loss of ductility.

10. **Creep and stress-rupture failures** were introduced in Section 6.5. Failure of this type can occur near room temperature for many polymers and certain low-melting-point metals, such as lead, but may occur above 1,000°C in many ceramics and certain high-melting-point metals, such as the superalloys.

11. **Complex failures** are those in which the failure occurs by the sequential operation of two distinct fracture mechanisms. An example would be initial cracking due to stress-corrosion cracking and, then, ultimate failure by fatigue after a cyclic load is introduced simultaneously with the removal of the corrosive environment.

A systematic sequence of procedures has been developed for the analysis of the failure of an engineering material. Although the specific methodology will vary with the specific failure, the principal components of the investigation and analysis are given in Table 8.7. With regard to failure analysis, *fracture mechanics* (Section 8.2) has provided a quantitative framework for evaluating structural reliability. It is worth noting that values of fracture toughness for various metals and alloys range between 20 and 200 MPa$\sqrt{m}$. Values of fracture toughness for

TABLE 8.7

Principal Components of Failure-Analysis Methodology

Collection of background data and samples
Preliminary examination of the failed part
Nondestructive testing
Mechanical testing
Selection, preservation, and cleaning of fracture surfaces
Macroscopic (1 to 100 ×) examination of fracture surfaces
Microscopic (> 100×) examination of fracture surfaces
Application of fracture mechanics
Simulated-service testing
Analysis of the evidence, formulation of conclusions, and report writing

Source: After *ASM Handbook*, Vol. 11: *Failure Analysis and Prevention,* ASM International, Materials Park, OH, 1986.

ceramics and glass are typically in the range of 1 to 9 MPa$\sqrt{m}$, values for polymers are typically 1 to 4 MPa$\sqrt{m}$, and values for composites are typically 10 to 60 MPa$\sqrt{m}$.

Finally, it is important to note that engineering designs can be improved by applying the concepts of failure analysis to failure prevention. An important example of this approach is to use designs without structural discontinuities that can serve as stress concentrators.

Summary

In discussing the mechanical and thermal behavior of materials in Chapters 6 and 7, we found numerous examples of failure. In this chapter, we have systematically surveyed a variety of additional examples. The impact test of a notched test specimen provides a measure of impact energy, an analog of the toughness (the area under the stress-versus-strain curve). Monitoring the impact energy over a range of environmental temperatures helps to identify the ductile-to-brittle transition temperature of bcc metal alloys, including common structural steels.

An introduction to fracture mechanics provides an especially useful and quantitative material parameter, the fracture toughness. For design applications of metal alloys, the fracture toughness helps to define the critical flaw size at the boundary between more desirable, general yielding and catastrophic, fast fracture. Transformation toughening techniques involve engineering design at the microstructural level to improve the fracture toughness of traditionally brittle ceramics.

Metal alloys and engineering polymers exhibit fatigue, a drop in strength as a result of cyclic loading. Fracture mechanics can provide a quantitative treatment of the approach to failure, as cracks grow in length with these repetitive stress cycles. Certain ceramics and glasses exhibit static fatigue, which is due to a chemical reaction with atmospheric moisture rather than cyclic loading.

Nondestructive testing, the evaluation of materials without impairing their usefulness, is critically important in identifying potentially critical flaws in engineering materials. Many techniques are available, but x-radiography and ultrasonic testing are primary examples.

Overall, failure analysis is the systematic methodology for determining which of a wide range of failure modes are operating for a specific material. Failure prevention uses the understanding provided by failure analysis to avoid future disasters.

Key Terms

brittle fracture (250)
Charpy test (228)
complex failure (252)
corrosion-fatigue failure (251)
creep and stress-rupture failures (252)
ductile fracture (250)
ductile-to-brittle transition temperature (232)
failure analysis (249)
failure prevention (249)
fast fracture (234)

fatigue (237)
fatigue curve (238)
fatigue failure (251)
fatigue strength (endurance limit) (240)
flaw-induced fracture (233)
fracture mechanics (233)
fracture toughness (233)
general yielding (233)
hydrogen embrittlement (252)
impact energy (228)
Izod test (229)

liquid-erosion failure (252)
liquid-metal embrittlement (252)
nondestructive testing (246)
static fatigue (241)
stress-corrosion cracking (251)
stress intensity factor (239)
transformation toughening (234)
ultrasonic testing (247)
wear failure (252)
x-radiography (246)

References

Ashby, **M. F.**, and **D. R. H. Jones**, *Engineering Materials— 1 An Introduction to Properties, Applications, and Design*, 3rd ed., Butterworth-Heinemann, Boston, 2005.

ASM Handbook, Vol. 11: *Failure Analysis and Prevention*, ASM International, Materials Park, OH, 2002.

ASM Handbook, Vol. 17: *Nondestructive Evaluation and*

Quality Control, ASM International, Materials Park, OH, 1989.

Chiang, **Y.**, **D. P. Birnie III**, and **W. D. Kingery**, *Physical Ceramics*, John Wiley & Sons, Inc., New York, 1997.

Engineered Materials Handbook, Desk Ed., ASM International, Materials Park, OH, 1995.

Problems

Section 8.1 • Impact Energy

8.1. Which of the alloys in Table 8.1 would you expect to exhibit ductile-to-brittle transition behavior? (State the basis of your selection.)

D 8.2. The relationship between service temperature and alloy selection in engineering design can be illustrated by the following case. **(a)** For the Fe–Mn–0.05 C alloys of Figure 8.3b, plot the ductile-to-brittle transition temperature (indicated by the sharp vertical rise in impact energy) against percentage Mn. **(b)** Using the plot from (a), estimate the percentage Mn level (to the nearest 0.1%) necessary to produce a ductile-to-brittle transition temperature of precisely 0°C.

D 8.3. As in Problem 8.2, estimate the percentage Mn level (to the nearest 0.1%) necessary to produce a ductile-to-brittle transition temperature of −25°C in the Fe–Mn–0.05 C alloy series of Figure 8.3b.

8.4. The following data were obtained from a series of Charpy impact tests on a particular structural steel. **(a)** Plot the data as impact energy versus temperature. **(b)** What is the ductile-to-brittle transition temperature, determined as corresponding to the average of the maximum and minimum impact energies?

Temperature (°C)	Impact energy (J)	Temperature (°C)	Impact energy (J)
100	84.9	−60	50.8
60	82.8	−80	36.0
20	81.2	−100	28.1
0	77.9	−140	25.6
−20	74.5	−180	25.1
−40	67.8		

Section 8.2 • Fracture Toughness

8.5. Calculate the specimen thickness necessary to make the plane–strain assumption used in Example 8.2 valid.

D 8.6. Generate a design plot similar to that shown in Figure 8.6 for a pressure-vessel steel with Y.S. = 1,000 MPa and $K_{IC} = 170$ MPa $\sqrt{m}$. For convenience, use the logarithmic scale for flaw size and cover a range of flaw sizes from 0.1 to 100 mm.

D 8.7. Repeat Problem 8.6 for an aluminum alloy with Y.S. = 400 MPa and $K_{IC} = 25$ MPa $\sqrt{m}$.

8.8. Critical flaw size corresponds to the transition between general yielding and fast fracture. If the fracture toughness of a high-strength steel can be increased by 50% (from 100 to 150 MPa $\sqrt{m}$) without changing its yield strength of 1,250 MPa, by what percentage is its critical flaw size changed?

8.9. A nondestructive testing program for a design component using the 1040 steel of Table 6.1 can ensure that no flaw greater than 1 mm will exist. If this steel has a fracture toughness of 120 MPa $\sqrt{m}$, can this inspection program prevent the occurrence of fast fracture?

8.10. Would the nondestructive testing program described in Problem 8.9 be adequate for the cast-iron alloy labeled 6(b) in Table 6.1, given a fracture toughness of 15 MPa $\sqrt{m}$?

8.11. A silicon-nitride turbine rotor fractures at a stress level of 300 MPa. Estimate the flaw size responsible for this failure.

8.12. Estimate the flaw size responsible for the failure of a turbine rotor made from alumina that fractures at a stress level of 300 MPa.

8.13. Estimate the flaw size responsible for the failure of a turbine rotor made from partially stabilized zirconia that fractures at a stress level of 300 MPa.

8.14. Plot the breaking stress for MgO as a function of flaw size, a, on a logarithmic scale using Equation 8.1 and taking $Y = 1$. Cover a range of a from 1 to 100 mm. (See Table 8.3 for fracture-toughness data and note the appearance of Figure 8.6.)

8.15. To appreciate the relatively low values of fracture toughness for traditional ceramics, plot, on a single graph, breaking stress versus flaw size, a, for an aluminum alloy with a K_{IC} of 30 MPa $\sqrt{m}$ and silicon carbide (see Table 8.3). Use Equation 8.1 and take $Y = 1$. Cover a range of a from 1 to 100 mm on a logarithmic scale. (Note also Problem 8.14.)

8.16. To appreciate the improved fracture toughness of the new generation of structural ceramics, superimpose a plot of breaking stress versus flaw size for partially stabilized zirconia on the result for Problem 8.15.

8.17. Some fracture mechanics data are given in Table 8.3. Plot the breaking stress for low-density polyethylene as a function of flaw size, a (on a logarithmic scale), using Equation 8.1 and taking $Y = 1$. Cover a range of a from 1 to 100 mm. (Note Figure 8.6.)

8.18. Superimpose a breaking stress plot for high-density polyethylene and ABS polymer on the result for Problem 8.17.

8.19. Calculate the breaking stress for a rod of ABS with a surface flaw size of 100 μm.

8.20. A nondestructive testing program can ensure that a thermoplastic polyester part will have no flaws greater than 0.1 mm in size. Calculate the maximum service stress available with this engineering polymer.

Section 8.3 • Fatigue

8.21. In Problem 6.41, a ductile iron was evaluated for a pressure-vessel application. For that alloy, determine the maximum pressure to which the vessel can be repeatedly pressurized without producing a fatigue failure.

8.22. Repeat Problem 8.21 for the ductile iron of Problem 6.42.

8.23. A structural steel with a fracture toughness of 60 MPa $\sqrt{m}$ has no surface crack larger than 3 mm in length. By how much (in %) would this largest surface crack have to grow before the system would experience fast fracture under an applied stress of 500 MPa?

8.24. For the conditions given in Problem 8.23, calculate the % increase in crack size if the applied stress in the structural design is 600 MPa.

8.25. The application of a C11000 copper wire in a control-circuit design will involve cyclic loading for extended periods at the elevated temperatures of a production plant. Use the data of Figure 8.15b to specify an upper temperature limit to ensure a fatigue strength of at least 100 MPa for a stress life of 10^7 cycles.

8.26. (a) The landing gear on a commercial aircraft experiences an impulse load upon landing. Assuming six such landings per day on average, how long would it take before the landing gear has been subjected to 10^8 load cycles? (b) The crankshaft in a given automobile rotates, on average, at 2,000 revolutions per minute for a period of 2 h per day. How long would it take before the crankshaft has been subjected to 10^8 load cycles?

8.27. In analyzing a hip-implant material for potential fatigue damage, it is important to note that an average person takes 4,800 steps on an average day. How many stress cycles would this average person produce in (a) 1 year and in (b) 10 years? (Note Problem 6.7.)

8.28. In analyzing a hip-implant material for potential fatigue damage when used by an active athlete, we find that he takes a total of 10,000 steps and/or strides on an average day. How many stress cycles would this active person produce in (a) 1 year and in (b) 10 years? (Note Problem 6.8.)

8.29. The time to fracture for a vitreous silica-glass fiber at $+50°C$ is 10^4 s. What will be the time to fracture at room temperature (25°C)? Assume the same activation energy as given in Example 8.5.

8.30. To illustrate the very rapid nature of the water reaction with silicate glasses above 150°C, calculate the time to fracture for the vitreous silica fiber of Problem 8.29 at 200°C.

D 8.31. A small pressure vessel is fabricated from an acetal polymer. The stress in the vessel wall is

$$\sigma = \frac{pr}{2t},$$

where p is the internal pressure, r is the outer radius of the sphere, and t is the wall thickness. For the vessel in question, $r = 30$ mm and $t = 2$ mm. What is the maximum permissible internal pressure for this design if the application is at room temperature and the wall stress is only tensile (due to internal pressurizations that will occur no more than 10^6 times)? (See Figure 8.21 for relevant data.)

D 8.32. Calculate the maximum permissible internal pressure for the design in Problem 8.31 if all conditions are the same except that you are certain there will be no more than 10,000 pressurizations.

Section 8.4 • Nondestructive Testing

8.33. In doing x-radiography of steel, assume the film can detect a variation in radiation intensity represented by $\Delta I/I_0 = 0.001$. What thickness variation could be detected using this system for the inspection of a 12.5-mm-thick plate of steel using a 100-keV beam?

8.34. A good rule of thumb for doing x-radiographic inspections is that the test piece should be five to eight times the half-value thickness $(t_{1/2})$ of the material, where $t_{1/2}$ is defined as the thickness value corresponding to an I/I_0 value of 0.5. Calculate the appropriate test-piece thickness range for titanium inspected by a 100-keV beam.

8.35. Using the background of Problem 8.34, calculate the appropriate test-piece thickness range for tungsten inspected by a 100-keV beam.

8.36. Using the background of Problem 8.34, calculate the appropriate test-piece thickness range for iron inspected by **(a)** a 100-keV beam and **(b)** a 1-MeV beam.

8.37. For an x-radiographic inspection of the plate in Figure 8.23 using a 100-keV beam, calculate the % change in I/I_0 between the flawed and flaw-free areas given that the effect of the flaw is to reduce the effective thickness of the plate by 10 μm. Assume the sample is aluminum with a flaw depth at 10 mm and an overall sample thickness of 20 mm.

8.38. Given the result of Problem 8.37, comment on the relative advantage of ultrasonic testing for the inspection of the flaw in that sample.

CHAPTER 9
Phase Diagrams—Equilibrium Microstructural Development

9.1 The Phase Rule

9.2 The Phase Diagram

9.3 The Lever Rule

9.4 Microstructural Development During Slow Cooling

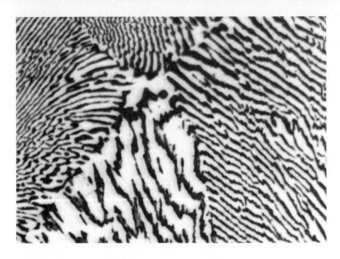

The microstructure of a slowly cooled "eutectic" soft solder (≈38 wt % Pb–62 wt % Sn) consists of a lamellar structure of tin-rich solid solution (white) and lead-rich solid solution (dark), 375X. (From ASM Handbook, Vol. 3: *Alloy Phase Diagrams, ASM International, Materials Park, OH, 1992.)*

From the beginning of this book we have seen that a fundamental concept of materials science is that the properties of materials follow from their atomic and microscopic structures. The dependence of transport and mechanical properties on atomic-scale structure was seen in Chapters 5 and 6. To appreciate fully the nature of the many microstructure-sensitive properties of engineering materials, we must spend some time exploring the ways in which microstructure is developed. An important tool in this exploration is the *phase diagram*, which is a map that will guide us in answering the general question: What microstructure should exist at a given temperature for a given material composition? This is a question with a specific answer based in part on the equilibrium nature of the material. Closely related is the next chapter, which deals with the heat treatment of materials. Still other related questions to be addressed in Chapter 10 will be "How fast will the microstructure form at a given temperature?" and "What temperature versus time history will result in an optimal microstructure?"

The discussion of microstructural development via phase diagrams begins with the *phase rule*, which identifies the number of microscopic phases associated with a given *state condition*, a set of values for temperature, pressure, and other variables that describe the nature of the material. We shall then describe the various characteristic phase diagrams for typical material systems. The *lever rule* will be used to quantify our interpretation of these phase diagrams. We shall specifically want to identify the composition and amount of each phase present. With these tools in hand, we can illustrate typical cases of microstructural development. Phase diagrams for several commercially important engineering materials are presented in this chapter. The most detailed discussion is reserved for the Fe–Fe₃C diagram, which is the foundation for much of the iron and steel industry.

9.1 | The Phase Rule

In this chapter we shall be quantifying the nature of microstructures. We begin with definitions of terms you need in order to understand the following discussion.

A **phase** is a chemically and structurally homogeneous portion of the microstructure. A single-phase microstructure can be polycrystalline (e.g., Figure 9.1), but each crystal grain differs only in crystalline orientation, not in chemical composition. Phase must be distinguished from **component**, which is a distinct chemical substance from which the phase is formed. For instance, we found in Section 4.1 that copper and nickel are so similar in nature that they are completely soluble in each other in any alloy proportions (e.g., Figure 4.2). For such a system, there is a single phase (a solid solution) and two components (Cu and Ni). For material systems involving compounds rather than elements, the compounds can be components. For example, MgO and NiO form solid solutions in a way similar to that for Cu and Ni (see Figure 4.5). In this case, the two components are MgO and NiO. As pointed out in Section 4.1, solid solubility is limited for many material systems. For certain compositions, the result is two phases, each richer in a different component. A classic example is the pearlite structure shown in Figure 9.2,

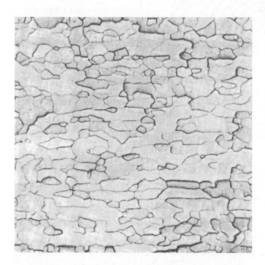

FIGURE 9.1 *Single-phase microstructure of commercially pure molybdenum, 200×. Although there are many grains in this microstructure, each grain has the same uniform composition. (From* ASM Handbook, *Vol. 9:* Metallography and Microstructures, *ASM International, Materials Park, OH, 2004.)*

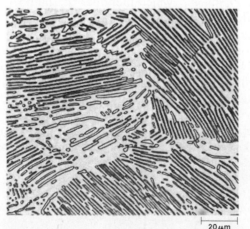

FIGURE 9.2 *Two-phase microstructure of pearlite found in a steel with 0.8 wt % C, 650 ×. This carbon content is an average of the carbon content in each of the alternating layers of ferrite (with <0.02 wt % C) and cementite (a compound, Fe_3C, which contains 6.7 wt % C). The narrower layers are the cementite phase. (From* ASM Handbook, *Vol. 9:* Metallography and Microstructures, *ASM International, Materials Park, OH, 2004.)*

which consists of alternating layers of ferrite and cementite. The ferrite is α-Fe with a small amount of cementite in solid solution. The cementite is nearly pure Fe_3C. The components, then, are Fe and Fe_3C.

Describing the ferrite phase as α-Fe with cementite in solid solution is appropriate in terms of our definition of the components for this system. However, on the atomic scale, the solid solution consists of carbon atoms dissolved interstitially in the α-Fe crystal lattice. The component Fe_3C does not dissolve as a discrete molecular unit, which is generally true for compounds in solid solution.

A third term can be defined relative to *phase* and *component*. The **degrees of freedom** are the number of independent variables available to the system. For example, a pure metal at precisely its melting point has no degrees of freedom. At this condition, or **state**, the metal exists in two phases in equilibrium (i.e., in solid and liquid phases simultaneously). Any increase in temperature will change the state of the microstructure. (All of the solid phase will melt and become part of the liquid phase.) Similarly, even a slight reduction in temperature will completely solidify the material. The important **state variables** over which the materials engineer has control in establishing microstructure are temperature, pressure, and composition.

The general relationship between microstructure and these state variables is given by the **Gibbs* phase rule**, which, without derivation, can be stated as

$$F = C - P + 2, \tag{9.1}$$

where F is the number of degrees of freedom, C is the number of components, and P is the number of phases.[†] The 2 in Equation 9.1 comes from limiting the state variables to two (temperature and pressure). For most routine materials processing involving condensed systems, the effect of pressure is slight, and we can consider pressure to be fixed at 1 atm. In this case, the phase rule can be rewritten to reflect one less degree of freedom:

$$F = C - P + 1. \tag{9.2}$$

For the case of the pure metal at its melting point, $C = 1$ and $P = 2$ (solid + liquid), giving $F = 1 - 2 + 1 = 0$, as we had noted previously. For a metal with a single impurity (i.e., with two components), solid and liquid phases can usually coexist over a range of temperatures (i.e., $F = 2 - 2 + 1 = 1$). The single degree of freedom means simply that we can maintain this two-phase microstructure while we vary the temperature of the material. However, we have only one independent variable ($F = 1$). By varying temperature, we indirectly vary the compositions of the individual phases. Composition is, then, a dependent variable. Such information that can be obtained from the Gibbs phase rule is most useful, but is also difficult

---

*Josiah Willard Gibbs (1839–1903), American physicist. As a professor of mathematical physics at Yale University, Gibbs was known as a quiet individual who made a profound contribution to modern science by almost single-handedly developing the field of thermodynamics. His phase rule was a cornerstone of this achievement.

[†]This derivation is an elementary one in the field of thermodynamics. Most students of this course will probably have a first course in thermodynamics 1 or 2 years hence. For now, we can take the Gibbs phase rule as a highly useful experimental fact. For those wishing to follow the derivation at this point, it is provided in the thermodynamics chapter of the related Web site.

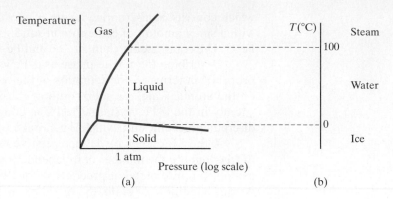

(a) (b)

FIGURE 9.3 *(a) Schematic representation of the one-component phase diagram for H_2O. (b) A projection of the phase diagram information at 1 atm generates a temperature scale labeled with the familiar transformation temperatures for H_2O (melting at $0°C$ and boiling at $100°C$).*

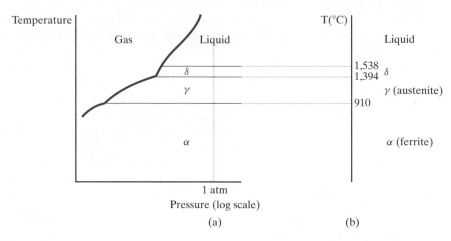

(a) (b)

FIGURE 9.4 *(a) Schematic representation of the one-component phase diagram for pure iron. (b) A projection of the phase diagram information at 1 atm generates a temperature scale labeled with important transformation temperatures for iron. This projection will become one end of important binary diagrams, such as that shown in Figure 9.19.*

to appreciate without the visual aid of phase diagrams. So, now we proceed to introduce these fundamentally important maps.

Our first example of a simple phase diagram is given in Figure 9.3. The one-component phase diagram (Figure 9.3a) summarizes the phases present for H_2O as a function of temperature and pressure. For the fixed pressure of 1 atm, we find a single, vertical temperature scale (Figure 9.3b) labeled with the appropriate transformation temperatures that summarize our common experience that solid H_2O (ice) transforms to liquid H_2O (water) at $0°C$ and that water transforms to gaseous H_2O (steam) at $100°C$. More relevant to materials engineering, Figure 9.4 provides a similar illustration for pure iron. As practical engineering materials are typically impure, we shall next discuss phase diagrams, in the more general sense, for the case of more than one component.

EXAMPLE 9.1

At 200°C, a 50:50 Pb–Sn solder alloy exists as two phases, a lead-rich solid and a tin-rich liquid. Calculate the degrees of freedom for this alloy and comment on its practical significance.

SOLUTION
Using Equation 9.2 (i.e., assuming a constant pressure of 1 atm above the alloy), we obtain

$$F = C - P + 1.$$

There are two components (Pb and Sn) and two phases (solid and liquid), giving

$$F = 2 - 2 + 1 = 1.$$

As a practical matter, we may retain this two-phase microstructure upon heating or cooling. However, such a temperature change exhausts the "freedom" of the system and must be accompanied by changes in composition. (The nature of the composition change will be illustrated by the Pb–Sn phase diagram shown in Figure 9.16.)

PRACTICE PROBLEM 9.1

Calculate the degrees of freedom at a constant pressure of 1 atm for **(a)** a single-phase solid solution of Sn dissolved in the solvent Pb, **(b)** pure Pb below its melting point, and **(c)** pure Pb at its melting point. (See Example 9.1.) **(Note that the solutions to all practice problems are provided on the related Web site.)**

9.2 | The Phase Diagram

A phase diagram experiment is available in a laboratory manual on the related Web site, along with phase diagram data.

A **phase diagram** is any graphical representation of the state variables associated with microstructures through the Gibbs phase rule. As a practical matter, phase diagrams in wide use by materials engineers are **binary diagrams**, which represent two-component systems ($C = 2$ in the Gibbs phase rule), and ternary diagrams, which represent three-component systems ($C = 3$). In this book, our discussion will be restricted to the binary diagrams. There are an abundant number of important binary systems that will give us full appreciation of the power of the phase rule, and at the same time, we avoid the complexities involved in extracting quantitative information from ternary diagrams.

In the following examples, keep in mind that phase diagrams are maps. Specifically, binary diagrams are maps of the equilibrium phases associated with various combinations of temperature and composition. Our concern will be to illustrate the change in phases and associated microstructure that follows from changes in the state variables (temperature and composition).

COMPLETE SOLID SOLUTION

Probably the simplest type of phase diagram is that associated with binary systems in which the two components exhibit **complete solid solution** in each other in both the solid and the liquid states. There have been previous references to such completely miscible behavior for Cu and Ni and for MgO and NiO. Figure 9.5 shows a typical phase diagram for such a system. Note that the diagram shows temperature as the variable on the vertical scale and composition as the horizontal variable. The melting points of pure components A and B are indicated. For relatively high temperatures, any composition will have melted completely to give a liquid **phase field**, the region of the phase diagram that corresponds to the existence of a liquid and that is labeled L. In other words, A and B are completely soluble in each other in the liquid state. What is unusual about this system is that A and B are also completely soluble in the solid state. The Hume-Rothery rules (see Section 4.1) state the criteria for this phenomenon in metal systems. In this book, we shall generally encounter complete miscibility in the liquid state (e.g., the L field in Figure 9.5). However, there are some systems in which liquid immiscibility occurs. Oil and water is a commonplace example. More relevant to materials engineering is the combination of various silicate liquids.

At relatively low temperatures, there is a single, solid-solution phase field labeled SS. Between the two single-phase fields is a two-phase region labeled L + SS. The upper boundary of the two-phase region is called the **liquidus** (i.e., the line above which a single liquid phase will be present). The lower boundary of the two-phase region is called the **solidus** and is the line below which the system has completely solidified. At a given **state point** (a pair of temperature and composition values) within the two-phase region, an A-rich liquid exists in equilibrium with a B-rich solid solution. The composition of each phase is established, as shown in Figure 9.6. The horizontal (constant-temperature) line passing through the state point cuts across both the liquidus and solidus lines. The composition

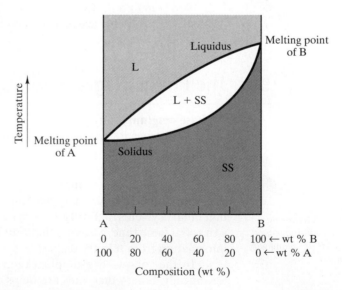

FIGURE 9.5 *Binary phase diagram showing complete solid solution. The liquid-phase field is labeled L, and the solid solution is designated SS. Note the two-phase region labeled L + SS.*

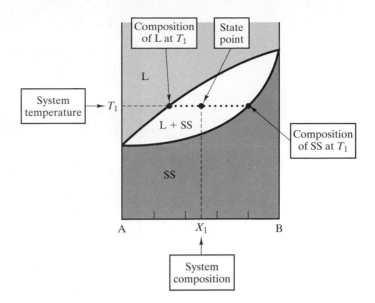

FIGURE 9.6 *The compositions of the phases in a two-phase region of the phase diagram are determined by a tie line (the horizontal line connecting the phase compositions at the system temperature).*

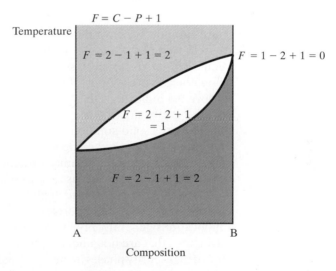

FIGURE 9.7 *Application of Gibbs phase rule (Equation 9.2) to various points in the phase diagram of Figure 9.5.*

of the liquid phase is given by the intersection point with the liquidus. Similarly, the solid-solution composition is established by the point of intersection with the solidus. This horizontal line connecting the two phase compositions is termed a **tie line**. This construction will prove even more useful in Section 9.3 when we calculate the relative amounts of the two phases by the lever rule.

Figure 9.7 shows the application of the Gibbs phase rule (Equation 9.2) to various points in this phase diagram. The discussions in Section 9.1 can now be appreciated in terms of the graphical summary provided by the phase diagram. For example, an **invariant point** (where $F = 0$) occurs at the melting point of

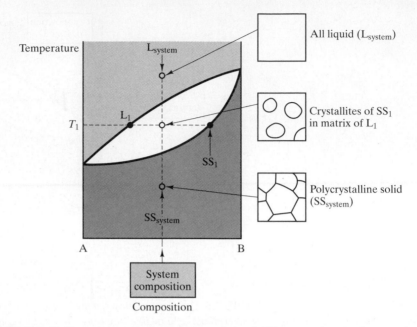

FIGURE 9.8 *Various microstructures characteristic of different regions in the complete solid-solution phase diagram.*

pure component B. At this limiting case, the material becomes a one-component system, and any change of temperature changes the microstructure to either all liquid (for heating) or all solid (for cooling). Within the two-phase (L + SS) region, there is one degree of freedom. A change of temperature is possible, but as Figure 9.6 indicates, the phase compositions are not independent. Instead, they will be established by the tie line associated with a given temperature. In the single-phase solid-solution region, there are two degrees of freedom; that is, both temperature and composition can be varied independently without changing the basic nature of the microstructure. Figure 9.8 summarizes microstructures characteristic of the various regions of this phase diagram.

By far the greatest use of phase diagrams in materials engineering is for the inorganic materials of importance to the metals and ceramics industries. Polymer applications generally involve single-component systems and/or nonequilibrium structures that are not amenable to presentation as phase diagrams. The common use of high-purity phases in the semiconductor industry also makes phase diagrams of limited use. The Cu–Ni system in Figure 9.9 is the classic example of a binary diagram with complete solid solution. A variety of commercial copper and nickel alloys falls within this system, including a superalloy called Monel.

The NiO–MgO system (Figure 9.10) is a ceramic system analogous to the Cu–Ni system (i.e., exhibiting complete solid solution). While the Hume–Rothery rules (Section 4.1) addressed solid solution in metals, the requirement of similarity of cations is a comparable basis for solid solution in this oxide structure (see Figure 4.5).

Note that the composition axis for the NiO–MgO phase diagram (and the other ceramic phase diagrams) is expressed in mole percent rather than weight percent. The use of mole percent has no effect on the validity of the lever-rule calculations that we will make in Section 9.3. The only effect on the results is that

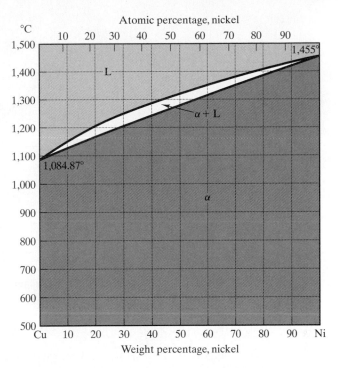

FIGURE **9.9** *Cu–Ni phase diagram. (From* Metals Handbook, *8th ed., Vol. 8:* Metallography, Structures, and Phase Diagrams, *American Society for Metals, Metals Park, OH, 1973, and* Binary Alloy Phase Diagrams, *Vol. 1, T. B. Massalski, ed., American Society for Metals, Metals Park, OH, 1986.)*

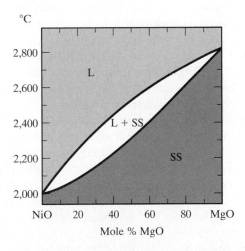

FIGURE **9.10** *NiO–MgO phase diagram. (After* Phase Diagrams for Ceramists, *Vol. 1, American Ceramic Society, Columbus, OH, 1964.)*

the answers obtained from such a calculation are in mole fractions rather than weight fractions.

EUTECTIC DIAGRAM WITH NO SOLID SOLUTION

We now turn to a binary system that is the opposite of the one just discussed. Some components are so dissimilar that their solubility in each other is nearly negligible.

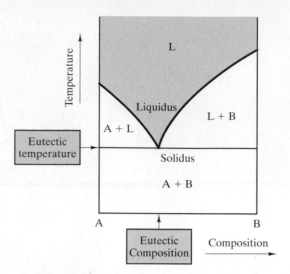

FIGURE 9.11 *Binary eutectic phase diagram showing no solid solution. This general appearance can be contrasted to the opposite case of complete solid solution illustrated in Figure 9.5.*

Figure 9.11 illustrates the characteristic phase diagram for such a system. Several features distinguish this diagram from the type characteristic of complete solid solubility. First is the fact that, at relatively low temperatures, there is a two-phase field for pure solids A and B, consistent with our observation that the two components (A and B) cannot dissolve in each other. Second, the solidus is a horizontal line that corresponds to the **eutectic temperature**. This name comes from the Greek word *eutektos*, meaning "easily melted." In this case, the material with the **eutectic composition** is fully melted at the eutectic temperature. Any composition other than the eutectic will not fully melt at the eutectic temperature. Instead, such a material must be heated further through a two-phase region to the liquidus line. This situation is analogous to the two-phase region (L+SS) found in Figure 9.5. Figure 9.11 differs in that we have two such two-phase regions (A + L and B + L) in the binary eutectic diagram.

Some representative microstructures for the binary **eutectic diagram** are shown in Figure 9.12. The liquid and the liquid + solid microstructures are comparable to cases found in Figure 9.8. However, a fundamental difference exists in the microstructure of the fully solid system. In Figure 9.12, we find a fine-grained eutectic microstructure in which there are alternating layers of the components, pure A and pure B. A fuller discussion of solid-state microstructures will be appropriate after the lever rule has been introduced in Section 9.3. For now, we can emphasize that the sharp solidification point of the eutectic composition generally leads to the fine-grained nature of the eutectic microstructure. Even during slow cooling of the eutectic composition through the eutectic temperature, the system must transform from the liquid state to the solid state relatively quickly. The limited time available prevents a significant amount of diffusion (Section 5.3). The segregation of A and B atoms (which were randomly mixed in the liquid state) into separate solid phases must be done on a small scale. Various morphologies occur for various eutectic systems. But whether lamellar, nodular, or other morphologies are stable, these various eutectic microstructures are commonly fine-grained.

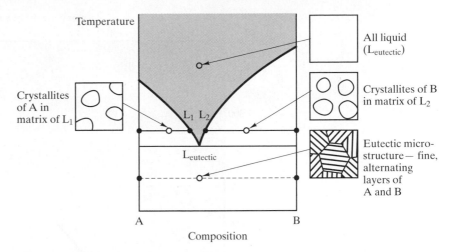

FIGURE 9.12 *Various microstructures characteristic of different regions in a binary eutectic phase diagram with no solid solution.*

The simple eutectic system Al–Si (Figure 9.13) is a close approximation to Figure 9.11, although a small amount of solid solubility does exist. The aluminum-rich side of the diagram describes the behavior of some important aluminum alloys. Although we are not dwelling on semiconductor-related examples, the silicon-rich side illustrates the limit of aluminum doping in producing *p*-type semiconductors (see Chapter 13).

EUTECTIC DIAGRAM WITH LIMITED SOLID SOLUTION

For many binary systems, the two components are partially soluble in each other. The result is a phase diagram intermediate between the two cases we have treated so far. Figure 9.14 shows a eutectic diagram with limited solid solution. It generally looks like Figure 9.11 except for the solid-solution regions near each edge. These single-phase regions are comparable to the SS region in Figure 9.5 except for the fact that the components in Figure 9.14 do not exist in a single solid solution near the middle of the composition range. As a result, the two solid-solution phases, α and β, are distinguishable, and they frequently have different crystal structures. In any case, the crystal structure of α will be that of A, and the crystal structure of β will be that of B because each component serves as a solvent for the other, "impurity" component (e.g., α consists of B atoms in solid solution in the crystal lattice of A). The use of tie lines to determine the compositions of α and β in the two-phase regions is identical to the diagram shown in Figure 9.6, and examples are shown in Figure 9.15 together with representative microstructures.

The Pb–Sn system (Figure 9.16) is a good example of a binary eutectic with limited solid solution. Common solder alloys fall within this system. Their low melting ranges allow for joining of most metals by convenient heating methods, with low risk of damage to heat-sensitive parts. Solders with less than 5 wt % tin are used for sealing containers and coating and joining metals, and are also used for applications with service temperatures that exceed 120°C. Solders with between 10 and 20 wt % tin are used for sealing cellular automobile radiators and filling seams and dents in automobile bodies. General-purpose solders are generally 40 or

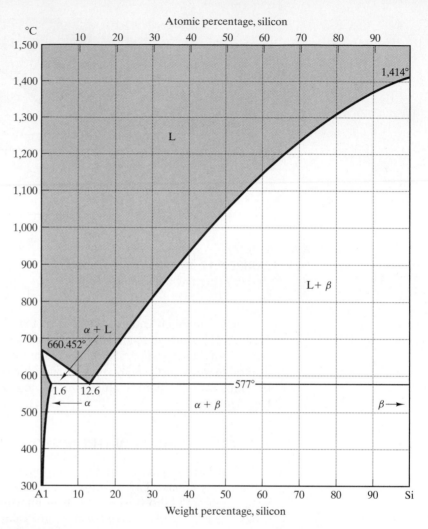

FIGURE 9.13 *Al–Si phase diagram. (After* Binary Alloy Phase Diagrams, *Vol. 1, T. B. Massalski, Ed., American Society for Metals, Metals Park, OH, 1986.)*

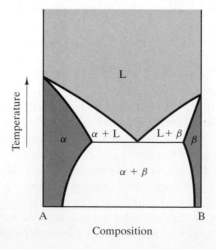

FIGURE 9.14 *Binary eutectic phase diagram with limited solid solution. The only difference between this diagram and the one shown in Figure 9.11 is the presence of solid-solution regions α and β.*

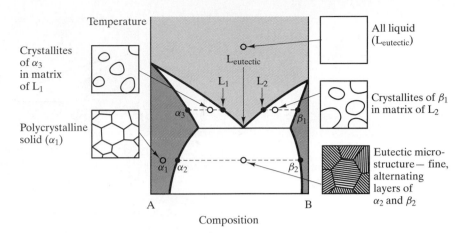

FIGURE 9.15 *Various microstructures characteristic of different regions in the binary eutectic phase diagram with limited solid solution. This illustration is essentially equivalent to the illustration shown in Figure 9.12, except that the solid phases are now solid solutions (α and β) rather than pure components (A and B).*

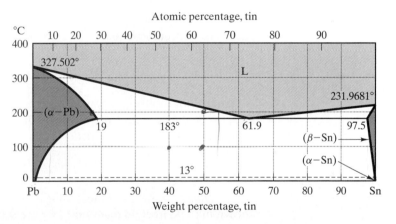

FIGURE 9.16 *Pb–Sn phase diagram. (After* Metals Handbook, *8th ed., Vol. 8:* Metallography, Structures, and Phase Diagrams, *American Society for Metals, Metals Park, Ohio, 1973, and* Binary Alloy Phase Diagrams, *Vol. 2, T. B. Massalski, Ed., American Society for Metals, Metals Park, OH, 1986.)*

50 wt % tin. These solders have a characteristic pastelike consistency during application, associated with the two-phase liquid plus solid region just above the eutectic temperature. Their wide range of applications includes well-known examples from plumbing to electronics. Solders near the eutectic composition (approximately 60 wt % tin) are used for heat-sensitive electronic components that require minimum heat application.

EUTECTOID DIAGRAM

The transformation of eutectic liquid to a relatively fine-grained microstructure of two solid phases upon cooling can be described as a special type of chemical

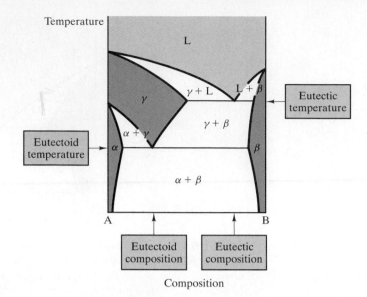

FIGURE 9.17 *This eutectoid phase diagram contains both a eutectic reaction (Equation 9.3) and its solid-state analog, a eutectoid reaction (Equation 9.4).*

reaction. This **eutectic reaction** can be written as

$$\text{L (eutectic)} \xrightarrow{\text{cooling}} \alpha + \beta, \tag{9.3}$$

where the notation corresponds to the phase labels from Figure 9.14. Some binary systems contain a solid-state analog of the eutectic reaction. Figure 9.17 illustrates such a case. The *eutectoid reaction* is

$$\gamma \text{(eutectoid)} \xrightarrow{\text{cooling}} \alpha + \beta, \tag{9.4}$$

where *eutectoid* means "eutectic-like." Some representative microstructures are shown in the **eutectoid diagram** of Figure 9.18. The different morphologies of the eutectic and eutectoid microstructures emphasize our previous point that although the specific nature of these diffusion-limited structures will vary, they will generally be relatively fine grained. A eutectoid reaction plays a fundamental role in the technology of steelmaking.

The Fe–Fe$_3$C system (Figure 9.19) is, by far, the most important commercial phase diagram we shall encounter. It provides the major scientific basis for the iron and steel industries. In Chapter 11, the boundary between irons and steels will be identified as a carbon content of 2.0 wt %. This point roughly corresponds to the carbon solubility limit in the **austenite*** (γ) phase of Figure 9.19. In addition,

---

*William Chandler Roberts-Austen (1843–1902), English metallurgist. Young William Roberts set out to be a mining engineer, but his opportunities led to an appointment in 1882 as "chemist and assayer of the mint," a position he held until his death. His varied studies of the technology of coin making led to his appointment as a professor of metallurgy at the Royal School of Mines. He was a great success in both his government and academic posts. His textbook, *Introduction to the Study of Metallurgy*, was published in six editions between 1891 and 1908. In 1885, he adopted the additional surname in honor of his uncle (Nathaniel Austen).

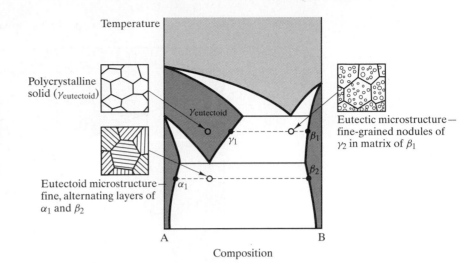

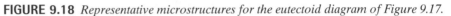

FIGURE 9.18 *Representative microstructures for the eutectoid diagram of Figure 9.17.*

Micrometer scale structures such as these developed during the slow cooling of commercial alloys play a central role in contemporary industries. Among the most important examples are Figures 9.19 and 9.20 relative to the iron and steel industries. More detailed examples are illustrated in Section 9.4.

this diagram is representative of microstructural development in many related systems with three or more components (e.g., some stainless steels that include large amounts of chromium). Although Fe_3C, and not carbon, is a component in this system, the composition axis is customarily given in weight percent carbon. The important areas of interest on this diagram are around the eutectic and the eutectoid reactions. The reaction near 1,500°C is of no practical consequence.

A final note of some irony is that the Fe–Fe_3C diagram is not a true equilibrium diagram. The Fe–C system (Figure 9.20) represents true equilibrium. Although graphite (C) is a more stable precipitate than Fe_3C, the rate of graphite precipitation is enormously slower than that of Fe_3C. The result is that in common steels (and many cast irons) the Fe_3C phase is **metastable**; that is, for all practical purposes it is stable with time and conforms to the Gibbs phase rule.

As just noted, the Fe–C system (Figure 9.20) is fundamentally more stable, but less common than the Fe–Fe_3C system because of slow *kinetics* (the subject of Chapter 10). Extremely slow cooling rates can produce the results indicated on the Fe–C diagram. The more practical method is to promote graphite precipitation by a small addition of a third component, such as silicon. Typically, silicon additions of 2 to 3 wt % are used to stabilize the graphite precipitation. This third component is not acknowledged in Figure 9.20. The result, however, is that the figure does describe microstructural development for some practical systems. An example will be given in Section 9.4.

PERITECTIC DIAGRAM

In all the binary diagrams inspected to this point, the pure components have had distinct melting points. In some systems, however, the components will form stable compounds that may not have such a distinct melting point. An example is illustrated in Figure 9.21. In this simple example, A and B form the stable compound AB, which does not melt at a single temperature as do components A and B. An additional simplification used here is to ignore the possibility of some

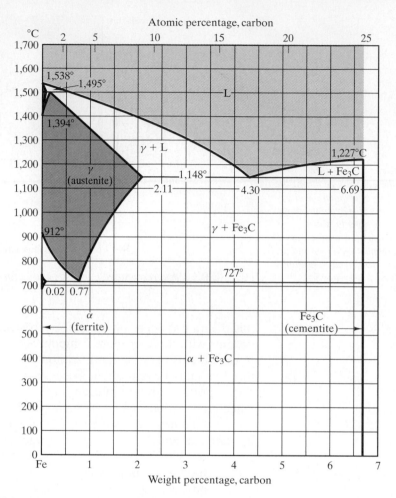

FIGURE 9.19 *Fe–Fe₃C phase diagram. Note that the composition axis is given in weight percent carbon even though Fe₃C, and not carbon, is a component. (After* Metals Handbook, *8th ed., Vol. 8:* Metallography, Structures, and Phase Diagrams, *American Society for Metals, Metals Park, OH, 1973, and* Binary Alloy Phase Diagrams, *Vol. 1, T. B. Massalski, Ed., American Society for Metals, Metals Park, OH, 1986.)*

solid solution for the components and intermediate compound. The components are said to undergo **congruent melting**; that is, the liquid formed upon melting has the same composition as the solid from which it was formed. On the other hand, compound AB (which is 50 mol % A plus 50 mol % B) is said to undergo **incongruent melting**; that is, the liquid formed upon melting has a composition other than AB. The term *peritectic* is used to describe this incongruent melting phenomenon. *Peritectic* comes from the Greek phrase meaning to "melt nearby." The **peritectic reaction** can be written as

$$AB \xrightarrow{\text{heating}} L + B, \qquad (9.5)$$

where the liquid composition is noted in the **peritectic diagram** of Figure 9.21. Some representative microstructures are shown in Figure 9.22. The Al₂O₃–SiO₂ phase diagram (Figure 9.23) is a classic example of a peritectic diagram.

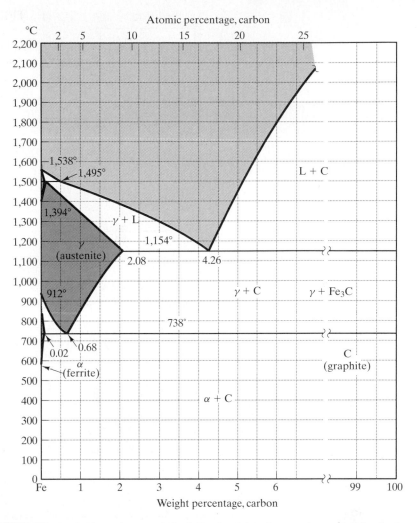

FIGURE 9.20 *Fe–C phase diagram. The left side of this diagram is nearly identical to the left side of the Fe–Fe₃C diagram (Figure 9.19). In this case, however, the intermediate compound Fe₃C does not exist. (After* Metals Handbook, *8th ed., Vol. 8:* Metallography, Structures, and Phase Diagrams, *American Society for Metals, Metals Park, OH, 1973, and* Binary Alloy Phase Diagrams, *Vol. 1, T. B. Massalski, Ed., American Society for Metals, Metals Park, OH, 1986.)*

The Al_2O_3–SiO_2 binary diagram is as important to the ceramic industry as the Fe–Fe₃C diagram is to the steel industry. Several important ceramics fall within this system. Refractory silica bricks are nearly pure SiO_2, with 0.2 to 1.0 wt % (0.1 to 0.6 mol %) Al_2O_3. For silica bricks required to operate at temperatures above 1,600°C, it is obviously important to keep the Al_2O_3 content as low as possible (by careful raw material selection) to minimize the amount of liquid phase. A small amount of liquid is tolerable. Common fireclay refractories are located in the range of 25 to 45 wt % (16 to 32 mol %) Al_2O_3. Their utility as structural elements in furnace designs is limited by the solidus (eutectic) temperature of 1,587°C. A dramatic increase in *refractoriness*, or temperature resistance, occurs at the composition of the incongruently melting compound mullite ($3Al_2O_3 \cdot 2SiO_2$).

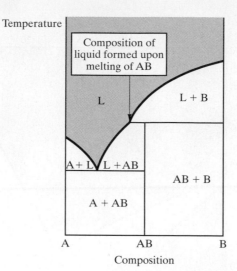

FIGURE 9.21 *Peritectic phase diagram showing a peritectic reaction (Equation 9.5). For simplicity, no solid solution is shown.*

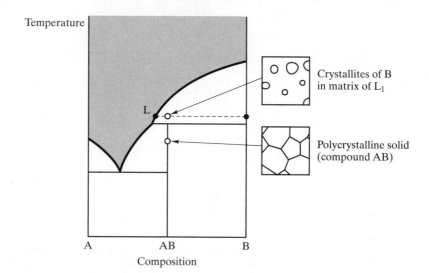

FIGURE 9.22 *Representative microstructures for the peritectic diagram of Figure 9.21.*

A controversy over the nature of the melting for mullite has continued for several decades. The peritectic reaction shown in Figure 9.23 is now widely accepted. The debate about such a commercially important system illustrates a significant point. Establishing equilibrium in high-temperature ceramic systems is not easy. Silicate glasses are similar examples of this point. Phase diagrams for this text represent our best understanding at this time, but we must remain open to refined experimental results in the future.

Care is exercised in producing mullite refractories to ensure that the overall composition is greater than 72 wt % (60 mol %) Al_2O_3 to avoid the two-phase region (mullite + liquid). By so doing, the refractory remains completely solid to the peritectic temperature of 1,890°C. So-called high-alumina refractories fall within the composition range 60 to 90 wt % (46 to 84 mol %) Al_2O_3. Nearly pure Al_2O_3

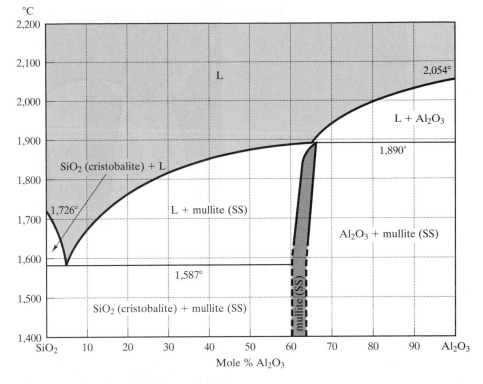

FIGURE 9.23 *Al_2O_3–SiO_2 phase diagram. Mullite is an intermediate compound with ideal stoichiometry $3Al_2O_3 \cdot 2SiO_2$. [After F. J. Klug, S. Prochazka, and R. H. Doremus,* J. Am. Ceram. Soc. 70, *750 (1987).]*

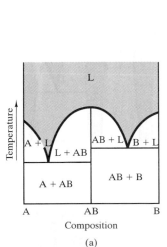

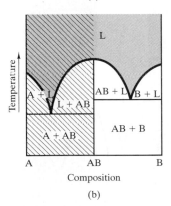

FIGURE 9.24 *(a) Binary phase diagram with a congruently melting intermediate compound, AB. This diagram is equivalent to two simple binary eutectic diagrams (the A–AB and AB–B systems). (b) For analysis of microstructure for an overall composition in the AB–B system, only that binary eutectic diagram need be considered.*

represents the highest refractoriness (temperature resistance) of commercial materials in the Al_2O_3–SiO_2 system. These materials are used in such demanding applications as refractories in glass manufacturing and laboratory crucibles.

GENERAL BINARY DIAGRAMS

The peritectic diagram was our first example of a binary system with an **intermediate compound**, a chemical compound formed between two components in a binary system. In fact, the formation of intermediate compounds is a relatively common occurrence. Of course, such compounds are not associated solely with the peritectic reaction. Figure 9.24a shows the case of an intermediate compound, AB, which melts congruently. An important point about this system is that it is equivalent to two adjacent binary eutectic diagrams of the type first introduced in Figure 9.11. Again, for simplicity, we are ignoring solid solubility. This is our first encounter with what can be termed a **general diagram**, a composite of two or more of the types covered in this section. The approach to analyzing these more complex systems is straightforward: Simply deal with the smallest binary system associated with the overall composition and ignore all others. This procedure is illustrated in Figure 9.24b, which shows that for an overall composition between AB and B, we can treat the diagram as a simple binary eutectic of AB and B. For all practical purposes, the A–AB binary does not exist for the overall composition

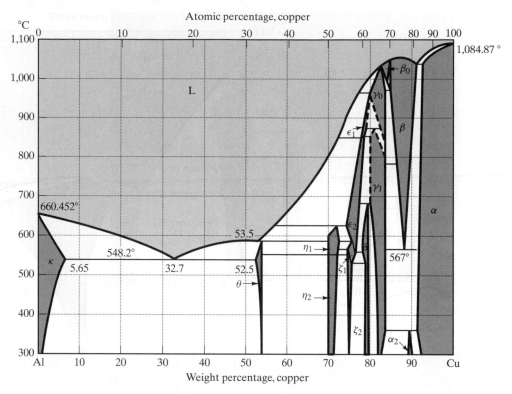

FIGURE 9.27 *Al–Cu phase diagram. (After* Binary Alloy Phase Diagrams, *Vol. 1, T. B. Massalski, Ed., American Society for Metals, Metals Park, OH, 1986.)*

The $CaO–ZrO_2$ phase diagram (Figure 9.29) is an example of a general diagram of a ceramic system. ZrO_2 has become an important refractory material through the use of stabilizing additions such as CaO. As seen in the phase diagram (Figure 9.29), pure ZrO_2 has a phase transformation at 1,000°C in which the crystal structure changes from monoclinic to tetragonal upon heating. This transformation involves a substantial volume change that is structurally catastrophic to the brittle ceramic. Cycling the pure material through the transformation temperature will effectively reduce it to powder. As the phase diagram also shows, the addition of approximately 10 wt % (20 mol %) CaO produces a solid-solution phase with a cubic crystal structure from room temperature to the melting point (near 2,500°C). This "stabilized zirconia" is a practical, and obviously very refractory, structural material. Various other ceramic components, such as Y_2O_3, serve as stabilizing components and have phase diagrams with ZrO_2 that look quite similar to the diagram shown in Figure 9.29.

EXAMPLE 9.2

An alloy in the A–B system described by Figure 9.25 is formed by melting equal parts of A and A_2B. Qualitatively describe the microstructural development that will occur upon slow cooling of this melt.

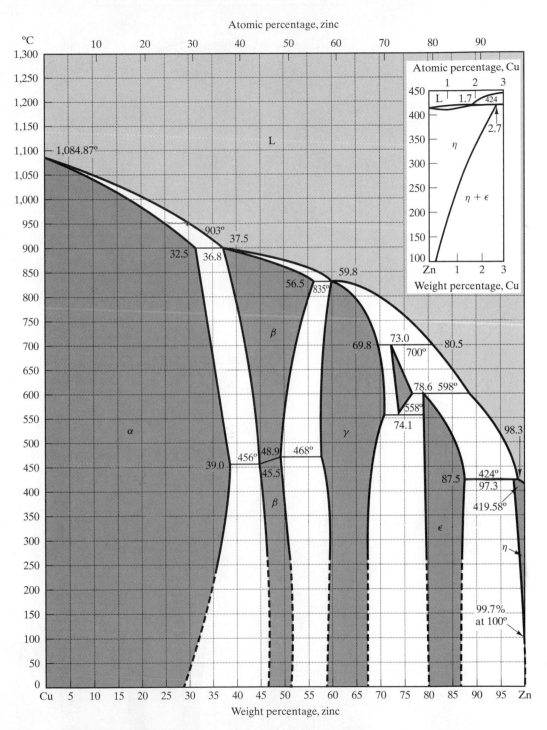

FIGURE 9.28 *Cu–Zn phase diagram. (After* Metals Handbook, *8th ed., Vol. 8:* Metallography, Structures, and Phase Diagrams, *American Society for Metals, Metals Park, OH, 1973, and* Binary Alloy Phase Diagrams, *Vol. 1, T. B. Massalski, Ed., American Society for Metals, Metals Park, OH, 1986.)*

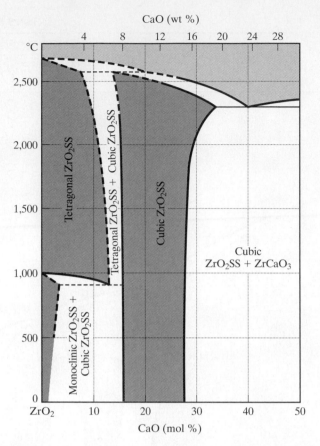

CaO (wt %)

FIGURE 9.29 *CaO–ZrO₂ phase diagram. The dashed lines represent tentative results. (After* Phase Diagrams for Ceramists, *Vol. 1, American Ceramic Society, Columbus, OH, 1964.)*

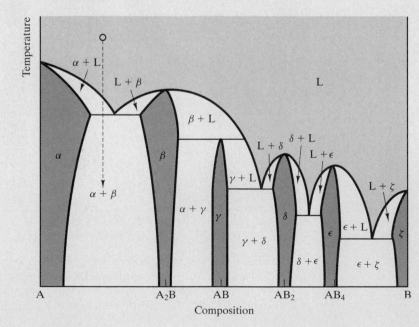

SOLUTION

A 50:50 combination of A and A_2B will produce an overall composition midway between A and A_2B. The cooling path is illustrated as follows:

The first solid to precipitate from the liquid is the A-rich solid solution
α. At the A–A$_2$B eutectic temperature, complete solidification occurs,
giving a two-phase microstructure of solid solutions α and β.

PRACTICE PROBLEM 9.2

Qualitatively describe the microstructural development that will
occur upon slow cooling of a melt of an alloy of equal parts of A$_2$B
and AB.

9.3 | The Lever Rule

In Section 9.2, we surveyed the use of phase diagrams to determine the phases
present at equilibrium in a given system and their corresponding microstructure.
The tie line (e.g., Figure 9.6) gives the composition of each phase in a two-phase
region. We now extend this analysis to determine the amount of each phase in the
two-phase region. We should first note that, for single-phase regions, the analysis
is trivial. By definition, the microstructure is 100% of the single phase. In the
two-phase regions, the analysis is not trivial, but is, nonetheless, simple.

The relative amounts of the two phases in a microstructure are easily calculated from a mass balance. Let us consider again the case of the binary diagram
for complete solid solution. The diagram shown in Figure 9.30 is equivalent to that
shown in Figure 9.6 and, again, shows a tie line that gives the composition of the
two phases associated with a state point in the L + SS region. In addition, the compositions of each phase and of the overall system are indicated. An overall **mass
balance** requires that the sum of the two phases equal the total system. Assuming
a total mass of 100 g gives an expression

$$m_L + m_{SS} = 100 \text{ g.} \qquad (9.6)$$

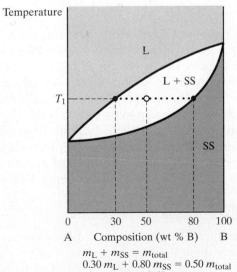

FIGURE 9.30 *A more quantitative
treatment of the tie line introduced in
Figure 9.6 allows the amount of each phase
(L and SS) to be calculated by means of a
mass balance (Equations 9.6 and 9.7).*

$m_L + m_{SS} = m_{total}$
$0.30\, m_L + 0.80\, m_{SS} = 0.50\, m_{total}$
$\rightarrow m_L = 0.60\, m_{total}$
$m_{SS} = 0.40\, m_{total}$

We can also compute an independent mass balance on either of the two components. For example, the amount of B in the liquid phase plus that in solid solution must equal the total amount of B in the overall composition. Noting in Figure 9.6 that (for temperature, T_1) L contains 30% B, SS 80% B, and the overall system 50% B, we can write

$$0.30 \, m_L + 0.80 \, m_{SS} = 0.50(100 \text{ g}) = 50 \text{ g}. \qquad (9.7)$$

Equations 9.6 and 9.7 are two equations with two unknowns, allowing us to solve for the amounts of each phase:

$$m_L = 60 \text{ g}$$

and

$$m_{SS} = 40 \text{ g}.$$

This material-balance calculation is convenient, but an even more streamlined version can be generated. To obtain this calculation, we can compute the mass balance in general terms. For two phases, α and β, the general mass balance will be

$$x_\alpha m_\alpha + x_\beta m_\beta = x(m_\alpha + m_\beta), \qquad (9.8)$$

where x_α and x_β are the compositions of the two phases and x is the overall composition. This expression can be rearranged to give the relative amount of each phase in terms of compositions:

$$\frac{m_\alpha}{m_\alpha + m_\beta} = \frac{x_\beta - x}{x_\beta - x_\alpha} \qquad (9.9)$$

and

$$\frac{m_\beta}{m_\alpha + m_\beta} = \frac{x - x_\alpha}{x_\beta - x_\alpha}. \qquad (9.10)$$

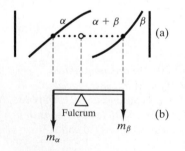

FIGURE 9.31 *The lever rule is a mechanical analogy to the mass-balance calculation. The (a) tie line in the two-phase region is analogous to (b) a lever balanced on a fulcrum.*

Together, Equations 9.9 and 9.10 constitute the **lever rule**. This mechanical analogy to the mass-balance calculation is illustrated in Figure 9.31. Its utility is due largely to the fact that it can be visualized so easily in terms of the phase diagram. The overall composition corresponds to the fulcrum of a lever with length corresponding to the tie line. The mass of each phase is suspended from the end of the lever corresponding to its composition. The relative amount of phase α is directly proportional to the length of the "opposite lever arm" ($= x_\beta - x$). It is this relationship that allows the relative amounts of phases to be determined by a simple visual inspection. With this final quantitative tool in hand, we can now proceed to the step-by-step analysis of microstructural development.

EXAMPLE 9.3

The temperature of 1 kg of the alloy shown in Figure 9.30 is lowered slowly until the liquid-solution composition is 18 wt % B and the solid-solution composition is 66 wt % B. Calculate the amount of each phase.

SOLUTION
Using Equations 9.9 and 9.10, we obtain

$$m_L = \frac{x_{SS} - x}{x_{SS} - x_L}(1\text{ kg}) = \frac{66 - 50}{66 - 18}(1\text{ kg})$$

$$= 0.333\text{ kg} = 333\text{ g}$$

and

$$m_{SS} = \frac{x - x_L}{x_{SS} - x_L}(1\text{ kg}) = \frac{50 - 18}{66 - 18}(1\text{ kg})$$

$$= 0.667\text{ kg} = 667\text{ g}.$$

Note. We can also calculate m_{SS} more swiftly by simply noting that $m_{SS} = 1{,}000\text{ g} - m_L = (1{,}000 - 333)\text{ g} = 667\text{ g}$. However, we shall continue to use both Equations 9.9 and 9.10 in the example problems in this chapter for the sake of practice and as a cross-check.

EXAMPLE 9.4

For 1 kg of eutectoid steel at room temperature, calculate the amount of each phase (α and Fe_3C) present.

SOLUTION
Using Equations 9.9 and 9.10 and Figure 9.19, we have

$$m_\alpha = \frac{x_{Fe_3C} - x}{x_{Fe_3C} - x_\alpha}(1\text{ kg}) = \frac{6.69 - 0.77}{6.69 - 0}(1\text{ kg})$$

$$= 0.885\text{ kg} = 885\text{ g}$$

and

$$m_{Fe_3C} = \frac{x - x_\alpha}{x_{Fe_3C} - x_\alpha}(1\text{ kg}) = \frac{0.77 - 0}{6.69 - 0}(1\text{ kg})$$

$$= 0.115\text{ kg} = 115\text{ g}.$$

EXAMPLE 9.5

A partially stabilized zirconia is composed of 4 wt % CaO. This product contains some monoclinic phase together with the cubic phase, which is the basis of fully stabilized zirconia. Estimate the mole percent of each phase present at room temperature.

SOLUTION
Noting that 4 wt % CaO = 8 mol % CaO and assuming that the solubility limits shown in Figure 9.29 do not change significantly below

500°C, we can use Equations 9.9 and 9.10:

$$\text{mol \% monoclinic} = \frac{x_{\text{cub}} - x}{x_{\text{cub}} - x_{\text{mono}}} \times 100\%$$

$$= \frac{15 - 8}{15 - 2} \times 100\% = 53.8 \text{ mol \%}$$

and

$$\text{mol \% cubic} = \frac{x - x_{\text{mono}}}{x_{\text{cub}} - x_{\text{mono}}} \times 100\%$$

$$= \frac{8 - 2}{15 - 2} \times 100\% = 46.2 \text{ mol \%}.$$

THE MATERIAL WORLD

Purifying Semiconductors by Zone Refining

The feature box in Chapter 3 gave us an appreciation of how semiconductors are produced with a high degree of structural perfection. In Chapter 13, we will see that solid-state electronics also require that semiconductors have a high degree of chemical purity. This chemical perfection is due to a special process prior to the crystal-growing step. This process is, in fact, a creative use of phase diagrams.

As seen in the following illustration, a bar of material (e.g., silicon) with a modest level of impurities is purified by the process of **zone refining**. In this technique, an induction coil produces a local molten "zone." As the coil passes along the length of the bar, the zone follows along. The molten material solidifies as soon as the induction coil moves away.

The following phase diagram illustrates that the impurity content in the liquid is substantially greater than in the solid. As a result, a single pass of the heating coil along the bar "sweeps" the impurities along with the liquid zone to one end. Multiple passes lead to substantial purification. Eventually, substantial levels of contamination will be swept to one end of the bar, which is simply sawed off and discarded. For the bulk of the bar, impurity levels in the parts per billion (ppb) range are practical and, in fact, were necessary to allow the development of solid-state electronics as we know it today.

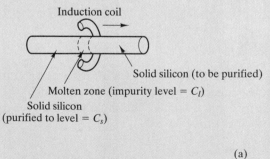

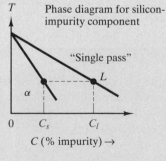

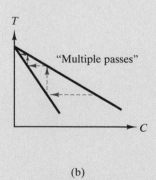

(a)   (b)

In zone refining, (a) a single pass of the molten "zone" through the bar leads to the concentration of impurities in the liquid. This is illustrated by the nature of the phase diagram. (b) Multiple passes of the molten zone increases purification of the solid.

PRACTICE PROBLEM 9.3

Suppose the alloy in Example 9.3 is reheated to a temperature at which the liquid composition is 48 wt % B and the solid-solution composition is 90 wt % B. Calculate the amount of each phase.

PRACTICE PROBLEM 9.4

In Example 9.4, we found the amount of each phase in a eutectoid steel at room temperature. Repeat this calculation for a steel with an overall composition of 1.13 wt % C.

PRACTICE PROBLEM 9.5

In Example 9.5, the phase distribution in a partially stabilized zirconia is calculated. Repeat this calculation for a zirconia with 5 wt % CaO.

9.4 | Microstructural Development During Slow Cooling

We are now in a position to follow closely the **microstructural development** in various binary systems. In all cases, we shall assume the common situation of cooling a given composition from a single-phase melt. Microstructure is developed in the process of solidification. We consider only the case of *slow* cooling; that is, equilibrium is essentially maintained at all points along the cooling path. The effect of more rapid temperature changes is the subject of Chapter 10, which deals with time-dependent microstructures developed during heat treatment.

Let us return to the simplest of binary diagrams, the case of complete solubility in both the liquid and solid phases. Figure 9.32 shows the gradual solidification of the 50% A–50% B composition treated previously (Figures 9.6, 9.8, and 9.30). The lever rule (Figure 9.31) is applied at three different temperatures in the two-phase (L + SS) region. It is important to appreciate that the appearance of the microstructures in Figure 9.32 corresponds directly with the relative position of the overall system composition along the tie line. At higher temperatures (e.g., T_1), the overall composition is near the liquid-phase boundary, and the microstructure is predominantly liquid. At lower temperatures (e.g., T_3), the overall composition is near the solid-phase boundary, and the microstructure is predominantly solid. Of course, the compositions of the liquid and solid phases change continuously during cooling through the two-phase region. At any temperature, however, the relative amounts of each phase are such that the overall composition is 50% A and 50% B, which is a direct manifestation of the lever rule as defined by the mass balance of Equation 9.8.

The understanding of microstructural development in the binary eutectic is greatly aided by the lever rule. The case for the eutectic composition itself is straightforward and was illustrated previously (Figures 9.12 and 9.15). Figure 9.33 repeats those cases in slightly greater detail. An additional comment is that the composition of each solid-solution phase (α and β) and their relative

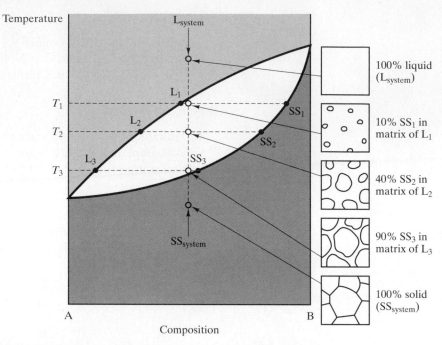

FIGURE 9.32 *Microstructural development during the slow cooling of a 50% A–50% B composition in a phase diagram with complete solid solution. At each temperature, the amounts of the phases in the microstructure correspond to a lever-rule calculation. The microstructure at T_2 corresponds to the calculation in Figure 9.30.*

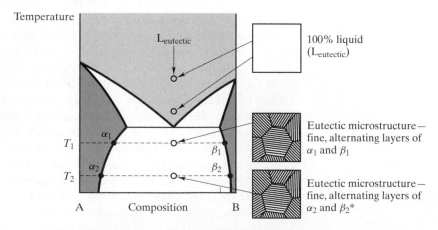

*The only differences between this structure and the T_1 microstructure are the phase compositions and the relative amounts of each phase. For example, the amount of b will be proportional to

$$\frac{x_{\text{eutectic}} - x_\alpha}{x_\beta - x_\alpha}.$$

FIGURE 9.33 *Microstructural development during the slow cooling of a eutectic composition.*

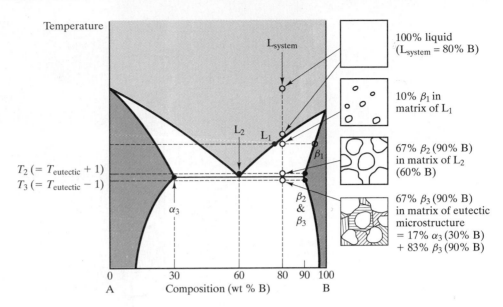

FIGURE 9.34 *Microstructural development during the slow cooling of a hypereutectic composition.*

amounts will change slightly with temperature below the eutectic temperature. The microstructural effect (corresponding to this compositional adjustment due to solid-state diffusion) is generally minor.

Microstructural development for a noneutectic composition is more complex. Figure 9.34 illustrates the microstructural development for a **hypereutectic composition** (composition greater than that of the eutectic). The gradual growth of β crystals above the eutectic temperature is comparable to the process found in Figure 9.32 for the complete solid-solution diagram. The one difference is that, in Figure 9.34, the crystallite growth stops at the eutectic temperature with only 67% of the microstructure solidified. Final solidification occurs when the remaining liquid (with the eutectic composition) transforms suddenly to the eutectic microstructure upon cooling through the eutectic temperature. In a sense, the 33% of the microstructure that is liquid just above the eutectic temperature undergoes the eutectic reaction illustrated in Figure 9.33. A lever-rule calculation just below the eutectic temperature (T_3 in Figure 9.34) indicates correctly that the microstructure is 17% α_3 and 83% β_3. However, following the entire cooling path has indicated that the β phase is present in two forms. The large grains produced during the slow cooling through the two-phase (L + β) region are termed **proeutectic** β; that is, they appear "before the eutectic." The finer β in the lamellar eutectic is appropriately termed *eutectic* β.

Figure 9.35 shows a similar situation that develops for a **hypoeutectic composition** (composition less than that of the eutectic). This case is analogous to that for the hypereutectic composition. In Figure 9.35, we see the development of large grains of procutectic α along with the eutectic microstructure of α and β layers. Two other types of microstructural development are illustrated in Figure 9.36. For an overall composition of 10% B, the situation is quite similar to that for the complete solid-solution binary in Figure 9.32. The solidification leads to a

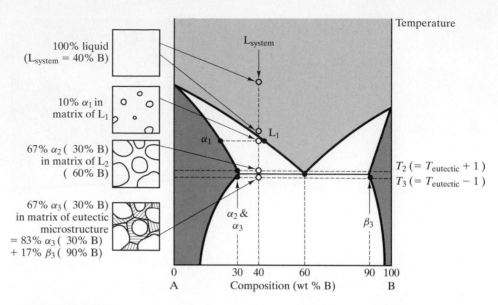

FIGURE 9.35 *Microstructural development during the slow cooling of a hypoeutectic composition.*

single-phase solid solution that remains stable upon cooling to low temperatures. The 20% B composition behaves in a similar fashion except that, upon cooling, the α phase becomes saturated with B atoms. Further cooling leads to precipitation of a small amount of β phase. In Figure 9.36b, this precipitation is shown occurring along grain boundaries. In some systems, the second phase precipitates within grains. For a given system, the morphology of the second phase can be a sensitive function of the rate of cooling. In Section 10.4, we shall encounter such a case for the Al–Cu system, in which precipitation hardening is an important example of heat treatment.

    With the variety of cases encountered in this section, we are now in a position to treat any composition in any of the binary systems presented in this chapter, including the general diagrams represented by Figure 9.25b.

    The cooling path for a **white cast iron** (see also Chapter 11) is shown in Figure 9.37. The schematic microstructure can be compared with a micrograph in Figure 11.1a. The eutectoid reaction to produce pearlite is shown in Figure 9.38. This composition (0.77 wt % C) is close to that for a 1080 plain-carbon steel (see Table 11.1). Many phase diagrams for the Fe–Fe_3C system give the eutectoid composition rounded off to 0.8 wt % C. As a practical matter, any composition near 0.77 wt % C will give a microstructure that is predominantly eutectoid. The actual pearlite microstructure is shown in the micrograph of Figure 9.2. A **hypereutectoid composition** (composition greater than the eutectoid composition of 0.77 wt % C) is treated in Figure 9.39. This case is similar in many ways to the hypereutectic path shown in Figure 9.34. A fundamental difference is that the **proeutectoid** cementite (Fe_3C) is the matrix in the final microstructure, whereas the proeutectic phase in Figure 9.34 was the isolated phase. The formation of the proeutectoid matrix occurs because the precipitation of proeutectoid cementite is a solid-state transformation and is favored at grain boundaries. Figure 9.40

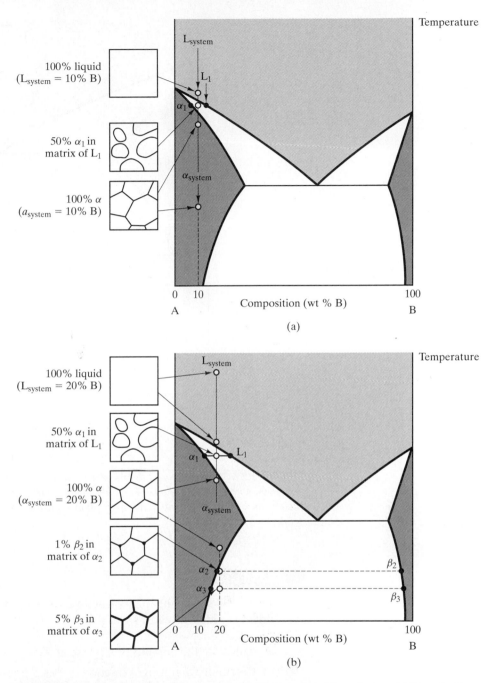

FIGURE 9.36 *Microstructural development for two compositions that avoid the eutectic reaction.*

illustrates the development of microstructure for a **hypoeutectoid composition** (less than 0.77 wt % C).

The Fe–C system (Figure 9.20) provides an illustration of the development of the microstructure of **gray cast iron** (Figure 9.41). This sketch can be compared with the micrograph in Figure 11.1b.

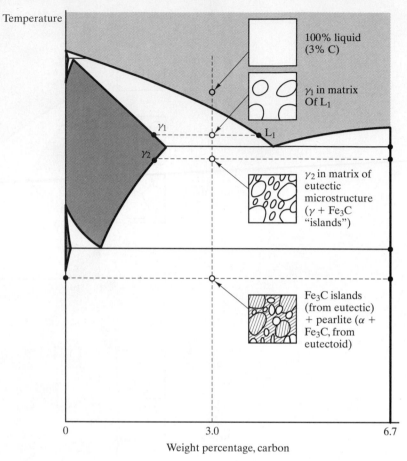

FIGURE 9.37 *Microstructural development for white cast iron (of composition 3.0 wt % C) shown with the aid of the Fe–Fe₃C phase diagram. The resulting (low-temperature) sketch can be compared with a micrograph in Figure 11.1a.*

EXAMPLE 9.6

Figure 9.34 shows the microstructural development for an 80 wt % B alloy. Consider instead 1 kg of a 70 wt % B alloy.

(a) Calculate the amount of β phase at T_3.

(b) Calculate what weight fraction of this β phase at T_3 is proeutectic.

SOLUTION

(a) Using Equation 9.10 gives us

$$m_{\beta, T_3} = \frac{x - x_\alpha}{x_\beta - x_\alpha}(1 \text{ kg}) = \frac{70 - 30}{90 - 30}(1 \text{ kg})$$

$$= 0.667 \text{ kg} = 667 \text{ g}.$$

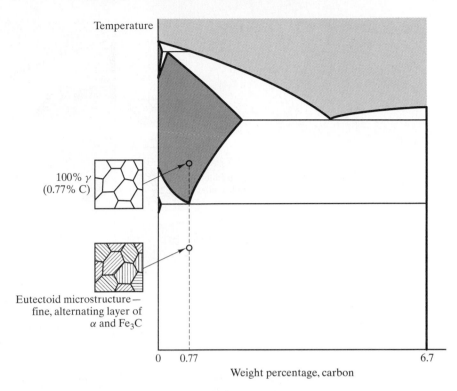

FIGURE 9.38 *Microstructural development for eutectoid steel (of composition 0.77 wt % C). The resulting (low-temperature) sketch can be compared with the micrograph in Figure 9.2.*

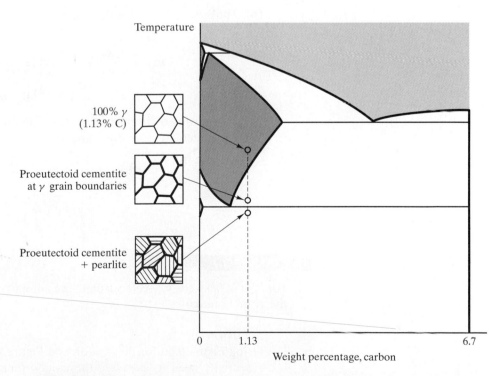

FIGURE 9.39 *Microstructural development for a slowly cooled hypereutectoid steel (of composition 1.13 wt % C).*

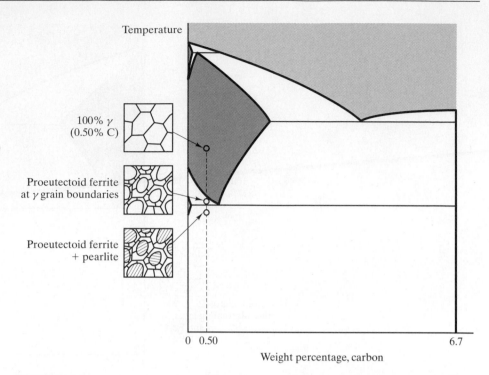

FIGURE 9.40 *Microstructural development for a slowly cooled hypoeutectoid steel (of composition 0.50 wt % C).*

(b) The proeutectic β was that which was present in the microstructure at T_2:

$$m_{\beta, T_2} = \frac{x - x_L}{x_\beta - x_L}(1 \text{ kg}) = \frac{70 - 60}{90 - 60}(1 \text{ kg})$$
$$= 0.333 \text{ kg} = 333 \text{ g}.$$

This portion of the microstructure is retained upon cooling through the eutectic temperature, giving

$$\text{fraction proeutectic} = \frac{\text{proeutectic } \beta}{\text{total } \beta}$$
$$= \frac{333 \text{ g}}{667 \text{ g}} = 0.50.$$

EXAMPLE 9.7

For 1 kg of 0.5 wt % C steel, calculate the amount of proeutectoid α at the grain boundaries.

SOLUTION
Using Figure 9.40 for illustration and Figure 9.19 for calculation, we essentially need to calculate the equilibrium amount of α at 728°C

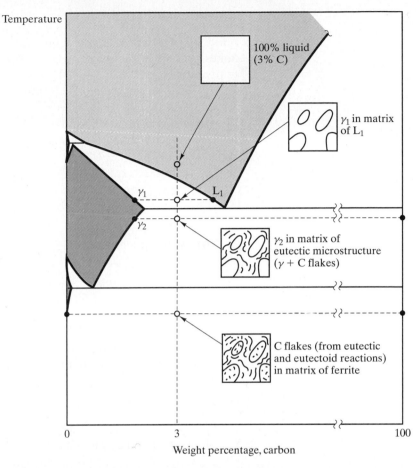

FIGURE 9.41 *Microstructural development for gray cast iron (of composition 3.0 wt % C) shown on the Fe–C phase diagram. The resulting low-temperature sketch can be compared with the micrograph in Figure 11.1b. A dramatic difference is that, in the actual microstructure, a substantial amount of metastable pearlite was formed at the eutectoid temperature. It is also interesting to compare this sketch with that for white cast iron in Figure 9.37. The small amount of silicon added to promote graphite precipitation is not shown in this two-component diagram.*

(i.e., 1 degree above the eutectoid temperature). Using Equation 9.9, we have

$$m_\alpha = \frac{x_\gamma - x}{x_\gamma - x_\alpha}(1\ \text{kg}) = \frac{0.77 - 0.50}{0.77 - 0.02}(1\ \text{kg})$$

$$= 0.360\ \text{kg} = 360\ \text{g}.$$

Note. You might have noticed that this calculation near the eutectoid composition used a value of x_α representative of the maximum solubility of carbon in α-Fe (0.02 wt %). At room temperature (see Example 9.4), this solubility goes to nearly zero.

For 1 kg of 3 wt % C gray iron, calculate the amount of graphite flakes present in the microstructure (a) at 1,153°C and (b) at room temperature.

(a) Using Figures 9.20 and 9.41, we note that 1,153°C is just below the eutectic temperature. Using Equation 9.10 gives us

$$m_C = \frac{x - x_\gamma}{x_C - x_\gamma}(1 \text{ kg}) = \frac{3.00 - 2.08}{100 - 2.08}(1 \text{ kg})$$

$$= 0.00940 \text{ kg} = 9.40 \text{ g}.$$

(b) At room temperature, we obtain

$$m_C = \frac{x - x_\alpha}{x_C - x_\alpha}(1 \text{ kg}) = \frac{3.00 - 0}{100 - 0}(1 \text{ kg})$$

$$= 0.030 \text{ kg} = 30.0 \text{ g}.$$

This calculation follows the ideal system of Figure 9.41 and ignores the possibility of any metastable pearlite being formed.

Consider 1 kg of an aluminum casting alloy with 10 wt % silicon.

(a) Upon cooling, at what temperature would the first solid appear?
(b) What is the first solid phase, and what is its composition?
(c) At what temperature will the alloy completely solidify?
(d) How much proeutectic phase will be found in the microstructure?
(e) How is the silicon distributed in the microstructure at 576°C?

We follow this microstructural development with the aid of Figure 9.13.

(a) For this composition, the liquidus is at ~595°C.
(b) It is solid solution α with a composition of ~1 wt % Si.
(c) At the eutectic temperature, 577°C.
(d) Practically all of the proeutectic α will have developed by 578°C. Using Equation 9.9, we obtain

$$m_\alpha = \frac{x_L - x}{x_L - x_\alpha}(1 \text{ kg}) = \frac{12.6 - 10}{12.6 - 1.6}(1 \text{ kg})$$

$$= 0.236 \text{ kg} = 236 \text{ g}.$$

(e) At $576°C$, the overall microstructure is $\alpha + \beta$. The amounts of each are

$$m_\alpha = \frac{x_\beta - x}{x_\beta - x_\alpha}(1 \text{ kg}) = \frac{100 - 10}{100 - 1.6}(1 \text{ kg})$$

$$= 0.915 \text{ kg} = 915 \text{ g}$$

and

$$m_\beta = \frac{x - x_\alpha}{x_\beta - x_\alpha}(1 \text{ kg}) = \frac{10 - 1.6}{100 - 1.6}(1 \text{ kg})$$

$$= 0.085 \text{ kg} = 85 \text{ g}.$$

However, we found in (d) that 236 g of the α is in the form of relatively large grains of proeutectic phase, giving

$$\alpha_{\text{eutectic}} = \alpha_{\text{total}} - \alpha_{\text{proeutectic}}$$

$$= 915 \text{ g} - 236 \text{ g} = 679 \text{ g}.$$

The silicon distribution is then given by multiplying its weight fraction in each microstructural region by the amount of that region:

$$\text{Si in proeutectic } \alpha = (0.016)(236 \text{ g}) = 3.8 \text{ g},$$

$$\text{Si in eutectic } \alpha = (0.016)(679 \text{ g}) = 10.9 \text{ g},$$

and

$$\text{Si in eutectic } \beta = (1.000)(85 \text{ g}) = 85.0 \text{ g}.$$

Finally, note that the total mass of silicon in the three regions sums to 99.7 g rather than 100.0 g ($= 10$ wt % of the total alloy) due to round-off errors.

EXAMPLE 9.10

The solubility of copper in aluminum drops to nearly zero at $100°C$. What is the maximum amount of θ phase that will precipitate out in a 4.5 wt % copper alloy quenched and aged at $100°C$? Express your answer in weight percent.

SOLUTION

As indicated in Figure 9.27, the solubility limit of the θ phase is essentially unchanging with temperature below $\sim400°C$ and is near a composition of 53 wt % copper. Using Equation 9.10 gives us

$$\text{wt \% } \theta = \frac{x - x_\kappa}{x_\theta - x_\kappa} \times 100\% = \frac{4.5 - 0}{53 - 0} \times 100\%$$

$$= 8.49\%.$$

EXAMPLE 9.11

In Example 9.1, we considered a 50:50 Pb–Sn solder.

(a) For a temperature of 200°C, determine (i) the phases present, (ii) their compositions, and (iii) their relative amounts (expressed in weight percent).

(b) Repeat part (a) for 100°C.

SOLUTION

(a) Using Figure 9.16, we find the following results at 200°C:

  i. The phases are α and liquid.
  ii. The composition of α is ~18 wt % Sn and of L is ~54 wt % Sn.
  iii. Using Equations 9.9 and 9.10, we have

$$\text{wt \% } \alpha = \frac{x_L - x}{x_L - x_\alpha} \times 100\% = \frac{54 - 50}{54 - 18} \times 100\%$$

$$= 11.1\%$$

and

$$\text{wt \% L} = \frac{x - x_\alpha}{x_L - x_\alpha} \times 100\% = \frac{50 - 18}{54 - 18} \times 100\%$$

$$= 88.9\%.$$

(b) Similarly, at 100°C, we obtain

  i. α and β.
  ii. α is ~5 wt % Sn and β is ~99 wt % Sn.
  iii. $\text{wt \% } \alpha = \dfrac{x_\beta - x}{x_\beta - x_\alpha} \times 100\% = \dfrac{99 - 50}{99 - 5} \times 100\% = 52.1\%$
  and

$$\text{wt \% } \beta = \frac{x - x_\alpha}{x_\beta - x_\alpha} \times 100\% = \frac{50 - 5}{99 - 5} \times 100\% = 47.9\%.$$

EXAMPLE 9.12

A fireclay refractory ceramic can be made by heating the raw material kaolinite, $Al_2(Si_2O_5)(OH)_4$, thus driving off the waters of hydration. Determine the phases present, their compositions, and their amounts for the resulting microstructure (below the eutectic temperature).

SOLUTION

A modest rearrangement of the kaolinite formula helps to clarify the production of this ceramic product:

$$Al_2(Si_2O_5)(OH)_4 = Al_2O_3 \cdot 2SiO_2 \cdot 2H_2O.$$

The firing operation yields

$$Al_2O_3 \cdot 2SiO_2 \cdot 2H_2O \xrightarrow{\text{heat}} Al_2O_3 \cdot 2SiO_2 + 2H_2O \uparrow.$$

The remaining solid, then, has an overall composition of

$$\text{mol \% } Al_2O_3 = \frac{\text{mol } Al_2O_3}{\text{mol } Al_2O_3 + \text{mol } SiO_2} \times 100\%$$

$$= \frac{1}{1+2} \times 100\% = 33.3\%.$$

Using Figure 9.23, we see that the overall composition falls in the SiO_2 + mullite two-phase region below the eutectic temperature. The SiO_2 composition is 0 mol % Al_2O_3 (i.e., 100% SiO_2). The composition of mullite is 60 mol % Al_2O_3.

Using Equations 9.9 and 9.10 yields

$$\text{mol \% } SiO_2 = \frac{x_{\text{mullite}} - x}{x_{\text{mullite}} - x_{SiO_2}} \times 100\% = \frac{60 - 33.3}{60 - 0} \times 100\%$$

$$= 44.5 \text{ mol \%}$$

and

$$\text{mol \% mullite} = \frac{x - x_{SiO_2}}{x_{\text{mullite}} - x_{SiO_2}} \times 100\% = \frac{33.3 - 0}{60 - 0} \times 100\%$$

$$= 55.5 \text{ mol \%}.$$

Note. Because the Al_2O_3–SiO_2 phase diagram is presented in mole percent, we have made our calculations in a consistent system. It would be a minor task to convert results to weight percent using data from Appendix 1.

PRACTICE PROBLEM 9.6

In Example 9.6, we calculate microstructural information about the β phase for the 70 wt % B alloy in Figure 9.34. In a similar way, calculate **(a)** the amount of α phase at T_3 for 1 kg of a 50 wt % B alloy and **(b)** the weight fraction of this α phase at T_3, which is proeutectic. (See also Figure 9.35.)

Calculate the amount of proeutectoid cementite at the grain boundaries in 1 kg of the 1.13 wt % C hypereutectoid steel illustrated in Figure 9.39. (See Example 9.7.)

In Example 9.8, the amount of carbon in 1 kg of a 3 wt % C gray iron is calculated at two temperatures. Plot the amount as a function of temperature over the entire temperature range of 1,135°C to room temperature.

In Example 9.9, we monitor the microstructural development for 1 kg of a 10 wt % Si–90 wt % Al alloy. Repeat this problem for a 20 wt % Si–80 wt % Al alloy.

In Example 9.10, we calculate the weight percent of θ phase at room temperature in a 95.5 Al–4.5 Cu alloy. Plot the weight percent of θ (as a function of temperature) that would occur upon slow cooling over a temperature range of 548°C to room temperature.

Calculate microstructures for **(a)** a 40:60 Pb–Sn solder and **(b)** a 60:40 Pb–Sn solder at 200°C and 100°C. (See Example 9.11.)

In the note at the end of Example 9.12, the point is made that the results can be easily converted to weight percent. Make these conversions.

Summary

The development of microstructure during slow cooling of materials from the liquid state can be analyzed by using phase diagrams. These "maps" identify the amounts and compositions of phases that are stable at given temperatures. Phase diagrams can be thought of as visual displays of the Gibbs phase rule. In this chapter, we restricted our discussion to binary diagrams, which represent phases present at various temperatures and compositions (with pressure fixed at 1 atm) in

systems with two components; the components can be elements or compounds.

Several types of binary diagrams are commonly encountered. For very similar components, complete solid solution can occur in the solid state as well as in the liquid state. In the two-phase (liquid solution + solid solution) region, the composition of each phase is indicated by a tie line. Many binary systems exhibit a eutectic reaction in which a low melting point (eutectic) composition produces a fine-grained, two-phase microstructure. Such eutectic diagrams are associated with limited solid solution. The completely solid-state analogy to the eutectic reaction is the eutectoid reaction, in which a single solid phase transforms upon cooling to a fine-grained microstructure of two other solid phases. The peritectic reaction represents the incongruent melting of a solid compound. Upon melting, the compound transforms to a liquid and another solid, each of composition different from the original compound. Many binary diagrams include various intermediate compounds, leading to a relatively complex appearance. However, such general binary diagrams can always be reduced to a simple binary diagram associated with the overall composition of interest.

The tie line that identifies the compositions of the phases in a two-phase region can also be used to calculate the amount of each phase. This calculation is done using the lever rule, in which the tie line is treated as a lever with its fulcrum located at the overall composition. The amounts of the two phases are such that they "balance the lever." The lever rule is, of course, a mechanical analog, but it follows directly from a mass balance for the two-phase system. The lever rule can be used to follow microstructural development as an overall composition is slowly cooled from the melt, which is especially helpful in understanding the microstructure that results in a composition near a eutectic composition. Several binary diagrams of importance to the metals and ceramics industries were given in this chapter. Special emphasis was given to the Fe–Fe_3C system, which provides the major scientific basis for the iron and steel industries.

Key Terms

austenite (270)
binary diagram (261)
complete solid solution (262)
component (258)
congruent melting (272)
degrees of freedom (259)
eutectic composition (266)
eutectic diagram (266)
eutectic reaction (270)
eutectic temperature (266)
eutectoid diagram (270)
general diagram (275)
Gibbs phase rule (259)
gray cast iron (289)

hypereutectic composition (287)
hypereutectoid composition (288)
hypoeutectic composition (287)
hypoeutectoid composition (289)
incongruent melting (272)
intermediate compound (275)
invariant point (263)
lever rule (282)
liquidus (262)
mass balance (281)
metastable (271)
microstructural development (285)
peritectic diagram (272)
peritectic reaction (272)

phase (258)
phase diagram (261)
phase field (262)
proeutectic (287)
proeutectoid (288)
solidus (262)
state (259)
state point (262)
state variables (259)
tie line (263)
white cast iron (288)
zone refining (284)

References

ASM Handbook, Vol. 3: *Alloy Phase Diagrams*, ASM International, Materials Park, OH, 1992.

Binary Alloy Phase Diagrams, 2nd ed., Vols. 1–3, T. B. Massalski, et al., Eds., ASM International, Materials Park, OH, 1990. The result of a cooperative program between ASM International and the National Institute of Standards and Technology for the critical review of 4,700 phase-diagram systems.

Phase Equilibria Diagrams, Vols. 1–14, American Ceramic Society, Westerville, OH, 1964–2005.

Problems

9.1 • The Phase Rule

9.1. Apply the Gibbs phase rule to the various points in the one-component H_2O phase diagram (Figure 9.3).

9.2. Apply the Gibbs phase rule to the various points in the one-component iron phase diagram (Figure 9.4).

9.3. Calculate the degrees of freedom for a 50:50 copper–nickel alloy at **(a)** $1,400°C$, where it exists as a single, liquid phase; **(b)** $1,300°C$, where it exists as a two-phase mixture of liquid and solid solutions; and **(c)** $1,200°C$, where it exists as a single, solid-solution phase. Assume a constant pressure of 1 atm above the alloy in each case.

9.4. In Figure 9.7, the Gibbs phase rule was applied to a hypothetical phase diagram. In a similar way, apply the phase rule to a sketch of the Pb–Sn phase diagram (Figure 9.16).

9.5. Apply the Gibbs phase rule to a sketch of the MgO–Al_2O_3 phase diagram (Figure 9.26).

9.6. Apply the Gibbs phase rule to the various points in the Al_2O_3–SiO_2 phase diagram (Figure 9.23).

9.2 • The Phase Diagram

9.7. Describe qualitatively the microstructural development that will occur upon slow cooling of a melt of equal parts (by weight) of copper and nickel (see Figure 9.9).

9.8. Describe qualitatively the microstructural development that will occur upon slow cooling of a melt composed of 50 wt % Al and 50 wt % Si (see Figure 9.13).

9.9. Describe qualitatively the microstructural development that will occur upon slow cooling of a melt composed of 87.4 wt % Al and 12.6 wt % Si (see Figure 9.13).

9.10. Describe qualitatively the microstructural development during the slow cooling of a melt composed of **(a)** 10 wt % Pb–90 wt % Sn, **(b)** 40 wt % Pb–60 wt % Sn, and **(c)** 50 wt % Pb–50 wt % Sn (see Figure 9.16).

9.11. Repeat Problem 9.10 for a melt composed of 38.1 wt % Pb–61.9 wt % Sn.

9.12. Describe qualitatively the microstructural development that will occur upon slow cooling of an alloy with equal parts (by weight) of aluminum and θ phase (Al_2Cu) (see Figure 9.27).

9.13. Describe qualitatively the microstructural development that will occur upon slow cooling of a melt composed of 20 wt % Cu, 80 wt % Al (see Figure 9.27).

9.14. Describe qualitatively the microstructural development during the slow cooling of a 30:70 brass (Cu with 30 wt % Zn). See Figure 9.28 for the Cu–Zn phase diagram.

9.15. Repeat Problem 9.14 for a 35:65 brass.

9.16. Describe qualitatively the microstructural development during the slow cooling of **(a)** a 50 mol % Al_2O_3–50 mol % SiO_2 ceramic and **(b)** a 70 mol % Al_2O_3–30 mol % SiO_2 ceramic (see Figure 9.23).

9.3 • The Lever Rule

9.17. Calculate the amount of each phase present in 1 kg of a 50 wt % Ni–50 wt % Cu alloy at **(a)** $1,400°C$, **(b)** $1,300°C$, and **(c)** $1,200°C$ (see Figure 9.9).

9.18. Calculate the amount of each phase present in 1 kg of a 50 wt % Pb–50 wt % Sn solder alloy at **(a)** $300°C$, **(b)** $200°C$, **(c)** $100°C$, and **(d)** $0°C$ (see Figure 9.16).

9.19. Repeat Problem 9.18 for a 60 wt % Pb–40 wt % Sn solder alloy.

9.20. Repeat Problem 9.18 for an 80 wt % Pb–20 wt % Sn solder alloy.

9.21. Calculate the amount of each phase present in 50 kg of a brass with composition 35 wt % Zn–65 wt % Cu at **(a)** $1,000°C$, **(b)** $900°C$, **(c)** $800°C$, **(d)** $700°C$, **(e)** $100°C$, and **(f)** $0°C$ (see Figure 9.28).

9.22. Some aluminum from a "metallization" layer on a solid-state electronic device has diffused into the silicon substrate. Near the surface, the silicon has an overall concentration of 1.0 wt % Al. In this region, what percentage of the microstructure would be composed of α-phase precipitates, assuming equilibrium? (See Figure 9.13 and assume the phase boundaries at $300°C$ will be essentially unchanged to room temperature.)

9.23. Calculate the amount of proeutectoid α present at the grain boundaries in 1 kg of a common 1020 structural steel (0.20 wt % C). (See Figure 9.19.)

9.24. Repeat Problem 9.23 for a 1040 structural steel (0.40 wt % C).

***9.25.** Suppose that you have a crucible containing 1 kg of an alloy of composition 90 wt % Sn–10 wt % Pb at a temperature of 184°C. How much Sn would you have to add to the crucible to completely solidify the alloy without changing the system temperature? (See Figure 9.16.)

9.26. Determine the phases present, their compositions, and their amounts (below the eutectic temperature) for a refractory made from equal molar fractions of kaolinite and mullite ($3Al_2O_3 \cdot 2SiO_2$). (See Figure 9.23.)

9.27. Repeat Problem 9.26 for a refractory made from equal molar fractions of kaolinite and silica (SiO_2).

***9.28.** You have supplies of kaolinite, silica, and mullite as raw materials. Using kaolinite plus either silica or mullite, calculate the batch composition (in weight percent) necessary to produce a final microstructure that is equimolar in silica and mullite. (See Figure 9.23.)

9.29. Calculate the phases present, their compositions, and their amounts (in weight percent) for the microstructure at 1,000°C for **(a)** a spinel ($MgO \cdot Al_2O_3$) refractory with 1 wt % excess MgO (i.e., 1 g MgO per 99 g $MgO \cdot Al_2O_3$) and **(b)** a spinel refractory with 1 wt % excess Al_2O_3. (See Figure 9.26.)

9.30. A partially stabilized zirconia (for a novel structural application) is desired to have an equimolar microstructure of tetragonal and cubic zirconia at an operating temperature of 1,500°C. Calculate the proper CaO content (in weight percent) for this structural ceramic. (See Figure 9.29.)

9.31. Repeat Problem 9.30 for a microstructure with equal weight fractions of tetragonal and cubic zirconia.

9.32. Calculate the amount of each phase present in a 1-kg alumina refractory with composition 70 mol % Al_2O_3– 30 mol % SiO_2 at **(a)** 2,000°C, **(b)** 1,900°C, and **(c)** 1,800°C. (See Figure 9.23.)

9.33. In a test laboratory, quantitative x-ray diffraction determines that a refractory brick has 25 wt % alumina phase and 75 wt % mullite solid solution. What is the overall SiO_2 content (in wt %) of this material? (See Figure 9.23.)

9.34. An important structural ceramic is PSZ, which has a composition that lies in the two-phase ZrO_2-cubic ZrO_2 (ss) region. Use Figure 9.29 to calculate the amount of each phase present in a 10 mol % CaO PSZ at 500°C.

***9.35.** In a materials laboratory experiment, a student sketches a microstructure observed under an optical microscope. The sketch appears as

The phase diagram for this alloy system is

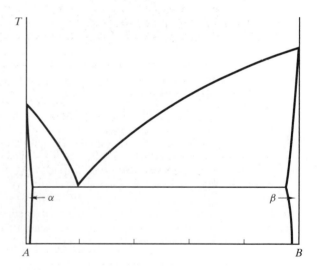

Determine **(a)** whether the black regions in the sketch represent α or β phase and **(b)** the approximate alloy composition.

9.4 • Microstructural Development During Slow Cooling

9.36. Calculate **(a)** the weight fraction of the α phase that is proeutectic in a 10 wt % Si–90 wt % Al alloy at 576°C and **(b)** the weight fraction of the β phase that is proeutectic in a 20 wt % Si–80 wt % Al alloy at 576°C.

9.37. Plot the weight percent of phases present as a function of temperature for a 10 wt % Si–90 wt % Al alloy slowly cooled from 700 to 300°C.

9.38. Plot the weight percent of phases present as a function of temperature for a 20 wt % Si–80 wt % Al alloy slowly cooled from 800 to 300°C.

9.39. Calculate the weight fraction of mullite that is proeutectic in a slowly cooled 20 mol % Al_2O_3–80 mol % SiO_2 refractory cooled to room temperature.

9.40. Microstructural analysis of a slowly cooled Al–Si alloy indicates there is a 5 volume % silicon-rich proeutectic phase. Calculate the overall alloy composition (in weight percent).

9.41. Repeat Problem 9.40 for a 10 volume % silicon-rich proeutectic phase.

9.42. Calculate the amount of proeutectic γ that has formed at 1,149°C in the slow cooling of the 3.0 wt % C white cast iron illustrated in Figure 9.37. Assume a total of 100 kg of cast iron.

9.43. Plot the weight percent of phases present as a function of temperature for the 3.0 wt % C white cast iron illustrated in Figure 9.37 slowly cooled from 1,400 to 0°C.

9.44. Plot the weight percent of phases present as a function of temperature from 1,000 to 0°C for the 0.77 wt % C eutectoid steel illustrated in Figure 9.38.

9.45. Plot the weight percent of phases present as a function of temperature from 1,000 to 0°C for the 1.13 wt % C hypereutectoid steel illustrated in Figure 9.39.

9.46. Plot the weight percent of phases present as a function of temperature from 1,000 to 0°C for a common 1020 structural steel (0.20 wt % C).

9.47. Repeat Problem 9.46 for a 1040 structural steel (0.40 wt % C).

9.48. Plot the weight percent of phases present as a function of temperature from 1,000 to 0°C for the 0.50 wt % C hypoeutectoid steel illustrated in Figure 9.40.

9.49. Plot the weight percent of phases present as a function of temperature from 1,400 to 0°C for a white cast iron with an overall composition of 2.5 wt % C.

9.50. Plot the weight percent of all phases present as a function of temperature from 1,400 to 0°C for a gray cast iron with an overall composition of 3.0 wt % C.

9.51. Repeat Problem 9.50 for a gray cast iron with an overall composition of 2.5 wt % C.

9.52. In comparing the equilibrium schematic microstructure in Figure 9.41 with the actual, room-temperature microstructure shown in Figure 11.1b, it is apparent that metastable pearlite can form at the eutectoid temperature (due to insufficient time for the more stable, but slower, graphite formation). Assuming that Figures 9.20 and 9.41 are accurate for 100 kg of a gray cast iron (3.0 wt % C) down to 738°C, but that pearlite forms upon cooling through the eutectoid temperature, calculate the amount of pearlite to be expected in the room-temperature microstructure.

9.53. For the assumptions in Problem 9.52, calculate the amount of flake graphite in the room-temperature microstructure.

9.54. Plot the weight percent of phases present as a function of temperature from 800 to 300°C for a 95 Al–5 Cu alloy.

9.55. Consider 1 kg of a brass with composition 35 wt % Zn–65 wt % Cu. **(a)** Upon cooling, at what temperature would the first solid appear? **(b)** What is the first solid phase to appear, and what is its composition? **(c)** At what temperature will the alloy completely solidify? **(d)** Over what temperature range will the microstructure be completely in the α-phase?

9.56. Repeat Problem 9.55 for 1 kg of a brass with composition 30 wt % Zn–70 wt % Cu.

9.57. Plot the weight percent of phases present as a function of temperature from 1,000 to 0°C for a 35 wt % Zn–65 wt % Cu brass.

9.58. Repeat Problem 9.57 for a 30 wt % Zn–70 wt % Cu brass.

9.59. Repeat Problem 9.57 for 1 kg of brass with a composition of 15 wt % Zn–85 wt % Cu.

9.60. For a 15 wt % Zn–85 wt % Cu brass, plot the weight percent of phases present as a function of temperature from 1,100°C to 0°C.

9.61. A solder batch is made by melting together 64 g of a 40:60 Pb–Sn alloy with 53 g of a 60:40 Pb–Sn alloy. Calculate the amounts of α and β phase that would be present in the overall alloy, assuming it is slowly cooled to room temperature, 25°C.

9.62. Plot the weight percent of phases present as a function of temperature from 400 to 0°C for a slowly cooled 50:50 Pb–Sn solder.

9.63. Plot the phases present (in mole percent) as a function of temperature for the heating of a refractory with the

composition 60 mol % Al$_2$O$_3$–40 mol % MgO from 1,000 to 2,500°C.

9.64. Plot the phases present (in mole percent) as a function of temperature for the heating of a partially stabilized zirconia with 10 mol % CaO from room temperature to 2,800°C.

CHAPTER 10
Kinetics—Heat Treatment

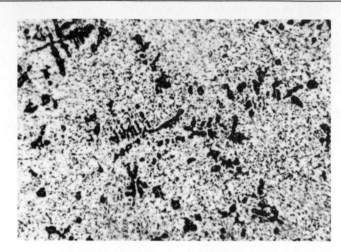

The microstructure of a rapidly cooled eutectic soft solder (≈38 wt % Pb −62 wt % Sn) consists of globules of lead-rich solid solution (dark) in a matrix of tin-rich solid solution (white), 375X. The contrast to the slowly cooled microstructure at the opening of Chapter 9 illustrates the effect of time on microstructural development. (From ASM Handbook, Vol. 3: Alloy Phase Diagrams, *ASM International, Materials Park, OH, 1992.)*

Chapter 9 introduced the powerful tool of phase diagrams for describing equilibrium microstructural development during slow cooling from the melt. Throughout that chapter, however, we were cautioned that phase diagrams represent microstructure that "should" develop, assuming that temperature is changed slowly enough to maintain equilibrium at all times. In practice, materials processing, like so much of daily life, is rushed, and time becomes an important factor. The practical aspect of this concept is **heat treatment**, the temperature versus time history necessary to generate a desired microstructure. The fundamental basis for heat treatment is **kinetics**, which we shall define as the science of time-dependent phase transformations.

We begin by adding a time scale to phase diagrams to show the approach to equilibrium. A systematic treatment of this kind generates a *TTT diagram*, which summarizes, for a given composition, the percentage completion of a given phase transformation on temperature–time axes (giving the three "T's" of temperature, time, and transformation). Such diagrams are maps in the same sense that phase diagrams are maps. TTT diagrams can include descriptions of transformations that involve time-dependent solid-state diffusion and of transformations that occur by a rapid, shearing mechanism, essentially independent of time. As with phase diagrams, some of our best illustrations of TTT diagrams will involve ferrous alloys. We shall explore some of the basic considerations in the heat treatment of steel. Related to this concept is the characterization of *hardenability*. *Precipitation hardening* is an important heat treatment illustrated by some nonferrous alloys. *Annealing* is a heat treatment that leads to reduced hardness by means of successive stages of *recovery*, *recrystallization*, and *grain growth*. Heat treatment is not a topic isolated to metallurgy. To illustrate this fact, we conclude this chapter with a discussion of some important phase transformations in nonmetallic systems.

10.1 | Time—The Third Dimension

Time did not appear in any quantitative way in the discussion of phase diagrams in Chapter 9. Aside from requiring temperature changes to occur relatively slowly, we did not consider time as a factor at all. Phase diagrams summarized equilibrium

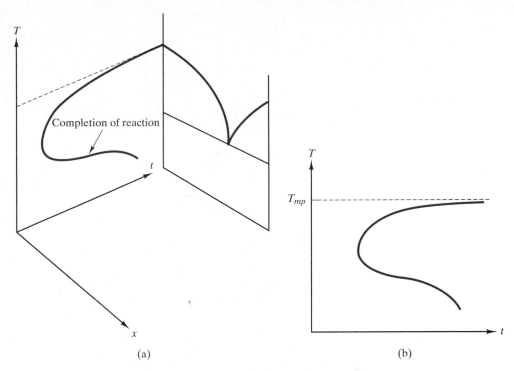

(a)   (b)

FIGURE 10.1 *Schematic illustration of the approach to equilibrium. (a) The time for solidification to go to completion is a strong function of temperature, with the minimum time occurring for a temperature considerably below the melting point. (b) The temperature–time plane with a transformation curve. We shall see later that the time axis is often plotted on a logarithmic scale.*

states and, as such, those states (and associated microstructures) should be stable and unchanging with time. However, these equilibrium structures take time to develop, and the approach to equilibrium can be mapped on a time scale.* A simple illustration of this concept is given in Figure 10.1, which shows a time axis perpendicular to the temperature–composition plane of a phase diagram. For component A, the phase diagram indicates that solid A should exist at any temperature below the melting point. However, Figure 10.1 indicates that the time required for the liquid phase to transform to the solid phase is a strong function of temperature.

Another way of stating this idea is that the time necessary for the solidification reaction to go to completion varies with temperature. In order to compare the reaction times in a consistent way, Figure 10.1 represents the rather ideal case of quenching the liquid from the melting point instantaneously to some lower temperature and then measuring the time for solidification to go to completion at that temperature. At first glance, the nature of the plot in Figure 10.1 may seem surprising. The reaction proceeds slowly near the melting point and at relatively low temperatures. The reaction is fastest at some intermediate temperature. To understand this "knee-shaped" transformation curve, we must explore some fundamental concepts of kinetics theory.

*The relationship between thermodynamics and kinetics is explored in the thermodynamics chapter on the related Web site.

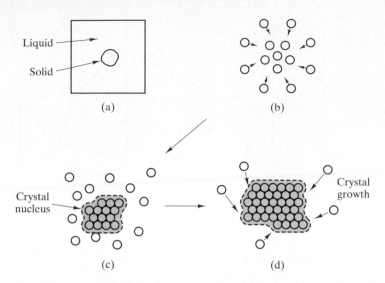

FIGURE 10.2 *(a) On a microscopic scale, a solid precipitate in a liquid matrix. The precipitation process is seen on the atomic scale as (b) a clustering of adjacent atoms to form (c) a crystalline nucleus followed by (d) the growth of the crystalline phase.*

For this discussion, we focus more closely on the precipitation of a single-phase solid within a liquid matrix (Figure 10.2). This process is an example of **homogeneous nucleation**, meaning that the precipitation occurs within a completely homogeneous medium. The more common case is **heterogeneous nucleation**, in which the precipitation occurs at some structural imperfection such as a foreign surface. The imperfection reduces the surface energy associated with forming the new phase.

Even homogeneous nucleation is rather involved. The precipitation process actually occurs in two stages. The first stage is **nucleation**. The new phase, which forms because it is more stable, first appears as small nuclei. These nuclei result from local atomic fluctuations and are, typically, only a few hundred atoms in size. This initial stage involves the random production of many nuclei. Only those larger than a given size are stable and can continue to grow. These critical-size nuclei must be large enough to offset the energy of formation for the solid–liquid interface. The rate of nucleation (i.e., the rate at which nuclei of critical size or larger appear) is the result of two competing factors. At the precise transformation temperature (in this case, the melting point), the solid and liquid phases are in equilibrium, and there is no net driving force for the transformation to occur. As the liquid is cooled below the transformation temperature, it becomes increasingly unstable. The classical theory of nucleation is based on an energy balance between the nucleus and its surrounding liquid. The key principle is that a small cluster of atoms (the nucleus) will be stable only if further growth reduces the net energy of the system. Taking the nucleus in Figure 10.2a as spherical, the energy balance can be illustrated as shown in Figure 10.3, demonstrating that the nucleus will be stable if its radius, r, is greater than a critical value, r_c.

The driving force for solidification increases with decreasing temperature, and the rate of nucleation increases sharply. This increase cannot continue indefinitely. The clustering of atoms to form a nucleus is a local-scale diffusion process. As such, this step will decrease in rate with decreasing temperature. This rate

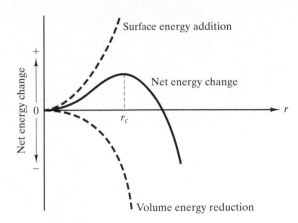

FIGURE 10.3 *Classical nucleation theory involves an energy balance between the nucleus and its surrounding liquid. A nucleus (cluster of atoms) as shown in Figure 10.2(c) will be stable only if further growth reduces the net energy of the system. An ideally spherical nucleus will be stable if its radius, r, is greater than a critical value, r_c.*

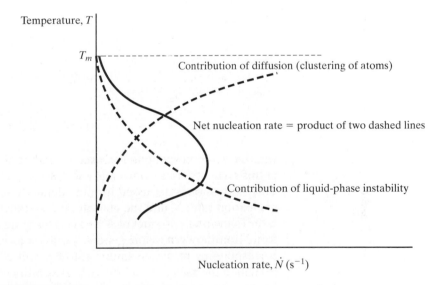

FIGURE 10.4 *The rate of nucleation is a product of two curves that represent two opposing factors (instability and diffusivity).*

decrease is exponential in nature and is another example of Arrhenius behavior (see Section 5.1). The overall nucleation rate reflects these two factors by increasing from zero at the transformation temperature (T_m) to a maximum value somewhere below T_m and then decreasing with further decreases in temperature (Figure 10.4). In a preliminary way, we now have an explanation for the shape of the curve in Figure 10.1. The time for reaction is long just below the transformation temperature because the driving force for reaction is small and the reaction rate is therefore small. The time for reaction is again long at low temperatures because the diffusion rate is small. In general, the time axis in Figure 10.1 is the inverse of the rate axis in Figure 10.4.

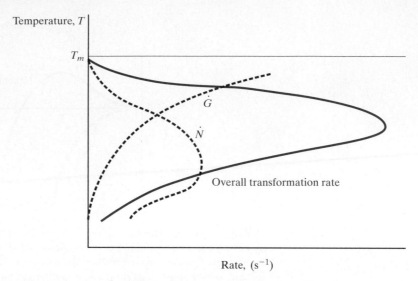

FIGURE 10.5 *The overall transformation rate is the product of the nucleation rate, $\dot{N}$, (from Figure 10.4) and the growth rate, $\dot{G}$ (given in Equation 10.1).*

Our explanation of Figure 10.1 using Figure 10.4 is preliminary because we have not yet included the growth step (see Figure 10.2). This process, like the initial clustering of atoms in nucleation, is diffusional in nature. As such, the growth rate, $\dot{G}$, is an Arrhenius expression,

$$\dot{G} = Ce^{-Q/RT}, \tag{10.1}$$

where C is a preexponential constant, Q is the activation energy for self-diffusion in this system, R is the universal gas constant, and T is the absolute temperature. This expression is discussed in some detail in Section 5.1. Figure 10.5 shows the nucleation rate, $\dot{N}$, and the growth rate, $\dot{G}$, together. The overall transformation rate is shown as a product of $\dot{N}$ and $\dot{G}$. This more complete picture of phase transformation shows the same general behavior as nucleation rate. The temperature corresponding to the maximum rate has shifted, but the general argument has remained the same. The maximum rate occurs in a temperature range where the driving forces for solidification and diffusion rates are both significant. Although this principle explains these knee-shaped curves in a qualitative way, we must acknowledge that transformation curves for many practical engineering materials frequently include additional factors, such as multiple diffusion mechanisms and mechanical strains associated with solid-state transformations.

EXAMPLE 10.1

At 900°C, growth rate, $\dot{G}$, is a dominant term in the crystallization of a copper alloy. By dropping the system temperature to 400°C, the growth rate drops six orders of magnitude and effectively reduces the crystallization rate to zero. Calculate the activation energy for self-diffusion in this alloy system.

SOLUTION

The following is a direct application of Equation 10.1:

$$\dot{G} = Ce^{-Q/RT}.$$

Considering two different temperatures yields

$$\frac{\dot{G}_{900^\circ C}}{\dot{G}_{400^\circ C}} = \frac{Ce^{-Q/R(900+273)K}}{Ce^{-Q/R(400+273)K}}$$

$$= e^{-Q/R(1/1,173-1/673)K^{-1}},$$

which gives

$$Q = -\frac{R \ln (\dot{G}_{900^\circ C}/\dot{G}_{400^\circ C})}{(1/1,173 - 1/673)\ K^{-1}}$$

$$= -\frac{[8.314\ J/(mol \cdot K)]\ \ln 10^6}{(1/1,173 - 1/673)\ K^{-1}} = 181\ kJ/mol.$$

Note. Because the crystallization rate is so high at elevated temperatures, it is not possible to suppress crystallization completely unless the cooling is accomplished at exceptionally high quench rates. The results in those special cases are the interesting amorphous metals (see Section 4.5).

PRACTICE PROBLEM 10.1

In Example 10.1, the activation energy for crystal growth in a copper alloy is calculated. Using that result, calculate the temperature at which the growth rate would have dropped three orders of magnitude relative to the rate at 900°C. **(Note that the solutions to all practice problems are provided on the related Web site.)**

10.2 | The TTT Diagram

The preceding section introduced time as an axis in monitoring microstructural development. The general term for a plot of the type shown in Figure 10.1 is a **TTT diagram**, where the letters stand for temperature, time, and (percent) transformation. This plot is also known as an **isothermal transformation diagram**. In the case of Figure 10.1, the time necessary for 100% completion of transformation was plotted. Figure 10.6 shows how the progress of the transformation can be traced with a family of curves showing different percentages of completion. Using the industrially important eutectoid transformation in steels as an example, we can now discuss in further detail the nature of **diffusional transformations** in solids (a change in structure due to the long-range migration of atoms). In addition, we shall find that some **diffusionless transformations** play an important role in microstructural development and can be superimposed on the TTT diagrams.

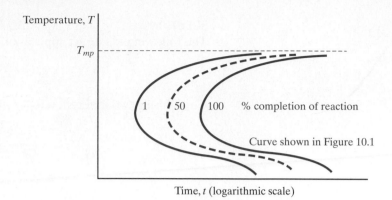

FIGURE 10.6 *A time–temperature–transformation diagram for the solidification reaction of Figure 10.1 with various percent completion curves illustrated.*

DIFFUSIONAL TRANSFORMATIONS

Diffusional transformations involve a change of structure due to the long-range migration of atoms. The development of microstructure during the slow cooling of eutectoid steel (Fe with 0.77 wt % C) was shown in Figure 9.38. A TTT diagram for this composition is shown in Figure 10.7. It is quite similar to the schematic for solidification shown in Figure 10.1. The most important new information provided in Figure 10.7 is that **pearlite** is not the only microstructure that can develop from the cooling of **austenite**. In fact, various types of pearlite are noted at various transformation temperatures. The slow cooling path assumed in Chapter 9 is illustrated in Figure 10.8 and clearly leads to the development of a coarse pearlite. Here, all references to size are relative. In Chapter 9, we made an issue of the fact that eutectic and eutectoid structures are generally fine grained. Figure 10.7 indicates that the pearlite produced near the eutectoid temperature is not as fine grained as that produced at slightly lower temperatures. The reason for this trend can be appreciated by studying Figure 10.5. Low nucleation rates and high diffusion rates near the eutectoid temperature lead to a relatively coarse structure. The increasingly fine pearlite formed at lower temperatures is eventually beyond the resolution of optical microscopes (approximately 0.25 μm features observable at about 2,000× magnification). Such fine structure can be observed with electron microscopy because electrons have effective wavelengths much smaller than those in the visible light range.

Pearlite formation is found from the eutectoid temperature (727°C) down to about 400°C. Below 400°C, the pearlite microstructure is no longer formed. The ferrite and carbide form as extremely fine needles in a microstructure known as **bainite**[*] (Figure 10.9), which represents an even finer distribution of ferrite and carbide than in fine pearlite. Although a different morphology is found in bainite, the general trend of finer structure with decreasing temperature is continued. It is important to note that the variety of morphologies that develops over the range of temperatures shown in Figure 10.7 all represent the same phase

---

[*]Edgar Collins Bain (1891–1971), American metallurgist, discovered the microstructure that now bears his name. His many achievements in the study of steels made him one of the most honored metallurgists of his generation.

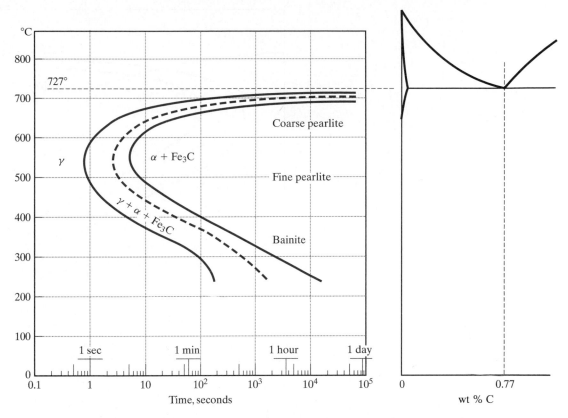

FIGURE 10.7 *TTT diagram for eutectoid steel shown in relation to the Fe–Fe₃ C phase diagram (see Figure 9.38). This diagram shows that, for certain transformation temperatures, bainite rather than pearlite is formed. In general, the transformed microstructure is increasingly fine grained as the transformation temperature is decreased. Nucleation rate increases and diffusivity decreases as temperature decreases. The solid curve on the left represents the onset of transformation (~1% completion). The dashed curve represents 50% completion. The solid curve on the right represents the effective (~99%) completion of transformation. This convention is used in subsequent TTT diagrams. (TTT diagram after* Atlas of Isothermal Transformation and Cooling Transformation Diagrams, *American Society for Metals, Metals Park, OH, 1977.)*

compositions and relative amounts of each phase. These terms all derive from the equilibrium calculations (using the tie line and lever rule) of Chapter 9. It is equally important to note that TTT diagrams represent specific thermal histories and are not state diagrams in the way that phase diagrams are. For instance, coarse pearlite is more stable than fine pearlite or bainite because it has less total interfacial boundary area (a high-energy region as discussed in Section 4.4). As a result, coarse pearlite, once formed, remains upon cooling, as illustrated in Figure 10.10.

DIFFUSIONLESS (MARTENSITIC) TRANSFORMATIONS

The eutectoid reactions in Figure 10.7 are all diffusional in nature. But close inspection of that TTT diagram indicates that no information is given below about 250°C. Figure 10.11 shows that a very different process occurs at lower

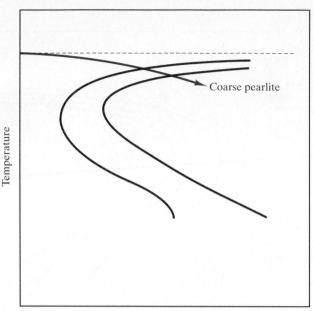

Time (logarithmic scale)

FIGURE 10.8 *A slow cooling path that leads to coarse pearlite formation is superimposed on the TTT diagram for eutectoid steel. This type of thermal history was assumed, in general, throughout Chapter 9.*

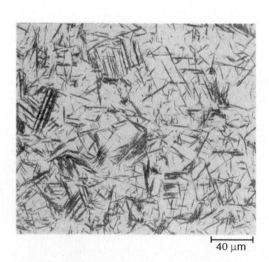

40 μm

FIGURE 10.9 *The microstructure of bainite involves extremely fine needles of ferrite and carbide, in contrast to the lamellar structure of pearlite (see Figure 9.2), 250×. (From* ASM Handbook, *Vol. 9:* Metallography and Microstructures, *ASM International, Materials Park, OH, 2004.)*

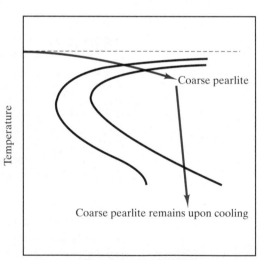

Time (logarithmic scale)

FIGURE 10.10 *The interpretation of TTT diagrams requires consideration of the thermal history "path." For example, coarse pearlite, once formed, remains stable upon cooling. The finer-grain structures are less stable because of the energy associated with the grain-boundary area. (By contrast, phase diagrams represent equilibrium and identify stable phases independent of the path used to reach a given state point.)*

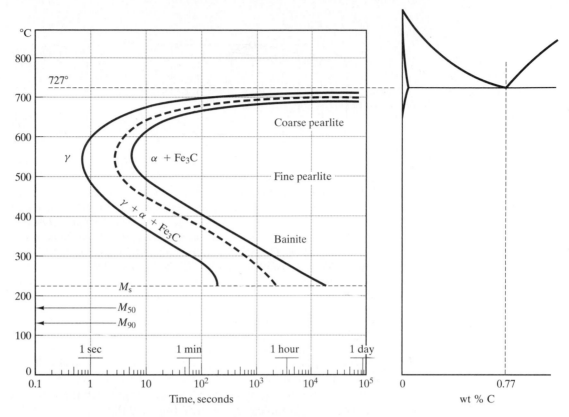

FIGURE 10.11 *A more complete TTT diagram for eutectoid steel than was given in Figure 10.7. The various stages of the time-independent (or diffusionless) martensitic transformation are shown as horizontal lines. M_s represents the start, M_{50} represents 50% transformation, and M_{90} represents 90% transformation. One hundred percent transformation to martensite is not complete until a final temperature (M_f) of $-46°C$.*

temperatures. Two horizontal lines are added to represent the occurrence of a *diffusionless* process known as the **martensitic* transformation**. This generic term refers to a broad family of diffusionless transformations in metals and non-metals alike. The most common example is the specific transformation in eutectoid steels. In this system, the product formed from the quenched austenite is termed **martensite**. In effect, the quenching of austenite rapidly enough to bypass the pearlite "knee" at approximately 550°C allows any diffusional transformation to be suppressed. However, there is a price to pay for avoiding the diffusional process. The austenite phase is still unstable and is, in fact, increasingly unstable with decreasing temperature. At approximately 215°C, the instability of austenite is so great that a small fraction (less than 1%) of the material transforms spontaneously to martensite. Instead of the diffusional migration of carbon atoms to

---

*Adolf Martens (1850–1914), German metallurgist, was originally trained as a mechanical engineer. Early in his career, he became involved in the developing field of testing materials for construction. He was a pioneer in using the microscope as a practical analytical tool for metals. Later, in an academic post, he produced the highly regarded *Handbuch der Materialienkunde* (1899).

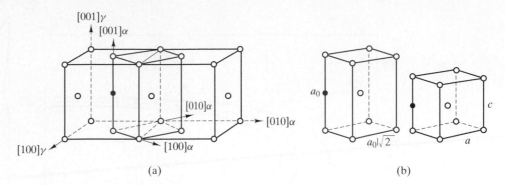

(a)   (b)

FIGURE 10.12 *For steels, the martensitic transformation involves the sudden reorientation of C and Fe atoms from the fcc solid solution of γ-Fe (austenite) to a body-centered tetragonal (bct) solid solution (martensite). In (a), the bct unit cell is shown relative to the fcc lattice by the ⟨100⟩ₐ axes. In (b), the bct unit cell is shown before (left) and after (right) the transformation. The open circles represent iron atoms. The solid circle represents an interstitially dissolved carbon atom. This illustration of the martensitic transformation was first presented by Bain in 1924, and while subsequent study has refined the details of the transformation mechanism, this diagram remains a useful and popular schematic. (After J. W. Christian, in* Principles of Heat Treatment of Steel, *G. Krauss, Ed., American Society for Metals, Metals Park, OH, 1980.)*

produce separate α and Fe_3C phases, the martensite transformation involves the sudden reorientation of C and Fe atoms from the fcc solid solution of γ-Fe to a body-centered tetragonal (bct) solid solution, which is martensite (Figure 10.12). The relatively complex crystal structure and the supersaturated concentration of carbon atoms in the martensite lead to a characteristically brittle nature. The start of the martensitic transformation is labeled M_s and is shown as a horizontal line (i.e., time independent) in Figure 10.11. If the quench of austenite proceeds below M_s, the austenite phase is increasingly unstable, and a larger fraction of the system is transformed to martensite. Various stages of the martensitic transformation are noted in Figure 10.11. Quenching to $-46°C$ or below leads to the complete transformation to martensite. The acicular, or needlelike, microstructure of martensite is shown in Figure 10.13. Martensite is a **metastable** phase; that is, it is stable with time, but upon reheating it will decompose into the even more stable phases of α and Fe_3C. The careful control of the proportions of these various phases is the subject of heat treatment, which is discussed in the next section.

As one might expect, the complex set of factors (discussed in Section 10.1) that determine transformation rates requires the TTT diagram to be defined in terms of a specific thermal history. The TTT diagrams in this chapter are generally *isothermal;* that is, the transformation time at a given temperature represents the time for transformation at the fixed temperature following an instantaneous quench. Figure 10.8 and several subsequent diagrams will superimpose cooling or heating paths on these diagrams. Such paths can affect the time at which the transformation will have occurred at a given temperature. In other words, the positions of transformation curves are shifted slightly downward and toward the right for nonisothermal conditions. Such a **continuous cooling transformation** (CCT) **diagram** is shown in Figure 10.14. For the purpose of illustration, we shall not generally make this refinement in this book. The principles demonstrated are, nonetheless, valid.

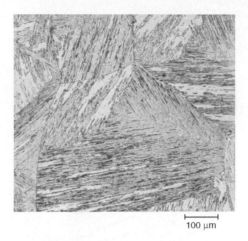

FIGURE 10.13 *Acicular, or needlelike, microstructure of martensite 100×. (From* ASM Handbook, *Vol. 9:* Metallography and Microstructures, *ASM International, Materials Park, OH, 2004.)*

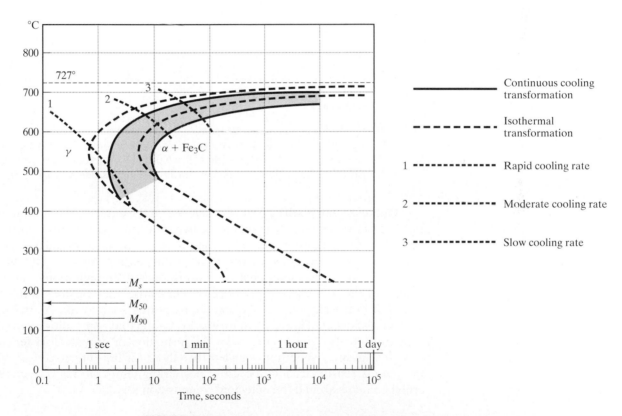

FIGURE 10.14 *A continuous cooling transformation (CCT) diagram is shown superimposed on the isothermal transformation diagram of Figure 10.11. The general effect of continuous cooling is to shift the transformation curves downward and toward the right. (After* Atlas of Isothermal Transformation and Cooling Transformation Diagrams, *American Society for Metals, Metals Park, OH, 1977.)*

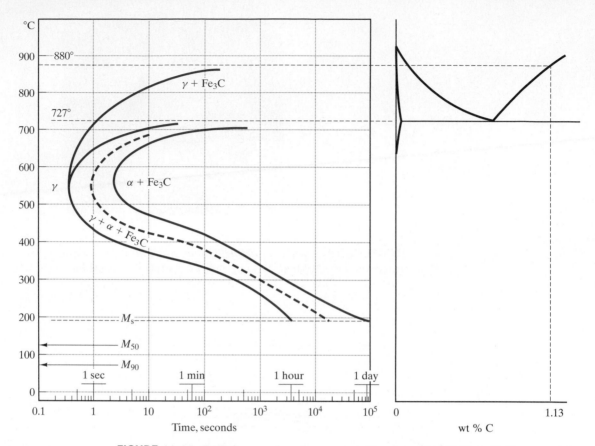

FIGURE 10.15 *TTT diagram for a hypereutectoid composition (1.13 wt % C) compared with the Fe–Fe$_3$ C phase diagram. Microstructural development for the slow cooling of this alloy was shown in Figure 9.39. (TTT diagram after* Atlas of Isothermal Transformation and Cooling Transformation Diagrams, *American Society for Metals, Metals Park, OH, 1977.)*

Our discussion to this point has centered on the eutectoid composition. Figure 10.15 shows the TTT diagram for the hypereutectoid composition introduced in Figure 9.39. The most obvious difference between this diagram and the eutectoid is the additional curved line extending from the pearlite "knee" to the horizontal line at 880°C. This additional line corresponds to the additional diffusional process for the formation of proeutectoid cementite. Less obvious is the downward shift in the martensitic reaction temperatures, such as M_s. A similar TTT diagram is shown in Figure 10.16 for the hypoeutectoid composition introduced in Figure 9.40. This diagram includes the formation of proeutectoid ferrite and shows martensitic temperatures higher than those for the eutectoid steel. In general, the martensitic reaction occurs at decreasing temperatures with increasing carbon contents around the eutectoid composition region.

HEAT TREATMENT OF STEEL

With the principles of TTT diagrams now available, we can illustrate some of the basic principles of the heat treatment of steels. This is a large field in itself with enormous, commercial significance. We can, of course, only touch on some

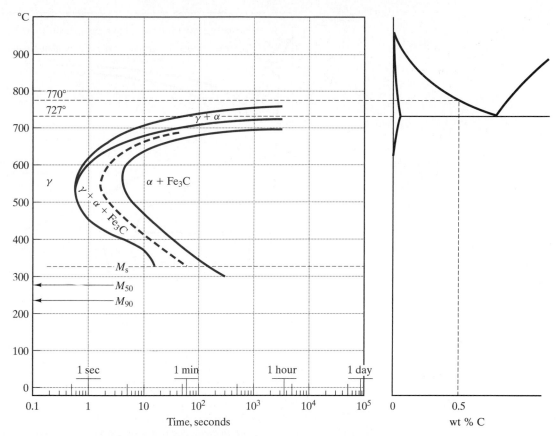

FIGURE 10.16 *TTT diagram for a hypoeutectoid composition (0.5 wt % C) compared with the Fe–Fe$_3$C phase diagram. Microstructural development for the slow cooling of this alloy was shown in Figure 9.40. By comparing Figures 10.11, 10.15, and 10.16, one will note that the martensitic transformation occurs at decreasing temperatures with increasing carbon content in the region of the eutectoid composition. (TTT diagrams after* Atlas of Isothermal Transformation and Cooling Transformation Diagrams, *American Society for Metals, Metals Park, OH, 1977.)*

elementary examples in this introductory textbook. For illustration, we shall select the eutectoid composition.

As discussed previously, martensite is a brittle phase. In fact, it is so brittle that a product of 100% martensite would be impractical, akin to a glass hammer. A common approach to fine-tuning the mechanical properties of a steel is to first form a completely martensitic material by rapid quenching. Then, this steel can be made less brittle by a careful reheat to a temperature where transformation to the equilibrium phases of α and Fe$_3$C is possible. By reheating for a short time at a moderate temperature, a high-strength, low-ductility product is obtained. By reheating for longer times, greater ductility occurs (due to less martensite). Figure 10.17 shows a thermal history $[T = fn(t)]$ superimposed on a TTT diagram that represents this conventional process, known as **tempering**. (It is important to recall that superimposing heating and cooling curves on an isothermal TTT diagram is a schematic illustration.) The α + Fe$_3$C microstructure produced by tempering is different from both pearlite and bainite, which is not surprising in

The tempering of steel is an excellent example of how the heat treatment of a material can lead to profound changes in mechanical behavior. The microstructure associated with this process is shown in Figure 10.18.

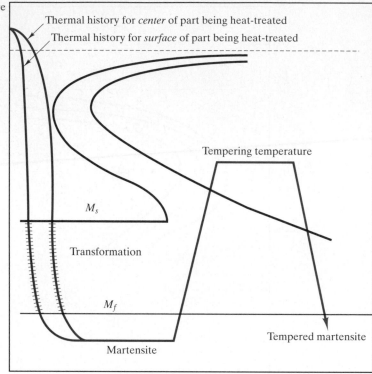

FIGURE 10.17 *Tempering is a thermal history $[T = fn(t)]$ in which martensite, formed by quenching austenite, is reheated. The resulting tempered martensite consists of the equilibrium phase of α-Fe and Fe_3 C, but in a microstructure different from both pearlite and bainite (note Figure 10.18). (After* Metals Handbook, *8th ed., Vol. 2, American Society for Metals, Metals Park, OH, 1964. It should be noted that the TTT diagram is, for simplicity, that of eutectoid steel. As a practical matter, tempering is generally done in steels with slower diffusional reactions that permit less-severe quenches.)*

light of the fundamentally different paths involved. Pearlite and bainite are formed by the cooling of austenite, a face-centered cubic solid solution. The microstructure known as **tempered martensite** (Figure 10.18) is formed by the heating of martensite, a body-centered tetragonal solid solution of Fe and C. The morphology in Figure 10.18 shows that the carbide has coalesced into isolated particles in a matrix of ferrite.

A possible problem with conventional quenching and tempering is that the part can be distorted and cracked due to uneven cooling during the quench step. The exterior will cool fastest and, therefore, transform to martensite before the interior. During the brief period of time in which the exterior and interior have different crystal structures, significant stresses can occur. The region that has the martensite structure is, of course, highly brittle and susceptible to cracking. A simple solution to this problem is a heat treatment known as **martempering** (or *marquenching*), illustrated in Figure 10.19. By stopping the quench above M_s, the entire piece can be brought to the same temperature by a brief isothermal step. Then, a slow cool allows the martensitic transformation to occur evenly through the piece. Again, ductility is produced by a final tempering step.

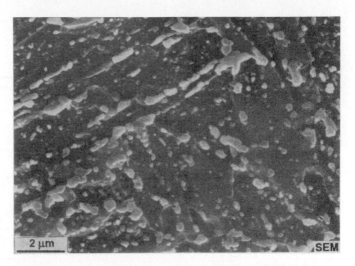

FIGURE 10.18 *The microstructure of tempered martensite, although an equilibrium mixture of α-Fe and Fe₃C, differs from those for pearlite (Figure 9.2) and bainite (Figure 10.9). This micrograph produced in a scanning electron microscope (SEM) shows carbide clusters in relief above an etched ferrite. (From* ASM Handbook, *Vol. 9:* Metallography and Microstructures, *ASM International, Materials Park, OH, 2004.)*

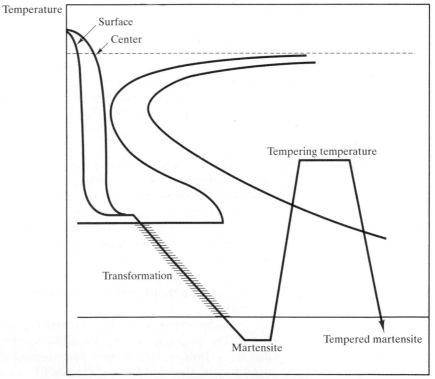

FIGURE 10.19 *In martempering, the quench is stopped just above M_S. Slow cooling through the martensitic transformation range reduces stresses associated with the crystallographic change. The final reheat step is equivalent to that in conventional tempering. (After* Metals Handbook, *8th ed., Vol. 2, American Society for Metals, Metals Park, OH, 1964.)*

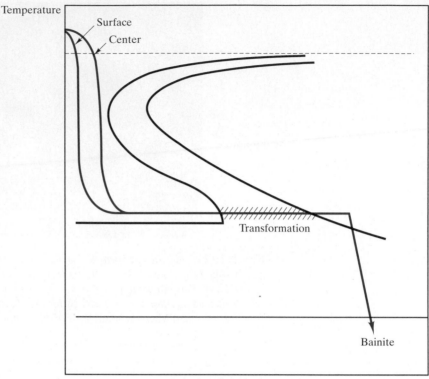

FIGURE 10.20 *As with martempering, austempering avoids the distortion and cracking associated with quenching through the martensitic transformation range. In this case, the alloy is held long enough just above M_s to allow full transformation to bainite. (After* Metals Handbook, *8th ed., Vol. 2, American Society for Metals, Metals Park, OH, 1964.)*

An alternative method to avoid the distortion and cracking of conventional tempering is the heat treatment known as **austempering**, illustrated in Figure 10.20. Austempering has the advantage of completely avoiding the costly reheating step. As with martempering, the quench is stopped just above M_s. In austempering, the isothermal step is extended until complete transformation to bainite occurs. Since this microstructure ($\alpha + Fe_3C$) is more stable than martensite, further cooling produces no martensite. Control of hardness is obtained by careful choice of the bainite transformation temperature. Hardness increases with decreasing transformation temperature due to the increasingly fine-grained structure.

A final comment on these schematic illustrations of heat treatment is in order. The principles were adequately shown using the simple eutectoid TTT diagram. However, the various heat treatments are similarly applied to a wide variety of steel compositions that have TTT diagrams that can differ substantially from that for the eutectoid. As an example, austempering is not practical for some alloy steels because the alloy addition substantially increases the time for bainite transformation. In addition, tempering of eutectoid steel, as illustrated, is of limited practicality due to the high quench rate needed to avoid the pearlite "knee."

EXAMPLE 10.2

(a) How much time is required for austenite to transform to 50% pearlite at 600°C?

(b) How much time is required for austenite to transform to 50% bainite at 300°C?

SOLUTION

(a) This problem is a direct application of Figure 10.7. The dotted line denotes the halfway point in the $\gamma \rightarrow \alpha + Fe_3C$ transformation. At 600°C, the time to reach that line is $\sim 3\frac{1}{2}$ s.

(b) At 300°C, the time is $\sim$480 s, or 8 min.

EXAMPLE 10.3

(a) Calculate the microstructure of a 0.77 wt % C steel that has the following heat treatment: (i) instantly quenched from the γ region to 500°C, (ii) held for 5 s, and (iii) quenched instantly to 250°C.

(b) What will happen if the resulting microstructure is held for 1 day at 250°C and then cooled to room temperature?

(c) What will happen if the resulting microstructure from part (a) is quenched directly to room temperature?

(d) Sketch the various thermal histories.

SOLUTION

(a) By having ideally fast quenches, we can solve this problem precisely in terms of Figure 10.7. The first two parts of the heat treatment lead to $\sim$70% transformation to fine pearlite. The final quench will retain this state:

$$30\% \ \gamma + 70\% \text{ fine pearlite } (\alpha + Fe_3C).$$

(b) The pearlite remains stable, but the retained γ will have time to transform to bainite, giving a final state:

$$30\% \text{ bainite } (\alpha + Fe_3C) + 70\% \text{ fine pearlite } (\alpha + Fe_3C).$$

(c) Again, the pearlite remains stable, but most of the retained γ will become unstable. For this case, we must consider the martensitic transformation data in Figure 10.11. The resulting microstructure will be

$$70\% \text{ fine pearlite } (\alpha + Fe_3C) + \sim 30\% \text{ martensite.}$$

(Because the martensitic transformation is not complete until -46°C, a small amount of untransformed γ will remain at room temperature.)

(d)

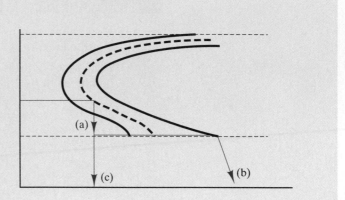

EXAMPLE 10.4

Estimate the quench rate needed to avoid pearlite formation in

(a) 0.5 wt % C steel,

(b) 0.77 wt % C steel, and

(c) 1.13 wt % C steel.

SOLUTION
In each case, we are looking at the rate of temperature drop needed to avoid the pearlite "knee":

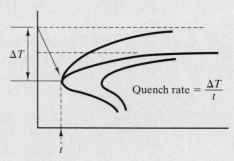

Note. This isothermal transformation diagram is used to illustrate a continuous cooling process. Precise calculation would require a true continuous cooling transformation curve.

(a) From Figure 10.16 for a 0.5 wt % C steel, we must quench from the austenite boundary (770°C) to ~520°C in ~0.6 s, giving

$$\frac{\Delta T}{t} = \frac{(770 - 520)°C}{0.6 \text{ s}} = 420°C/s.$$

(b) From Figure 10.11 for a 0.77 wt % C steel, we quench from the eutectoid temperature (727°C) to ~550°C in ~0.7 s, giving

$$\frac{\Delta T}{t} = \frac{(727 - 550)°C}{0.7 \text{ s}} = 250°C/s.$$

(c) From Figure 10.15 for a 1.13 wt % C steel, we quench from the austenite boundary (880°C) to ~550°C in ~3.5 s, giving

$$\frac{\Delta T}{t} = \frac{(880 - 550)°C}{0.35\ s} = 940°C/s.$$

EXAMPLE 10.5

Calculate the time required for austempering at 5°C above the M_s temperature for

(a) 0.5 wt % C steel,

(b) 0.77 wt % C steel, and

(c) 1.13 wt % C steel.

SOLUTION

(a) Figure 10.16 for 0.5 wt % C steel indicates that complete bainite formation will have occurred 5°C above M_s by

$$\sim180\ s \times 1\ m/60\ s = 3\ min.$$

(b) Similarly, Figure 10.11 for 0.77 wt % C steel gives a time of

$$\sim\frac{1.9 \times 10^4\ s}{3,600\ s/h} = 5.3\ h.$$

(c) Finally, Figure 10.15 for 1.13 wt % C steel gives an austempering time of

$$\sim1\ day.$$

PRACTICE PROBLEM 10.2

In Example 10.2, we use Figure 10.7 to determine the time for 50% transformation to pearlite and bainite at 600 and 300°C, respectively. Repeat these calculations for (a) 1% transformation and (b) 99% transformation.

PRACTICE PROBLEM 10.3

A detailed thermal history is outlined in Example 10.3. Answer all of the questions in that problem if only one change is made in the history; namely, step (i) is an instantaneous quench to 400°C (not 500°C).

PRACTICE PROBLEM 10.4

In Example 10.4, we estimate quench rates necessary to retain austenite below the pearlite "knee." What would be the percentage of martensite formed in each of the alloys if these quenches were continued to 200°C?

The time necessary for austempering is calculated for three alloys in Example 10.5. In order to do martempering (Figure 10.19), it is necessary to cool the alloy before bainite formation begins. How long can the alloy be held at $5°$ above M_s before bainite formation begins in **(a)** 0.5 wt % C steel, **(b)** 0.77 wt % C steel, and **(c)** 1.13 wt % C steel?

10.3 | Hardenability

There are software and representative data for the Jominy end-quench test on the related Web site.

In the remainder of this chapter, we shall encounter several heat treatments in which the primary purpose is to affect the **hardness** of a metal alloy. In Section 6.4, hardness was defined by the degree of indentation produced in a standard test. The indentation decreases with increasing hardness. An important feature of the hardness measurement is its direct correlation with strength. We shall now concentrate on heat treatments, with hardness serving to monitor the effect of the thermal history on alloy strength.

Our experience with TTT diagrams has shown a general trend. For a given steel, hardness is increased with increasing quench rates. However, a systematic comparison of the behavior of different steels must take into account the enormous range of commercial steel compositions. The relative ability of a steel to be hardened by quenching is termed **hardenability**. Fortunately, a relatively simple experiment has become standardized for industry to provide such a systematic comparison. The **Jominy** end-quench test is illustrated in Figure 10.21. A standard-size steel bar (25 mm in diameter by 100 mm long) is taken to the austenizing temperature, and then one end is subjected to a water spray. For virtually all carbon and low-alloy steels, this standard quench process produces a common cooling-rate gradient along the Jominy bar because the thermal properties (e.g., thermal conductivity) are nearly identical for these various alloys. (See Chapter 11 for other members of the steel family. The carbon and low-alloy steels are the most commonly used ones for quenching-induced hardness, for which the Jominy test is so useful.)

Figure 10.22 shows how the cooling rate varies along the Jominy bar. An end-quench test of the type illustrated in Figure 10.21 and 10.22 is the basis of the continuous cooling diagram of Figure 10.14. Of course, the cooling rate is greatest near the end subjected to the water spray. The resulting variation in hardness along a typical steel bar is illustrated in Figure 10.23. A similar plot comparing various steels is given in Figure 10.24. Here, comparisons of hardenability can be made, where hardenability corresponds to the relative magnitude of hardness along the Jominy bar.

The hardenability information from the end-quench test can be used in two complementary ways. If the quench rate for a given part is known, the Jominy data can predict the hardness of that part. Conversely, hardness measurements on various areas of a large part (which may have experienced uneven cooling) can identify different quench rates.

---

*Walter Jominy (1893–1976), American metallurgist. A contemporary of E. C. Bain, Jominy was a similarly productive researcher in the field of ferrous metallurgy. He held important appointments in industrial, government, and university laboratories.

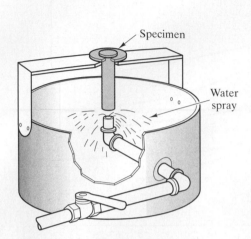

FIGURE 10.21 *Schematic illustration of the Jominy end-quench test for hardenability. (After W. T. Lankford et al., Eds.,* The Making, Shaping, and Treating of Steel, *10th ed., United States Steel, Pittsburgh, PA, 1985. Copyright 1985 by United States Steel Corporation.)*

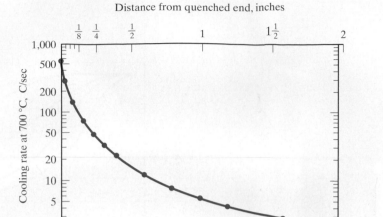

FIGURE 10.22 *The cooling rate for the Jominy bar (see Figure 10.21) varies along its length. This curve applies to virtually all carbon and low-alloy steels. (After L. H. Van Vlack,* Elements of Materials Science and Engineering, *4th ed., Addison-Wesley Publishing Co., Inc., Reading, MA, 1980.)*

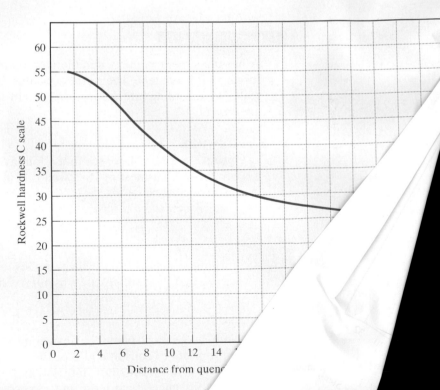

FIGURE 10.23 *Variation in hardness al... et al., Eds.,* The Making, Shaping, and Treat... *Pittsburgh, PA, 1985. Copyright 1985 by United ...*

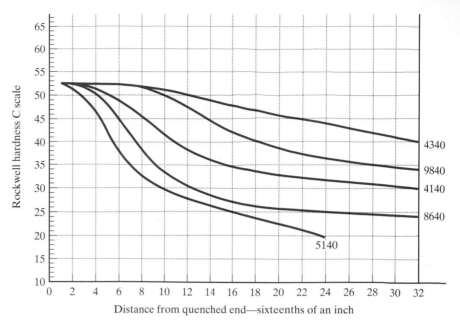

FIGURE 10.24 *Hardenability curves for various steels with the same carbon content (0.40 wt %) and various alloy contents. The codes designating the alloy compositions are defined in Table 11.1. (From W. T. Lankford et al., Eds.,* The Making, Shaping, and Treating of Steel, *10th ed., United States Steel, Pittsburgh, PA, 1985. Copyright 1985 by United States Steel Corporation.)*

EXAMPLE 10.6

A hardness measurement is made at a critical point on a trailer-axle forging of 4340 steel. The hardness value is 45 on the Rockwell C scale. What cooling rate was experienced by the forging at the point in question?

SOLUTION

Using Figure 10.24, we see that a Jominy end-quench test on this alloy produces a hardness of Rockwell C45 at 22/16 in. from the quenched end, which is equal to

$$D_{qe} = \frac{22}{16} \text{ in.} \times 25.4 \text{ mm/in.} = 35 \text{ mm.}$$

Turning to Figure 10.22, which applies to carbon and low-alloy steels, we see that the cooling rate was approximately

$$4°C/s \text{ (at } 700°C).$$

Note. To be more precise in answering a question such as this one, it is appropriate to consult a plot of the "hardness band" from a series of Jominy tests on the alloy in question. For most alloys, there is a considerable range of hardness that can occur at a given point along D_{qe}.

EXAMPLE 10.7

Estimate the hardness that would be found at the critical point on the axle discussed in Example 10.6 if that part were fabricated from 4140 steel rather than 4340 steel.

SOLUTION

This problem is straightforward in that Figure 10.22 shows us that the cooling behavior of various carbon and low-alloy steel is essentially the same. We can read the 4140 hardness from the plot in Figure 10.24 at the same D_{qe} as that calculated in Example 10.6 (i.e., at $\frac{22}{16}$ of an inch). The result is a hardness of Rockwell C32.5.

Note. The comment in Example 10.6 applies here also; that is, there are "uncertainty bars" associated with the Jominy data for any given alloy. However, Figure 10.24 is still very useful for indicating that the 4340 alloy is significantly more hardenable and, for a given quench rate, can be expected to yield a higher hardness part.

PRACTICE PROBLEM 10.6

In Example 10.6, we are able to estimate a quench rate that leads to a hardness of Rockwell C45 in a 4340 steel. What quench rate would be necessary to produce a hardness of **(a)** C50 and **(b)** C40?

PRACTICE PROBLEM 10.7

In Example 10.7, we find that the hardness of a 4140 steel is lower than that for a 4340 steel (given equal quench rates). Determine the corresponding hardness for **(a)** a 9840 steel, **(b)** an 8640 steel, and **(c)** a 5140 steel.

10.4 | Precipitation Hardening

In Section 6.3, we found that small obstacles to dislocation motion can strengthen (or harden) a metal (e.g., Figure 6.26). Small, second-phase precipitates are effective in this way. In Chapter 9, we found that cooling paths for certain alloy compositions lead to second-phase precipitation (e.g., Figure 9.36b). Many alloy systems use such **precipitation hardening**. The most common illustration is found in the Al–Cu system. Figure 10.25 shows the aluminum-rich end of the Al–Cu phase diagram, together with the microstructure that develops upon slow cooling. As the precipitates are relatively coarse and isolated at grain boundaries, little hardening is produced by the presence of the second phase. A substantially different thermal history is shown in Figure 10.26. Here, the coarse microstructure is first reheated to the single-phase (κ) region, which is appropriately termed a **solution treatment**. Then, the single-phase structure is quenched to room temperature, where the precipitation is quite slow, and the supersaturated solid solution remains a metastable phase. Upon reheating to some intermediate temperature, the solid-state diffusion of copper atoms in aluminum is sufficiently rapid to allow a

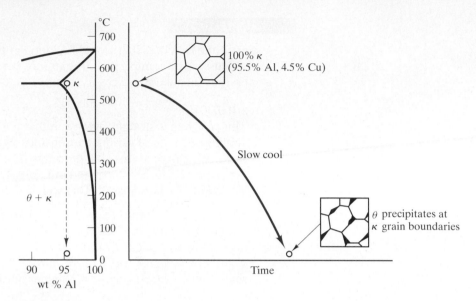

FIGURE 10.25 *Coarse precipitates form at grain boundaries in an Al–Cu (4.5 wt %) alloy when slowly cooled from the single-phase (κ) region of the phase diagram to the two-phase (θ + κ) region. These isolated precipitates do little to affect alloy hardness.*

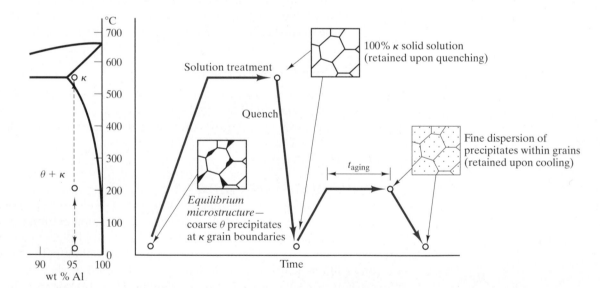

FIGURE 10.26 *By quenching and then reheating an Al–Cu (4.5 wt %) alloy, a fine dispersion of precipitates forms within the κ grains. These precipitates are effective in hindering dislocation motion and, consequently, increasing alloy hardness (and strength). This process is known as precipitation hardening, or age hardening.*

Precipitation hardening is another excellent example of how the heat treatment of a material can lead to significant changes to the microscale structure and resulting, profound changes in mechanical behavior.

fine dispersion of precipitates to form. These precipitates are effective dislocation barriers and lead to a substantial hardening of the alloy. Because this precipitation takes time, this process is also termed **age hardening**. Figure 10.27 illustrates **over-aging**, in which the precipitation process is continued so long that the precipitates have an opportunity to coalesce into a more coarse dispersion. This dispersion is

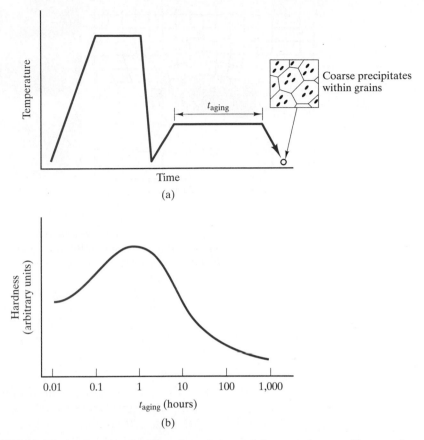

FIGURE 10.27 *(a) By extending the reheat step, precipitates coalesce and become less effective in hardening the alloy. The result is referred to as* overaging. *(b) The variation in hardness with the length of the reheat step (*aging time*).*

less effective as a dislocation barrier. Figure 10.28 shows the structure (formed during the early stages of precipitation), which is so effective as a dislocation barrier. These precipitates are referred to as **Guinier–Preston*** (or *G.P.*) **zones** and are distinguished by **coherent interfaces** at which the crystal structures of the matrix and precipitate maintain registry. This coherency is lost in the larger precipitates formed as overaging occurs.

EXAMPLE 10.8

(a) Calculate the amount of θ phase that would precipitate at the grain boundaries in the equilibrium microstructure shown in Figure 10.25.

(b) What is the maximum amount of Guinier–Preston zones to be expected in a 4.5 wt % Cu alloy?

---

*Andre Guinier (1911–2000), French physicist, and George Dawson Preston (1896–1972), English physicist. The detailed atomic structure (Figure 10.28) was determined in the 1930s by these physicists using the powerful tool of x-ray diffraction (see Section 3.7).

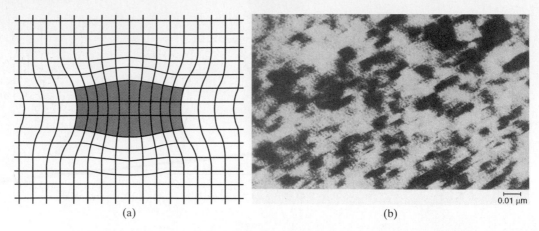

0.01 μm

(a)   (b)

FIGURE 10.28 *(a) Schematic illustration of the crystalline geometry of a Guinier–Preston (G.P.) zone. This structure is most effective for precipitation hardening and is the structure developed at the hardness maximum shown in Figure 10.27b. Note the coherent interfaces lengthwise along the precipitate. The precipitate is approximately 15 nm × 150 nm. (From H. W. Hayden, W. G. Moffatt, and J. Wulff,* The Structure and Properties of Materials, *Vol. 3:* Mechanical Behavior, *John Wiley & Sons, Inc., NY, 1965.) (b) Transmission electron micrograph of G.P. zones at 720,000×. (From* ASM Handbook, *Vol. 9:* Metallography and Microstructures, *ASM International, Materials Park, OH, 2004.)*

SOLUTION

(a) This is an equilibrium question that returns us to the concept of phase diagrams from Chapter 9. Using the Al–Cu phase diagram (Figure 9.27) and Equation 9.10, we obtain

$$\text{wt \% } \theta = \frac{x - x_\kappa}{x_\theta - x_\kappa} \times 100\% = \frac{4.5 - 0}{53 - 0} \times 100\%$$

$$= 8.49\%.$$

(b) As the G.P. zones are precursors to the equilibrium precipitation, the maximum amount would be 8.49%.

Note.   This calculation was made in a similar case treated in Example 9.10.

PRACTICE PROBLEM 10.8

The nature of precipitation in a 95.5 Al–4.5 Cu alloy is considered in Example 10.8. Repeat these calculations for a 96 Al–4 Cu alloy.

10.5 | Annealing

One of the more important heat treatments introduced in this chapter (in Section 10.2) is tempering, in which a material (martensite) is softened by high temperature for an appropriate time. **Annealing** is a comparable heat treatment in which the hardness of a mechanically deformed microstructure is reduced at high temperatures. In order to appreciate the details of this microstructural development, we need to explore four terms: *cold work*, *recovery*, *recrystallization*, and *grain growth*.

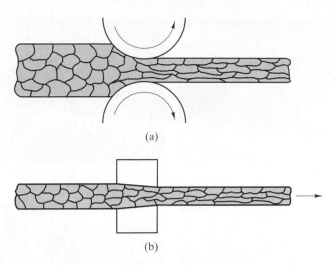

FIGURE 10.29 *Examples of cold-working operations: (a) cold-rolling a bar or sheet and (b) cold-drawing a wire. Note in these schematic illustrations that the reduction in area caused by the cold-working operation is associated with a preferred orientation of the grain structure.*

COLD WORK

Cold work means to mechanically deform a metal at relatively low temperatures. This concept was introduced in Section 6.3 in relating dislocation motion to mechanical deformation. The amount of cold work is defined relative to the reduction in cross-sectional area of the alloy by processes such as rolling or drawing (Figure 10.29). The percent cold work is given by

$$\% \text{ CW} = \frac{A_0 - A_f}{A_0} \times 100\%, \tag{10.2}$$

where A_0 is the original cross-sectional area and A_f is the final cross-sectional area after cold working. The hardness and strength of alloys are increased with increasing % CW, a process termed *strain hardening*. The relationship of mechanical properties to % CW of brass is illustrated in Figure 11.7, relative to a discussion of design specifications. The mechanism for this hardening is the resistance to plastic deformation caused by the high density of dislocations produced in the cold working. (Recall the discussion in Section 6.3.) The density of dislocations can be expressed as the length of dislocation lines per unit volume (e.g., m/m^3 or net units of m^{-2}). An annealed alloy can have a dislocation density as low as 10^{10} m^{-2}, with a correspondingly low hardness. A heavily cold-worked alloy can have a dislocation density as high as 10^{16} m^{-2}, with a significantly higher hardness (and strength).

A cold-worked microstructure is shown in Figure 10.30a. The severely distorted grains are quite unstable. By taking the microstructure to higher temperatures where sufficient atom mobility is available, the material can be softened and a new microstructure can emerge.

RECOVERY

An experiment on recovery, recrystallization, and grain growth, along with representative data, is provided on the related Web site.

The most subtle stage of annealing is **recovery**. No gross microstructural change occurs. However, atomic mobility is sufficient to diminish the concentration of point defects within grains and, in some cases, to allow dislocations to move to

(a)   (b)   (c)

(d)   (e)

FIGURE 10.30 *Annealing can involve the complete recrystallization and subsequent grain growth of a cold-worked microstructure. (a) A cold-worked brass (deformed through rollers such that the cross-sectional area of the part was reduced by one-third). (b) After 3 s at 580°C, new grains appear. (c) After 4 s at 580°C, many more new grains are present. (d) After 8 s at 580°C, complete recrystallization has occurred. (e) After 1 h at 580°C, substantial grain growth has occurred. The driving force for this growth is the reduction of high-energy grain boundaries. The predominant reduction in hardness for this overall process had occurred by step (d). All micrographs have a magnification of 75×. (Courtesy of J. E. Burke, General Electric Company, Schenectady, NY.)*

lower-energy positions. This process yields a modest decrease in hardness and can occur at temperatures just below those needed to produce significant microstructural change. Although the structural effect of recovery (primarily a reduced number of point defects) produces a modest effect on mechanical behavior, electrical conductivity does increase significantly. (The relationship between conductivity and structural regularity is explored further in Section 13.3.)

RECRYSTALLIZATION

In Section 6.3, we stated an important concept: "The temperature at which atomic mobility is sufficient to affect mechanical properties is approximately one-third to one-half times the absolute melting point, T_m." A microstructural result of exposure to such temperatures is termed **recrystallization** and is illustrated dramatically in Figures 10.30a–d. New equi-axed, stress-free grains nucleate at high-stress regions in the cold-worked microstructure (Figure 10.30b). These grains then grow together until they constitute the entire microstructure (Figure 10.30c and d). As the nucleation step occurs in order to stabilize the system, it is not surprising that the concentration of new grain nuclei increases with the degree of cold work. As a result, the grain size of the recrystallized microstructure decreases with the

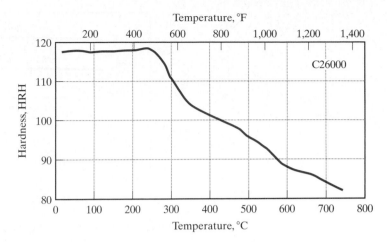

FIGURE 10.31 *The sharp drop in hardness identifies the recrystallization temperature as ~290°C for the alloy C26000, "cartridge brass." (From* Metals Handbook, *9th ed., Vol. 4, American Society for Metals, Metals Park, OH, 1981.)*

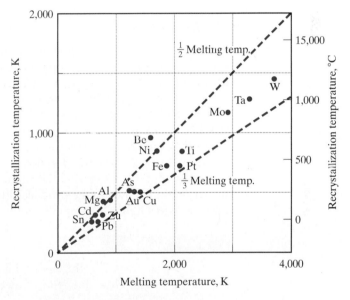

FIGURE 10.32 *Recrystallization temperature versus melting points for various metals. This plot is a graphic demonstration of the rule of thumb that atomic mobility is sufficient to affect mechanical properties above approximately $\frac{1}{3}$ to $\frac{1}{2}T_m$ on an absolute temperature scale. (From L. H. Van Vlack,* Elements of Materials Science and Engineering, *3rd ed., Addison-Wesley Publishing Co., Inc., Reading, MA, 1975.)*

degree of cold work. The decrease in hardness due to annealing is substantial, as indicated by Figure 10.31. Finally, the rule of thumb quoted at the beginning of this discussion of recrystallization effectively defines the **recrystallization temperature** (Figure 10.32). For a given alloy composition, the precise recrystallization temperature will depend slightly on the percentage of cold work. Higher values of % CW correspond to higher degrees of strain hardening and a correspondingly lower recrystallization temperature; that is, less thermal energy input is required to initiate the reformation of the microstructure (Figure 10.33).

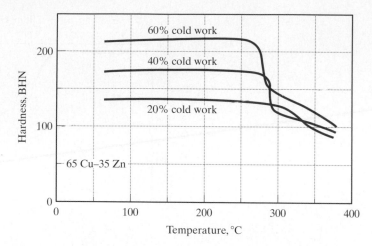

FIGURE 10.33 *For this cold-worked brass alloy, the recrystallization temperature drops slightly with increasing degrees of cold work. (From L. H. Van Vlack,* Elements of Materials Science and Engineering, *4th ed., Addison-Wesley Publishing Co., Inc., Reading, MA, 1980.)*

GRAIN GROWTH

The microstructure developed during recrystallization (Figure 10.30d) occurred spontaneously. It is stable compared with the original cold-worked structure (Figure 10.30a). However, the recrystallized microstructure contains a large concentration of grain boundaries. We have noted frequently since Chapter 4 that the reduction of these high-energy interfaces is a method of stabilizing a system further. The stability of coarse pearlite (Figure 10.10) was such an example. The coarsening of annealed microstructures by grain growth is another. Figure 10.30e illustrates **grain growth**, which is not dissimilar to the coalescence of soap bubbles, a process similarly driven by the reduction of surface area. Figure 10.34 shows that this grain-growth stage produces little additional softening of the alloy. That effect is associated predominantly with recrystallization.

EXAMPLE 10.9

Cartridge brass has the approximate composition of 70 wt % Cu, 30 wt % Zn. How does this alloy compare with the trend shown in Figure 10.32?

SOLUTION
The recrystallization temperature is indicated by Figure 10.31 as ~290°C. The melting point for this composition is indicated by the Cu–Zn phase diagram (Figure 9.28) as ~920°C (the solidus temperature). The ratio of recrystallization temperature to melting point is then

$$\frac{T_R}{T_m} = \frac{(290 + 273)\ \mathrm{K}}{(920 + 273)\ \mathrm{K}} = 0.47,$$

which is within the range of one-third to one-half indicated by Figure 10.32.

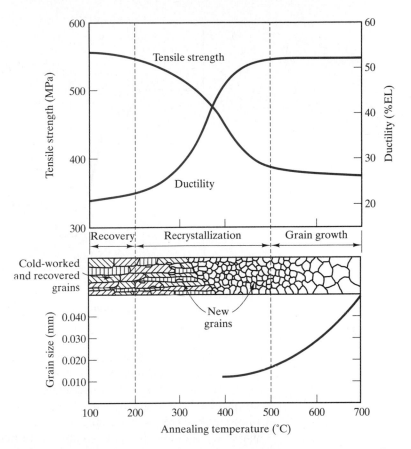

FIGURE 10.34 *Schematic illustration of the effect of annealing temperature on the strength and ductility of a brass alloy shows that most of the softening of the alloy occurs during the recrystallization stage. (After G. Sachs and K. R. Van Horn,* Practical Metallurgy: Applied Physical Metallurgy and the Industrial Processing of Ferrous and Nonferrous Metals and Alloys, *American Society for Metals, Cleveland, OH, 1940.)*

PRACTICE PROBLEM 10.9

Noting the result of Example 10.9, plot the estimated temperature range for recrystallization of Cu–Zn alloys as a function of composition over the entire range from pure Cu to pure Zn.

10.6 | The Kinetics of Phase Transformations for Nonmetals

As with phase diagrams in Chapter 9, our discussion of the kinetics of phase transformations has dwelled on metallic materials. The rates at which phase transformations occur in nonmetallic systems are, of course, also important to the processing of those materials. The crystallization of some single-component polymers is a model example of the nucleation and growth kinetics illustrated in Figure 10.5. Careful control of the rate of melting and solidification of silicon is critical to the growth and subsequent purification of large, single crystals, which are the foundation of the semiconductor industry. As with phase diagrams, ceramics

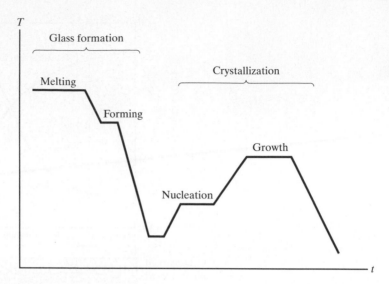

FIGURE 10.35 *Typical thermal history for producing a glass ceramic by the controlled nucleation and growth of crystalline grains.*

(rather than polymers or semiconductors) provide the closest analogy to the treatment of kinetics for metals.

TTT diagrams are not widely used for nonmetallic materials, but some examples have been generated, especially in systems where transformation rates play a critical role in processing, such as the crystallization of glass. In this regard, glass-ceramics are closely associated with nucleation and growth kinetics. These products are formed as glasses and are then carefully crystallized to produce a polycrystalline product. The result can be relatively strong ceramics formed in complex shapes for modest cost. A typical temperature–time schedule for producing a glass-ceramic is shown in Figure 10.35. These interesting materials are presented further in Chapter 11.

The CaO–ZrO$_2$ phase diagram was presented in Figure 9.29. The production of stabilized zirconia was shown to be the result of adding sufficient CaO (around 20 mol %) to ZrO$_2$ to form the cubic-zirconia solid-solution phase. As a practical matter, PSZ with a composition in the two-phase (monoclinic zirconia + cubic zirconia) region exhibits mechanical properties superior to those of the fully stabilized material, especially thermal-shock resistance (see Section 7.4). Electron microscopy has revealed that, in cooling the partially stabilized material, some monoclinic precipitates form within the cubic phase. This precipitation strengthening is the analog of precipitation hardening in metals. Electron microscopy has also revealed that the tetragonal-to-monoclinic phase transformation in pure zirconia is a martensitic-type transformation (Figure 10.36).

The subject of grain growth has played an especially important role in the development of ceramic processing in recent decades. Our first example of a microstructurally sensitive property was the transparency of a polycrystalline Al$_2$O$_3$ ceramic (see the feature box in Chapter 1). The material can be nearly transparent if it is essentially pore-free. However, it is formed by the densification of a powder. The bonding of powder particles occurs by solid-state diffusion. In the course of this densification stage, the pores between adjacent particles steadily shrink. This overall process is known as **sintering**, which refers to any process for forming a

FIGURE 10.36 *Transmission electron micrograph of monoclinic zirconia showing a microstructure characteristic of a martensitic transformation. Included in the evidence are twins labeled T. See Figure 4.15 for an atomic-scale schematic of a twin boundary, and see Figure 10.13 for the microstructure of martensitic steel. (Courtesy of Arthur H. Heuer.)*

dense mass by heating, but without melting. This rather unusual term comes from the Greek word *sintar*, meaning "slag" or "ash." It shares this source with the more common term *cinder*. The mechanism of shrinkage is the diffusion of atoms away from the grain boundary (between adjacent particles) to the pore. In effect, the pore is filled in by diffusing material (Figure 10.37). Unfortunately, grain growth can begin long before the pore shrinkage is complete. The result is that some pores become trapped within grains. The diffusion path from grain boundary to pore is too long to allow further pore elimination (Figure 10.38). A microstructure for this case was shown in the feature box in Chapter 1. The solution to this problem is to add a small amount (about 0.1 wt %) of MgO, which severely retards grain growth and allows pore shrinkage to go to completion. The resulting microstructure was also shown in the feature box in Chapter 1. The mechanism for the retardation of grain growth appears to be associated with the effect of Mg^{2+} ions on the retarding of grain-boundary mobility by a solid-solution pinning mechanism. The result is a substantial branch of ceramic technology in which nearly transparent polycrystalline ceramics can be reliably fabricated.

EXAMPLE 10.10

Calculate the maximum amount of monoclinic phase you would expect to find in the microstructure of a partially stabilized zirconia with an overall composition of 3.4 wt % CaO.

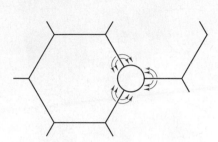

FIGURE 10.37 *An illustration of the sintering mechanism for shrinkage of a powder compact is the diffusion of atoms away from the grain boundary to the pore, thereby "filling in" the pore. Each grain in the microstructure was originally a separate powder particle in the initial compact.*

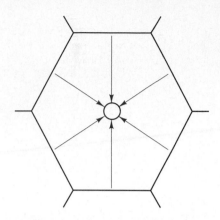

FIGURE 10.38 *Grain growth hinders the densification of a powder compact. The diffusion path from grain boundary to pore (now isolated within a large grain) is prohibitively long.*

SOLUTION

This is another example of the important relationship between phase diagrams and kinetics. We can make this prediction by calculating the equilibrium concentration of monoclinic phase at room temperature in the CaO–ZrO$_2$ system (Figure 9.29). A composition of 3.4 wt % CaO is approximately 7 mol % CaO (as shown by the upper and lower composition scales in Figure 9.29). Extrapolating the phase boundaries in the monoclinic + cubic two-phase region to room temperature gives a monoclinic phase composition of ~2 mol % CaO and a cubic phase composition of ~15 mol % CaO. Using Equation 9.9, we get

$$\text{mol \% monoclinic} = \frac{x_{\text{cubic}} - x}{x_{\text{cubic}} - x_{\text{mono}}} \times 100\%$$

$$= \frac{15 - 7}{15 - 2} \times 100\%$$

$$= 62 \text{ mol \%.}$$

Note. We have made this calculation in terms of mole percent because of the nature of the plot in Figure 9.29. It would be a straightforward matter to convert the results to weight percent.

PRACTICE PROBLEM 10.10

Convert the 62 mol % from Example 10.10 to weight percent.

THE MATERIAL WORLD

Field–Assisted Sintering Technique (FAST)

In this chapter, we are introduced to *sintering*, an important processing technique for materials with relatively high melting points. Sintering involves densification of a powder by bonding individual powder particles via solid-state diffusion. This densification can be done at substantially lower temperatures than required in casting, in which a molten liquid is solidified. In our introduction to sintering, however, we also find an interesting challenge involving kinetics. A sufficiently high temperature that can allow particle bonding by solid-state diffusion can also allow grain growth (illustrated in Figure 10.38).

A creative approach to minimizing the "kinetics challenge" of grain growth is provided by the **field-assisted sintering technique** (FAST). The FAST technique is illustrated as follows. The key distinction of FAST is that the sample is heated by the direct application of an external electrical current. (By contrast, conventional sintering involves heating the sample in a chamber in which coils

surrounding the sample are heated by an electrical current.) For nonconductive samples in FAST, heating is provided by transfer from the resistively heated die and punch.

The sintering equipment consists of a mechanical device (including die and punch) to provide pressure application to the powder and electrical components, which in turn provide pulsed and steady direct current (DC). A typical pulse discharge is achieved by applying a low voltage ($\approx$30 V) and a high current ($\approx$1,000 A). The duration of the short pulse can vary between 1 and 300 ms. These short pulses can be applied throughout the sintering process or prior to the application of a steady DC.

The great benefit of FAST is that the sintering process can be carried out at substantially reduced temperatures in significantly reduced times. For example, fully dense aluminum nitride (AlN) ceramics can be prepared by FAST within 5 minutes at 2,000 K, while conventional sintering at 2,220 K requires 30 hours to achieve 95% density. By

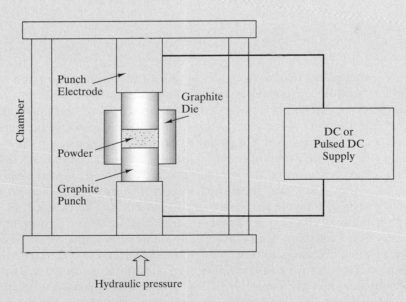

Schematic of the field-assisted sintering technique (FAST) system. (Courtesy of J. R. Groza, University of California, Davis.)

(Continued)

minimizing grain growth in conjunction with these reduced sintering times and temperatures, FAST can be an important technique for producing grain structures in the nanometer range. These *nanostructures*, as opposed to microstructures, can take advantage of the benefits of nanotechnology introduced in the feature box in Chapter 4.

Finally, we should emphasize that FAST is an example of technology leading science. FAST takes advantage of the electrical discharge among the powder particles under an applied pressure. Successful sintering techniques have been developed with substantial trial and error, while the precise mechanism of the electrical discharge is not fully understood. The technique has become widely described as **spark plasma sintering** (SPS), however, there is some doubt that either electrical sparking or plasma formation truly occurs. Nonetheless, a wide variety of materials has been produced in this way, with most being ceramics [e.g., Al_2O_3, AlN, Si_3N_4, TiN, TiO_2, Al_2TiO_5, and hydroxyapatite, $Ca_{10}(PO_4)_6(OH)_2$].

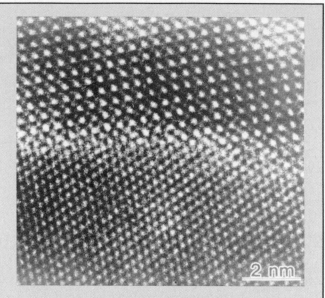

High-resolution transmission electron micrograph illustrates a high-quality grain boundary in fully dense AlN produced by the field-assisted sintering technique (FAST). The crystalline order of each grain extends fully to the boundary region. (Courtesy of J. R. Groza, University of California, Davis.)

Summary

In Chapter 9, we presented phase diagrams as two-dimensional maps of temperature and composition (at a fixed pressure of 1 atm). In this chapter, we added time as a third dimension to monitor the kinetics of microstructural development. Time-dependent phase transformations begin with the nucleation of the new phase followed by its subsequent growth. The overall transformation rate is a maximum at some temperature below the equilibrium transformation temperature because of a competition between the degree of instability (which increases with decreasing temperature) and atomic diffusivity (which decreases with decreasing temperature). A plot of percentage transformation on a temperature-versus-time plot is known as a TTT diagram. A prime example of the TTT diagram is given for the eutectoid steel composition (Fe with 0.77 wt % C). Pearlite and bainite are the products of diffu-

sional transformation formed by cooling austenite. If the austenite is cooled in a sufficiently rapid quench, a diffusionless, or martensitic, transformation occurs. The product is a metastable, supersaturated solid solution of C in Fe known as martensite. The careful control of diffusional and diffusionless transformations is the basis of the heat treatment of steel. As an example, tempering involves a quench to produce a non-diffusional martensitic transformation followed by a reheat to produce a diffusional transformation of martensite into the equilibrium phases of α-Fe and Fe_3C. This tempered martensite is different from both pearlite and bainite. Martempering is a slightly different heat treatment in which the steel is quenched to just above the starting temperature for the martensitic transformation and then cooled slowly through the martensitic range. This process reduces the stresses produced by quenching during the martensitic transformation. The reheat to produce tempered martensite is the same as before. A third type of heat treatment is austempering. Again, the quench stops short of the martensitic range, but in this case the quenched austenite is held long enough for bainite to form. As a result, the reheat step is unnecessary.

The ability of steel to be hardened by quenching is termed hardenability. This characteristic is evaluated in a straightforward manner by the Jominy end-quench test. High-aluminum alloys in the Al–Cu system are excellent examples of precipitation hardening. By careful cooling from a single phase to a two-phase region of the phase diagram, it is possible to generate a fine dispersion of second-phase precipitates that are dislocation barriers and the source of hardening (and strengthening). Annealing is a heat treatment that reduces hardness in cold-worked alloys. Cold working involves the mechanical deformation of an alloy at a relatively low temperature. In combination with annealing, cold working provides for wide-

ranging control of mechanical behavior. A subtle form of annealing is recovery, in which a slight reduction in hardness occurs due to the effect of atomic mobility on structural defects. Elevated temperature is required to permit the atomic mobility. At slightly higher temperatures, a more dramatic reduction in hardness occurs due to recrystallization. The entire cold-worked microstructure is transformed to a set of small, stress-free grains. With longer time, grain growth occurs in the new microstructure.

Time also plays a crucial role in the development of microstructure in nonmetallic materials. Examples are given for polymers and ceramics. The production of glass ceramics is a classic example of nucleation and growth theory as applied to the crystallization of glass. Structural zirconia ceramics exhibit microstructures produced by both diffusional and diffusionless transformations. The production of transparent, polycrystalline ceramics has been the result of control of grain growth in the densification of compressed powders (sintering).

Key Terms

age hardening (328)
annealing (330)
austempering (320)
austenite (310)
bainite (310)
coherent interface (329)
cold work (331)
continuous cooling transformation (CCT) diagram (314)
diffusional transformation (309)
diffusionless transformation (309)
field-assisted sintering technique (FAST) (339)
grain growth (334)

Guinier–Preston zone (329)
hardenability (324)
hardness (324)
heat treatment (304)
heterogeneous nucleation (306)
homogeneous nucleation (306)
isothermal transformation diagram (309)
Jominy end-quench test (324)
kinetics (304)
martempering (318)
martensite (313)
martensitic transformation (313)
metastable (314)

nucleation (306)
overaging (328)
pearlite (310)
precipitation hardening (327)
recovery (331)
recrystallization (332)
recrystallization temperature (333)
sintering (336)
solution treatment (327)
spark plasma sintering (SPS) (340)
tempered martensite (318)
tempering (317)
TTT diagram (309)

References

ASM Handbook, Vol. 4: *Heat Treating*, ASM International, Materials Park, OH, 1991.

Chiang, **Y.**, **D. P. Birnie III**, and **W. D. Kingery**, *Physical Ceramics*, John Wiley & Sons, Inc., New York, 1997.

Problems

10.1 • Time—The Third Dimension

10.1. For an aluminum alloy, the activation energy for crystal growth is 120 kJ/mol. By what factor would the rate of crystal growth change by dropping the alloy temperature from 500°C to room temperature (25°C)?

10.2. Repeat Problem 10.1 for a copper alloy for which the activation energy for crystal growth is 195 kJ/mol.

10.3. Although Section 10.1 concentrates on crystal nucleation and growth from a liquid, similar kinetics laws apply to solid-state transformations. For example, Equation 10.1 can be used to describe the rate of β precipitation upon cooling supersaturated α phase in a 10 wt % Sn–90 wt % Pb alloy. Given precipitation rates of 3.77×10^3 s^{-1} and 1.40×10^3 s^{-1} at 20°C and 0°C, respectively, calculate the activation energy for this process.

10.4. Use the result of Problem 10.3 to calculate the precipitation rate at room temperature, 25°C.

10.5. Relative to Figure 10.3, derive an expression for r_c as a function of σ, the surface energy per unit area of the nucleus, and ΔG_v, the volume energy reduction per unit volume. Recall that the area of a sphere is $4\pi r^2$, and its volume is $\left(\frac{4}{3}\right)\pi r^3$.

*10.6. The work of formation, W, for a stable nucleus is the maximum value of the net energy change (occurring at r_c) in Figure 10.3. Derive an expression for W in terms of σ and ΔG_v (defined in Problem 10.5).

*10.7. A theoretical expression for the rate of pearlite growth from austenite is

$$\dot{R} = Ce^{-Q/RT}(T_E - T)^2,$$

where C is a constant, Q is the activation energy for carbon diffusion in austenite, R is the gas constant, and T is an absolute temperature below the equilibrium-transformation temperature, T_E. Derive an expression for the temperature, T_M, that corresponds to the maximum growth rate (i.e., the "knee" of the transformation curve).

10.8. Use the result of Problem 10.7 to calculate T_M (in °C). (Recall that the activation energy is given in Chapter 5, and the transformation temperature is given in Chapter 9.)

10.2 • The TTT Diagram

10.9. (a) A 1050 steel (iron with 0.5 wt % C) is rapidly quenched to 330°C, held for 10 minutes, and then cooled to room temperature. What is the resulting microstructure? (b) What is a name for this heat treatment?

10.10. (a) A eutectoid steel is (i) quenched instantaneously to 500°C, (ii) held for 5 seconds, (iii) quenched instantaneously to room temperature, (iv) reheated to 300°C for 1 hour, and (v) cooled to room temperature. What is the final microstructure? (b) A carbon steel with 1.13 wt % C is given exactly the same heat treatment described in part (a). What is the resulting microstructure in this case?

10.11. (a) A carbon steel with 1.13 wt % C is given the following heat treatment: (i) instantaneously quenched to 200°C, (ii) held for 1 day, and (iii) cooled slowly to room temperature. What is the resulting microstructure? (b) What microstructure would result if a carbon steel with 0.5 wt % C were given exactly the same heat treatment?

10.12. Three different eutectoid steels are given the following heat treatments: (a) instantaneously quenched to 600°C, held for 2 minutes, then cooled to room temperature; (b) instantaneously quenched to 400°C, held for 2 minutes, then cooled to room temperature; and (c) instantaneously quenched to 100°C, held for 2 minutes, then cooled to room temperature. List these heat treatments in order of decreasing hardness of the final product. Briefly explain your answer.

10.13. (a) A eutectoid steel is cooled at a steady rate from 727 to 200°C in exactly 1 day. Superimpose this cooling curve on the TTT diagram of Figure 10.11. (b) From the result of your plot for part (a), determine at what temperature a phase transformation would first be observed. (c) What would be the first phase to be observed? (Recall the approximate nature of using an isothermal diagram to represent a continuous cooling process.)

10.14. Repeat Problem 10.13 for a steady rate of cooling in exactly 1 minute.

10.15. Repeat Problem 10.13 for a steady rate of cooling in exactly 1 second.

10.16. (a) Using Figures 10.11, 10.15, and 10.16 as data sources, plot M_s, the temperature at which the martensitic transformation begins, as a function of carbon content. **(b)** Repeat part (a) for M_{50}, the temperature at which the martensitic transformation is 50% complete. **(c)** Repeat part (a) for M_{90}, the temperature at which the martensitic transformation is 90% complete.

10.17. Using the trends of Figures 10.11, 10.15, and 10.16, sketch as specifically as you can a TTT diagram for a hypoeutectoid steel that has 0.6 wt % C. (Note Problem 10.16 for one specific compositional trend.)

10.18. Repeat Problem 10.17 for a hypereutectoid steel with 0.9 wt % C.

10.19. What is the final microstructure of a 0.6 wt % C hypoeutectoid steel given the following heat treatment: (i) instantaneously quenched to 500°C, (ii) held for 10 seconds, and (iii) instantaneously quenched to room temperature. (Note the TTT diagram developed in Problem 10.17.)

10.20. Repeat Problem 10.19 for a 0.9 wt % C hypereutectoid steel. (Note the TTT diagram developed in Problem 10.18.)

10.21. It is worth noting that in TTT diagrams such as Figure 10.7, the 50% completion (dashed-line) curve lies roughly midway between the onset (1%) curve and completion (99%) curve. It is also worth noting that the progress of transformation is not linear, but instead is sigmoidal (s-shaped) in nature. For example, careful observation of Figure 10.7 at 500°C shows that 1%, 50%, and 99% completion occur at 0.9 s, 3.0 s, and 9.0 s, respectively. Intermediate completion data, however, can be given as follows:

% completion	$t(s)$
20	2.3
40	2.9
60	3.2
80	3.8

Plot the % completion at 500°C versus log t to illustrate the sigmoidal nature of the transformation.

10.22. (a) Using the result of Problem 10.21, determine t for 25% and 75% completion. **(b)** Superimpose the result of (a) on a sketch of Figure 10.7 to illustrate that the 25% and 75% completion lines are much closer to the 50% line than to either the 1% or 99% completion lines.

10.3 • Hardenability

10.23. (a) Specify a quench rate necessary to ensure a hardness of at least Rockwell C40 in a 4140 steel. **(b)** Specify a quench rate necessary to ensure a hardness of no more than Rockwell C40 in the same alloy.

10.24. The surface of a forging made from a 4340 steel is unexpectedly subjected to a quench rate of 100°C/s (at 700°C). The forging is specified to have a hardness between Rockwell C46 and C48. Is the forging within specifications? Briefly explain your answer.

D 10.25. A flywheel shaft is made of 8640 steel. The surface hardness is found to be Rockwell C35. By what percentage would the cooling rate at the point in question have to change in order for the hardness to be increased to a more desirable value of Rockwell C45?

D 10.26. Repeat Problem 10.25 assuming that the shaft is made of 9840 steel.

10.27. Quenching a bar of 4140 steel at 700°C into a stirred water bath produces an instantaneous quench rate at the surface of 100°C/s. Use Jominy test data to predict the surface hardness resulting from this quench. (Note: The quench described is not in the Jominy configuration. Nonetheless, Jominy data provide general information on hardness as a function of quench rate.)

10.28. Repeat Problem 10.27 for a stirred oil bath that produces a surface quench rate of 20°C/s.

10.29. A bar of steel quenched into a stirred liquid will cool much more slowly in the center than at the surface in contact with the liquid. The following limited data represent the initial quench rates at various points across the diameter of a bar of 4140 steel (initially at 700°C):

Position	Quench rate (°C/s)
center	35
15 mm from center	55
30 mm from center (at surface)	200

(a) Plot the quench rate profile across the diameter of the bar. (Assume the profile to be symmetrical.) **(b)** Use Jominy test data to plot the resulting hardness profile across the diameter of the bar. (*Note:* The quench rates described are not in the Jominy configuration. Nonetheless, Jominy data provide general information on hardness as a function of quench rate.)

10.30. Repeat Problem 10.29b for a bar of 5140 steel that would have the same quench-rate profile.

D 10.31. In heat-treating a complex-shaped part made from a 5140 steel, a final quench in stirred oil leads to a hardness of Rockwell C30 at 3 mm beneath the surface. This hardness is unacceptable, as design specifications require a hardness of Rockwell C45 at that point. Select an alloy substitution to provide this hardness, assuming the heat treatment must remain the same.

10.32. In general, hardenability decreases with decreasing alloy additions. Illustrate this concept by superimposing the following data for a plain carbon 1040 steel on the plot for other xx40 steels in Figure 10.24:

Distance from quenched end (in sixteenths of an inch)	Rockwell C hardness
2	44
4	27
6	22
8	18
10	16
12	13
14	12
16	11

10.4 • Precipitation Hardening

10.33. Specify an aging temperature for a 95 Al–5 Cu alloy that will produce a maximum of 5 wt % θ precipitate.

10.34. Specify an aging temperature for a 96 Al–4 Cu alloy that will produce a maximum of 5 wt % θ precipitate.

10.35. As second-phase precipitation is a thermally activated process, an Arrhenius expression can be used to estimate the time required to reach maximum hardness (see Figure 10.27b). As a first approximation, you can treat t_{max}^{-1} as a "rate," where t_{max} is the time to reach maximum hardness. For a given aluminum alloy, t_{max} is 40 hours at 150°C and only 4 hours at 190°C. Use Equation 5.1 to calculate the activation energy for this precipitation process.

10.36. Estimate the time to reach maximum hardness, t_{max}, at 250°C for the aluminum alloy of Problem 10.35.

10.5 • Annealing

10.37. A 90:10 Ni–Cu alloy is heavily cold-worked. It will be used in a structural design that is occasionally subjected to 200°C temperatures for as much as 1 hour. Do you expect annealing effects to occur?

10.38. Repeat Problem 10.37 for a 90:10 Cu–Ni alloy.

10.39. A 12.5-mm-diameter rod of steel is drawn through a 10-mm-diameter die. What is the resulting percentage of cold work?

*10.40. An annealed copper-alloy sheet is cold-worked by rolling. The x-ray diffraction pattern of the original annealed sheet with an fcc crystal structure is represented schematically by Figure 3.33. Given that the rolling operation tends to produce the preferred orientation of the (220) planes parallel to the sheet surface, sketch the x-ray diffraction pattern you would expect for the rolled sheet. (To locate the 2θ positions for the alloy, assume pure copper and note the calculations for Problem 3.75.)

10.41. Recrystallization is a thermally activated process and, as such, can be characterized by the Arrhenius expression (Equation 5.1). As a first approximation, we can treat t_R^{-1} as a "rate," where t_R is the time necessary to fully recrystallize the microstructure. For a 75% cold-worked aluminum alloy, t_R is 100 hours at 256°C and only 10 hours at 283°C. Calculate the activation energy for this recrystallization process. (Note Problem 10.35, in which a similar method was applied to the case of precipitation hardening.)

10.42. Calculate the temperature at which complete recrystallization would occur for the aluminum alloy of Problem 10.41 within 1 hour.

10.6 • The Kinetics of Phase Transformations for Nonmetals

10.43. The sintering of ceramic powders is a thermally activated process and shares the "rule of thumb" about temperature with diffusion and recrystallization. Estimate a minimum sintering temperature for **(a)** pure Al_2O_3, **(b)** pure mullite, and **(c)** pure spinel. (See Figures 9.23 and 9.26.)

10.44. Four ceramic phase diagrams were presented in Figures 9.10, 9.23, 9.26, and 9.29. In which systems would you expect precipitation hardening to be a possible heat treatment? Briefly explain your answer.

10.45. The total sintering rate for $BaTiO_3$ increases by a factor of 10 between 750°C and 794°C. Calculate the activation energy for sintering in $BaTiO_3$.

10.46. Given the data in Problem 10.45, predict the temperature at which the initial sintering rate for $BaTiO_3$ would have increased **(a)** by a factor of 50 and **(b)** by a factor of 100 compared to 750°C.

10.47. At room temperature, moisture absorption will occur slowly in parts made of nylon, a common engineering polymer. Such absorption will increase dimensions and lower strength. To stabilize dimensions, nylon products are sometimes given a preliminary "moisture conditioning" by immersion in hot or boiling water. At 60°C, the time to condition a 5-mm-thick nylon part to a 2.5% moisture content is 20 hours. At 77°C, the time for the same size part is 7 hours. As the conditioning is diffusional in nature, the time required for other temperatures can be estimated from the Arrhenius expression (Equation 5.1). Calculate the activation energy for this moisture-conditioning process. (Note the method used for similar kinetics examples in Problems 10.35 and 10.41.)

10.48. Estimate the conditioning time in boiling water (100°C) for the nylon part discussed in Problem 10.47.

PART II
Materials and Their Applications

Times change. The first transistor radio had four discrete transistor devices. The contemporary integrated circuit chip resting on top of that antique has over 800 million transistors. (Photograph courtesy of Intel Corporation).

A modern building is a sophisticated collection of all categories of structural materials. (Courtesy of the University of California, Davis.)

The materials science fundamentals from Part I provide a menu of materials for structural applications. We recall that metallic, ionic, and covalent bonding roughly corresponds to the categories of metals, ceramics/glasses, and polymers. Chapters 11 and 12 identify these four categories of *structural materials* (metals, ceramics, and glasses in Chapter 11 and polymers in Chapter 12). In Chapter 12, we include a fifth category of composites (microscopic combinations of the four other categories). Chapters 11 and 12 also survey numerous processing techniques for the diverse menu of structural materials available to the engineering profession.

Materials selection is not limited to structural applications. We often choose materials because of their performance in electronic applications. Although the menu of structural materials can apply in many of these circumstances, Chapter 13 shows that a classification of materials based on electrical conductivity rather than chemical bonding produces an additional, sixth category, viz. semiconductors. These materials have values of conductivity intermediate between the generally conductive metals and the generally insulating ceramics, glasses, and polymers. Superconductors are dramatic exceptions, in that certain metals and ceramics have no electrical resistivity and provide intriguing design possibilities. Chapter 13 also reviews the wide range of semiconductor materials and the unique processing technologies that have evolved in the manufacturing of solid-state electronics.

We conclude our exploration of the field of *materials science and engineering* by focusing on *materials engineering*. Chapter 14 explores the theme of "materials in engineering design," including *materials selection* indicating how the material properties introduced throughout the book are parameters in the overall engineering design process. Materials selection is illustrated by case studies for both structural and electronic materials. A discussion of the engineering design process is not complete without acknowledging the various environmental aspects of this design process and the need to recycle materials where possible. Various forms of *environmental degradation* can limit material applications in engineering designs.

CHAPTER 11
Structural Materials—Metals, Ceramics, and Glasses

11.1 Metals

11.2 Ceramics and Glasses

11.3 Processing the Structural Materials

Reducing weight and subsequently fuel costs is a strong driving force in the selection of structural materials for modern transportation systems. These automobile steering wheels are made of a magnesium alloy that combines high strength with low weight. (Photograph courtesy of TRW.)

The materials science fundamentals from the previous ten chapters provide a menu of materials for structural applications. We recall that metallic, ionic, and covalent bonding roughly corresponds to the categories of metals, ceramics, and polymers. Chapters 11 and 12 identify, in fact, four categories of *structural materials*, with composites (combinations of the three primary categories) being the fourth. In Chapter 11, we cover metals and ceramics, along with the closely related glasses. Metals are especially versatile candidates for a wide range of structural applications. This chapter details the large family of *metal* alloys along with the processing techniques used to produce them. *Ceramics* and *glasses* are chemically similar to each other but are distinguished by their atomic-scale structure. The processing of this diverse family of materials reflects the distinctive character of the crystalline ceramics and the noncrystalline glasses. Glass-ceramics are sophisticated materials that are processed as a glass and then carefully crystallized into a strong and fracture-resistant crystalline ceramic. Together, metals, ceramics, and glasses represent the inorganic materials used in structural applications. The processing techniques for these materials are as diverse as the materials themselves.

11.1 | Metals

The related Web site contains an article "Steel Bars for Automotive Applications" that appeared in the journal Advanced Materials and Processes, *published by ASM International. Used by permission.*

As discussed in Chapter 1, probably no materials are more closely associated with the engineering profession than metals such as structural steel. In this chapter, we shall explore in greater detail the wide variety of engineering metals. We begin with the dominant examples: the iron-based or **ferrous alloys**, which include carbon and alloy steels and the cast irons. The *nonferrous alloys* are all other metals that do not contain iron as the major constituent. We shall look specifically at alloys based on aluminum, magnesium, titanium, copper, nickel, zinc, and lead, as well as the refractory and precious metals.

FERROUS ALLOYS

More than 90% by weight of the metallic materials used by human beings are ferrous alloys, which represent an immense family of engineering materials with a wide range of microstructures and related properties. The majority of engineering designs that require structural-load support or power transmission involve ferrous alloys. As a practical matter, these alloys fall into two broad categories based on the amount of carbon in the alloy composition. **Steel** generally contains between 0.05 and 2.0 wt % C. The *cast irons* generally contain between 2.0 and 4.5 wt % C. Within the steel category, we shall distinguish whether or not a significant amount of alloying elements other than carbon is used. A composition of 5 wt % total noncarbon additions will serve as an arbitrary boundary between **low-alloy** and **high-alloy steels**. These alloy additions are chosen carefully because they invariably bring with them sharply increased material costs. They are justified only by essential improvements in properties such as higher strength or improved corrosion resistance.

The majority of ferrous alloys are **carbon steels** and low-alloy steels. The reasons for this are straightforward. Such alloys are moderately priced due to the absence of large amounts of alloying elements, and they are sufficiently ductile to be readily formed. The final product is strong and durable. These eminently practical materials find applications from ball bearings to metal sheet formed into automobile bodies. A convenient designation system* for these useful alloys is given in Table 11.1. In this American Iron and Steel Institute–Society of Automotive Engineers (AISI–SAE) system, the first two numbers give a code designating the type of alloy additions and the last two or three numbers give the average carbon content in hundredths of a weight percent. As an example, a plain-carbon steel with 0.40 wt % C is a 1040 steel, whereas a steel with 1.45 wt % Cr and 1.50 wt % C is a 52150 steel. One should keep in mind that chemical compositions quoted in alloy designations such as those shown in Table 11.1 are approximate and will vary slightly from product to product within acceptable limits of industrial quality control.

An interesting class of alloys known as **high-strength, low-alloy** (HSLA) **steels** has emerged in response to requirements for weight reduction of vehicles. The compositions of many commercial HSLA steels are proprietary and they are specified by mechanical 40 properties rather than composition. A typical example, though, might contain 0.2 wt % C and about 1 wt % or less of such elements as Mn, P, Si, Cr, Ni, or Mo. The high strength of HSLA steels is the result of optimal alloy selection and carefully controlled processing, such as hot rolling (deformation at temperatures sufficiently elevated to allow some stress relief).

As mentioned earlier in this section, alloy additions must be made with care and justification because they are expensive. We shall now look at three cases in which engineering design requirements justify high-alloy compositions (i.e., total

*Alloy designations are convenient but arbitrary listings usually standardized by professional organizations such as AISI and SAE. These traditional designations tend to be as varied as the alloys themselves. In this chapter, we shall attempt, where possible, to use a more general Unified Numbering System (UNS) together with the traditional designation. A student confronted with a mysterious designation not included in this introductory text might consult *Metals and Alloys in the Unified Numbering System*, 10th ed., Society of Automotive Engineers, Warrendale, PA, and American Society for Testing and Materials, Philadelphia, PA, 2004.

TABLE 11.1

AISI–SAE Designation System for Carbon and Low-Alloy Steels

Numerals and digits[a]	Type of steel and nominal alloy content[b]	Numerals and digits	Type of steel and nominal alloy content	Numerals and digits	Type of steel and nominal alloy content
Carbon Steels		**Nickel–Chromium–Molybdenum Steels**		**Chromium Steels**	
10XX(a)	Plain carbon (Mn 1.00% max)	43XX	Ni 1.82; Cr 0.50 and 0.80; Mo 0.25	50XXX	Cr 0.50
11XX	Resulfurized	43BVXX	Ni 1.82; Cr 0.50; Mo 0.12 and 0.25; V 0.03 min	51XXX	Cr 1.02 $\;\}$ C 1.00 min
12XX	Resulfurized and rephosphorized	47XX	Ni 1.05; Cr 0.45; Mo 0.20 and 0.35	52XXX	Cr 1.45
15XX	Plain carbon (max Mn range—1.00 to 1.65%)	81XX	Ni 0.30; Cr 0.40; Mo 0.12	**Chromium–Vanadium Steels**	
Manganese Steels		86XX	Ni 0.55; Cr 0.50; Mo 0.20	61XX	Cr 0.60, 0.80. and 0.95; V 0.10 and 0.15 min
13XX	Mn 1.75	87XX	Ni 0.55; Cr 0.50; Mo 0.25	**Tungsten–Chromium Steel**	
Nickel Steels		88XX	Ni 0.55; Cr 0.50; Mo 0.35	72XX	W 1.75; Cr 0.75
23XX	Ni 3.50	93XX	Ni 3.25; Cr 1.20; Mo 0.12	**Silicon–Manganese Steels**	
25XX	Ni 5.00	94XX	Ni 0.45; Cr 0.40; Mo 0.12	92XX	Si 1.40 and 2.00; Mn 0.65, 0.82, and 0.85; Cr 0.00 and 0.65
Nickel–Chromium Steels		97XX	Ni 0.55; Cr 0.20; Mo 0.20	**High-Strength Low-Alloy Steels**	
31XX	Ni 1.25; Cr 0.65 and 0.80	98XX	Ni 1.00; Cr 0.80; Mo 0.25	9XX	Various SAE grades
32XX	Ni 1.75; Cr 1.07	**Nickel–Molybdenum Steels**		**Boron Steels**	
33XX	Ni 3.50; Cr 1.50 and 1.57	46XX	Ni 0.85 and 1.82; Mo 0.20 and 0.25	XXBXX	B denotes boron steel
34XX	Ni 3.00; Cr 0.77	48XX	Ni 3.50; Mo 0.25	**Leaded Steels**	
Molybdenum Steels		**Chromium Steels**		XXLXX	L denotes leaded steel
40XX	Mo 0.20 and 0.25	50XX	Cr 0.27, 0.40, 0.50, and 0.65		
44XX	Mo 0.40 and 0.52	51XX	Cr 0.80, 0.87, 0.92, 0.95, 1.00, and 1.05		
Chromium–Molybdenum Steels					
41XX	Cr 0.50, 0.80, and 0.95; Mo 0.12, 0.20, 0.25, and 0.30				

[a] XX or XXX in the last two or three digits of these designations indicates that the carbon content (in hundredths of a weight percent) is to be inserted.

[b] All alloy contents are expressed in weight percent.

Source: Metals Handbook, 9th ed., Vol. 1, American Society for Metals, Metals Park, OH, 1978.

noncarbon additions greater than 5 wt %). Stainless steels require alloy additions to prevent damage from a corrosive atmosphere. Tool steels require alloy additions to obtain sufficient hardness for machining applications. So-called superalloys require alloy additions to provide stability in high-temperature applications such as turbine blades.

Stainless steels are more resistant to rusting and staining than carbon and low-alloy steels, due primarily to the addition of chromium. The amount of chromium is at least 4 wt % and is usually above 10 wt %. Levels as high as 30 wt % Cr are sometimes used. Several of the common stainless steels fall into four main categories:

1. The **austenitic stainless steels** have the austenite structure retained at room temperature. As discussed in Section 3.2, γ-Fe, or austenite, has the fcc structure and is stable above 910°C. This structure can occur at room temperature when it is stabilized by an appropriate alloy addition such as nickel. While the bcc structure is energetically more stable than the fcc structure for pure iron at room temperature, the opposite is true for iron containing a significant number of nickel atoms in substitutional solid solution.

2. Without the high nickel content, the bcc structure is stable, as seen in the **ferritic stainless steels**. For many applications not requiring the high corrosion resistance of austenitic stainless steels, these lower-alloy (and less expensive) ferritic stainless steels are quite serviceable.

3. A rapid-quench heat treatment discussed in Chapter 10 allows the formation of a more complex body-centered tetragonal crystal structure called martensite. Consistent with our discussion in Section 6.3, this crystal structure yields high strength and low ductility. As a result, these **martensitic stainless steels** are excellent for applications such as cutlery and springs.

4. Precipitation hardening is another heat treatment covered in Chapter 10. Essentially, it involves producing a multiphase microstructure from a single-phase microstructure. The result is increased resistance to dislocation motion and, thereby, greater strength or hardness. **Precipitation-hardening stainless steels** can be found in applications such as corrosion-resistant structural members.

In this chapter, these four basic types of stainless steel are categorized. We leave the discussion of mechanisms of corrosion protection to Chapter 14.

Tool steels are used for cutting, forming, or otherwise shaping another material. The principal types include a plain-carbon steel (designated as W1). For shaping operations that are not too demanding, such a material is adequate. In fact, tool steels were historically of the plain-carbon variety until the mid-19th century. Now, high-alloy additions are common. Their advantage is that they can provide the necessary hardness with simpler heat treatments and retain that hardness at higher operating temperatures. The primary alloying elements used in these materials are tungsten, molybdenum, and chromium.

Superalloys include a broad class of metals with especially high strength at elevated temperatures (even above 1,000°C). Many of the stainless steels serve a dual role as heat-resistant alloys. These steels are iron-based superalloys. However, there are also cobalt- and nickel-based alloys. Most contain chromium additions for oxidation and corrosion resistance. These materials are expensive, in some cases extremely so, but the severe requirements of modern technology frequently justify such costs. Between 1950 and 1980, for example, the use of superalloys in aircraft turbojet engines rose from 10% to 50% by weight.

At this point, our discussion of steels has taken us into closely related non-ferrous alloys. Before going on to the general area of all other nonferrous alloys, we must discuss the traditional and important ferrous system, the cast irons, and the less traditional category of rapidly solidified alloys.

As stated earlier, we define **cast irons** as the ferrous alloys with greater than 2 wt % carbon. They also generally contain up to 3 wt % silicon for control of carbide formation kinetics. Cast irons have relatively low melting temperatures and liquid-phase viscosities, do not form undesirable surface films when poured, and undergo moderate shrinkage during solidification and cooling. The cast irons must balance good formability of complex shapes against inferior mechanical properties compared with those of wrought alloys.

A cast iron is formed into a final shape by pouring molten metal into a mold. The shape of the mold is retained by the solidified metal. Inferior mechanical properties result from a less uniform microstructure, including some porosity. **Wrought alloys** are initially cast but are rolled or forged into final, relatively simple shapes. (In fact, *wrought* simply means "worked.")

There are four general types of cast irons:

1. **White iron** has a characteristic white, crystalline fracture surface. Large amounts of Fe_3C are formed during casting, giving a hard, brittle material.

2. **Gray iron** has a gray fracture surface with a finely faceted structure. A significant silicon content (2 to 3 wt %) promotes graphite (C) precipitation rather than cementite (Fe_3C). The sharp, pointed graphite flakes contribute to characteristic brittleness in gray iron.

3. By adding a small amount (0.05 wt %) of magnesium to the molten metal of the gray-iron composition, spheroidal graphite precipitates rather than flakes are produced. This resulting **ductile iron** derives its name from the improved mechanical properties. Ductility is increased by a factor of 20, and strength is doubled.

4. A more traditional form of cast iron with reasonable ductility is **malleable iron**, which is first cast as white iron and then heat-treated to produce nodular graphite precipitates.

Figure 11.1 shows typical microstructures of these four cast irons.

In Section 4.5, the relatively new technology of amorphous metals was introduced. Various ferrous alloys in this category have been commercially produced. Alloy design for these systems originally involved searching for eutectic compositions, which permitted cooling to a glass transition temperature at a practical quench rate (10^5 to $10^{6\circ}$C/s). Refined alloy design has included optimizing mismatch in the size of solvent and solute atoms. Boron, rather than carbon, has been a primary alloying element for amorphous ferrous alloys. Iron–silicon alloys have been a primary example of successful commercialization of this technology. The absence of grain boundaries in these alloys helps make them among the most easily magnetized materials and especially attractive as soft-magnet transformer cores.

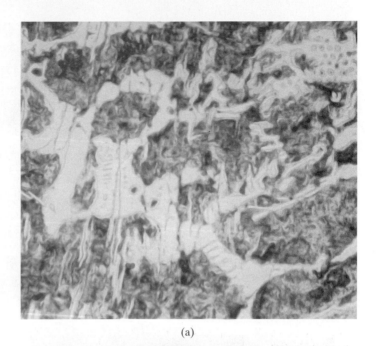

(a)

The very different microstructures of these four common cast irons correlate to their very different characteristics, appreciated in terms of the fundamental concepts introduced in the first ten chapters.

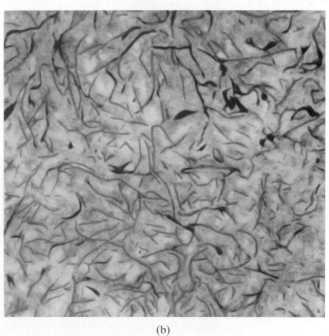

(b)

FIGURE 11.1 *Typical microstructures of (a) white iron (400 ×), eutectic carbide (light constituent) plus pearlite (dark constituent); (b) gray iron (100 ×), graphite flakes in a matrix of 20% free ferrite (light constituent) and 80% pearlite (dark constituent). (Continued)*

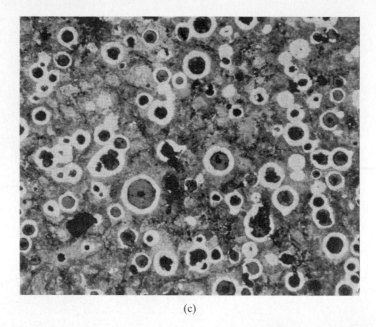

(c)

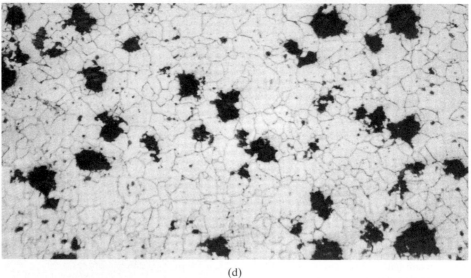

(d)

FIGURE 11.1 *(Continued) (c) ductile iron (100 ×), graphite nodules (spherulites) encased in envelopes of free ferrite, all in a matrix of pearlite; and (d) malleable iron (100 ×), graphite nodules in a matrix of ferrite. (From* Metals Handbook, *9th ed., Vol. 1, American Society for Metals, Metals Park, OH, 1978.)*

In addition to superior magnetic properties, amorphous metals have the potential for exceptional strength, toughness, and corrosion resistance. All these advantages can be related to the absence of microstructural inhomogeneities, especially grain boundaries. Rapid solidification methods do not, in all cases, produce a truly amorphous, or noncrystalline, product. Nonetheless, some unique and attractive materials have resulted as a by-product of the development of amorphous metals. An appropriate label for these novel materials (both crystalline and noncrystalline) is **rapidly solidified alloys**. Although rapid solidification may not produce a noncrystalline state for many alloy compositions, microstructures of rapidly solidified crystalline alloys are characteristically fine grained (e.g., 0.5 μm compared to 50 μm for a traditional alloy). In some cases, grain sizes less than 0.1 μm (= 100 nm) are possible and are appropriately labeled nanoscale structures. In addition, rapid solidification can produce metastable phases and novel precipitate morphologies. Correspondingly novel alloy properties are the focus of active research and development.

NONFERROUS ALLOYS

Although ferrous alloys are used in the majority of metallic applications in current engineering designs, **nonferrous alloys** play a large and indispensable role in our technology, and the list of nonferrous alloys is long and complex. We shall briefly list the major families of nonferrous alloys and their key attributes.

Aluminum alloys are best known for low density and corrosion resistance. Electrical conductivity, ease of fabrication, and appearance are also attractive features. Because of these features, the world production of aluminum roughly doubled between the 1960s and the 1970s. During the 1980s and 1990s, demand for aluminum and other metals tapered off due to increasing competition from ceramics, polymers, and composites. However, the importance of aluminum within the metals family has increased due to its low density, which is also a key factor in the increased popularity of nonmetallic materials. For example, the total mass of a new American automobile dropped by 16% between 1976 and 1986, from 1,705 kg to 1,438 kg. In large part, this was the result of a 29% decrease in the use of conventional steels (from 941 kg to 667 kg) and a 63% increase in the use of aluminum alloys (from 39 kg to 63 kg), as well as a 33% increase in the use of polymers and composites (from 74 kg to 98 kg). The total aluminum content in American automobiles further increased by 102% in the 1990s. Ore reserves for aluminum are large (representing 8% of the earth's crust), and aluminum can easily be recycled. The alloy designation system for wrought aluminum alloys is summarized in Table 11.2.

TABLE 11.2

Alloy Designation System for Aluminum Alloys	
Numerals	**Major alloying element(s)**
1XXX	None ($\geq$ 99.00% Al)
2XXX	Cu
3XXX	Mn
4XXX	Si
5XXX	Mg
6XXX	Mg and Si
7XXX	Zn
8XXX	Other elements

Source: Data from *Metals Handbook*, 9th ed., Vol. 2, American Society for Metals, Metals Park, OH, 1979.

TABLE 11.3

Temper Designation System for Aluminum Alloys[a]

Temper	Definition
F	As fabricated
O	Annealed
H1	Strain-hardened only
H2	Strain-hardened and partially annealed
H3	Strain-hardened and stabilized (mechanical properties stabilized by low-temperature thermal treatment)
T1	Cooled from an elevated-temperature shaping process and naturally aged to a substantially stable condition
T2	Cooled from an elevated-temperature shaping process, cold-worked, and naturally aged to a substantially stable condition
T3	Solution heat-treated, cold-worked, and naturally aged to a substantially stable condition
T4	Solution heat-treated and naturally aged to a substantially stable condition
T5	Cooled from an elevated-temperature shaping process and artificially aged
T6	Solution heat-treated and artificially aged
T7	Solution heat-treated and stabilized
T8	Solution heat-treated, cold-worked, and artificially aged
T9	Solution heat-treated, artificially aged, and cold-worked
T10	Cooled from an elevated-temperature shaping process, cold-worked, and artificially aged

Note: A more complete listing and more detailed descriptions are given on pages 24–27 of *Metals Handbook*, 9th ed., Vol. 2, American Society for Metals, Metals Park, OH, 1979.

[a]General alloy designation: XXXX-Temper, where XXXX is the alloy numeral from Table 11.7 (e.g., 6061–T6).

One of the most active areas of development in aluminum metallurgy is in the 8XXX-series, involving Li as the major alloying element. The Al–Li alloys provide especially low density, as well as increased stiffness. The increased cost of Li (compared to the cost of traditional alloying elements) and controlled atmosphere processing (due to lithium's reactivity) appear justified for several advanced aircraft applications.

In Chapter 10, we discussed a wide variety of heat treatments for alloys. For some alloy systems, standard heat treatments are given code numbers and become an integral part of the alloy designations. The temper designations for aluminum alloys in Table 11.3 are good examples.

Magnesium alloys have even lower density than aluminum and, as a result, appear in numerous structural applications, such as aerospace designs. Magnesium's density of 1.74 Mg/m^3 is, in fact, the lowest of any of the common structural metals. Extruded magnesium alloys have found a wide range of applications in consumer products, from tennis rackets to suitcase frames. These structural components exhibit especially high strength-to-density ratios. This is an appropriate time to recall Figure 6.24, which indicates the basis for the marked difference in characteristic mechanical behavior of fcc and hcp alloys. Aluminum is an fcc material and therefore has numerous (12) slip systems, leading to good ductility. By contrast, magnesium is hcp with only three slip systems and characteristic brittleness.

Titanium alloys have become widely used since World War II. Before that time, a practical method of separating titanium metal from reactive oxides and nitrides was not available. Once formed, titanium's reactivity works to its advantage. A thin, tenacious oxide coating forms on its surface, giving excellent resistance to corrosion. This *passivation* will be discussed further in Chapter 14. Titanium alloys, like Al and Mg, are of lower density than iron. Although more dense than Al or Mg, titanium alloys have a distinct advantage of retaining strength at moderate service temperatures (e.g., skin temperatures of high-speed aircraft), leading to numerous aerospace design applications. Titanium shares the hcp structure with magnesium, leading to characteristically low ductility. However, a high-temperature bcc structure can be stabilized at room temperature by addition of certain alloying elements, such as vanadium.

Copper alloys possess a number of superior properties. Their excellent electrical conductivity makes copper alloys the leading material for electrical wiring. Their excellent thermal conductivity leads to applications for radiators and heat exchangers. Superior corrosion resistance is exhibited in marine and other corrosive environments. The fcc structure contributes to their generally high ductility and formability. Their coloration is frequently used for architectural appearance. Widespread uses of copper alloys through history have led to a somewhat confusing collection of descriptive terms. The **brasses**, in which zinc is a predominant substitutional solute, are the most common copper alloys. Examples include yellow, naval, and cartridge brass. Applications include automotive radiators, coins, cartridge casings, musical instruments, and jewelry. The **bronzes**, copper alloys involving elements such as tin, aluminum, silicon, and nickel, provide a high degree of corrosion resistance associated with brasses, but a somewhat higher strength. The mechanical properties of copper alloys rival the steels in their variability. High-purity copper is an exceptionally soft material. The addition of 2 wt % beryllium followed by a heat treatment to produce CuBe precipitates is sufficient to push the tensile strength beyond 10^3 MPa.

Nickel alloys have much in common with copper alloys. We have already used the Cu–Ni system as the classic example of complete solid solubility (Section 4.1). *Monel* is the name given to commercial alloys with Ni–Cu ratios of roughly 2:1 by weight. These alloys are good examples of **solution hardening**, in which the alloys are strengthened by the restriction of plastic deformation due to solid-solution formation. (Recall the discussion relative to Figure 6.26.) Nickel is harder than copper, but Monel is harder than nickel. The effect of solute atoms on dislocation motion and plastic deformation was illustrated in Section 6.3. Nickel exhibits excellent corrosion resistance and high-temperature strength.

The superalloys *Inconel* (nickel–chromium–iron) and *Hastelloy* (nickel–molybdenum–iron–chromium) are important examples of superalloys. Widely used in jet engines and developed over a period of roughly 70 years, nickel-based superalloys often contain precipitates with a composition of Ni_3Al. This intermetallic compound has the same crystal structure options shown for Cu_3Au in Figure 4.3, with Ni corresponding to Cu and Al to Au. Most common is the *gamma-prime phase* precipitate with the ordered crystal structure of Figure 4.3b. The magnetic properties of various nickel alloys are covered in Chapter 13D on the related Web site.

Zinc alloys are ideally suited for die castings due to their low melting point and lack of corrosive reaction with steel crucibles and dies. Automobile parts and hardware are typical structural applications, although the extent of use in this industry is steadily diminishing for the sake of weight savings. Zinc coatings on ferrous alloys are important means of corrosion protection. This method, termed *galvanization*, will be discussed in Chapter 14.

Lead alloys are durable and versatile materials. The lead pipes installed by the Romans at the public baths in Bath, England, nearly 2,000 years ago are still in use. Lead's high density and deformability, combined with a low melting point, add to its versatility. Lead alloys find use in battery grids (alloyed with calcium or antimony), solders (alloyed with tin), radiation shielding, and sound-control structures. The toxicity of lead restricts design applications and the handling of its alloys. (Environmental pressures to identify lead-free solders are discussed in the feature box in Chapter 14.)

The **refractory metals** include molybdenum, niobium, rhenium, tantalum, and tungsten. They are, even more than the superalloys, especially resistant to high temperatures. However, their general reactivity with oxygen requires the high-temperature service to be in a controlled atmosphere or with protective coatings.

The **precious metals** include gold, iridium, osmium, palladium, platinum, rhodium, ruthenium, and silver. Excellent corrosion resistance, combined with various inherent properties, justifies the many costly applications of these metals and alloys. Gold circuitry in the electronics industry, various dental alloys, and platinum coatings for catalytic converters are a few of the better known examples. Platinum-based intermetallics, such as Pt_3Al, which, along with Ni_3Al, has a crystal structure comparable to that of Cu_3Au in Figure 4.3, are promising candidates for the next generation of jet-engine materials due to their high melting points.

Rapidly solidified ferrous alloys were introduced earlier. Research and development in this area is equally active for nonferrous alloys. Various amorphous Ni-based alloys have been developed for superior magnetic properties. Rapidly solidified crystalline aluminum and titanium alloys have demonstrated superior mechanical properties at elevated temperatures. The control of fine-grained precipitates by rapid solidification is an important factor for both of these alloy systems of importance to the aerospace industry. The interesting quasi-crystal structures discussed in the chapter on advanced structural topics on the related Web site were first produced by rapid solidification. By adding multiple alloying elements, the kinetics of crystallization can be sufficiently slowed so that **bulk amorphous alloys** can be produced. Such titanium- and zirconium-based alloys have been produced in large enough sizes to be fabricated as golf-club heads.

EXAMPLE 11.1

In redesigning an automobile for a new model year, 25 kg of conventional steel parts are replaced by aluminum alloys of the same dimensions. Calculate the resulting mass savings for the new model, approximating the alloy densities by those for pure Fe and Al, respectively.

SOLUTION

From Appendix 1, $\rho_{Fe} = 7.87$ Mg/m^3 and $\rho_{Al} = 2.70$ Mg/m^3. The volume of steel parts replaced would be

$$V = \frac{m_{Fe}}{\rho_{Fe}} = \frac{25 \text{ kg}}{7.87 \text{ Mg/m}^3} \times \frac{1 \text{ Mg}}{10^3 \text{ kg}} = 3.21 \times 10^{-3} \text{ m}^3.$$

The mass of new aluminum parts would be

$$m_{Al} = \rho_{Al} V_{Al} = 2.70 \text{ Mg/m}^3 \times 3.21 \times 10^{-3} \text{ m}^3 \times \frac{10^3 \text{ kg}}{1 \text{ Mg}} = 8.65 \text{ kg}.$$

The resulting mass savings is then

$$m_{Fe} - m_{Al} = 25 \text{ kg} - 8.65 \text{ kg} = 16.3 \text{ kg}.$$

PRACTICE PROBLEM 11.1

A common basis for selecting nonferrous alloys is their low density as compared to the density of structural steels. Alloy density can be approximated as a weighted average of the densities of the constituent elements. In this way, calculate the densities of the aluminum alloys given in Table 6.1, given that the composition of 3003-H14 is 1.25 wt.% Mn, 1.0 wt.% Mg, balance Al, and the composition of 2048 plate is 0.40 wt.% Mn, 3.3 wt.% Cu, 1.5 wt.% Mg, balance Al. **(Note that the solutions to all practice problems are provided on the related Web site.)**

11.2 | Ceramics and Glasses

A link to the American Ceramic Society Web site on the related Web site opens a vast portal to the ceramics and glass industries.

Ceramics and glasses represent some of the earliest and most environmentally durable materials for engineering. They also represent some of the most advanced materials being developed for the aerospace and electronics industries. In this section, we divide this highly diverse collection of engineering materials into three main categories. **Crystalline ceramics** include the traditional silicates and the many oxide and nonoxide compounds widely used in both traditional and advanced technologies. *Glasses* are noncrystalline solids with compositions comparable to the crystalline ceramics. The absence of crystallinity, which results from specific processing techniques, gives a unique set of mechanical and optical properties. Chemically, the glasses are conveniently subdivided as silicates and nonsilicates. *Glass-ceramics*, the third category, are another type of crystalline ceramics that are initially formed as glasses and then crystallized in a carefully controlled way. Rather specific compositions lend themselves to this technique, with the Li$_2$–Al$_2$O$_3$–SiO$_2$ system being the most important commercial example.

TABLE 11.4

Compositions[a] of Some Silicate Ceramics

Ceramic	Composition (wt %)					
	SiO_2	Al_2O_3	K_2O	MgO	CaO	Others
Silica refractory	96					4
Fireclay refractory	50–70	45–25				5
Mullite refractory	28	72				—
Electrical porcelain	61	32	6			1
Steatite porcelain	64	5		30		1
Portland cement	25	9			64	2

[a]These are approximate compositions, indicating primary components. Impurity levels can vary significantly from product to product.

CERAMICS—CRYSTALLINE MATERIALS

It is appropriate to begin our discussion of crystalline ceramics by looking at the SiO_2–based **silicates**. Since silicon and oxygen are the two most abundant elements in the earth's crust, these materials are abundant and economical. Many of the traditional ceramics that we use fall into this category. One of the best tools for characterizing early civilizations is *pottery*, burnt clayware that has been a commercial product since roughly 4000 B.C. Pottery is a part of the category of ceramics known as **whitewares**, which are commercial fired ceramics with a typically white and fine-grained microstructure. A well-known example is the translucent porcelain of which fine china is composed. In addition to pottery and related whiteware ceramics, **clay** is the basis of *structural clay products*, such as brick, tile, and sewer pipe. The range of silicate ceramics reflects the diversity of the silicate minerals that are usually available to local manufacturing plants. Table 11.4 summarizes the general compositions of some common examples. This listing includes refractories based on fired clay. **Refractories** are high-temperature-resistant structural materials that play crucial roles in industry (e.g., in the steelmaking process). About 40% of the refractories' industry output consists of the clay-based silicates. Also listed in Table 11.4 is a representative of the cement industry. The example is portland cement, a complex mixture that can be described overall as a calcium aluminosilicate.

Nonsilicate oxide ceramics include some traditional materials such as alumina (Al_2O_3), a widely used structural ceramic introduced in Chapter 1 and magnesia (MgO), a refractory widely used in the steel industry, as well as many of the more advanced ceramic materials. **Pure oxides** are compounds with impurity levels sometimes less than 1 wt % and, in some cases, impurity levels in the ppm range. The expense of chemical separation and subsequent processing of these materials is a sharp contrast to the economy of silicate ceramics made from locally available and generally impure minerals. These materials find many uses in areas such as the electronics industry, where demanding specifications are required. However, many of these products with one predominant oxide compound may contain several percent oxide additions and impurities. UO_2 is our best example of a **nuclear ceramic**. This compound contains radioactive uranium and is a widely used reactor fuel. **Partially stabilized zirconia** (ZrO_2) (PSZ) is a primary candidate for advanced structural applications, including many traditionally filled by metals. A key to the potential for metals substitution is the mechanism of transformation toughening, which was discussed in Section 8.2. **Electronic ceramics**, such as $BaTiO_3$, and the **magnetic ceramics**, such as $NiFe_2O_4$ (nickel ferrite), represent the largest part of the industrial ceramics market; they are discussed in Chapters 13B and 13D, respectively on the related Web site.

Nonoxide ceramics, such as silicon carbide, have been common industrial materials for several decades. Silicon carbide has served as a furnace heating element and as an abrasive material. Silicon nitride and related materials (e.g., the oxygen-containing SiAlON, pronounced "sigh-a-lon") have been, along with partially stabilized zirconia, the focus of substantial research and development for more than three decades for the purpose of producing superior gas-turbine components. The development of a *ceramic engine* for automobiles has been an attractive, but generally elusive goal. Nonetheless, silicon carbide-matrix ceramic composites are still attractive candidates for ultra-high-temperature jet engines.

GLASSES—NONCRYSTALLINE MATERIALS

The concept of the noncrystalline solid or **glass** was discussed in Section 4.5. As shown there, the traditional examples of this type of material are the **silicate glasses**. As with crystalline silicates, these glasses are generally moderate in cost due to the abundance of elemental Si and O in the earth's crust. For much of routine glass manufacturing, SiO_2 is readily available in local sand deposits of adequate purity. In fact, the manufacturing of various glass products accounts for a much larger tonnage than that involved in producing crystalline ceramics. Table 11.5 lists key examples of commercial silicate glassware. We can interpret the significance of the compositions in Table 11.5 by distinguishing the oxides that are network formers, modifiers, and intermediates.

Network formers include oxides such as B_2O_3 that form oxide polyhedra, with low coordination numbers. These polyhedra can connect with the network of SiO_4^{4-} tetrahedra associated with vitreous SiO_2. Alkali and alkaline earth oxides such as Na_2O and CaO do not form such polyhedra in the glass structure, but, instead, tend to break up the continuity of the polymer-like SiO_2 network. One might refer back to the schematic of an alkali–silicate glass structure in

TABLE 11.5

Compositions of Some Silicate Glasses									
Glass	**Composition (wt %)**								
	SiO_2	**B_2O_3**	**Al_2O_3**	**Na_2O**	**CaO**	**MgO**	**K_2O**	**ZnO**	**PbO**	**Others**
Vitreous silica	100									—
Borosilicate	76	13	4	5	1					1
Window	72		1	14	8	4				1
Container	73		2	14	10					1
Fiber (E-glass)	54	8	15		22					1
Bristol glaze	60		16		7		11	6		—
Copper enamel	34	3	4				17		42	—

Figure 4.23. The breaking of the network leads to the term **network modifier**. These modifiers make the glass article easier to form at a given temperature but increase its chemical reactivity in service environments. Some oxides such as Al_2O_3 and ZrO_2 are not, in themselves, glass formers, but the cation (Al^{3+} or Zr^{4+}) may substitute for the Si^{4+} ion in a network tetrahedron, thereby contributing to the stability of the network. Such oxides, which are neither formers nor modifiers, are referred to as **intermediates**.

Returning to Table 11.5, we can consider the nature of the major commercial silicate glasses:

1. **Vitreous silica** is high-purity SiO_2. *Vitreous* means "glassy" and is generally used interchangeably with *amorphous* and *noncrystalline*. With the absence of any significant network modifiers, vitreous silica can withstand service temperatures in excess of $1,000°C$. High-temperature crucibles and furnace windows are typical applications.

2. **Borosilicate glasses** involve a combination of BO_3^{3-} triangular polyhedra and SiO_4^{4-} tetrahedra in the glass-former network. About 5 wt % Na_2O provides good formability of the glassware without sacrificing the durability associated with the glass-forming oxides. The borosilicates are widely used for this durability in applications such as chemical labware and cooking ware. The great bulk of the glass industry is centered around a **soda–lime–silica** composition of approximately 15 wt % Na_2O, 10 wt % CaO, and 70 wt % SiO_2. The majority of common **window glasses** and **glass containers** can be found within a moderate range of composition.

3. The **E-glass** composition in Table 11.5 represents one of the most common glass fibers. It will be a central example of the fiber-reinforcing composite systems in Section 12.2.

4. **Glazes** are glass coatings applied to ceramics such as clayware pottery. The glaze generally provides a substantially more impervious surface than the unglazed material alone. A wide control of surface appearance is possible, as is discussed in Chapter 13C on optical properties on the related Web site.

5. **Enamels** are glass coatings applied to metals. This term must be distinguished from *enamel* as applied to polymer-based paints. Frequently more important than the surface appearance provided by the enamel is the protective barrier it provides against environments corrosive to the metal. This corrosion-prevention system will be discussed further in Chapter 14. Table 11.5 lists the composition of a typical glaze and a typical enamel.

There are also many **nonsilicate glasses**. The nonsilicate oxide glasses such as B_2O_3 are generally of little commercial value because of their reactivity with typical environments such as water vapor. However, they can be useful additions to silicate glasses (e.g., the common borosilicate glasses). Some of the nonoxide glasses have become commercially significant. For example, chalcogenide glasses are frequently semiconductors and are discussed in Chapter 13. The term *chalcogenide* comes from the Greek word *chalco*, meaning "copper" and is associated with compounds of S, Se, and Te. All three elements form strong compounds with copper, as well as with many other metal ions. Zirconium tetrafluoride (ZrF_4) glass fibers have proven to have superior light transmission properties in the infrared region compared with traditional silicates.

GLASS–CERAMICS

Among the most sophisticated ceramic materials are the **glass-ceramics**, which combine the nature of crystalline ceramics with glass. The result is a product with especially attractive qualities (Figure 11.2). Glass-ceramics begin as relatively ordinary glassware. A significant advantage is their ability to be formed into a product shape as economically and precisely as glasses. By a carefully controlled heat treatment, more than 90% of the glassy material crystallizes (see Figure 10.35). The final crystallite grain sizes are generally between 0.1 and 1 μm. The small amount of residual glass phase effectively fills the grain-boundary volume, creating a pore-free structure. The final glass-ceramic product is characterized by mechanical and thermal-shock resistance far superior to that of conventional ceramics. In Chapter 6, the sensitivity of ceramic materials to brittle failure was discussed. The resistance of glass-ceramics to mechanical shock is largely due to the elimination of stress-concentrating pores. The resistance to thermal shock results from characteristically low thermal-expansion coefficients of these materials. The significance of this concept was demonstrated in Section 7.4.

We have alluded to the importance of a carefully controlled heat treatment to produce the uniformly fine-grained microstructure of a glass-ceramic. The theory of heat treatment (the kinetics of solid-state reactions) was dealt with in Chapter 10. For now, we need to recall that the crystallization of a glass is a stabilizing process. Such a transformation begins (or is nucleated) at some impurity phase boundary. For an ordinary glass in the molten state, crystallization will tend to nucleate at a few isolated spots along the surface of the melt container. This process is followed by the growth of a few large crystals. The resulting microstructure is coarse and nonuniform. Glass-ceramics differ in that they contain several weight percent of a nucleating agent such as TiO_2. A fine dispersion of small TiO_2 particles gives a nuclei density as high as 10^{12} per cubic millimeter. There is some controversy about the exact role of nucleating agents such as TiO_2. In some cases, it appears that the TiO_2 contributes to a finely dispersed second phase of TiO_2–SiO_2 glass, which is unstable and crystallizes, thereby initiating the crystallization of the entire system. For a given composition, optimum temperatures exist for nucleating and growing the small crystallites.

FIGURE 11.2 *Cookware made of a glass-ceramic provides good mechanical and thermal properties. The casserole dish can withstand the thermal shock of simultaneous high temperature (the torch flame) and low temperature (the block of ice). (Courtesy of Corning Glass Works.)*

TABLE 11.6

Compositions of Some Glass–Ceramics								
	Composition (wt %)							
Glass-ceramic	**SiO_2**	**Li_2O**	**Al_2O_3**	**MgO**	**ZnO**	**B_2O_3**	**TiO_2^a**	**$P_2O_5^a$**
Li_2O–Al_2O_3–SiO_2 system	74	4	16				6	
MgO–Al_2O_3–SiO_2 system	65		19	9			7	
Li_2O–MgO–SiO_2 system	73	11		7		6		3
Li_2O–ZnO–SiO_2 system	58	23			16			3

[a]Nucleating agents.

Source: Data from P. W. McMillan, *Glass-Ceramics*, 2nd ed., Academic Press, Inc., New York, 1979.

Table 11.6 lists the principal commercial glass-ceramics. By far the most important example is the Li_2O–Al_2O_3–SiO_2 system. Various commercial materials in this composition range exhibit excellent thermal-shock resistance due to the low thermal-expansion coefficient of the crystallized ceramic. Examples are Corning's Corning Ware and Schott Glaswerke's Ceran. Contributing to the low expansion coefficient is the presence of crystallites of β-spodumene ($Li_2O\cdot Al_2O_3\cdot 4SiO_2$), which has a characteristically small expansion coefficient, or β-eucryptite ($Li_2O\cdot Al_2O_3\cdot SiO_2$), which actually has a negative expansion coefficient.

EXAMPLE 11.2

Common soda–lime–silica glass is made by melting together Na_2CO_3, $CaCO_3$, and SiO_2. The carbonates break down, liberating CO_2 gas bubbles that help to mix the molten glass. For 1,000 kg of container glass (15 wt % Na_2O, 10 wt % CaO, 75 wt % SiO_2), what is the raw material batch formula (weight percent of Na_2CO_3, $CaCO_3$, and SiO_2)?

SOLUTION

1,000 kg of glass consists of 150 kg of Na_2O, 100 kg of CaO, and 750 kg of SiO_2.

Using data from Appendix 1 gives us

$$\text{mol wt } Na_2O = 2(22.99) + 16.00$$

$$= 61.98 \text{ amu,}$$

$$\text{mol wt } Na_2CO_3 = 2(22.99) + 12.00 + 3(16.00)$$

$$= 105.98 \text{ amu,}$$

$$\text{mol wt } CaO = 40.08 + 16.00$$

$$= 56.08 \text{ amu,}$$

and

$$\text{mol wt } CaCO_3 = 40.08 + 12.00 + 3(16.00)$$
$$= 100.08 \text{ amu.}$$

$$Na_2CO_3 \text{ required} = 150 \text{ kg} \times \frac{105.98}{61.98} = 256 \text{ kg,}$$

$$CaCO_3 \text{ required} = 100 \text{ kg} \times \frac{100.08}{56.08} = 178 \text{ kg,}$$

and

$$SiO_2 \text{ required} = 750 \text{ kg.}$$

The batch formula is

$$\frac{256 \text{ kg}}{(256 + 178 + 750) \text{ kg}} \times 100 = 21.6 \text{ wt \% } Na_2CO_3,$$

$$\frac{178 \text{ kg}}{(256 + 178 + 750) \text{ kg}} \times 100 = 15.0 \text{ wt \% } CaCO_3,$$

and

$$\frac{750 \text{ kg}}{(256 + 178 + 750) \text{ kg}} \times 100 = 63.3 \text{ wt \% } SiO_2.$$

PRACTICE PROBLEM 11.2

In Example 11.2, we calculated a batch formula for a common soda–lime–silica glass. To improve chemical resistance and working properties, Al_2O_3 is often added to the glass by adding soda feldspar (albite), $Na(AlSi_3)O_8$, to the batch formula. Calculate the formula of the glass produced when 2,000 kg of the batch formula is supplemented with 100 kg of this feldspar.

11.3 | Processing the Structural Materials

Finally, we shall consider how various metals, ceramics, and glasses are made into forms convenient for engineering applications. This topic of *processing* is the subject of more specialized courses that may be available to many students. However, even an introductory course in materials requires a brief discussion of how those materials are produced. (Additional examples of processing will be found in Chapters 11A–B on the related Web site.) This discussion serves two functions. First, it provides a fuller understanding of the nature of each material. Second, and more important, it provides an appreciation of the effects of processing history on properties.

TABLE 11.7

Major Processing Methods for Metals	
Wrought processing	Joining
Rolling	Welding
Extrusion	Brazing
Forming	Soldering
Stamping	Powder metallurgy
Forging	Hot isostatic pressing
Drawing	Superplastic forming
Casting	Rapid solidification

PROCESSING OF METALS

Table 11.7 summarizes some of the major **processing** techniques for metals. An especially comprehensive example is the general production of steel by the **wrought process**. While the range of wrought products is large, there is a common processing history. Raw materials are combined and melted, leading eventually to a rough cast form. The casting is then worked to final product shapes. A potential problem with **casting**, as illustrated in Figure 11.3, is the presence of residual porosity (Figure 11.4). The use of mechanical deformation to form the final product shape in the wrought process largely eliminates this porosity.

Complex structural designs are generally not fabricated in a single-step process. Instead, relatively simple forms produced by wrought or casting processes are joined together. Joining technology is a broad field in itself. Our most common example is **welding**, in which the metal parts being joined are partially melted in the vicinity of the join. Welding frequently involves a metal weld rod, which is also melted. In **brazing**, the braze metal melts, but the parts being joined may not. The bond is more often formed by the solid-state diffusion of that braze metal into the joined parts. In **soldering**, neither melting nor solid-state diffusion is required. The join is usually produced by the adhesion of the melted solder to the surface of each metal part.

Figure 11.5 shows a solid-state alternative to the more conventional processing techniques. **Powder metallurgy** involves the solid-state bonding of a fine-grained powder into a polycrystalline product. Each grain in the original powder roughly corresponds to a grain in the final polycrystalline microstructure. Sufficient solid-state diffusion can lead to a fully dense product, but some residual porosity is common. This processing technique is advantageous for high-melting-point alloys and products of intricate shape. The discussion of the sintering of ceramics in Section 10.6 is relevant here also. An advance in the field of powder metallurgy is the technique of **hot isostatic pressing** (HIP), in which a uniform pressure is applied to the part using a high-temperature, inert gas. **Superplastic forming** is an economical technique developed for forming complex shapes. This process is closely associated with creep deformation. Certain fine-grained alloys can exhibit several thousand percent elongations, making the forming of complex product shapes possible. The rapid solidification of alloys was discussed earlier in this chapter in conjunction with the recent development of amorphous metals and a variety of novel crystalline microstructures. The quasi-crystals introduced in the chapter on advanced structural topics on the related Web site were originally produced as a by-product of research on rapid solidification.

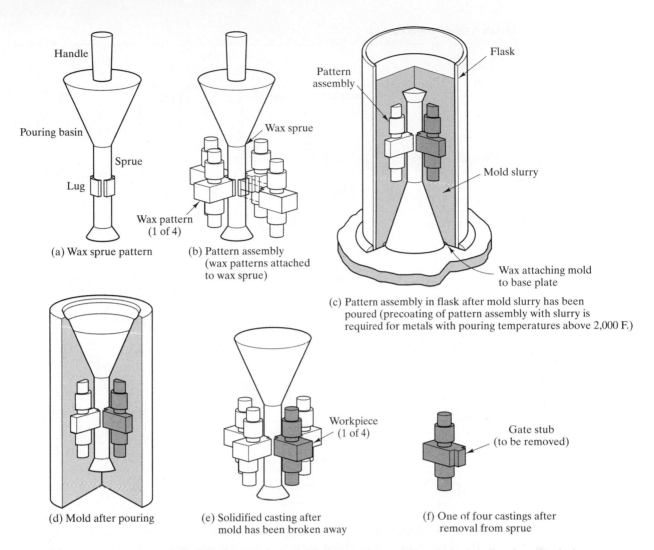

(a) Wax sprue pattern

Handle

Pouring basin

Sprue

Lug

(b) Pattern assembly
(wax patterns attached
to wax sprue)

Wax sprue

Wax pattern
(1 of 4)

Pattern
assembly

Flask

Mold slurry

Wax attaching mold
to base plate

(c) Pattern assembly in flask after mold slurry has been
poured (precoating of pattern assembly with slurry is
required for metals with pouring temperatures above 2,000 F.)

(d) Mold after pouring

(e) Solidified casting after
mold has been broken away

Workpiece
(1 of 4)

(f) One of four castings after
removal from sprue

Gate stub
(to be removed)

FIGURE 11.3 *Schematic illustration of the casting of a metal alloy form by the* investment molding *process. (From* Metals Handbook, *8th ed., Vol. 5:* Forging and Casting, *American Society for Metals, Metals Park, OH, 1970.)*

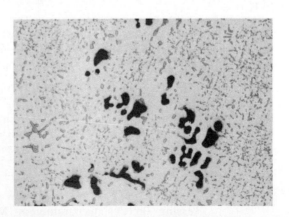

FIGURE 11.4 *Microstructure of a cast alloy (354–T4 aluminum), 50×. The black spots are voids, and gray particles are a silicon-rich phase. (From* Metals Handbook, *9th ed., Vol. 9:* Metallography and Microstructures, *American Society for Metals, Metals Park, OH, 1985.)*

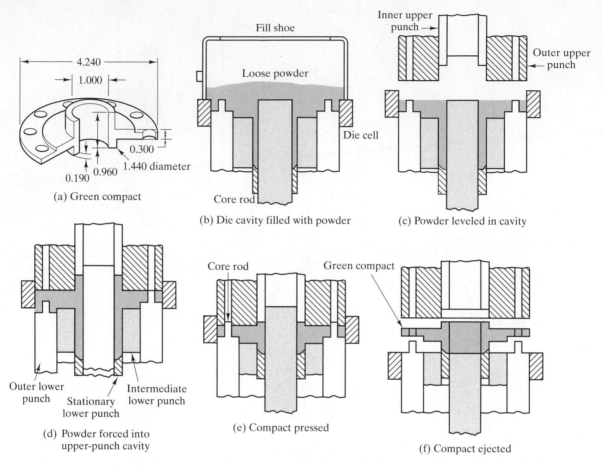

FIGURE 11.5 *Schematic illustration of powder metallurgy. The green, or unfired, compact is subsequently heated to a sufficiently high temperature to produce a strong piece by solid-state diffusion between the adjacent powder particles. (From* Metals Handbook, *8th ed., Vol. 4:* Forming, *American Society for Metals, Metals Park, OH, 1969.)*

TABLE 11.8

Some General Effects of Processing on the Properties of Metals	
Strengthening by	**Weakening by**
Cold working	Porosity (produced by casting, welding, or powder metallurgy)
Alloying (e.g., solution hardening)	Annealing
Phase transformations (e.g., martensitic)	Hot-working
	Heat-affected zone (welding)
	Phase transformations (e.g., tempered martensite)

Table 11.8 summarizes a few rules of thumb about the effects of metals processing on design parameters. Metals exhibit an especially wide range of behavior as a function of processing. As with any generalizations, we must be alert to exceptions. In addition to the fundamental processing topics discussed in this section, Table 11.8 refers to issues of microstructural development and heat

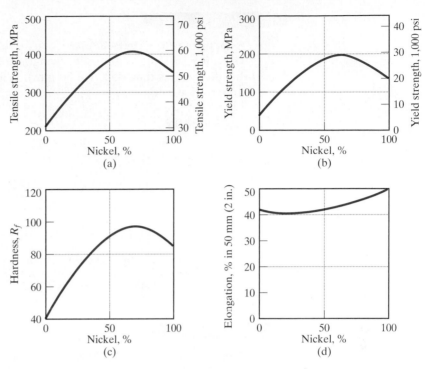

FIGURE 11.6 *Variation of mechanical properties of copper–nickel alloys with composition. Recall that copper and nickel form a complete solid-solution phase diagram (Figure 9.9). (From L. H. Van Vlack,* Elements of Materials Science and Engineering, *4th ed., Addison-Wesley Publishing Co., Inc., Reading, MA, 1980.)*

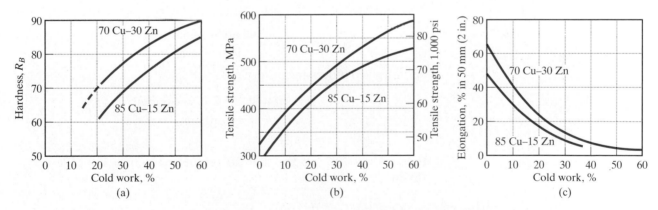

FIGURE 11.7 *Variation of mechanical properties of two brass alloys with degree of cold work. (From L. H. Van Vlack,* Elements of Materials Science and Engineering, *4th ed., Addison-Wesley Publishing Co., Inc., Reading, MA, 1980.)*

treatment discussed in Chapters 9 and 10. A specific example is given in Figure 11.6, which shows how strength, hardness, and ductility vary with alloy composition in the Cu–Ni system. Similarly, Figure 11.7 shows how these mechanical properties vary with mechanical history for a given alloy, in this case, brass. Variations in alloy chemistry and thermomechanical history allow considerable fine-tuning of structural design parameters.

D   **EXAMPLE 11.3**

A copper–nickel alloy is required for a particular structural application. The alloy must have a tensile strength greater than 400 MPa and a ductility of less than 45% (in 50 mm). What is the permissible alloy composition range?

SOLUTION

Using Figure 11.6, we can determine a "window" corresponding to the given property ranges:

$$\text{Tensile strength} > 400 \text{ MPa} : 59 < \%\text{Ni} < 79$$

and

$$\text{Elongation} < 45\% \qquad : 0 < \%\text{Ni} < 79,$$

$$\text{giving a net window of} \qquad :$$

$$\text{permissible alloy range} \quad : 59 < \%\text{Ni} < 79.$$

EXAMPLE 11.4

A bar of annealed 70 Cu–30 Zn brass (10-mm diameter) is cold-drawn through a die with a diameter of 8 mm. What is **(a)** the tensile strength and **(b)** the ductility of the resulting bar?

SOLUTION

The results are available from Figure 11.7 once the percentage of cold work is determined. This percentage is given by

$$\% \text{ cold work} = \frac{\text{initial area} - \text{final area}}{\text{initial area}} \times 100\%.$$

For the given processing history,

$$\% \text{ cold work} = \frac{\pi/4(10 \text{ mm})^2 - \pi/4(8 \text{ mm})^2}{\pi/4(10 \text{ mm})^2} \times 100\%$$

$$= 36\%.$$

From Figure 11.7, we see that **(a)** tensile strength $= 520$ MPa and **(b)** ductility (elongation) $= 9\%$.

D   **PRACTICE PROBLEM 11.3**

(a) In Example 11.3, we determine a range of copper–nickel alloy compositions that meet structural requirements for strength and ductility. Make a similar determination for the following specifications: hardness greater than 80 R_F and ductility less than 45%. **(b)** For the range of copper–nickel alloy compositions determined in part (a), which specific alloy would be preferred on a cost basis, given that the cost of copper is approximately \$5.00/kg, and the cost of nickel \$21.00/kg?

THE MATERIAL WORLD

The Foundry of the Future

Over the past 2 decades, more than 2,000 foundries across the United States have closed. These difficult economic decisions have often been based on increasingly stringent environmental health and safety laws. Concerns about the potential economic impact of a continuing trend of this sort has led to the Casting Emissions Reduction Program (CERP), a collaborative effort between government and private industry, located at McClellan (near Sacramento), California. The CERP Foundry of the Future is a 5,600-square-meter facility that can produce aluminum and gray-iron castings as large as engine blocks. The CERP mission is to help the metal-casting industry remain competitive with international foundries while meeting federal clean-air standards. The overall objective is to maintain manufacturing jobs while contributing to a clean environment. A specific objective is to help reduce emissions

from the foundry industry, which employs more than 200,000 workers in the United States alone.

The CERP foundry produces high-quality castings using environmentally sound technologies and new materials. The casting equipment within the Foundry of the Future represents the state of the art, enhanced with specialized ventilation and monitoring systems to capture air emissions. Collaboration with the American Industry/Government Emission Research (AIGER) project involves research on emission-measurement technologies for both stationary (e.g., foundry) and mobile (e.g., automobile) sources. The resulting information and research data are shared with the general foundry industry. CERP is operated under contract by Technikon, LLC, and is managed by the U.S. Army Industrial Ecology Center (IEC).

[Courtesy of the Casting Emissions Reduction Program (CERP).]

PRACTICE PROBLEM 11.4

In Example 11.4, we calculate the tensile strength and ductility for a cold-worked bar of 70 Cu–30 Zn brass. **(a)** What percent increase does that tensile strength represent compared to that for the annealed bar? **(b)** What percent decrease does that ductility represent compared to that for the annealed bar?

TABLE 11.9

Some Major Processing Methods for Ceramics and Glasses
Fusion casting
Slip casting
Sintering
Hot isostatic pressing (HIP)
Glass forming
Controlled devitrification
Sol–gel processing
Biomimetic processing
Self-propagating high temperature synthesis (SHS)

PROCESSING OF CERAMICS AND GLASSES

Table 11.9 summarizes some of the major processing techniques for ceramics and glasses. Many of these techniques are direct analogs of metallic processing introduced in Table 11.7. However, wrought processing does not exist per se for ceramics. The deformation forming of ceramics is limited by their inherent brittleness. Although cold working and hot working are not practical, a wider variety of casting techniques are available. **Fusion casting** refers to a process equivalent to metal casting. This technique is not a predominant one for ceramics because of their generally high melting points. Some low-porosity refractories are formed in this way, but at a relatively high cost. **Slip casting**, shown in Figure 11.8, is a more typical ceramic processing technique. Here, the casting is done at room temperature. The "slip" is a powder–water mixture that is poured into a porous mold. Much of the water is absorbed by the mold, leaving a relatively rigid powder form that can be removed from the mold. To develop a strong product, the piece must be heated. Initially, the remaining absorbed water is driven off. **Firing** is done at higher temperatures, typically above $1,000°C$. As in powder metallurgy, much of the strength of the fired piece is due to solid-state diffusion. For many ceramics, especially clayware, additional high-temperature reactions are involved. Chemically combined water can be driven off, various phase transformations can take place, and substantial glassy phases, such as silicates, can be formed. Sintering is the direct analog of powder metallurgy. This subject was introduced in Section 10.6. The high melting points of common ceramics make sintering a widespread processing technique. As with powder metallurgy, hot isostatic pressing is finding increased applications in ceramics, especially in providing fully dense products with superior mechanical properties. A typical **glass-forming** process is shown in Figure 11.9. The viscous nature of the glassy state plays a central role in this processing (see also Section 6.6). **Controlled devitrification** (i.e., crystallization) leads to the formation of glass-ceramics. This topic was raised earlier (see Sections 10.6 and earlier in this section). **Sol–gel processing** is among the more rapidly developing technologies for fabricating ceramics and glasses. For ceramics, the method provides for the formation of uniform, fine particulates of high purity at relatively low temperatures. Such powders can subsequently be sintered to high density with correspondingly good mechanical properties. In such techniques, the essential feature is the formation of an organometallic solution. The dispersed phase "sol" is then converted

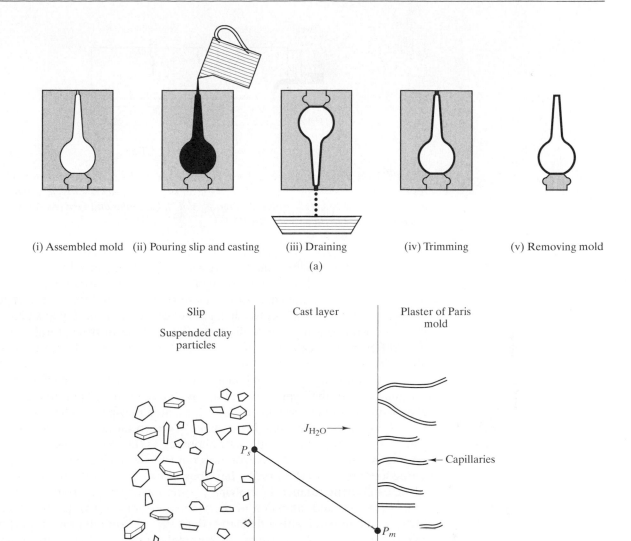

(i) Assembled mold   (ii) Pouring slip and casting   (iii) Draining   (iv) Trimming   (v) Removing mold

(a)

FIGURE 11.8 *(a) Schematic illustration of the slip casting of ceramics. The slip is a powder–water mixture. (After F. H. Norton,* Elements of Ceramics, *2nd ed., Addison-Wesley Publishing Co., Inc., Reading, MA, 1974.) (b) Much of that water is absorbed into the porous mold. The final form must be fired at elevated temperatures to produce a structurally strong piece. (From W. D. Kingery, H. K. Bowen, and D. R. Uhlmann,* Introduction to Ceramics, *2nd ed., John Wiley & Sons, Inc., New York, 1976.)*

into a rigid "gel," which, in turn, is reduced to a final composition by various thermal treatments. A key advantage of the sol–gel process is that the product formed initially through this liquid phase route can be fired at lower temperatures, compared with the temperatures used in conventional ceramic processes. The cost savings of the lower firing temperatures can be significant.

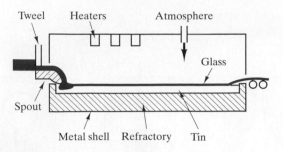

FIGURE 11.9 *The high degree of flatness achieved in modern architectural plate glass is the result of the float glass process in which the layer of glass is drawn across a bath of molten tin. (After* Engineered Materials Handbook, *Vol. 4, Ceramics and Glasses, ASM International, Materials Park, OH, 1991.)*

Recently, ceramic engineers have realized that certain natural ceramic fabrication processes, such as the formation of sea shells, take the liquid phase processing route to its ultimate conclusion. Scientists have noted that the formation of an abalone shell that takes place in an aqueous medium entirely at ambient temperature, with no firing step at all. Attractive features of this natural bioceramic, in addition to fabrication at ambient conditions from readily available materials, include a final microstructure that is fine grained with an absence of porosity and microcracks and a resulting high strength and fracture toughness. Such bioceramics are normally produced at a slow buildup rate from a limited range of compositions, usually calcium carbonate, calcium phosphate, silica, or iron oxide.

Biomimetic processing is the name given to fabrication strategies for engineering ceramics that imitate natural processes (i.e., low-temperature, aqueous syntheses of oxides, sulfides, and other ceramics by adapting biological principles). Three key aspects of this process have been identified: (1) the occurrence within specific microenvironments (implying stimulation of crystal production at certain functional sites and inhibition of the process at other sites); (2) the production of a specific mineral with a defined crystal size and orientation; and (3) macroscopic growth by packaging many incremental units together (resulting in a unique composite structure and accommodating later stages of growth and repair). This natural process occurs for bone and dental enamel, as well as the shells. Biomimetic engineering processes have not yet been able to duplicate the sophisticated level of control exhibited in the natural materials. Nonetheless, promising results have come from efforts in this direction. A simple example is the addition of water-soluble polymers to portland cement mixes, thereby reducing freeze-thaw damage by inhibiting the growth of large ice crystals. The ceramic-like cement particles resemble biological hard tissue. The polymer addition can change hardening reactions, microstructure, and properties of cement products in the same way that extracellular biopolymers contribute to the properties of bones and shells. Current research in biomimetic processing is centered on the control of crystal nucleation and growth in advanced ceramics using inorganic and organic polymers, as well as biopolymers.

A final note of importance with regard to biomimetic processing is that an additional attractive feature of the natural formation of bones and shells is that it represents **net-shape processing** (i.e., the "product," once formed, does not require a final shaping operation). Much of the effort in recent years for both engineered metals and ceramics can be described as **near net-shape processing**, in which the goal is to minimize any final shaping operation. The superplastic forming of metal alloys and the sol–gel processing of ceramics and glass are such examples. By contrast, biominerals are formed as relatively large, dense parts in a "moving front" process in which incremental matrix-defined units are sequentially mineralized. The resulting net-shape forming of a dense material represents an exceptional level of microstructural control.

A final type of ceramic processing, **self-propagating high-temperature synthesis** (SHS), is a novel technique that involves the use of the substantial heat evolved by certain chemical reactions, once initiated, to sustain the reaction and produce the final product. For example, igniting a reaction between titanium powder in a nitrogen gas atmosphere produces an initial amount of titanium nitride $[Ti(s) + 1/2\, N_2(g) = TiN(s)]$. The substantial heat evolved in this highly exothermic reaction can be enough to produce a combustion wave that passes through the remaining titanium powder, sustaining the reaction and converting all of the titanium to a TiN product. Although high temperatures are involved, SHS shares an advantage of the relatively low temperature sol–gel and biomimetic processing (i.e., energy savings, in that much of the high temperature in SHS comes from the self-sustaining reactions). In addition, SHS is a simple process that provides relatively pure products and the possibility of simultaneous formation and densification of the product. A wide variety of materials is formed in this way. These materials are generally produced in powder form, although dense products can be formed by subsequent sintering or the simultaneous application of pressure or casting operations during the combustion. In addition to forming ceramics, intermetallic compounds (e.g., TiNi) and composites can be produced by SHS.

EXAMPLE 11.5

In firing 5 kg of kaolinite, $Al_2(Si_2O_5)(OH)_4$, in a laboratory furnace to produce an aluminosilicate ceramic, how much H_2O is driven off?

SOLUTION
As in Example 9.12, note that

$$Al_2(Si_2O_5)(OH)_4 = Al_2O_3 \cdot 2SiO_2 \cdot 2H_2O$$

and

$$Al_2O_3 \cdot 2SiO_2 \cdot 2H_2O \xrightarrow{\text{heat}} Al_2O_3 \cdot 2SiO_2 + 2H_2O \uparrow.$$

Then,

$$1 \text{ mol } Al_2O_3 \cdot 2SiO_2 \cdot 2H_2O = [2(26.98) + 3(16.00)] \text{ amu}$$
$$+ 2[28.09 + 2(16.00)] \text{ amu}$$
$$+ 2[2(1.008) + 16.00] \text{ amu}$$
$$= 258.2 \text{ amu},$$

and

$$2 \text{ mol } H_2O = 2[2(1.008) + 16.00] \text{ amu} = 36.03 \text{ amu}.$$

As a result, the mass of H_2O driven off will be

$$m_{H_2O} = \frac{36.03 \text{ amu}}{258.2 \text{ amu}} \times 5 \text{ kg} = 0.698 \text{ kg} = 698 \text{ g}.$$

EXAMPLE 11.6

Assume a glass bottle is formed at a temperature of 680°C with a viscosity of 10^7 P. If the activation energy for viscous deformation for the glass is 460 kJ/mol, calculate the annealing range for this product.

SOLUTION

Following the methods described in Section 6.6, we have an application for Equation 6.20:

$$\eta = \eta_0 e^{+Q/RT}.$$

For 680°C = 953 K,

$$10^7 \text{P} = \eta_0 e^{+(460 \times 10^3 \text{ J/mol})/[8.314 \text{ J/(mol·K)}](953 \text{ K})},$$

or

$$\eta_0 = 6.11 \times 10^{-19} \text{ P}.$$

For the annealing range, $\eta = 10^{12.5}$ to $10^{13.5}$ P. For $\eta = 10^{12.5}$ P,

$$T = \frac{460 \times 10^3 \text{ J/mol}}{[8.314 \text{ J/(mol · K)}] \ln(10^{12.5}/6.11 \times 10^{-19})}$$

$$= 782 \text{ K} = 509°C.$$

For $\eta = 10^{13.5}$ P,

$$T = \frac{460 \times 10^3 \text{ J/mol}}{[8.314 \text{ J/(mol · K)}] \ln(10^{13.5}/6.11 \times 10^{-19})}$$

$$= 758 \text{ K} = 485°C.$$

Therefore,

$$\text{annealing range} = 485°C \text{ to } 509°C.$$

PRACTICE PROBLEM 11.5

In scaling up the laboratory firing operation of Example 11.5 to a production level, 6.05×10^3 kg of kaolinite are fired. How much H_2O is driven off in this case?

PRACTICE PROBLEM 11.6

For the glass-bottle production described in Example 11.6, calculate the melting range for this manufacturing process.

Summary

Metals play a major role in engineering design, especially as structural elements. Over 90% by weight of the materials used for engineering are iron-based, or ferrous, alloys, which include the steels (containing 0.05 to 2.0 wt % C) and the cast irons (with 2.0 to 4.5 wt % C). Nonferrous alloys include a wide range of materials with individual attributes. Aluminum, magnesium, and titanium alloys have found wide use as lightweight structural members. Copper and nickel alloys are especially attractive for chemical and temperature resistance and electrical and magnetic applications. Other important nonferrous alloys include the zinc and lead alloys and the refractory and precious metals.

Ceramics and glasses represent a diverse family of engineering materials. The term *ceramics* is associated with predominantly crystalline materials. Glasses are noncrystalline solids chemically similar to the crystalline ceramics. The glass-ceramics are distinctive products that are initially processed as glasses and then carefully crystallized to form a dense, fine-grained ceramic product with excellent mechanical and thermal shock resistance.

Many applications of materials in engineering design are dependent on the processing of the material. Many common metal alloys are produced in the wrought process, in which a simple cast form is mechanically worked into a final shape. Other alloys are produced directly by casting. More complex structural shapes depend on joining techniques such as welding. An alternative to wrought and casting processes is the entirely solid-state technique of powder metallurgy. Contemporary metal-forming techniques include hot isostatic pressing, superplastic forming, and rapid solidification. Ceramics can be formed by fusion casting, which is similar to metal casting. Slip casting is more common. Sintering and HIP are directly analogous to powder metallurgy methods. Glass forming involves careful control of the viscosity of the supercooled silicate liquid. Glass-ceramics require the additional step of controlled devitrification to form a fine-grained, fully crystalline product. Sol–gel processing, biomimetic processing, and SHS are all energy-saving processes for producing economical products of high purity.

Key Terms

Metals
austenitic stainless steel (351)
brass (357)
brazing (366)
bronze (357)
bulk amorphous alloy (358)
carbon steel (349)
casting (366)
cast iron (352)
ductile iron (352)
ferritic stainless steel (351)
ferrous alloy (348)
gray iron (352)
high-alloy steel (349)
high-strength low-alloy steel (349)

hot isostatic pressing (HIP) (366)
low-alloy steel (349)
malleable iron (352)
martensitic stainless steel (351)
nonferrous alloy (355)
powder metallurgy (366)
precious metal (358)
precipitation-hardening stainless
    steel (351)
processing (366)
rapidly solidified alloy (355)
refractory metal (358)
soldering (366)
solution hardening (357)
steel (349)

superalloy (351)
superplastic forming (366)
tool steel (351)
welding (366)
white iron (352)
wrought alloy (352)
wrought process (366)
Ceramics and Glasses
biomimetic processing (374)
borosilicate glass (362)
clay (360)
controlled devitrification (372)
crystalline ceramic (359)
E-glass (362)
electronic ceramic (361)

References

Metals

Ashby, **M. F.**, and **D. R. H. Jones**, *Engineering Materials 1—An Introduction to Their Properties, Applications, and Design*, 3rd ed., Butterworth-Heinemann, Boston, 2005.

Ashby, **M. F.**, and **D. R. H. Jones**, *Engineering Materials 2— An Introduction to Microstructures, Processing and Design*, 2nd ed., Butterworth-Heinemann, Boston, 2006.

Davis, **J. R.**, Ed., *Metals Handbook*, Desk Ed., 2nd ed., ASM International, Materials Park, OH, 1998. A one-volume summary of the extensive *Metals Handbook* series.

ASM Handbook, Vols. 1 (*Properties and Selection: Irons, Steels, and High-Performance Alloys*) and 2 (*Properties and Selection: Nonferrous Alloys and Special-Purpose Metals*), ASM International, Materials Park, OH, 1990 and 1991.

Ceramics and Glasses

Chiang, **Y.**, **D. P. Birnie III**, and **W. D. Kingery**, *Physical Ceramics*, John Wiley & Sons, Inc., New York, 1997.

Doremus, **R. H.**, *Glass Science*, 2nd ed., John Wiley & Sons, Inc., New York, 1994.

Engineered Materials Handbook, Vol. 4, *Ceramics and Glasses*, ASM International, Materials Park, OH, 1991.

Reed, **J. S.**, *Principles of Ceramic Processing*, 2nd ed., John Wiley & Sons, Inc., New York, 1995.

Problems

11.1 • Metals

11.1. **(a)** Estimate the density of 1040 carbon steel as the weighted average of the densities of constituent elements.

(b) The density of 1040 steel is what percentage of the density of pure Fe?

11.2. Repeat Problem 11.1 for the type 304 stainless steel in Table 6.1.

11.3. A prototype Al–Li alloy is being considered for replacement of a 7075 alloy in a commercial aircraft. The compositions are compared in the following table. **(a)** Assuming that the same volume of material is used, what percentage reduction in density would occur by this material substitution? **(b)** If a total mass of 75,000 kg of 7075 alloy is currently used on the aircraft, what net mass reduction would result from the substitution of the Al–Li alloy?

	Primary alloying elements (wt %)					
Alloy	**Li**	**Zn**	**Cu**	**Mg**	**Cr**	**Zr**
Al–Li	2.0		3.0			0.12
7075		5.6	1.6	2.5	0.23	

11.4. Estimate the alloy densities for **(a)** the magnesium alloys of Table 6.1 and **(b)** the titanium alloys of Table 6.1.

11.5. Between 1975 and 1985, the volume of all iron and steel in a given automobile model decreased from 0.162 m^3 to 0.116 m^3. In the same time frame, the volume of all aluminum alloys increased from 0.012 m^3 to 0.023 m^3. Using the densities of pure Fe and Al, estimate the mass reduction resulting from this trend in materials substitution.

11.6. For an automobile design equivalent to the model described in Problem 11.5, the volume of all iron and steel is further reduced to 0.082 m^3 by the year 2000. In the same time frame, the total volume of aluminum alloys increases to 0.034 m^3. Estimate the mass reduction (compared to 1975) resulting from this materials substitution.

11.7. Estimate the density of a cobalt–chrome alloy used for artificial-hip joints as a weighted average of the densities of constituent elements: 50 wt % Co, 20 wt % Cr, 15 wt % W, and 15 wt % Ni.

11.8. Consider an artificial-hip component made of the cobalt–chrome alloy of Problem 11.7 with a volume of 160×10^{-6} m^3. What mass savings would result from substituting a Ti–6 Al–4 V alloy with the same component shape and consequently the same volume?

11.2 • Ceramics and Glasses

11.9. As pointed out in the discussion relative to the Al_2O_3–SiO_2 phase diagram (Figure 9.23), an alumina-rich mullite is desirable to ensure a more refractory (temperature-resistant) product. Calculate the composition of a refractory made by adding 2.5 kg of Al_2O_3 to 100 kg of stoichiometric mullite.

11.10. A fireclay refractory of simple composition can be produced by heating the raw material kaolinite, $Al_2(Si_2O_5)(OH)_4$, driving off the waters of hydration. Calculate the composition (weight percent basis) for the resulting refractory. (Note that this process was introduced in Example 9.12 relative to the Al_2O_3–SiO_2 phase diagram.)

11.11. Using the results of Problem 11.10 and Example 9.12, calculate the weight percent of SiO_2 and mullite present in the final microstructure of a fireclay refractory made by heating kaolinite.

11.12. Estimate the density of **(a)** a PSZ (with 4 wt % CaO) as the weighted average of the densities of ZrO_2 ($= 5.60$ Mg/m^3) and CaO ($= 3.35$ Mg/m^3) and **(b)** a fully stabilized zirconia with 8 wt % CaO.

11.13. The primary reason for introducing ceramic components in automotive engine designs is the possibility of higher operating temperatures and, therefore, improved efficiencies. A by-product of this substitution, however, is mass reduction. For the case of 2 kg of cast iron (density $= 7.15$ Mg/m^3) being replaced by an equivalent volume of PSZ (density $= 5.50$ Mg/m^3), calculate the mass reduction.

11.14. Calculate the mass reduction attained if silicon nitride (density $= 3.18$ Mg/m^3) is used in place of 2 kg of cast iron (density $= 7.15$ Mg/m^3).

11.15. A batch formula for a window glass contains 400 kg Na_2CO_3, 300 kg $CaCO_3$, and 1,300 kg SiO_2. Calculate the resulting glass formula.

11.16. For the window glass in Problem 11.15, calculate the glass formula if the batch is supplemented by 100 kg of lime feldspar (anorthite), $Ca(Al_2Si_2)O_8$.

11.17. An economical substitute for vitreous silica is a high-silica glass made by leaching the B_2O_3-rich phase from a two-phase borosilicate glass. (The resulting porous microstructure is densified by heating.) A typical starting composition is 81 wt % SiO_2, 4 wt % Na_2O, 2 wt % Al_2O_3, and 13 wt % B_2O_3. A typical final composition is 96 wt % SiO_2, 1 wt % Al_2O_3, and 3 wt % B_2O_3. How much product (in kilograms) would be produced from 100 kg of starting material, assuming no SiO_2 is lost by leaching?

11.18. How much B_2O_3 (in kilograms) is removed by leaching in the glass-manufacturing process described in Problem 11.17?

11.19. A novel electronic material involves the dispersion of small silicon particles in a glass matrix. These *quantum dots* are discussed in Chapter 13. If 4.85×10^{16} particles of Si are dispersed per mm^3 of glass corresponding to a total content of 5 wt %, calculate the average particle size of the quantum dots. (Assume spherical particles and note that the density of the silicate glass matrix is 2.60 Mg/m^3.)

11.20. Calculate the average separation distance between the centers of adjacent Si particles in the quantum dot material of Problem 11.19. (For simplicity, assume a simple cubic array of dispersed particles.)

11.21. Assuming the TiO_2 in a Li_2O–Al_2O_3–SiO_2 glass-ceramic is uniformly distributed with a dispersion of 10^{12} particles per cubic millimeter and a total amount of 6 wt %, what is the average particle size of the TiO_2 particles? (Assume spherical particles. The density of the glass-ceramic is 2.85 Mg/m^3, and the density of the TiO_2 is 4.26 Mg/m^3.)

11.22. Repeat Problem 11.21 for a 3 wt % dispersion of P_2O_5 with a concentration of 10^{12} particles per cubic millimeter. (The density of P_2O_5 is 2.39 Mg/m^3.)

11.23. What is the overall volume percent of TiO_2 in the glass-ceramic described in Problem 11.21?

11.24. What is the overall volume percent of P_2O_5 in the glass-ceramic described in Problem 12.22?

11.25. Calculate the average separation distance between the centers of adjacent TiO_2 particles in the glass-ceramic described in Problems 11.21 and 11.23. (Note Problem 11.20.)

11.26. Calculate the average separation distance between the centers of adjacent P_2O_5 particles in the glass-ceramic described in Problems 11.22 and 11.24. (Note Problem 11.20.)

11.3 • Processing the Structural Materials

At the outset of the text, we established a policy of avoiding subjective homework problems. Questions about materials processing quickly turn to the subjective. As a result, Section 11.3 contains only a few problems that can be kept in the objective style used in preceding chapters.

11.27. A bar of annealed 85 Cu–15 Zn (12-mm diameter) is cold-drawn through a die with a diameter of 10 mm. What are **(a)** the tensile strength and **(b)** the ductility of the resulting bar?

11.28. For the bar analyzed in Problem 11.27, **(a)** what percentage does the tensile strength represent compared to that for the annealed bar, and **(b)** what percentage decrease does the ductility represent compared to that for the annealed bar?

D11.29. You are given a 2-mm-diameter wire of 85 Cu–15 Zn brass. It must be drawn down to a diameter of 1 mm. The final product must meet specifications of tensile strength greater than 375 MPa and a ductility of greater than 20%. Describe a processing history to provide this result.

D11.30. How would your answer to Problem 11.29 change if a 70 Cu–30 Zn brass wire were used instead of the 85 Cu–15 Zn material?

11.31. For the ceramic in Example 11.5, use the Al_2O_3–SiO_2 phase diagram in Chapter 9 to determine the maximum firing temperature to prevent formation of a silica-rich liquid.

11.32. How would your answer to Problem 11.31 change if the ceramic being fired was composed of two parts Al_2O_3 in combination with one part kaolin?

11.33. For simplified processing calculations, glass-bottle shaping temperature can be taken as the softening point (at which $\eta = 10^{7.6}$ P), and the subsequent annealing temperature can be taken as the annealing point (at which $\eta = 10^{13.4}$ P). If the glass-bottle processing sequence involves a shaping temperature of 700°C for a glass with an activation energy for viscous deformation of 475 kJ/mol, calculate the appropriate annealing temperature.

11.34. Using the approach of Problem 11.33, assume a change in raw-material suppliers changes the glass composition, thus reducing the softening point to 690°C and the activation energy to 470 kJ/mol. Recalculate the appropriate annealing temperature.

11.35. How many grams of N_2 gas are consumed in forming 100 g of TiN by SHS?

11.36. How many grams of Ti powder were initially required in the SHS process of Problem 11.35?

D11.37. Using the data on modulus of elasticity and strength (modulus of rupture) in Table 6.4, select the sintered ceramic that would meet the following design specifications for a furnace refractory application:

modulus of elasticity, E: $< 350 \times 10^3$ MPa

and

modulus of rupture, MOR: > 125 MPa.

D11.38. Repeat the materials-selection exercise of Problem 11.37 for any of the ceramics and glasses of Table 6.4 produced by any processing technique, including hot pressing and glass forming.

D11.39. In designing a bushing of TiN for an aerospace application, the design engineer determines that the part should have a mass of 78 g. How much starting Ti powder will she need in order to produce this part by SHS? (Note also Problem 11.36.)

D11.40. How much nitrogen gas is consumed in producing the bushing of TiN in Problem 11.39 by SHS?

D11.41. The bushing of TiN in Problem 11.39 needs to be redesigned with a resulting mass of 97 g. How much starting Ti powder would be required in this case?

D11.42. How much nitrogen gas is consumed in producing the bushing of TiN in Problem 11.41?

CHAPTER 12
Structural Materials—Polymers and Composites

12.1 Polymers

12.2 Composites

12.3 Processing the Structural Materials

This athlete is enjoying some of the latest advances in polymers and composites. These materials meet the demanding mechanical requirements of the inline skates while minimizing weight.

Chapter 11 covered the inorganic structural materials: metals, ceramics, and glasses. The *polymers* represent a large family of organic structural materials. The organic nature of these covalently bonded materials makes them attractive alternatives to traditional metal alloys and allows some unique processing techniques. The *composites* are defined as microscopic-scale combinations of metals, ceramics, glasses, and/or polymers. Many of our most common examples have a polymer matrix and can be considered predominantly organic materials also. Fiberglass is a traditional example, combining high-modulus glass fibers in a ductile polymer matrix. Some of the most sophisticated structural materials are advanced composites that provide especially impressive combinations of properties not available from individual components by themselves. Of course, not all composite systems involve an organic phase. Metal matrix composites with metal or ceramic dispersed phases and ceramic-matrix composites with ceramic dispersed phases are important examples. The processing techniques for composites represent the full spectrum of methods used to produce their individual components.

12.1 | Polymers

A link to the Society of Plastics Engineers Web site on the related Web site opens a vast portal to the field of engineering polymers.

We follow our discussions of metals and ceramics with a third category of structural materials, polymers. A common synonym for polymers is **plastics**, a name derived from the deformability associated with the fabrication of most polymeric products. To some critics, *plastic* is a synonym for modern culture. Accurate or not, it represents the impact that this complex family of engineering materials has had on our society. Polymers, or plastics, are available in a wide variety of commercial forms: fibers, thin films and sheets, foams, and in bulk.

The metals, ceramics, and glasses we have considered in previous chapters are inorganic materials. The polymers discussed in this chapter are organic. Our decision to limit the discussion to organic polymers is a common one, though it is somewhat arbitrary. Several inorganic materials have structures composed of building blocks connected in chain and network configurations. We occasionally

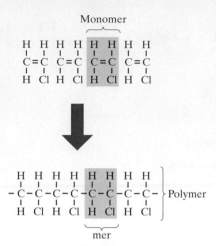

Monomer

FIGURE 12.1 *Polymerization is the joining of individual monomers (e.g., vinyl chloride, C_2H_3Cl) to form a polymer [$(C_2H_3Cl)_n$] consisting of many mers (again, C_2H_3Cl).*

point out that silicate ceramics and glasses are examples. (This section, together with Chapter 2, will provide any of the fundamentals of organic chemistry needed to appreciate the unique nature of polymeric materials.)

POLYMERIZATION

The term **polymer** simply means "many mers," where **mer** is the building block of the long-chain or network molecule. Figure 12.1 shows how a long-chain structure results from the joining together of many **monomers** by chemical reaction. **Polymerization**, the process by which long-chain or network molecules are made from relatively small organic molecules, can take place in two distinct ways. **Chain growth**, or **addition polymerization**, involves a rapid chain reaction of chemically activated monomers. **Step growth**, or **condensation polymerization**, involves individual chemical reactions between pairs of reactive monomers and is a much slower process. In either case, the critical feature of a monomer, which permits it to join with similar molecules and form a polymer, is the presence of reactive sites—*double bonds* in chain growth, or reactive functional groups in step growth. As discussed in Chapter 2, each covalent bond is a pair of electrons shared between adjacent atoms. The double bond is two such pairs. The chain growth reaction in Figure 12.1 converts the double bond in the monomer to a single bond in the mer. The remaining two electrons become parts of the single bonds joining adjacent mers.

Figure 12.2 illustrates the formation of polyethylene by the process of chain growth. The overall reaction can be expressed as

$$nC_2H_4 \rightarrow \{C_2H_4\}_n. \tag{12.1}$$

The process begins with an **initiator**—a hydroxyl *free radical* in this case. A free radical is a reactive atom or group of atoms containing an unpaired electron. The initiation reaction converts the double bond of one monomer into a single bond. Once completed, the one unsatisfied bonding electron (see step $1'$ of Figure 12.2) is free to react with the nearest ethylene monomer, extending the molecular chain by one unit (step 2). This chain reaction can continue in rapid succession limited only by the availability of unreacted ethylene monomers. The rapid progression

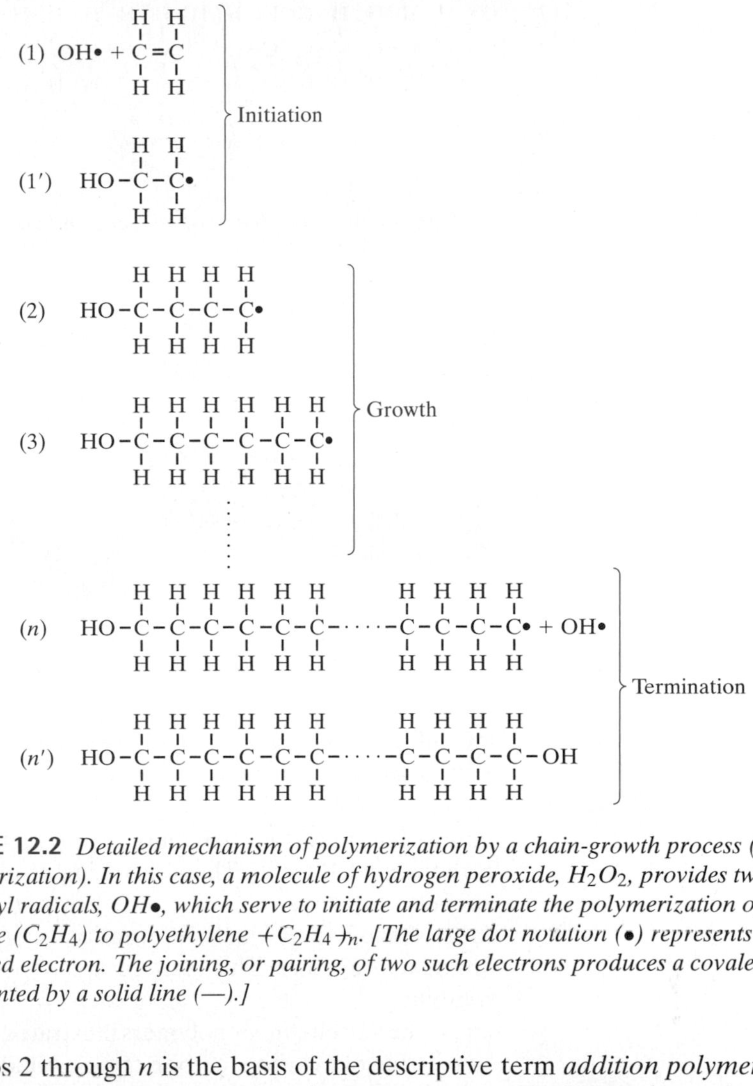

FIGURE 12.2 *Detailed mechanism of polymerization by a chain-growth process (addition polymerization). In this case, a molecule of hydrogen peroxide, H_2O_2, provides two hydroxyl radicals, OH•, which serve to initiate and terminate the polymerization of ethylene (C_2H_4) to polyethylene* $\text{+}C_2H_4\text{+}_n$. *[The large dot notation (•) represents an unpaired electron. The joining, or pairing, of two such electrons produces a covalent bond, represented by a solid line (—).]*

of steps 2 through *n* is the basis of the descriptive term *addition polymerization.* Eventually, another hydroxyl radical can act as a **terminator** (step *n*), giving a stable molecule with *n* mer units (step *n'*). For the specific case of hydroxyl groups as initiators and terminators, hydrogen peroxide is the source of the radicals:

$$H_2O_2 \rightarrow 2OH \bullet . \tag{12.2}$$

Each hydrogen-peroxide molecule provides an initiator–terminator pair for each polymeric molecule. The termination step in Figure 12.2 is termed *recombination.* Although simpler to illustrate, it is not the most common mechanism of termination. Both hydrogen abstraction and disproportionation are more common termination steps than recombination. *Hydrogen abstraction* involves obtaining a hydrogen atom (with unpaired electron) from an impurity hydrocarbon group. *Disproportionation* involves the formation of a monomer-like double bond.

If an intimate solution of different types of monomers is polymerized, the result is a **copolymer** (Figure 12.3), which is analogous to the solid-solution alloy of metallic systems (Figure 4.2). Figure 12.3 represents specifically a **block copolymer**;

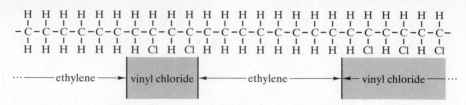

FIGURE 12.3 *A copolymer of ethylene and vinyl chloride is analogous to a solid-solution metal alloy.*

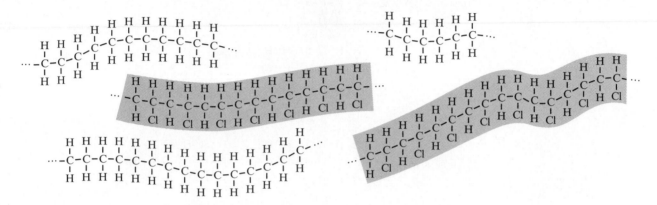

FIGURE 12.4 *A blend of polyethylene and polyvinyl chloride is analogous to a metal alloy with limited solid solution.*

that is, the individual polymeric components appear in "blocks" along a single carbon-bonded chain. The alternating arrangement of the different mers can be irregular (as shown in Figure 12.3) or regular. A **blend** (Figure 12.4) is another form of alloying in which different types of already formed polymeric molecules are mixed together. This blend is analogous to metallic alloys with limited solid solubility.

The various linear polymers illustrated in Figures 12.1 to 12.4 are based on the conversion of a carbon–carbon double bond into two carbon–carbon single bonds. It is also possible to convert the carbon–oxygen double bond in formaldehyde to single bonds. The overall reaction for this case can be expressed as

$$n\ CH_2O \rightarrow (CH_2O)_n \tag{12.3}$$

and is illustrated in Figure 12.5. The product is known by various names including polyformaldehyde, polyoxymethylene, and polyacetal. The important acetal group of engineering polymers is based on the reaction of Figure 12.5.

Figure 12.6 illustrates the formation of phenol-formaldehyde by the process of step growth. Only a single step is shown. The two phenol molecules are linked by the formaldehyde molecule in a reaction in which the phenols each give up a hydrogen atom and the formaldehyde gives up an oxygen atom to produce a water-molecule by-product (condensation product). Extensive polymerization requires this three-molecule reaction to be repeated for each unit increase in molecular length. The time required for this process is substantially greater than that for the chain reaction of Figure 12.2. The common occurrence of condensation

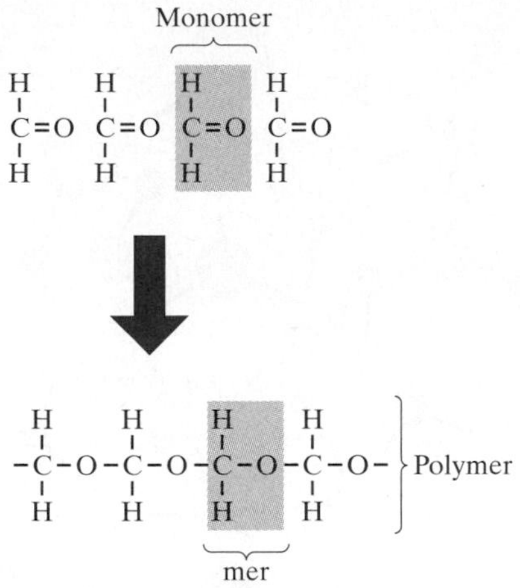

FIGURE 12.5 *The polymerization of formaldehyde to form polyacetal. (Compare with Figure 12.1.)*

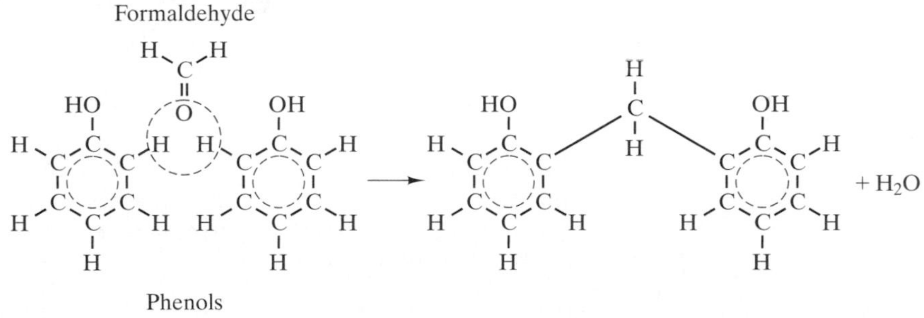

FIGURE 12.6 *Single first step in the formation of phenol-formaldehyde by a step-growth process (condensation polymerization). A water molecule is the condensation product.*

by-products in step-growth processes provides the descriptive term *condensation polymerization*. The polyethylene mer in Figure 12.1 has two points of contact with adjacent mers and is said to be **bifunctional**, which leads to a **linear molecular structure**. On the other hand, the phenol molecule in Figure 12.6 has several potential points of contact and is termed **polyfunctional**. In practice, there is room for no more than three CH_2 connections per phenol molecule, but this number is sufficient to generate a three-dimensional **network molecular structure**, as opposed to the linear structure of polyethylene. Figure 12.7 illustrates this network structure. The terminology here is reminiscent of inorganic glass structure, which was discussed in Section 11.2. The breaking up of network arrangements of silica tetrahedra by network modifiers produced substantially "softer" glass. Similarly, linear polymers are "softer" than network polymers. A key difference between silicate "polymers" and the organic materials of this chapter is that the silicates contain predominantly primary bonds that cause their viscous behavior to occur at substantially higher temperatures. It should be noted that a bifunctional monomer will produce a linear molecule by either chain-growth or step-growth processes, and a polyfunctional monomer will produce a network structure by either process.

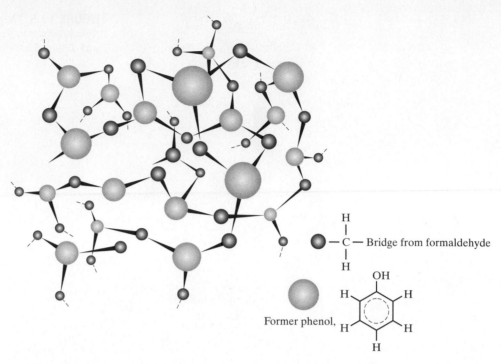

Former phenol,

—C— Bridge from formaldehyde

FIGURE 12.7 *After several reaction steps like those in Figure 12.6, polyfunctional mers form a three-dimensional network molecular structure. (From L. H. Van Vlack,* Elements of Materials Science and Engineering, *4th ed., Addison-Wesley Publishing Co., Inc., Reading, MA, 1980.)*

EXAMPLE 12.1

How much H_2O_2 must be added to ethylene to yield an average degree of polymerization of 750? Assume that all H_2O_2 dissociates to OH groups that serve as terminals for the molecules, and express the answer in weight percent.

SOLUTION
Referring to Figure 12.2, we note that there is one H_2O_2 molecule (= two OH groups) per polyethylene molecule. Thus,

$$\text{wt \% } H_2O_2 = \frac{\text{mol wt } H_2O_2}{750 \times (\text{mol wt } C_2H_4)} \times 100.$$

Using the data of Appendix 1 yields

$$\text{wt \% } H_2O_2 = \frac{2(1.008) + 2(16.00)}{750[2(12.01) + 4(1.008)]} \times 100$$

$$= 0.162 \text{ wt \%}.$$

EXAMPLE 12.2

Calculate the molecular weight of a polyacetal molecule with a degree of polymerization of 500.

SOLUTION

$$\text{mol wt } (CH_2O)_n = n(\text{mol wt } CH_2O).$$

Using the data of Appendix 1, we obtain

$$\text{mol wt } (CH_2O)_n = 500[12.01 + 2(1.008) + 16.00] \text{ amu}$$
$$= 15,010 \text{ amu}.$$

PRACTICE PROBLEM 12.1

How much H_2O_2 must be added to ethylene to yield an average degree of polymerization of **(a)** 500 and **(b)** 1,000? (See Example 12.1.)

PRACTICE PROBLEM 12.2

Calculate the degree of polymerization for a polyacetal molecule with a molecular weight of 25,000 amu. (See Example 12.2.)

STRUCTURAL FEATURES OF POLYMERS

The first aspect of polymer structure that needs to be specified is the length of the polymeric molecule. For example, how large is n in $(C_2H_4)_n$? In general, n is termed the **degree of polymerization** and is also designated $\overline{DP}$. It is usually determined from the measurement of physical properties, such as viscosity and light scattering. For typical commercial polymers, n can range from approximately 100 to 1,000, but for a given polymer, the degree of polymerization represents an average. As you might suspect from the nature of both chain-growth and step-growth mechanisms, the extent of the molecular-growth process varies from molecule to molecule. The result is a statistical distribution of molecular lengths around the average value. Directly related to molecular length is molecular weight, which is simply the degree of polymerization (n) times the molecular weight of the individual mer. Less simple is the concept of molecular length. For network structures, there is, by definition, no meaningful one-dimensional measure of length. For linear structures, there are two such parameters. First is the **root-mean-square length**, $\bar{L}$, given by

$$\bar{L} = l\sqrt{m}, \tag{12.4}$$

where l is the length of a single bond in the backbone of the hydrocarbon chain and m is the number of bonds. Equation 12.4 results from the statistical analysis of a freely kinked linear chain, as illustrated in Figure 12.8. Each bond angle between three adjacent C atoms is near 109.5° (as discussed in Chapter 2), but as seen in Figure 12.8, this angle can be rotated freely in space. The result is the kinked and coiled molecular configuration. The root-mean-square length represents the

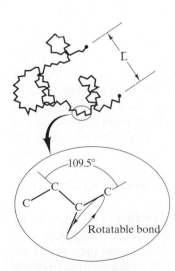

FIGURE 12.8 *The length of kinked molecular chain is given by Equation 12.4, due to the free rotation of the C—C—C bond angle of 109.5°.*

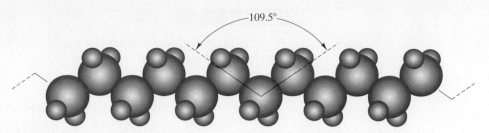

FIGURE 12.9 *"Sawtooth" geometry of a fully extended molecule. The relative sizes of carbon and hydrogen atoms are shown in the polyethylene configuration.*

effective length of the linear molecule as it would be present in the polymeric solid. The second length parameter is a hypothetical one in which the molecule is extended as straight as possible (without bond-angle distortion),

$$L_{\text{ext}} = ml \sin \frac{109.5^\circ}{2}, \qquad \textbf{(12.5)}$$

where L_{ext} is the **extended length**. The "sawtooth" geometry of the extended molecule is illustrated in Figure 12.9. For typical bifunctional, linear polymers such as polyethylene and PVC, there are two bond lengths per mer, or

$$m = 2n, \qquad \textbf{(12.6)}$$

where n is the degree of polymerization.

In general, the rigidity and melting point of polymers increase with the degree of polymerization. As with any generalization, there can be important exceptions. For example, the melting point of nylon does not change with degree of polymerization. This fact raises a useful rule of thumb; that is, rigidity and melting point increase as the complexity of the molecular structure increases. For example, the phenol-formaldehyde structure in Figure 12.6 produces a rigid, even brittle, polymer. By contrast, the linear polyethylene structure of Figure 12.2 produces a relatively soft material. Students of civil engineering should appreciate the rigidity of a structure with extensive cross members. The network structure has the strength of covalent bonds linking all adjacent mers. The linear structure has covalent bonding only along the backbone of the chain. Only weak secondary (van der Waals) bonding holds together adjacent molecules. Molecules are relatively free to slide past each other, corresponding to the relatively low values of elastic modulus discussed in Section 6.1. Now let us explore a series of structural features that add to the complexity of linear molecules and take them closer to the nature of the network structure.

We will begin with the ideally simple hydrocarbon chain of polyethylene (Figure 12.10a). By replacing some hydrogen atoms with large side groups (R), a less-symmetrical molecule results. The placement of the side groups can be regular and all along one side, or **isotactic** (Figure 12.10b), or the placement of the groups can be alternating along opposite sides, or **syndiotactic** (Figure 12.10c). An even less-symmetrical molecule is the **atactic** form (Figure 12.10d), in which the side groups are irregularly placed. For $R = CH_3$, Figure 12.10b to d represents polypropylene. As the side groups become larger and more irregular, rigidity and melting point tend to rise for two reasons. First, the side groups serve as hindrances to molecular sliding. By contrast, the polyethylene molecules (Figure 12.10a) can

```
    H H H H H H H H H H H H H H H H H H H H H H
    | | | | | | | | | | | | | | | | | | | | | |
····-C-C-C-C-C-C-C-C-C-C-C-C-C-C-C-C-C-C-C-C-C-C-····
    | | | | | | | | | | | | | | | | | | | | | |
    H H H H H H H H H H H H H H H H H H H H H H
                          (a)
```

```
    H H H H H H H H H H H H H H H H H H H H H H
    | | | | | | | | | | | | | | | | | | | | | |
····-C-C-C-C-C-C-C-C-C-C-C-C-C-C-C-C-C-C-C-C-C-C-····
    | H | H | H | H | H | H | H | H | H | H | H |
    R   R   R   R   R   R   R   R   R   R   R
                          (b)
```

```
    R       R       R       R       R
    | H H H | H H H | H H H | H H H | H H
    H | | | H | | | H | | | H | | | H | | |
····-C-C-C-C-C-C-C-C-C-C-C-C-C-C-C-C-C-C-C-C-C-····
    H H H | H H H | H H H | H H H | H H H |
          R       R       R       R       R
                          (c)
```

```
    R   R       R               R
    | H | H H H | H H H H H H H | H II II II
    H | | | | | H | | | | | | | H | | | | |
····-C-C-C-C-C-C-C-C-C-C-C-C-C-C-C-C-C-C-C-C-C-····
    H H H H | H H H | H | H | H H H | H |
    R       R       R   R       R   R
                          (d)
```

FIGURE 12.10 *(a) The symmetrical polyethylene molecule. (b) A less symmetrical molecule is produced by replacing one H in each mer with a large side group, R. The isotactic structure has all R along one side. (c) The syndiotactic structure has the R groups regularly alternating on opposite sides. (d) The least-symmetrical structure is the atactic, in which the side groups irregularly alternate on opposite sides. Increasing irregularity decreases crystallinity while increasing rigidity and melting point. When R = CH₃, parts (b)–(d) illustrate various forms of polypropylene. (One might note that these schematic illustrations can be thought of as "top views" of the more pictorial representations of Figure 12.9.)*

slide past each other readily under an applied stress. Second, increasing size and complexity of the side group lead to greater secondary bonding forces between adjacent molecules (see Chapter 2).

An extension of the concept of adding large side groups is to add a polymeric molecule to the side of the chain. This process, called *branching*, is illustrated in Figure 12.11. It can occur as a fluctuation in the chain-growth process illustrated by Equation 12.1 (in which a hydrogen further back on the chain is abstracted by a free radical) or as a result of an addition agent that strips away a hydrogen, allowing chain growth to commence at that site. The complete transition from linear to network structure is produced by *cross-linking*, as shown in Figure 6.47, which illustrates **vulcanization**. Rubbers are the most common examples of cross-linking. The bifunctional isoprene mer still contains a double bond after the initial polymerization, which permits covalent bonding of a sulfur atom to two adjacent mers. The extent of cross-linking is controlled by the amount of sulfur addition. This permits control of the rubber behavior from a gummy material to a tough, elastic one and finally, to a hard, brittle product as the sulfur content is increased.

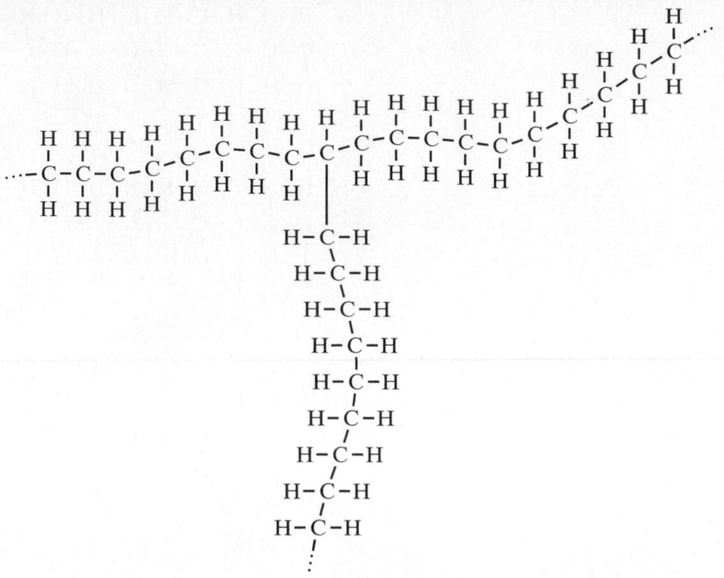

FIGURE 12.11 *Branching involves adding a polymeric molecule to the side of the main molecular chain.*

In Chapters 3 and 4, we pointed out that the complexity of long-chain molecular structures leads to complex crystalline structures and a significant degree of noncrystalline structure in commercial materials. We can now comment that the degree of crystallinity will decrease with the increasing structural complexity discussed in this section. For instance, branching in polyethylene can drop the crystallinity from 90% to 40%. An isotactic polypropylene can be 90% crystalline, while atactic polypropylene is nearly all noncrystalline. Control of polymeric structure has been an essential component of the development of polymers that are competitive with metals for various engineering design applications.

EXAMPLE 12.3

A sample of polyethylene has an average degree of polymerization of 750.

(a) What is the coiled length?

(b) Determine the extended length of an average molecule.

SOLUTION

(a) Using Equations 12.4 and 12.6, we have

$$\bar{L} = l\sqrt{2n}.$$

From Table 2.2, $l = 0.154$ nm, giving

$$\bar{L} = (0.154 \text{ nm})\sqrt{2(750)}$$

$$= 5.96 \text{ nm}.$$

(b) Using Equations 12.5 and 12.6, we obtain

$$L_{\text{ext}} = 2\,nl \sin \frac{109.5°}{2}$$

$$= 2(750)(0.154 \text{ nm}) \sin \frac{109.5°}{2}$$

$$= 189 \text{ nm}.$$

PRACTICE PROBLEM 12.3

In Example 12.3, coiled and extended molecular lengths are calculated for a polyethylene with a degree of polymerization of 750. If the degree of polymerization of this material is increased by one-third (to $n = 1{,}000$), by what percentage is **(a)** the coiled length and **(b)** the extended length increased?

THERMOPLASTIC POLYMERS

Thermoplastic polymers become soft and deformable upon heating, which is characteristic of linear polymeric molecules (including those that are branched but not cross-linked). The high-temperature plasticity is due to the ability of the molecules to slide past one another, which is another example of a thermally activated, or Arrhenius, process. In this sense, thermoplastic materials are similar to metals that gain ductility at high temperatures (e.g., creep deformation). It should be noted that, as with metals, the ductility of thermoplastic polymers is reduced upon cooling. The key distinction between thermoplastics and metals is what we mean by "high" temperatures. The secondary bonding, which must be overcome to deform thermoplastics, may allow substantial deformation around 100°C for common thermoplastics. However, metallic bonding generally restricts creep deformation to temperatures closer to 1,000°C in typical alloys.

Although polymers cannot, in general, be expected to duplicate fully the mechanical behavior of traditional metal alloys, a major effort is made to produce some polymers with sufficient strength and stiffness to be serious candidates for structural applications once dominated by metals. These polymers are called **engineering polymers**, which retain good strength and stiffness up to 150–175°C. The categories are, in fact, somewhat arbitrary. The "general-use" textile fiber nylon is also a pioneering example of an engineering polymer, and it continues to be the most important. It has been estimated that industry has developed more than half a million engineering polymer-part designs specifying nylon. The other members of the family of engineering polymers are part of a steadily expanding list. The importance of these materials to design engineers goes beyond their relatively small percentage of the total polymer market (less than 10% compared to approximately 30% for polyethylene alone). Nonetheless, the bulk of that market is devoted to the materials referred to as *general-use polymers,* including polyethylene. These polymers include the various films, fabrics, and packaging materials that are so much a part of everyday life. The market shares for nylon (1%) and polyester (5%) include their major uses as textile fibers, which is the reason for polyester's larger market share even though nylon is a more common metal substitute. General-use polymers include some of the more familiar product trade names (including Dacron and Mylar, trade names of Du Pont), as well as the chemical names of the polymers.

Polyethylene, as the most common thermoplastic, is subdivided into *low-density polyethylene* (LDPE), *high-density polyethylene* (HDPE), and *ultra-high molecular-weight polyethylene* (UHMWPE). LDPE has much more chain branching than HDPE, which is essentially linear. UHMWPE has very long linear chains. Greater chain linearity and chain length tend to increase the melting point and improve the physical and mechanical properties of the polymer due to the greater crystallinity possible within the polymer morphology. *Linear low-density polyethylene* (LLDPE) is a copolymer with α-olefins that has less chain branching and better properties than LDPE. HDPE and UHMWPE are two good examples of engineering polymers, although polyethylene overall is a general-use polymer. Note that acrylonitrile–butadiene–styrene (ABS) is an important example of a copolymer, as discussed earlier in this section. ABS is a **graft copolymer** as opposed to the block copolymer shown in Figure 12.3. Acrylonitrile and styrene chains are "grafted" onto the main polymeric chain composed of polybutadiene.

A third category of thermoplastic polymers is that of **thermoplastic elastomers**. **Elastomers** are polymers with mechanical behavior analogous to natural rubber. Elastomeric deformation was discussed in Section 6.6. Traditional synthetic rubbers have been, upon vulcanization, thermosetting polymers as discussed later in this section. The relatively novel thermoplastic elastomers are essentially composites of rigid elastomeric domains in a relatively soft matrix of a crystalline thermoplastic polymer. A key advantage of thermoplastic elastomers is the convenience of processing by traditional thermoplastic techniques, including being recyclable.

THERMOSETTING POLYMERS

Thermosetting polymers are the opposite of thermoplastics. They become hard and rigid upon heating. Unlike thermoplastic polymers, this phenomenon is not lost upon cooling, which is characteristic of network molecular structures formed by the step-growth mechanism. The chemical reaction steps are enhanced by higher temperatures and are irreversible; that is, the polymerization remains upon cooling. Thermosetting products can be removed from the mold at the fabrication temperature (typically 200 to 300°C). Phenol-formaldehyde is a common example. By contrast, thermoplastics must be cooled in the mold to prevent distortion. Common thermosetting polymers are subdivided into two categories: thermosets and elastomers. In this case, thermosets refers to materials that share with the engineering polymers significant strength and stiffness so as to be common metal substitutes. However, thermosets have the disadvantages of not being recyclable and, in general, having less-variable processing techniques. As noted in the discussion of thermoplastic polymers, traditional elastomers are thermosetting copolymers. Examples include some familiar trade names (such as Neoprene and Viton, trade names of Du Pont). In addition to the many applications found for thermoplastics, such as films, foams, and coatings, thermosets include the important application of **adhesives**. The adhesive serves to join the surfaces of two solids (adherends) by secondary forces similar to those between molecular chains in thermoplastics. If the adhesive layer is thin and continuous, the adherend material will often fail before the adhesive. Epoxies are common examples.

Finally, it might also be noted that **network copolymers** can be formed similar to the block and graft copolymers already discussed for thermoplastics. The network copolymer will result from polymerization of a combination of more than one species of polyfunctional monomers.

EXAMPLE 12.4

Metallurgical samples to be polished flat for optical microscopy are frequently mounted in a cylinder of phenol-formaldehyde, a thermosetting polymer. Because of the three-dimensional network structure, the polymer is essentially one large molecule. What would be the molecular weight of a 10-cm^3 cylinder of this polymer? (The density of the phenol-formaldehyde is 1.4 g/cm^3.)

SOLUTION

In general, the phenol molecule is trifunctional (i.e., a phenol is attached to three other phenols by three formaldehyde bridges). One such bridge is shown in Figure 12.6. A network of trifunctional bridges is shown in Figure 12.7.

As each formaldehyde bridge is shared by two phenols, the overall ratio of phenol to formaldehyde that must react to form the three-dimensional structure of Figure 12.6 is $1:\frac{3}{2}$ or 1:1.5. As each formaldehyde reaction produces one H_2O molecule, we can write that

$$1 \text{ phenol} + 1.5 \text{ formaldehyde} \rightarrow$$

$$1 \text{ phenol-formaldehyde mer } + 1.5H_2O \uparrow$$

In this way, we can calculate the mer molecular weight as

$$(\text{m.w.})_{mer} = (\text{m.w.})_{phenol} + 1.5(\text{m.w.})_{formaldehyde}$$

$$- 1.5(\text{m.w.})_{H_2O}$$

$$= [6(12.01) + 6(1.008) + 16.00]$$

$$+ 1.5[12.01 + 2(1.008) + 16.00]$$

$$- 1.5[2(1.008) + 16.00]$$

$$= 112.12 \text{ amu.}$$

The mass of the polymer in question is

$$m = \rho V = 1.4 \frac{\text{g}}{\text{cm}^3} \times 10 \text{ cm}^3$$

$$= 14 \text{ g.}$$

Therefore, the number of mers in the cylinder is

$$n = \frac{14 \text{ g}}{112.12 \text{ g}/0.6023 \times 10^{24} \text{ mers}}$$

$$= 7.52 \times 10^{22} \text{ mers,}$$

which gives a molecular weight of

$$\text{mol wt} = 7.52 \times 10^{22} \text{ mers} \times 112.12 \text{ amu/mer}$$
$$= 8.43 \times 10^{24} \text{ amu.}$$

PRACTICE PROBLEM 12.4

The molecular weight of a product of phenol-formaldehyde is calculated in Example 12.4. How much water by-product is produced in the polymerization of this product?

ADDITIVES

Copolymers and blends were discussed earlier in this section as analogs of metallic alloys. There are several other alloylike *additives* that have traditionally been used in polymer technology to provide specific characteristics to the polymers.

A **plasticizer** is added to soften a polymer. This addition is essentially blending with a low-molecular-weight (approximately 300 amu) polymer. Note that a large addition of plasticizer produces a liquid. Common paint is an example. The "drying" of the paint involves the plasticizer evaporating (usually accompanied by polymerization and cross-linking by oxygen).

A **filler**, on the other hand, can strengthen a polymer by restricting chain mobility. In general, fillers are largely used for volume replacement, providing dimensional stability and reduced cost. Relatively inert materials are used. Examples include short-fiber cellulose (an organic filler) and asbestos (an inorganic filler). Roughly one-third of the typical automobile tire is a filler (i.e., carbon black). **Reinforcements** such as glass fibers are also categorized as additives. These reinforcements are widely used in engineering polymers to enhance their strength and stiffness, thereby increasing their competitiveness as metal substitutes. Reinforcements are generally given surface treatments to ensure good interfacial bonding with the polymer and, thereby, maximum effectiveness in enhancing properties. The use of such additives up to a level of roughly 50 vol % produces a material generally still referred to as a polymer. For additions above roughly 50 vol %, the material is more properly referred to as a composite. A good example of a composite is fiberglass, which is discussed in detail in Section 12.2.

Stabilizers are additives used to reduce polymer degradation. They include a complex set of materials because of the large variety of degradation mechanisms (oxidation, thermal, and ultraviolet). As an example, polyisoprene can absorb up to 15% oxygen at room temperature, with its elastic properties being destroyed by the first 1%. Natural rubber latex contains complex phenol groups that retard the room-temperature oxidation reactions. However, these naturally occurring antioxidants are not effective at elevated temperatures. Therefore, additional stabilizers (e.g., other phenols, amines, or sulfur compounds) are added to rubber intended for tire applications.

Flame retardants are added to reduce the inherent combustibility of certain polymers such as polyethylene. Combustion is simply the reaction of a hydrocarbon with oxygen accompanied by substantial heat evolution. Many polymeric hydrocarbons exhibit combustibility. Others, such as PVC, exhibit reduced combustibility. The resistance of PVC to combustion appears to come from the evolution of the chlorine atoms from the polymeric chain. These halogens hinder the process of combustion by terminating free-radical chain reactions. Additives that provide this function for halogen-free polymers include chlorine-, bromine-, and phosphorus-containing reactants.

Colorants are additions used to provide color to a polymer where appearance is a factor in materials selection. Two types of colorants are used, pigments and dyes. **Pigments** are insoluble, colored materials added in powdered form. Typical examples are crystalline ceramics such as titanium oxide and aluminum silicate, although organic pigments are also available. **Dyes** are soluble, organic colorants that can provide transparent colors. The nature of color is discussed further in Chapter 13B on the related Web site.

EXAMPLE 12.5

A nylon 66 polymer is reinforced with 33 wt % glass fiber. Calculate the density of this engineering polymer. (The density of a nylon $66 = 1.14$ Mg/m^3, and the density of the reinforcing glass = 2.54 Mg/m^3.)

SOLUTION
For 1 kg of final product, there will be

$$0.33 \times 1 \text{ kg} = 0.33 \text{ kg glass}$$

and

$$1 \text{ kg} - 0.33 \text{ kg} = 0.67 \text{ kg nylon 66.}$$

The total volume of product will be

$$
\begin{aligned}
V_{\text{product}} &= V_{\text{nylon}} + V_{\text{glass}} \\
&= \frac{m_{\text{nylon}}}{\rho_{\text{nylon}}} + \frac{m_{\text{glass}}}{\rho_{\text{glass}}} \\
&= \left(\frac{0.67 \text{ kg}}{1.14 \text{ Mg/m}^3} + \frac{0.33 \text{ kg}}{2.54 \text{ Mg/m}^3} \right) \\
&\quad \times \frac{1 \text{ Mg}}{1{,}000 \text{ kg}} = 7.18 \times 10^{-4} \text{ m}^3.
\end{aligned}
$$

The overall density of the final product is then

$$\rho = \frac{1 \text{ kg}}{7.18 \times 10^{-4} \text{ m}^3} \times \frac{1 \text{ Mg}}{1{,}000 \text{ kg}} = 1.39 \text{ Mg/m}^3.$$

PRACTICE PROBLEM 12.5

Example 12.5 describes a high-strength and high-stiffness engineering polymer. Strength and stiffness can be further increased by a greater "loading" of glass fibers. Calculate the density of a nylon 66 with 43 wt % glass fibers.

12.2 | Composites

Our final category of structural engineering materials is that of composites. These materials involve some combination of two or more components from the fundamental structural material types: metals, ceramics, glasses, and polymers. A key philosophy in selecting composite materials is that they should provide the "best of both worlds" (i.e., attractive properties from each component). A classic example is fiberglass. The strength of small-diameter glass fibers is combined with the ductility of the polymeric **matrix**. The combination of these two components provides a product superior to either component alone. Many composites, such as fiberglass, involve combinations that cross over the boundaries set in the preceding three sections. Others, such as ceramic matrix composites, involve different components from within a single material type. In general, we shall use a fairly narrow definition of composites. We shall consider only those materials that combine different components on the microscopic (rather than macroscopic) scale. We shall not include multiphase alloys and ceramics, which are the result of routine processing discussed in Chapters 9 and 10. Similarly, the microcircuits to be discussed in Chapter 13 are not included because each component retains its distinctive character in these material systems. In spite of these restrictions, we shall find this category to include a tremendously diverse collection of materials, from the common to some of the most sophisticated. Fiberglass is among our most common construction materials. The aerospace industry has driven much of the development of our most sophisticated composite systems (e.g., "stealth" aircraft with high-performance, nonmetallic materials). Increasingly, these advanced materials are being used in civilian applications, such as improved strength-to-weight ratio bridges and more fuel-efficient automobiles.

We shall consider two broad categories of composite materials. Fiberglass, or glass fiber-reinforced polymer, is an excellent example of a synthetic fiber-reinforced composite. The fiber reinforcement is generally found in one of three primary configurations: aligned in a single direction, randomly chopped, or woven in a fabric that is laminated with the matrix. Cermets are examples of an aggregate composite, in which ceramic particles rather than fibers reinforce a metal matrix. Two common construction materials, wood and concrete, are also examples of fiber-reinforced and aggregate composites, respectively. These materials are covered in some detail in Chapter 12B on the related Web site.

The related Web site contains the chapter "Introduction to Composites" from ASM Handbook, *Vol. 21* (Composites), *ASM International, Materials Park, OH, 2001. Used by permission.*

FIBER–REINFORCED COMPOSITES

The most common examples of synthetic composite materials are those with micron-scale reinforcing fibers. Within this category are two distinct subgroups: (1) fiberglass generally using glass fibers with moderately high values of elastic modulus, and (2) advanced composites with even higher moduli fibers. We shall also compare these synthetic materials with an important natural, **fiber-reinforced composite**—wood.

Fiberglass is a classic example of a modern composite system. Reinforcing fibers are shown in Figure 12.12. A typical fracture surface of a composite (Figure 12.13) shows such fibers embedded in the polymeric matrix. Table 12.1 lists some common glass compositions used for fiber reinforcement. Each is the result of substantial development that has led to optimal suitability for specific

FIGURE 12.12 *Glass fibers to be used for reinforcement in a fiberglass composite. (Courtesy of Owens-Corning Fiberglas Corporation.)*

FIGURE 12.13 *The glass-fiber reinforcement in a fiberglass composite is clearly seen in a scanning electron microscope image of a fracture surface. (Courtesy of Owens-Corning Fiberglas Corporation.)*

applications. For example, the most generally used glass-fiber composition is **E-glass**, in which E stands for "electrical type." The low sodium content of E-glass is responsible for its especially low electrical conductivity and its attractiveness as a dielectric. Its popularity in structural composites is related to the chemical durability of the borosilicate composition. Table 12.2 lists some of the common polymeric matrix materials. Three common fiber configurations are illustrated

TABLE 12.1

Compositions of Glass-Reinforcing Fibers

Designation	Characteristic	Composition[a] (wt %)								
		SiO_2	$(Al_2O_3 + Fe_2O_3)$	CaO	MgO	Na_2O	K_2O	B_2O_3	TiO_2	ZrO_2
A-glass	Common soda–lime silica	72	<1	10		14				
AR-glass	Alkali resistant (for concrete reinforcement)	61	<1	5	<1	14	3		7	10
C-glass	Chemical corrosion resistant	65	4	13	3	8	2	5		
E-glass	Electrical composition	54	15	17	5	<1	<1	8		
S-glass	High strength and modulus	65	25		10					

[a]Approximate and not representing various impurities.

Source: Data from J. G. Mohr and W. P. Rowe, *Fiber Glass,* Van Nostrand Reinhold Company, Inc., New York, 1978.

TABLE 12.2

Polymeric Matrix Materials for Fiberglass	
Polymer	**Characteristics and applications**
Thermosetting	
Epoxies	High strength (for filament-wound vessels)
Polyesters	For general structures (usually fabric reinforced)
Phenolics	High-temperature applications
Silicones	Electrical applications (e.g., printed-circuit panels)
Thermoplastic	
Nylon 66	
Polycarbonate	Less common, especially good ductility
Polystyrene	

Source: Data from L. J. Broutman and R. H. Krock, Eds., *Modern Composite Materials,* Addison-Wesley Publishing Co., Inc., Reading, MA, 1967, Chapter 13.

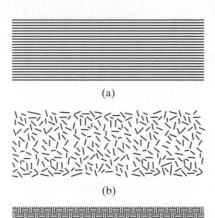

(a)

(b)

(c)

FIGURE 12.14 *Three common fiber configurations for composite reinforcement are (a) continuous fibers, (b) discrete (or chopped) fibers, and (c) woven fabric, which is used to make a laminated structure.*

in Figure 12.14. Parts (a) and (b) show the use of **continuous fibers** and **discrete (chopped) fibers**, respectively. Part (c) shows the **woven fabric** configuration, which is layered with the matrix polymer to form a **laminate**. The implications of these various geometries on mechanical properties will be covered in the discussion of property averaging. For now, we note that optimal strength is achieved by the aligned, continuous fiber reinforcement. Caution is necessary, however, in citing this strength because it is maximal only in the direction parallel to the fiber axes. In other words, the strength is highly **anisotropic**—it varies with direction.

Advanced composites include those systems in which reinforcing fibers have moduli higher than that of E-glass. For example, fiberglass used in most U.S. helicopter blades contains high modulus S-glass fibers (see Table 12.1). Advanced composites, however, generally involve fibers other than glass. Table 12.3 lists a variety of advanced composite systems. These systems include some of the most

TABLE **12.3**

Advanced Composite Systems Other Than Fiberglass	
Class	**Fiber/Matrix**
Polymer matrix	Para-aramid (Kevlar[a])/epoxy
	Para-aramid (Kevlar[a])/polyester
	C (graphite)/epoxy
	C (graphite)/polyester
	C (graphite)/polyetheretherketone (PEEK)
	C (graphite)/polyphenylene sulfide (PPS)
Metal matrix	B/Al
	C/Al
	Al_2O_3/Al
	Al_2O_3/Mg
	SiC/Al
	SiC/Ti (alloys)
Ceramic matrix	Nb/$MoSi_2$
	C/C
	C/SiC
	SiC/Al_2O_3
	SiC/SiC
	SiC/Si_3N_4
	SiC/Li–Al–silicate (glass-ceramic)

[a]Trade name, Du Pont.

Source: Data from K. K. Chawla, University of Alabama, Birmingham; A. K. Dhingra, the Du Pont Company; and A. J. Klein, ASM International.

sophisticated materials developed for some of the most demanding engineering applications. The growth of the advanced composites industry began with the materials advances of World War II and accelerated rapidly with the space race of the 1960s and with subsequent growth in demand for commercial aviation and high-performance leisure-time products such as golf clubs and tennis rackets.

Carbon and Kevlar fiber reinforcements represent advances over traditional glass fibers for **polymer–matrix composites**. Carbon fibers typically range in diameter between 4 and 10 μm, with the carbon being a combination of crystalline graphite and noncrystalline regions. Kevlar is a Du Pont trade name for poly *p*-phenyleneterephthalamide (PPD-T), a para-aramid with the formula

$$\left(HN-\bigcirc-N-\underset{\underset{O}{\|}}{C}-\bigcirc-CO\right)_n$$

(12.7)

Epoxies and polyesters (thermosetting polymers) are traditional matrices. Substantial progress has been made in developing thermoplastic polymer matrices, such as polyetheretherketone (PEEK) and polyphenylene sulfide (PPS). These materials have the advantages of increased toughness and recyclability. Carbon- and Kevlar-reinforced polymers are used in pressure vessels, and Kevlar reinforcement is widely used in tires. Carbon-reinforced PEEK and PPS demonstrate good temperature resistance and are, as a result, attractive for aerospace applications.

Metal–matrix composites have been developed for use in temperature, conductivity, and load conditions beyond the capability of polymer–matrix systems. For example, boron-reinforced aluminum is used in the Space Shuttle Orbiter, and carbon-reinforced aluminum is used in the Hubble Telescope. Alumina-reinforced aluminum is used in automobile engine components.

A primary driving force for the development of **ceramic–matrix composites** is superior high-temperature resistance. These composites, as opposed to traditional ceramics, represent the greatest promise to obtain the requisite toughness for structural applications such as high-efficiency jet-engine designs. An especially advanced composite system in this category is the **carbon–carbon composite**. This high-modulus and high-strength material is also quite expensive. The expense is significantly increased by the process of forming the large carbon chain molecules of the matrix by pyrolysis (heating in an inert atmosphere) of a polymeric hydrocarbon. Carbon–carbon composites are currently being used in high-performance automobiles as friction-resistant materials and in a variety of aerospace applications, such as ablative shields for reentry vehicles.

Metal fibers are frequently small-diameter wires. Especially high-strength reinforcement comes from **whiskers**, small, single-crystal fibers that can be grown with a nearly perfect crystalline structure. Unfortunately, whiskers cannot be grown as continuous filaments in the manner of glass fibers or metal wires. Figure 12.15 contrasts the wide range of cross-sectional geometries associated with reinforcing fibers.

During the 1980s, production of advanced composites in the United States doubled every 5 years. In the first three years of the 1990s, however, production suddenly declined by 20% due to the end of the Cold War and the resulting effect on defense budgets. Various trends have emerged in the advanced composites field in response to these changes. Emerging product applications include the marine market (e.g., high-performance powerboats), improved strength-to-weight ratio civil-engineering structures, and electric-car development. A fundamental challenge to the wider use of advanced composites in the general automotive industry is the need for reduced costs, which cannot occur until greater production capacity is in place. In turn, production capacity cannot be increased without greater demand in the automotive field.

Specific technological developments are occurring in response to the new trend toward nondefense applications. A major thrust is reduced production costs. For this reason, nonautoclave curing of thermosetting resins is being developed for bridge construction. Similarly, resin transfer molding (RTM) involving textile preforms substantially reduces cure times. (The related technique of transfer molding for polymers is illustrated in Figure 12.24.) Also, automated fiber-placement equipment produces a more rapid fabrication process. The addition of thermoplastics or elastomeric microspheres to thermosetting resins is among the various techniques being used to improve fracture toughness with a goal of reducing delamination and impact damage. Bismaleimide (BMI) resins are an advance over epoxies for heat resistance (over 300°C).

A single composite may contain various types of reinforcing fibers. **Hybrids** are woven fabrics consisting of two or more types of reinforcing fibers (e.g., carbon and glass or carbon and aramid). The combination is a design approach to optimize composite performance. For example, high-strength, noncarbon fibers can be added to carbon fibers to improve the impact resistance of the overall composite.

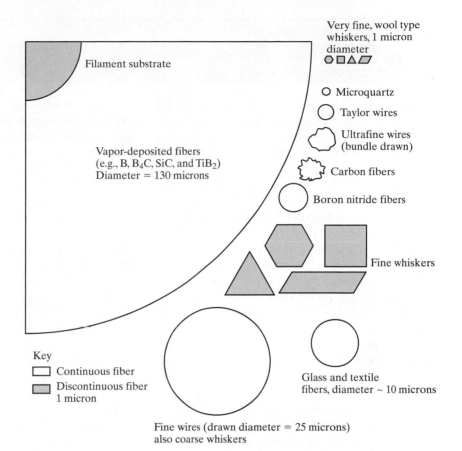

Very fine, wool type
whiskers, 1 micron
diameter

Filament substrate

Vapor-deposited fibers
(e.g., B, B₄C, SiC, and TiB₂)
Diameter = 130 microns

Microquartz

Taylor wires

Ultrafine wires
(bundle drawn)

Carbon fibers

Boron nitride fibers

Fine whiskers

Key

☐ Continuous fiber

▨ Discontinuous fiber
    1 micron

Glass and textile
fibers, diameter ~ 10 microns

Fine wires (drawn diameter = 25 microns)
also coarse whiskers

FIGURE 12.15 *Relative cross-sectional areas and shapes of a wide variety of reinforcing fibers. (After L. J. Broutman and R. H. Krock, eds.,* Modern Composite Materials, *Addison-Wesley Publishing Co., Inc., Reading, MA, 1967, Chapter 14.)*

EXAMPLE 12.6

A fiberglass composite contains 70 vol % E-glass fibers in an epoxy matrix.

(a) Calculate the weight percent glass fibers in the composite.

(b) Determine the density of the composite. The density of E-glass is 2.54 Mg/m³ (= g/cm³) and for epoxy is 1.1 Mg/m³.

SOLUTION

(a) For 1 m³ composite, we would have 0.70 m³ of E-glass and $(1.00 - 0.70)$ m³ $= 0.30$ m³ of epoxy. The mass of each component will be

$$m_{\text{E-glass}} = \frac{2.54 \text{ Mg}}{\text{m}^3} \times 0.70 \text{ m}^3 = 1.77 \text{ Mg}$$

THE MATERIAL WORLD

A Novel Composite for Synthetic Bone

A ceramic [hydroxyapatite (HA)] and a polymer (collagen) are major components of natural bone. With this microscopic-scale combination (43 wt % HA and 36 wt % collagen), bone is also a good example of a natural composite material. The balance of the composition of bone is composed of viscous liquids. Altogether, this composite can be considered one of nature's premier building materials.

Bone's adequate mechanical properties combined with its ability to repair and remodel itself make animal skeletons impressive structural systems. Occasional trauma, however, extends bone beyond its ability for complete self-repair. Orthopaedic surgeons use the term *large defects* to describe centimeter-scale gaps in the skeletal system. Historically, such defects were repaired by harvesting bone from another part of the body (autogenous bone grafting, or autografts) or by using cadaver bone (allografts). Each of these surgeries presents problems. Autografts have significant morbidity and cost, whereas allografts have significant risk of immunologic reaction and disease transmission.

An example of an engineered composite to fill these large gaps is the product *Collagraft* (see photo). In this system, millimeter-scale ceramic particles are embedded in a matrix of collagen (see micrograph). Each ceramic particle is, in fact, a fine mixture of micrometer-scale grains of HA and tricalcium phosphate (TCP), a material chemically similar to HA, but that reacts much more rapidly with the physiological environment. This combination is effective in that TCP is resorbed by the body in a few days, allowing rapid attachment of the implant to natural bone, while HA retains its structural integrity for a few months while new bone is growing into and filling the defect region. Collagen appears to facilitate new bone formation in a role similar to that in normal bone growth. The composite performance is maximized by adding bone marrow from the patient to the commercial material, thereby imitating the third component (viscous fluids) of natural bone.

Scanning electron microscope image of the aggregate composite composed of ceramic (hydroxyapatite plus tricalcium phosphate) granules in a polymeric (collagen) matrix. The image shows collagen as a darker gray. [From J. P. McIntyre, J. F. Shackelford, M. W. Chapman, and R. R. Pool, Bull. Amer. Ceram. Soc. 70 1499 (1991).]

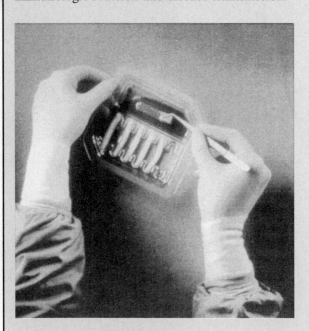

(Courtesy of Zimmer Corporation.)

and

$$m_{epoxy} = \frac{1.1\,\text{Mg}}{m^3} \times 0.30\,\text{m}^3 = 0.33\,\text{Mg},$$

giving

$$wt\ \%\ glass = \frac{1.77\,\text{Mg}}{(1.77 + 0.33)\,\text{Mg}} \times 100 = 84.3\%.$$

(b) The density will be given by

$$\rho = \frac{m}{V} = \frac{(1.77 + 0.33)\,\text{Mg}}{m^3} = 2.10\,\text{Mg/m}^3.$$

PRACTICE PROBLEM 12.6

In Example 12.6 we found the density of a typical fiberglass composite. Repeat the calculations for **(a)** 50 vol % and **(b)** 75 vol % E-glass fibers in an epoxy matrix.

AGGREGATE COMPOSITES

There are a variety of examples of **aggregate composites** in which particles reinforce a matrix. A large number of composite systems are based on particle reinforcement. Examples are listed in Table 12.4. As with Table 12.3, these systems include some of our most sophisticated engineering materials. Two groups of modern composites are identified in Table 12.4. **Particulate composites** refer specifically to systems in which the dispersed particles are relatively large (at least several micrometers in diameter) and are present in relatively high concentrations

TABLE 12.4

Aggregate Composite Systems
Particulate composites
Thermoplastic elastomer (elastomer in thermoplastic polymer)
SiC in Al
W in Cu
Mo in Cu
WC in Co
W in NiFe
Dispersion-strengthened metals
Al_2O_3 in Al
Al_2O_3 in Cu
Al_2O_3 in Fe
ThO_2 in Ni

Source: Data from L. J. Broutman and R. H. Krock, eds., *Modern Composite Materials,* Addison-Wesley Publishing Co., Inc., Reading, MA, 1967, Chapters 16 and 17; and K. K. Chawla, University of Alabama, Birmingham.

(greater than 25 vol % and frequently between 60 and 90 vol %). Earlier, we encountered a material system that can be included in this category—the polymers containing fillers discussed in Section 12.1. Remember that automobile tires are a rubber with roughly one-third carbon black particles.

A good example of a particulate composite is WC/Co, an excellent cutting-tool material. A high-hardness carbide in a ductile metal matrix is an important example of a **cermet**, a ceramic–metal composite. The carbide is capable of cutting hardened steel but needs the toughness provided by the ductile matrix, which also prevents crack propagation that would be caused by particle-to-particle contact of the brittle carbide phase. As both the ceramic and metal phases are relatively refractory, they can both withstand the high temperatures generated by the machining process. The **dispersion-strengthened metals** contain fairly small concentrations (less than 15 vol %) of small-diameter oxide particles (0.01 to 0.1 μm in diameter). The oxide particles strengthen the metal by serving as obstacles to dislocation motion. This concept can be appreciated from the discussion in Section 6.3 and Figure 6.26. Later in this section, we will find that a 10 vol % dispersion of Al_2O_3 in aluminum can increase tensile strength by as much as a factor of 4.

As wood is a common structural material and an example of a fiber-reinforced composite, concrete is an excellent example of an aggregate composite. Common concrete is rock and sand in a calcium aluminosilicate (cement) matrix. As with wood, this common construction material is used in staggering quantities. The weight of concrete used annually exceeds that of all metals combined. A detailed discussion of concrete is given in Chapter 12B on the related Web site.

EXAMPLE 12.7

A dispersion-strengthened aluminum contains 10 vol % Al_2O_3. Assuming that the metal phase is essentially pure aluminum, calculate the density of the composite. (The density of Al_2O_3 is 3.97 Mg/m^3.)

SOLUTION
From Appendix 1, we see that

$$\rho_{Al} = 2.70 \text{ Mg/m}^3.$$

For 1 m^3 of composite, we shall have 0.1 m^3 of Al_2O_3 and $1.0 - 0.1 = 0.9$ m^3 of Al. The mass of each component will be

$$m_{Al_2O_3} = 3.97 \frac{\text{Mg}}{\text{m}^3} \times 0.1 \text{ m}^3 = 0.40 \text{ Mg}$$

and

$$m_{Al} = 2.70 \frac{\text{Mg}}{\text{m}^3} \times 0.9 \text{ m}^3 = 2.43 \text{ Mg},$$

giving

$$\rho_{composite} = \frac{m}{V} = \frac{(0.40 + 2.43) \text{ Mg}}{1 \text{ m}^3}$$
$$= 2.83 \text{ Mg/m}^3.$$

Calculate the density of a particulate composite containing 50 vol % W particles in a copper matrix. (See Example 12.7.)

PROPERTY AVERAGING

It is obvious that the properties of composites must, in some way, represent an average of the properties of their individual components. However, the precise nature of the "average" is a sensitive function of microstructural geometry. Because of the wide variety of such geometries in modern composites, we must be cautious of generalities. However, we will identify one important example.

We will use the modulus of elasticity to illustrate **property averaging**, which is consistent with our emphasis on the structural applications of composites. The uniaxial stressing of a fiber-reinforced composite (stress parallel to the direction of fiber alignment) is shown in Figure 12.16. If the matrix is intimately bonded to the reinforcing fibers, the strain of both the matrix and the fibers must be the same. This **isostrain** condition is true even though the elastic moduli of the components will tend to be quite different. In other words,

$$\varepsilon_c = \frac{\sigma_c}{E_c} = \varepsilon_m = \frac{\sigma_m}{E_m} = \varepsilon_f = \frac{\sigma_f}{E_f}, \tag{12.8}$$

where all terms are defined in Figure 12.16. It is also apparent that the load carried by the composite, P_c, is the simple sum of loads carried by each component:

$$P_c = P_m + P_f. \tag{12.9}$$

Each load is equal, by definition, to a stress times an area; that is,

$$\sigma_c A_c = \sigma_m A_m + \sigma_f A_f, \tag{12.10}$$

where, again, terms are illustrated in Figure 12.16. Combining Equations 12.8 and 12.10 gives

$$E_c \varepsilon_c A_c = E_m \varepsilon_m A_m + E_f \varepsilon_f A_f. \tag{12.11}$$

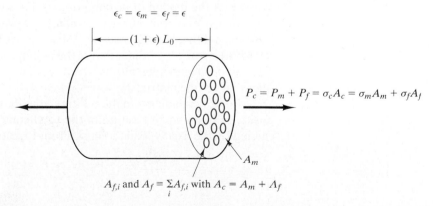

FIGURE 12.16 *Uniaxial stressing of a composite with continuous fiber reinforcement. The load is parallel to the reinforcing fibers. The terms in Equations 12.8 to 12.10 are illustrated.*

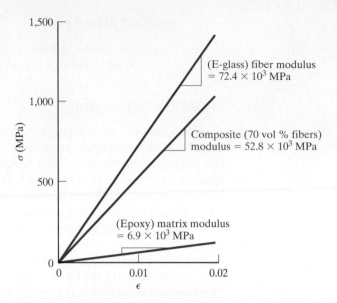

FIGURE 12.17 *Simple stress–strain plots for a composite and its fiber and matrix components. The slope of each plot gives the modulus of elasticity. The composite modulus is given by Equation 12.13.*

Let us note that $\varepsilon_c = \varepsilon_m = \varepsilon_f$ and divide both sides of Equation 12.10 by A_c:

$$E_c = E_m \frac{A_m}{A_c} + E_f \frac{A_f}{A_c}. \tag{12.12}$$

Because of the cylindrical geometry of Figure 12.16, the area fraction is also the volume fraction, or

$$E_c = v_m E_m + v_f E_f, \tag{12.13}$$

where v_m and v_f are the volume fractions of matrix and fibers, respectively. In this case, of course, $v_m + v_f$ must equal 1. Equation 12.13 is an important result. It identifies the modulus of a fibrous composite loaded axially as a simple, weighted average of the moduli of its components. Figure 12.17 shows the modulus as the slope of a stress–strain curve for a composite with 70 vol % reinforcing fibers. In this typical fiberglass (E-glass-reinforced epoxy), the glass-fiber modulus (72.4×10^3 MPa) is roughly 10 times that of the polymeric matrix modulus (6.9×10^3 MPa). The composite modulus, although not equal to that for glass, is substantially higher than that for the matrix.

Equally significant to the relative contribution of the glass fibers to the composite modulus is the fraction of the total composite load, P_c, in Equation 12.9, carried by the axially loaded fibers. From Equation 12.9, we note that

$$\frac{P_f}{P_c} = \frac{\sigma_f A_f}{\sigma_c A_c} = \frac{E_f \varepsilon_f A_f}{E_c \varepsilon_c A_c} = \frac{E_f}{E_c} v_f. \tag{12.14}$$

For the fiberglass example under discussion, $P_f/P_c = 0.96$; that is, nearly the entire uniaxial load is carried by the 70 vol % of high-modulus fibers. This geometry is an ideal application of a composite. The high modulus and strength of the fibers are effectively transmitted to the composite as a whole. At the same time, the ductility of the matrix is available to produce a substantially less brittle material than glass by itself.

The result of Equation 12.13 is not unique to the modulus of elasticity. A number of important properties exhibit this behavior, which is especially true of transport properties. In general, we can write

$$X_c = v_m X_m + v_f X_f, \tag{12.15}$$

where X can be diffusivity, D (see Section 5.3); thermal conductivity, k (see Section 7.3); or electrical conductivity, σ (see Section 13.1). The Poisson's ratio for loading parallel to reinforcing fibers can also be predicted from Equation 12.15.

It is also worth noting that a similar analysis can be done for stress loads perpendicular to the direction of fiber alignment, termed **isostress** loading. The resulting average is much lower than obtained under isostrain conditions. Thus, isostrain configurations are generally preferred in design applications. Furthermore, the modulus average obtained for the loading of aggregate composites is midway between the isostrain and isostress cases for fiber-reinforced composites. Detailed discussions of these other cases are given in Chapter 12B on the related Web site.

EXAMPLE 12.8

Calculate the composite modulus for polyester reinforced with 60 vol % E-glass under isostrain conditions, using data from Tables 12.5 and 12.6.

SOLUTION
This problem is a direct application of Equation 12.13:

$$E_c = v_m E_m + v_f E_f.$$

Data for elastic moduli of the components can be found in Tables 12.5 and 12.6:

$$E_{\text{polyester}} = 6.9 \times 10^3 \text{ MPa}$$

and

$$E_{\text{E-glass}} = 72.4 \times 10^3 \text{ MPa},$$

giving

$$E_c = (0.4)(6.9 \times 10^3 \text{ MPa}) + (0.6)(72.4 \times 10^3 \text{ MPa})$$
$$= 46.2 \times 10^3 \text{ MPa}.$$

TABLE 12.5

Mechanical Properties of Common Matrix Materials

Class	Example	E [MPa (ksi)]	T.S. [MPa (ksi)]	Flexural strength [MPa (ksi)]	Compressive strength (after 28 days) [MPa (ksi)]	Percent elongation at failure	K_{IC} (MPa $\sqrt{m}$)
Polymer[a]	Epoxy	6,900 (1,000)	69 (10)	—	—	0	0.3–0.5
	Polyester	6,900 (1,000)	28 (4)	—	—	0	—
Metal[b]	Al	69×10^3 (10×10^3)	76 (11)	—	—	—	—
	Cu	115×10^3 (17×10^3)	170 (25)	—	—	—	—
Ceramic[c]	Al$_2$O$_3$	—	—	550 (80)	—	—	4–5
	SiC	—	—	500 (73)	—	—	4.0
	Si$_3$N$_4$ (reaction bonded)	—	—	260 (38)	—	—	2–3

[a]From Tables 6.7 and 8.3.
[b]For high-purity alloys with no significant cold working from *Metals Handbook*, 9th ed., Vol. 2, American Society for Metals, Metals Park, OH, 1979.
Source: Data from A. J. Klein, *Advanced Materials and Processes 2*, 26 (1986).

TABLE 12.6

Mechanical Properties of Common Dispersed-Phase Materials

Group	Dispersed phase	E [MPa (ksi)]	T.S. [MPa (ksi)]	Compressive strength [MPa (ksi)]	Percent elongation at failure
Glass fiber[a]	C-glass	69×10^3 (10×10^3)	3,100 (450)	—	4.5
	E-glass	72.4×10^3 (10.5×10^3)	3,400 (500)	—	4.8
	S-glass	85.5×10^3 (12.4×10^3)	4,800 (700)	—	5.6
Ceramic fiber[a]	C (graphite)	$340–380 \times 10^3$ ($49–55 \times 10^3$)	2,200–2,400 (320–350)	—	—
	SiC	430×10^3 (62×10^3)	2,400 (350)	—	—
Ceramic whisker[a]	Al$_2$O$_3$	430×10^3 (62×10^3)	21×10^3 (3,000)	—	—
Polymer fiber[b]	Kevlar[c]	131×10^3 (19×10^3)	3,800 (550)	—	2.8
Metal filament[a]	Boron	410×10^3 (60×10^3)	3,400 (500)	—	—

[a]Source: Data from L. J. Broutman and R. H. Krock, Eds., *Modern Composite Materials*, Addison–Wesley Publishing Co. Inc., Reading, MA, 1967.
[b]Source: Data from A. K. Dhingra, Du Pont Company.
[c]Trade name, Du Pont.

EXAMPLE 12.9

Calculate the thermal conductivity parallel to continuous, reinforcing fibers in a composite with 60 vol % E-glass in a matrix of polyester. [The thermal conductivity of E-glass is 0.97 W/(m · K) and for polyester is 0.17 W/(m · K). Both of these values are for room temperature.]

SOLUTION

This problem is the analog of our calculation in Example 12.8. We use Equation 12.15, with X being the thermal conductivity k:

$$k_c = v_m k_m + v_f k_f$$

$$= (0.4)[0.17 \text{ W/(m} \cdot \text{K)}] + (0.6)[0.97 \text{ W/(m} \cdot \text{K)}]$$

$$= 0.65 \text{ W/(m} \cdot \text{K)}.$$

PRACTICE PROBLEM 12.8

Calculate the composite modulus for a composite with 50 vol % E-glass in a polyester matrix. (See Example 12.8.)

PRACTICE PROBLEM 12.9

The thermal conductivity of a particular fiberglass composite is calculated in Example 12.9. Repeat this calculation for a composite with 50 vol % E-glass in a polyester matrix.

The averaging of properties in a useful composite material can be represented by the typical examples just discussed. But before we leave this topic, we must note an important consideration so far taken for granted. That is, the interface between the matrix and discontinuous phase must be strong enough to transmit the stress or strain due to a mechanical load from one phase to the other. Without this strength, the dispersed phase can fail to "communicate" with the matrix. Rather than have the "best of both worlds" as implied in the introduction to this section, we may instead obtain the worst behavior of each component. Reinforcing fibers easily slipping out of a matrix can be an example. Figure 12.18 illustrates the contrasting microstructures of (a) poorly bonded and (b) well-bonded interfaces in a fiberglass composite. Substantial effort has been devoted to controlling **interfacial strength**. Surface treatment, chemistry, and temperature are a few considerations in the "art and science" of interfacial bonding. To summarize, some interfacial strength is required in all composites to ensure that property averaging is available at relatively low stress levels.

Under isostrain conditions, the axial loading of a reinforcing fiber of finite length leads to a constant shear stress, τ, along the fiber surface, which, in turn, leads to the buildup of tensile stress near the ends of the fiber. Figure 12.19 shows how the tensile stress, σ, varies along the fiber. A "long" fiber (Figure 12.19b) is one with length greater than a critical length, l_c, in which the middle of the fiber reaches a maximum, constant value that corresponds to fiber failure. For maximum efficiency in reinforcement, the fiber length should be much greater than l_c to ensure that the average tensile strength in the fiber is near σ_{critical}.

(a)   (b)

Microstructural images such as these can help engineers analyze the failure of structural components. The lack of bonding between fiber and matrix in part (a) is such evidence.

FIGURE 12.18 *The utility of a reinforcing phase in this polymer–matrix composite depends on the strength of the interfacial bond between the reinforcement and the matrix. These scanning electron micrographs contrast (a) poor bonding with (b) a well-bonded interface. In metal–matrix composites, high interfacial strength is also desirable to ensure high overall composite strength. (Courtesy of Owens-Corning Fiberglas Corporation.)*

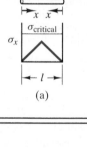

(a)

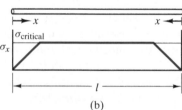

(b)

FIGURE 12.19 *(a) Plot of the tensile stress along a "short" fiber in which the buildup of stress near the fiber ends never exceeds the critical stress associated with fiber failure. (b) A similar plot for the case of a "long" fiber in which the stress in the middle of the fiber reaches the critical value.*

Two fundamentally different philosophies are applied relative to the behavior of fiber composites at relatively high stress levels. For polymer–matrix and metal–matrix composites, failure originates in or along the reinforcing fibers. As a result, a high interfacial strength is desirable to maximize the overall composite strength (Figure 12.18). In ceramic–matrix composites, failure generally originates in the matrix phase. To maximize the fracture toughness for these materials, it is desirable to have a relatively weak interfacial bond that allows fibers to pull out. As a result, a crack initiated in the matrix is deflected along the fiber–matrix interface. This increased crack-path length significantly improves fracture toughness. The mechanism of fiber pullout for improving fracture toughness is shown in Figure 12.20 and can be compared with the two mechanisms for unreinforced ceramics illustrated in Figure 8.7. In general, ceramic composites have achieved substantially higher fracture-toughness levels than the unreinforced ceramics, with values between 20 and 30 MPa$\sqrt{m}$ being common.

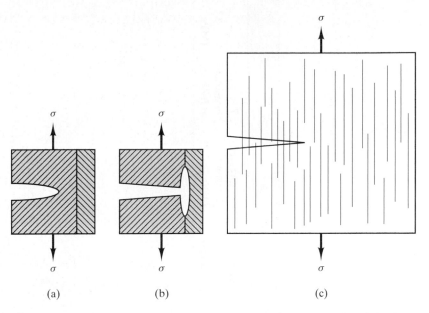

FIGURE 12.20 *For ceramic–matrix composites, low interfacial strength is desirable (in contrast to the case for ductile–matrix composites, such as those shown in Figure 12.18). We see that (a) a matrix crack approaching a fiber is (b) deflected along the fiber–matrix interface. For the overall composite (c), the increased crack-path length due to fiber pullout significantly improves fracture toughness. (Two toughening mechanisms for unreinforced ceramics are illustrated in Figure 8.7.)*

MECHANICAL PROPERTIES OF COMPOSITES

It is obvious from the discussion of property averaging that citing a single number for a given mechanical property of a given composite material is potentially misleading. The concentration and geometry of the discontinuous phase play an important role. Unless otherwise stated, one can assume that composite properties cited in this chapter correspond to optimal conditions (e.g., loading parallel to reinforcing fibers). It is also useful to have information about the component materials separate from the composite. Table 12.5 gives some key mechanical properties for some of the common matrix materials. Table 12.6 gives a similar list for some common dispersed-phase materials. Table 12.7 gives properties for various composite systems. In the preceding tables, the mechanical properties listed are those first defined in Section 6.1 for metals. Comparing the data in Table 12.7 with the data for metals, ceramics and glasses, and polymers in Tables 12.5 and 12.6 will give us some appreciation for the relative mechanical behavior of composites. The dramatic improvement in fracture toughness for ceramic–matrix composites compared with that for unreinforced ceramics is illustrated by Figure 12.21. A more dramatic comparison, which emphasizes the importance of composites in the aerospace field, is given by the list of **specific strength** (strength/density) values in Table 12.8. The specific strength is sometimes referred to as the **strength-to-weight ratio**. The key point is that the substantial cost associated with many "advanced composites" is justified not so much by their absolute strength, but by the fact that they can provide adequate strength in some very low density configurations. The savings in fuel costs alone can frequently justify the higher material

TABLE 12.7

Mechanical Properties of Common Composite Systems

Class	E [MPa (ksi)]	T.S. [MPa (ksi)]	Flexural strength [MPa (ksi)]	Compressive strength [MPa (ksi)]	Percent elongation at failure	K_{IC}[a] (MPa $\sqrt{m}$)
Polymer–matrix						
E-glass (73.3 vol %) in epoxy (parallel loading of continuous fibers)[b]	56×10^3 (8.1×10^3)	1,640 (238)	—	—	2.9	42–60
Al$_2$O$_3$ whiskers (14 vol %) in epoxy[b]	41×10^3 (6×10^3)	779 (113)	—	—	—	—
C (67 vol %) in epoxy (parallel loading)[c]	221×10^3 (32×10^3)	1,206 (175)	—	—	—	—
Kevlar[d] (82 vol %) in epoxy (parallel loading)[c]	86×10^3 (12×10^3)	1,517 (220)	—	—	—	—
B (70 vol %) in epoxy (parallel loading of continuous filaments)[b]	$210\text{–}280 \times 10^3$ ($30\text{–}40 \times 10^3$)[c]	1,400–2,100 (200–300)[c]	—	—	—	46
Metal matrix						
Al$_2$O$_3$ (10 vol %) dispersion-strengthened aluminum[b]	—	330 (48)	—	—	—	—
W (50 vol %) in copper (parallel loading of continuous filaments)[b]	260×10^3 (38×10^3)	1,100 (160)	—	—	—	—
W particles (50 vol %) in copper[b]	190×10^3 (27×10^3)	380 (55)	—	—	—	—
Ceramic–matrix						
SiC whiskers in Al$_2$O$_3$[e]	—	—	800 (116)	—	—	8.7
SiC fibers in SiC[e]	—	—	750 (109)	—	—	25.0
SiC whiskers in reaction-bonded Si$_3$N$_4$[e]	—	—	900 (131)	—	—	20.0

[a] Source: Data from M. F. Ashby and D. R. H. Jones, *Engineering Materials—An Introduction to Their Properties and Applications*, Pergamon Press, Inc., Elmsford, NY, 1980.

[b] L. J. Broutman and R. H. Krock, Eds., *Modern Composite Materials*, Addison-Wesley Publishing Co., Inc., Reading, MA, 1967.

[c] A. K. Dhingra, Du Pont Company.

[d] Trade name, Du Pont.

[e] A. J. Klein, *Advanced Materials and Processes 2*, 26 (1986).

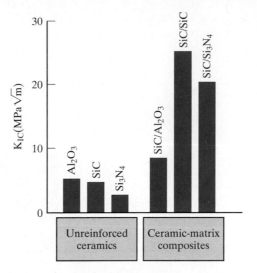

FIGURE 12.21 *The fracture toughness of these structural ceramics is substantially increased by the use of a reinforcing phase. (Note the toughening mechanism illustrated in Figure 12.20.)*

TABLE 12.8

Specific Strengths (Strength/Density)

Group	Material	Specific strength [mm (in.)]
Noncomposites	1040 steel[a]	9.9×10^6 (0.39×10^6)
	2048 plate aluminum[a]	16.9×10^6 (0.67×10^6)
	Ti–5Al–2.5Sn[a]	19.7×10^6 (0.78×10^6)
	Epoxy[b]	6.4×10^6 (0.25×10^6)
Composites	E-glass (73.3 vol %) in epoxy (parallel loading of continuous fibers)[c]	77.2×10^6 (3.04×10^6)
	Al_2O_3 whiskers (14 vol %) in epoxy[c]	48.8×10^6 (1.92×10^6)
	C (67 vol %) in epoxy (parallel loading)[d]	76.9×10^6 (3.03×10^6)
	Kevlar[e] (82 vol %) in epoxy (parallel loading)[d]	112×10^6 (4.42×10^6)
	Douglas fir, kiln-dried to 12% moisture (loaded in bending)[c]	18.3×10^6 (0.72×10^6)

[a]Sources: Data from *Metals Handbook*, 9th ed., Vols. 1–3, and 8th ed., Vol. 1, American Society for Metals, Metals Park, OH, 1961, 1978, 1979, and 1980.

[b]R. A. Flinn and P. K. Trojan, *Engineering Materials and Their Applications*, 2nd ed., Houghton Mifflin Company, Boston, 1981; and M. F. Ashby and D. R. H. Jones, *Engineering Materials*, Pergamon Press, Inc., Elmsford, NY, 1980.

[c]R. A. Flinn and P. K. Trojan, *Engineering Materials and Their Applications*, 2nd ed., Houghton Mifflin Company, Boston, 1981.

[d]A. K. Dhingra, Du Pont Company.

[e]Trade name, Du Pont.

costs. Figure 12.22 illustrates (with the data of Table 12.8) the distinct advantage of advanced composite systems in this regard. Finally, note that the higher material costs of advanced composites can be offset by reduced assembly costs (e.g., one-piece automobile frames), as well as by high specific strength values.

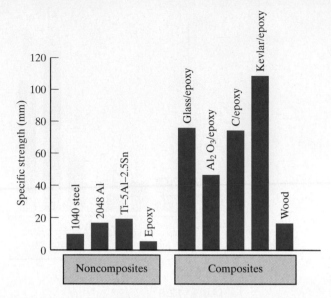

FIGURE 12.22 *A bar graph plot of the data of Table 12.8 illustrates the substantial increase in specific strength possible with composites.*

EXAMPLE 12.10

Calculate the isostrain modulus of epoxy reinforced with 73.3 vol % E-glass fibers and compare your result with the measured value given in Table 12.7.

SOLUTION

Using Equation 12.13 and data from Tables 12.5 and 12.6, we obtain

$$E_c = v_m E_m + v_f E_f$$
$$= (1.000 - 0.733)(6.9 \times 10^3 \text{ MPa}) + (0.733)(72.4 \times 10^3 \text{ MPa})$$
$$= 54.9 \times 10^3 \text{ MPa}.$$

Table 12.7 gives for this case $E_c = 56 \times 10^3$ MPa, or

$$\% \text{ error} = \frac{56 - 54.9}{56} \times 100 = 2.0\%.$$

The calculated value comes within 2% of the measured value.

EXAMPLE 12.11

The tensile strength of the dispersion-strengthened aluminum in Example 12.7 is 350 MPa. The tensile strength of the pure aluminum is 175 MPa. Calculate the specific strengths of these two materials.

SOLUTION

The specific strength is simply

$$\text{sp.str.} = \frac{\text{T.S.}}{\rho}.$$

Using the preceding strengths and densities from Example 12.7, we obtain

$$\text{sp. str., Al} = \frac{(175 \text{ MPa}) \times (1.02 \times 10^{-1} \text{ kg/mm}^2)/\text{MPa}}{(2.70 \text{ Mg/m}^3)(10^3 \text{ kg/Mg})(1 \text{ m}^3/10^9 \text{ mm}^3)}$$

$$= 6.61 \times 10^6 \text{ mm}$$

and

$$\text{sp. str., Al/10 vol \% Al}_2\text{O}_3 = \frac{(350)(1.02 \times 10^{-1})}{(2.83)(10^{-6})} \text{ mm}$$

$$= 12.6 \times 10^6 \text{ mm}.$$

Note. In following this example, you may have been disturbed by the rather casual use of units. We canceled kg in the strength term (numerator) with kg in the density term (denominator). This cancellation is, of course, not rigorously correct because the strength term uses kg force and the density term uses kg mass. However, this convention is commonly used and was, in fact, the basis of the numbers in Table 12.8. The important thing about specific strength is not the absolute number, but the relative values for competitive structural materials.

PRACTICE PROBLEM 12.10

In Example 12.10, the isostrain modulus for a fiberglass composite is shown to be close to a calculated value. Repeat this comparison for the isostrain modulus of B (70 vol %)/epoxy composite given in Table 12.7.

PRACTICE PROBLEM 12.11

In Example 12.11, we find that dispersion-strengthened aluminum has a substantially higher specific strength than pure aluminum. In a similar way, calculate the specific strength of the E-glass/epoxy composite of Table 12.7 compared to that of the pure epoxy of Table 12.5. For density information, refer to Example 12.6. (You may wish to compare your calculations with the values in Table 12.8.)

12.3 | Processing the Structural Materials

Finally, we consider the various ways in which polymers and composites are produced. (Additional examples of their processing will be found in Chapters 12A–B on the related Web site.)

PROCESSING OF POLYMERS

Table 12.9 summarizes some of the major processing techniques for polymers. For thermoplastics, **injection molding** and **extrusion molding** are the predominant processes. Figure 12.23 illustrates the extrusion-molding technique. Injection molding involves the melting of polymer powder prior to injection. Both injection and extrusion molding are similar to metallurgical processing, but are carried out at relatively low temperatures. **Blow molding** is a third major processing technique

TABLE 12.9

Some Major Processing Methods for Polymers	
Thermoplastics	**Thermosetting**
Injection molding	Compression molding
Extrusion molding	Transfer molding
Blow molding	

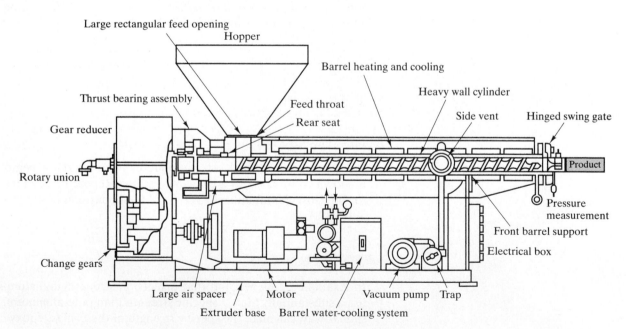

FIGURE 12.23 *Extrusion molding of a thermoplastic polymer. (After* Modern Plastics Encyclopedia, 1981–82, *Vol. 58, No. 10A, McGraw-Hill Book Company, New York, October 1981.)*

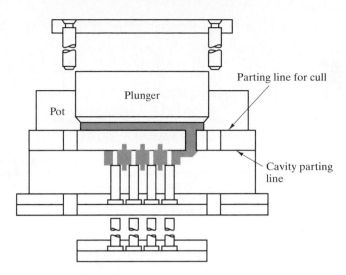

FIGURE 12.24 *Transfer molding of a thermosetting polymer. (After* Modern Plastics Encyclopedia, 1981–82, *Vol. 58, No. 10A, McGraw-Hill Book Company, New York, October 1981.)*

for thermoplastics. With this technique, the specific shaping process is quite similar to the glass-forming technique for bottles in which jets of air form the bottle shape into a mold, except that relatively low molding temperatures are required. As with glass-container manufacturing, blow molding is often used to produce polymeric containers. In addition, various commercial products, including automobile body parts, can be economically fabricated by this method. **Compression molding** and **transfer molding** are the predominant processes for thermosetting polymers. Transfer molding is illustrated in Figure 12.24. Compression molding is generally impractical for thermoplastics because the mold would have to be cooled to ensure that the part would not lose its shape upon ejection from the mold. In transfer molding, a partially polymerized material is forced into a closed mold, where final cross-linking occurs at elevated temperature and pressure.

In Chapter 6 and earlier in this chapter, there are numerous references to the effect of polymeric structure on mechanical behavior. Table 12.10 summarizes these various relations, along with corresponding references to the processing techniques that lead to the various structures.

PROCESSING OF COMPOSITES

Composites represent such a wide range of structural materials that a brief list of processing techniques cannot do justice to the full field. Table 12.11 is restricted to the key examples for fiberglass. Even this conventional material represents a diverse set of processing techniques. Figure 12.25 illustrates the fabrication of typical fiberglass configurations. These configurations are often standard polymer processing methods with glass fibers added at an appropriate point in the procedure. A major factor affecting properties is the orientation of the fibers. The issue of anisotropy of properties was discussed earlier in this section. Note that the open-mold processes include *pultrusion*, which is especially well suited for producing complex cross-section products continuously.

Open-mold processes

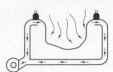

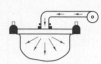

Contact Molding

Resin is in contact with air. Lay-up normally cures at room temperature. Heat may accelerate cure. A smoother exposed side may be achieved by wiping on cellophane.

Vacuum Bag

Cellophane or polyvinyl acetate is placed over lay-up. Joints are sealed with plastic; vacuum is drawn. Resultant atmospheric pressure eliminates voids and forces out entrapped air and excess resin.

Pressure Bag

Tailored bag—normally rubber sheeting—is placed against layup. Air or steam pressure up to 50 psi is applied between pressure plate and bag.

Autoclave

Modification of the pressure bag method: after lay-up, entire assembly is placed in steam autoclave at 50 to 100 psi. Additional pressure achieves higher glass loadings and improved removal of air.

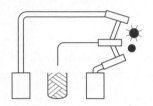

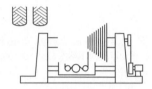

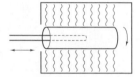

Spray-up

Roving is fed through a chopper and ejected into a resin stream, which is directed at the mold by either of two spray systems: (1) A gun carries resin premixed with catalyst, another gun carries resin premixed with accelerator. (2) Ingredients are fed into a single run mixing chamber ahead of the spray nozzle. By either method the resin mix precoats the strands and the merged spray is directed into the mold by the operator. The glass-resin mix is rolled by hand to remove air, lay down the fibers, and smooth the surface. Curing is similar to hand lay-up.

Filament Winding

Uses continuous reinforcement to achieve efficient utilization of glass fiber strength. Roving or single strands are fed from a creel through a bath of resin and wound on a mandrel. Preimpregnated roving is also used. Special lathes lay down glass in a predetermined pattern to give max. strength in the directions required. When the right number of layers have been applied, the wound mandrel is cured at room temperature or in an oven.

Centrifugal Casting

Round objects such as pipe can be formed using the centrifugal casting process. Chopped strand mat is positioned inside a hollow mandrel. The assembly is then placed in an oven and rotated. Resin mix is distributed uniformly throughout the glass reinforcement. Centrifugal action forces glass and resin against walls of rotating mandrel prior to and during the cure. To accelerate cure, hot air is passed through the oven.

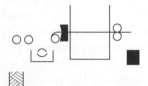

Continuous Pultrusion

Continuous strand—roving or other forms of reinforcement—is impregnated in a resin bath and drawn through a die which sets the shape of the stock and controls the resin content. Final cure is effected in an oven through which the stock is drawn by a suitable pulling device.

Encapsulation

Short chopped strands are combined with catalyzed resin and poured into open molds. Cure is at room temperature. A post-cure of 30 minutes at 200 °F is normal.

(a)

FIGURE 12.25 *Summary of the diverse methods of processing fiberglass products: (a) open-mold processes. (b) preforming methods, and (c) closed-mold processes. (After illustrations from Owens-Corning Fiberglas Corporation as abstracted in R. Nicholls,* Composite Construction Materials Handbook, *Prentice Hall, Inc., Englewood Cliffs, NJ, 1976.)*

Preforming methods

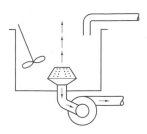

Directed Fiber

Roving is cut into 1- to 2-inch lengths of chopped strand which are blown through a flexible hose onto a rotating preform screen. Suction holds them in place while a binder is sprayed on the preform and cured in an oven. The operator controls both deposition of chopped strands and binder.

Plenum Chamber

Roving is fed into a cutter on top of plenum chamber. Chopped strands are directed onto a spinning fiber distributor to separate chopped strands and distribute strands uniformly in plenum chamber. Falling strands are sucked onto preform screen. Resinous binder is sprayed on. Preform is positioned in a curing oven. New screen is indexed in plenum chamber for repeat cycle.

Water Slurry

Chopped strands are pre-impregnated with pigmented polyester resin and blended with cellulosic fiber in a water slurry. Water is exhausted through a contoured, perforated screen and glass fibers and cellulosic material are deposited on the surface. The wet preform is transferred to an oven where hot air is sucked through the preform. When dry, the preform is sufficiently strong to be handled and molded.

(b)

Closed-mold processes

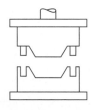

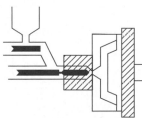

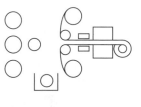

Premix/Molding Compound

Prior to molding, glass reinforcement, usually chopped spun roving, is thoroughly mixed with resin, pigment, filler, and catalyst. The premixed material can be extruded into a rope-like form for easy handling or may be used in bulk form.

The premix is formed into accurately weighed charges and placed in the mold cavity under heat and pressure. Amount of pressure varies from 100 to 1500 psi. Length of cycle depends on cure temperature, resin, and wall thickness. Cure temperatures range from 225 °F to 300 °F. Time varies from 30 seconds to 5 minutes.

Injection Molding

For use with thermoplastic materials. The glass and resin molding compound is introduced into a heating chamber where it softens. This mass is then injected into a mold cavity that is kept at a temperature below the softening point of the resin. The part then cools and solidifies.

Continuous Laminating

Fabric or mat is passed through a resin dip and brought together between cellophane covering sheets; the lay-up is passed through a heating zone and the resin is cured. Laminate thickness and resin content are controlled by squeeze rolls as the various plies are brought together.

(c)

FIGURE 12.25 *(Continued)*

TABLE 12.10

The Relationship of Processing, Molecular Structure, and Mechanical Behavior for Polymers			
Category	Processing technique	Molecular structure	Mechanical effect
Thermoplastic polymers	Addition agent	Branching	Increased strength and stiffness
	Vulcanization	Cross-linking	Increased strength and stiffness
	Crystallization	Increased crystallinity	Increased strength and stiffness
	Plasticizer	Decreased molecular weight	Decreased strength and stiffness
	Filler	Restricted chain mobility	Increased strength and stiffness
Thermosetting polymers	Setting at elevated temperatures	Network formation	Rigid (remaining upon cooling)

TABLE 12.11

Major Processing Methods for Fiberglass Composites	
Composite	Processing methods
Fiberglass	Open mold
	Preforming
	Closed mold

Summary

Polymers, or plastics, are organic materials composed of long-chain or network organic molecules formed from small molecules (monomers) by polymerization reactions. Thermoplastic polymers become softer upon heating due to thermal agitation of weak, secondary bonds between adjacent linear molecules. Thermoplastics include engineering polymers for metals substitution and thermoplastic elastomers, rubberlike materials with the processing convenience of traditional thermoplastics. Thermosetting polymers are network structures that form upon heating, resulting in greater rigidity, and include the traditional vulcanized elastomers. Additives are materials added to polymers to provide specific characteristics.

Composites bring together in a single material the benefits of various components discussed in Chapters 11 and 12. Fiberglass typifies synthetic fiber-reinforced composites. Glass fibers provide high strength and modulus in a polymeric matrix that provides ductility. Various fiber geometries are commonly used. In any case, properties tend to reflect the highly anisotropic geometry of the composite microstructure. To produce structural materials that have properties beyond the capability of polymer–matrix composites, substantial effort is under way in the development of new advanced composites, such as metal–matrix and ceramic–matrix composites. There are various examples of aggregate composites in which particles reinforce a matrix. The property averaging that results from combining more than one component in a composite material is highly dependent on the microstructural geometry of the composite. The major mechanical properties important to structural materials are summarized for various composites. An additional parameter of importance to aerospace applications (among others) is the specific strength or strength-to-weight ratio, which is characteristically large for many advanced composite systems.

Thermoplastic polymers are generally processed by injection molding, extrusion molding, or blow molding. Thermosetting polymers are generally formed by compression molding or transfer molding. The processing of composites involves a wide range of methods, representing the especially diverse nature of this family of materials. Only a few representative examples for fiberglass manufacturing have been discussed in this chapter.

Key Terms

Polymers
addition polymerization (382)
adhesive (392)
atactic (388)
bifunctional (385)
blend (384)
block copolymer (383)
blow molding (416)
chain growth (382)
colorant (395)
compression molding (417)
condensation polymerization (382)
copolymer (383)
degree of polymerization (387)
dye (395)
elastomer (392)
engineering polymer (391)
extended length (388)
extrusion molding (416)
filler (394)
flame retardant (394)
graft copolymer (392)
initiator (382)
injection molding (416)
isotactic (388)

linear molecular structure (385)
mer (382)
monomer (382)
network copolymer (393)
network molecular structure (385)
pigment (395)
plastics (381)
plasticizer (394)
polyfunctional (385)
polymer (382)
polymerization (382)
reinforcement (394)
root-mean-square length (387)
stabilizer (394)
step growth (382)
syndiotactic (388)
terminator (383)
thermoplastic elastomer (392)
thermoplastic polymers (391)
thermosetting polymers (392)
transfer molding (417)
vulcanization (389)
Composites
advanced composites (398)
aggregate composite (403)

anisotropic (398)
carbon–carbon composite (400)
ceramic–matrix composite (400)
cermet (404)
continuous fiber (398)
discrete (chopped) fiber (398)
dispersion-strengthened metal (404)
E-glass (397)
fiberglass (396)
fiber-reinforced composite (396)
hybrid (400)
interfacial strength (409)
isostrain (405)
isostress (407)
laminate (398)
matrix (396)
metal–matrix composite (400)
particulate composite (403)
polymer–matrix composite (399)
property averaging (405)
specific strength (411)
strength-to-weight ratio (411)
whisker (400)
woven fabric (398)

References

Polymers

Brandrup, **J.**, **E. H. Immergut**, and **E. A. Grulke**, Eds., *Polymer Handbook*, 4th ed., John Wiley & Sons, Inc., New York, 2003.

Engineered Materials Handbook, Vol. 2, *Engineering Plastics*, ASM International, Materials Park, OH, 1988.

Mark, **H. F.**, Ed., *Encyclopedia of Polymer Science and Engineering, Concise*, 3rd ed., John Wiley & Sons, Inc., New York, 2007.

Composites

Agarwal, **B. D.**, **L. J. Broutman,** and **K. Chandrashekhara**, *Analysis and Performance of Fiber Composites*, 3rd ed., John Wiley & Sons, Inc., New York, 2006.

Chawla, **K. K.**, *Composite Materials: Science and Engineering*, 3rd ed., Springer, New York, 2009.

ASM Handbook, Vol. 21, *Composites*, ASM International, Materials Park, OH, 2001.

Jones, **R. M.**, *Mechanics of Composite Materials*, 2nd ed., Taylor and Francis, Philadelphia, 1999.

Nicholls, **R.**, *Composite Construction Materials Handbook*, Prentice Hall, Inc., Englewood Cliffs, NJ, 1976.

Problems

12.1 • Polymers

12.1. What is the average molecular weight of a polypropylene, $(C_3H_6)_n$, with a degree of polymerization of 500?

12.2. What is the average molecular weight of a polystyrene, $(C_8H_8)_n$, with a degree of polymerization of 500?

12.3. How many grams of H_2O_2 would be needed to yield 1 kg of a polypropylene, $(C_3H_6)_n$, with an average degree of polymerization of 600? (Use the same assumptions given in Example 12.1.)

12.4. A blend of polyethylene and PVC (see Figure 12.4) contains 10 wt % PVC. What is the molecular percentage of PVC?

12.5. A blend of polyethylene and PVC (see Figure 12.4) contains 10 mol % PVC. What is the weight percentage of PVC?

12.6. Calculate the degree of polymerization for **(a)** a low-density polyethylene with a molecular weight of 20,000 amu, **(b)** a high-density polyethylene with a molecular weight of 300,000 amu, and **(c)** an ultra-high molecular weight polyethylene with a molecular weight of 4,000,000 amu.

12.7. A simplified mer formula for natural rubber (isoprene) is C_5H_8. Calculate the molecular weight for a molecule of isoprene with a degree of polymerization of 500.

12.8. Calculate the molecular weight for a molecule of chloroprene (a common synthetic rubber) with a degree of polymerization of 500. (The mer formula for chloroprene is C_4H_5Cl.)

12.9. The distribution of the degree of polymerization of a polymer can be represented in tabular form as follows:

n Range	n_i (Mid value)	Population fraction
1–100	50	—
101–200	150	—
201–300	250	0.01
301–400	350	0.10
401–500	450	0.21
501–600	550	0.22
601–700	650	0.18
701–800	750	0.12
801–900	850	0.07
901–1,000	950	0.05
1,001–1,100	1,050	0.02
1,101–1,200	1,150	0.01
1,201–1,300	1,250	0.01
		$\sum = 1.00$

Calculate the average degree of polymerization for this system.

12.10. If the polymer evaluated in Problem 12.9 is polypropylene, what would be the **(a)** coiled length and **(b)** extended length of the average molecule?

12.11. What would be the maximum fraction of cross-link sites that would be connected in 1 kg of chloroprene, C_4H_5Cl, with the addition of 250 g of sulfur?

12.12. Calculate the average molecular length (extended) for a polyethylene with a molecular weight of 20,000 amu.

12.13. Calculate the average molecular length (extended) for PVC with a molecular weight of 20,000 amu.

12.14. If 0.2 g of H_2O_2 is added to 100 g of ethylene to establish the degree of polymerization, what would be the resulting average molecular length (coiled)? (Use the assumptions of Example 12.1.)

12.15. What would be the extended length of the average molecule described in Problem 12.14?

*****12.16.** The acetal polymer in Figure 12.5 contains, of course, C–O bonds rather than C–C bonds along its molecular chain backbone. As a result, there are two types of bond angles to consider. The O–C–O bond angle is approximately the same as the C–C–C bond angle (109.5°) because of the tetrahedral bonding configuration in carbon (see Figure 2.19). However, the C–O–C bond is a flexible one with a possible bond angle ranging up to 180°. **(a)** Make a sketch similar to that shown in Figure 12.9 for a fully extended polyacetal molecule. **(b)** Calculate the extended length of a molecule with a degree of polymerization of 500. (Refer to Table 2.2 for bond length data.) **(c)** Calculate the coiled length of the molecule in part (b).

12.17. Calculate **(a)** the molecular weight, **(b)** coiled molecular length, and **(c)** extended molecular length for a polytetrafluoroethylene polymer with a degree of polymerization of 500.

12.18. Repeat Problem 12.17 for a polypropylene polymer with a degree of polymerization of 700.

12.19. Calculate the degree of polymerization of a polycarbonate polymer with a molecular weight of 100,000 amu.

12.20. Calculate the molecular weight for a polymethyl methacrylate polymer with a degree of polymerization of 500.

12.21. In Problem 11.5, the mass reduction in an automobile design was calculated based on trends in metal-alloy selection. A more complete picture is obtained by noting that, in the same 1975 to 1985 period, the volume of polymers used increased from 0.064 m³ to 0.100 m³. Estimate the mass reduction (compared to that in 1975) including this additional polymer data. (Approximate the polymer density as 1 Mg/m³.)

12.22. Repeat Problem 12.21 for the mass reduction in the year 2000 given the data in Problem 11.6 and the fact that the total volume of polymer used at that time is 0.122 m³.

12.23. What would be the molecular weight of a 50,000 mm³ plate made from urea-formaldehyde? (The density of the urea-formaldehyde is 1.50 Mg/m³.)

12.24. How much water by-product would be produced in the polymerization of the urea-formaldehyde product in Problem 12.23?

12.25. Polyisoprene loses its elastic properties with 1 wt % O_2 addition. If we assume that this loss is due to a cross-linking mechanism similar to that for sulfur, what fraction of the cross-link sites are occupied in this case?

12.26. Repeat the calculation of Problem 12.25 for the case of the oxidation of polychloroprene by 1 wt % O_2.

12.27. An epoxy (density = 1.1 Mg/m³) is reinforced with 25 vol % E-glass fibers (density = 2.54 Mg/m³). Calculate **(a)** the weight percent of E-glass fibers and **(b)** the density of the reinforced polymer.

12.28. Calculate the % mass savings that would occur if the reinforced polymer described in Problem 12.27 were used to replace a steel gear. (Assume the gear volume is the same for both materials, and approximate the density of steel by pure iron.)

12.29. Repeat Problem 12.28 if the polymer replaces an aluminum alloy. (Again, assume the gear volume is the same, and approximate the density of the alloy by pure aluminum.)

12.30. Some injection moldable nylon 66 contains 40 wt % glass spheres as a filler. Improved mechanical properties are the result. If the average glass sphere diameter is 100 μm, estimate the density of such particles per cubic millimeter.

12.31. Calculate the average separation distance between the centers of adjacent glass spheres in the nylon described in Problem 12.30. (Assume a simple cubic array of dispersed particles as in Problem 11.20.)

12.32. Repeat Problem 12.31 for the case of the same wt % spheres, but an average glass sphere diameter of 50 μm.

12.33. Bearings and other parts requiring exceptionally low friction and wear can be fabricated from an acetal polymer with an addition of polytetrafluoroethylene (PTFE) fibers. The densities of acetal and PTFE are 1.42 Mg/m³ and 2.15 Mg/m³, respectively. If the density of the polymer with additive is 1.54 Mg/m³, calculate the weight percent of the PTFE addition.

12.34. Repeat Problem 12.33 for the case of a polymer density of 1.48 Mg/m³.

12.2 • Composites

12.35. Calculate the density of a fiber-reinforced composite composed of 14 vol % Al_2O_3 whiskers in a matrix of epoxy. The density of Al_2O_3 is 3.97 Mg/m^3 and of epoxy is 1.1 Mg/m^3.

12.36. Calculate the density of a boron-filament-reinforced epoxy composite containing 70 vol % filaments. The density of epoxy is 1.1 Mg/m^3.

12.37. Using the information in Equation 12.7, calculate the molecular weight of an aramid polymer with an average degree of polymerization of 500.

12.38. Calculate the density of the Kevlar fiber-reinforced epoxy composite in Table 12.7. The density of Kevlar is 1.44 Mg/m^3 and of epoxy is 1.1 Mg/m^3.

12.39. In a contemporary commercial aircraft, a total of 0.25 m^3 of its exterior surface is constructed of a Kevlar/epoxy composite, rather than of a conventional aluminum alloy. Calculate the mass savings using the density calculated in Problem 12.38 and approximating the alloy density by that of pure aluminum.

12.40. What would be the mass savings (relative to the aluminum alloy) if a carbon/epoxy composite with a density of 1.5 Mg/m^3 is used rather than the Kevlar/epoxy composite of Problem 12.39?

12.41. Calculate the density of a particulate composite composed of 50 vol % Mo in a copper matrix.

12.42. Calculate the density of a dispersion-strengthened copper with 10 vol % Al_2O_3.

12.43. Calculate the composite modulus for epoxy reinforced with 70 vol % boron filaments under isostrain conditions.

12.44. Calculate the composite modulus for aluminum reinforced with 50 vol % boron filaments under isostrain conditions.

12.45. Calculate the modulus of elasticity of a metal–matrix composite under isostrain conditions. Assume an aluminum matrix is reinforced by 60 vol % SiC fibers.

12.46. Calculate the modulus of elasticity of a ceramic–matrix composite under isostrain conditions. Assume an Al_2O_3 matrix is reinforced by 60 vol % SiC fibers.

12.47. On a plot similar to that displayed in Figure 12.17, show the composite modulus for **(a)** 60 vol % fibers (the result of Example 12.8) and **(b)** 50 vol % fibers

(the result of Practice Problem 12.8). Include the individual glass and polymer plots.

12.48. On a plot similar to that displayed in Figure 12.17, show the composite modulus for an epoxy reinforced with 70 vol % carbon fibers under isostrain conditions. (Use the midrange value for carbon modulus in Table 12.6. Epoxy data are given in Table 12.5. For effectiveness of comparison with the case in Figure 12.17, use the same stress and strain scales. Include the individual matrix and fiber plots.)

12.49. Calculate the composite modulus for polyester reinforced with 10 vol % Al_2O_3 whiskers under isostrain conditions. (See Tables 12.5 and 12.6 for appropriate moduli.)

12.50. Plot Poisson's ratio as a function of reinforcing fiber content for an SiC fiber-reinforced Si_3N_4 composite system loaded parallel to the fiber direction and SiC contents between 50 and 75 vol %. (Note the discussion relative to Equation 12.15 and data in Table 6.5.)

*12.51. Consider further the discussion of interfacial strength relative to Figure 12.19. **(a)** Taking the tensile stress in the fiber (with radius r) at a distance x from either end of the fiber to be σ_x, use a force balance between tensile and shear components to derive an expression for σ_x in terms of the fiber geometry and the interfacial shear stress, τ (which is uniform along the entire interface). **(b)** Show how this expression produces the plot of the tensile stress in a short fiber (in which σ_x is always less than $\sigma_{critical}$, the failure stress of the fiber), as shown in Figure 12.19a.

*12.52. **(a)** Referring to Problem 12.51, show how this expression produces the sketch of the tensile stress distribution in a long fiber (in which the stress in the middle portion of the fiber reaches a maximum, constant value, corresponding to fiber failure), as shown in Figure 12.19b. **(b)** Using the result of Problem 12.51a, derive an expression for the critical stress transfer length, l_c, the minimum fiber length that must be exceeded if fiber failure is to occur (i.e., if σ_x is to reach $\sigma_{critical}$).

12.53. Compare the calculated value of the isostrain modulus of a W fiber (50 vol %)/copper composite with that given in Table 12.7. The modulus of tungsten is 407×10^3 MPa.

12.54. Determine the error made in Problem 12.43 in calculating the isostrain modulus of the B/epoxy composite of Table 12.7.

12.55. Calculate the error in assuming the isostrain modulus of an epoxy reinforced with 67 vol % C fibers is given by Equation 12.13. (Note Table 12.7 for experimental data.)

12.56. Calculate the specific strength of the Kevlar/epoxy composite in Table 12.7. (Note Problem 12.38.)

12.57. Calculate the specific strength of the B/epoxy composite in Table 12.7. (Note Problem 12.36.)

12.58. Calculate the specific strength of the W particles (50 vol %)/copper composite listed in Table 12.7. (See Practice Problem 12.7.)

12.59. Calculate the specific strength for the W fibers (50 vol %)/copper composite listed in Table 12.7.

12.60. Calculate the (flexural) specific strength of the SiC/Al_2O_3 ceramic–matrix composite in Table 12.7, assuming 50 vol % whiskers. (The density of SiC is 3.21 Mg/m^3. The density of Al_2O_3 is 3.97 Mg/m^3.)

12.61. To appreciate the relative toughness of (i) traditional ceramics, (ii) high-toughness, unreinforced ceramics, and (iii) ceramic–matrix composites, plot the breaking stress versus flaw size, a, for **(a)** silicon carbide, **(b)** partially stabilized zirconia, and **(c)** silicon carbide reinforced with SiC fibers. Use Equation 8.1 and take $Y=1$. Cover a range of a from 1 to 100 mm on a logarithmic scale. (See Tables 8.3 and 12.7 for data.)

D 12.62. In Problem 6.9, a competition among various metallic pressure-vessel materials was illustrated. We can expand the selection process by including some composites, as listed in the following table:

Material	ρ (Mg/m³)	Cost ($/kg)	Y.S. (MPa)
1040 carbon steel	7.80	0.63	
304 stainless steel	7.80	3.70	
3003–H14 aluminum	2.73	3.00	
Ti–5Al–2.5Sn	4.46	15.00	
Reinforced concrete	2.50	0.40	200
Fiberglass	1.80	3.30	200
Carbon fiber-reinforced polymer	1.50	70.00	600

(a) From this expanded list, select the material that will produce the lightest vessel. **(b)** Select the material that will produce the minimum-cost vessel.

12.3 • Processing the Structural Materials

As with the comments regarding problems for Section 11.3, Section 12.3 also contains only a few problems that can be kept in the objective style used in preceding chapters.

12.63. The heat evolved in the manufacturing of polyethylene sheet was evaluated in Problem 2.29. In a similar way, calculate the heat evolution occurring during a 24-hour period in which 864 km of a sheet 1 mil (25.4 μm) thick by 300 mm wide are manufactured. (The density of the sheet is 0.910 Mg/m^3.)

12.64. What would be the total heat evolution occurring in the manufacturing of the polyethylene sheet in Problem 12.63 if that production rate could be maintained for a full year?

D 12.65. Given data on modulus of elasticity (in tension) and tensile strength for various thermoplastic polymers in Table 6.6, select the polymers that would meet the following design specifications for a mechanical gear application:

modulus of elasticity, E:

$$2{,}000 \text{ MPa} < E < 3{,}000 \text{ MPa}$$

and

tensile strength (T.S.): > 50 MPa.

D 12.66. Using the design specifications given for Problem 12.65, consider the various thermosetting polymers of Table 6.7. How many of these polymers would meet the design specifications?

D 12.67. Consider the injection molding of low-cost casings using a polyethylene–clay particle composite system. The modulus of elasticity of the composite increases and the tensile strength of the composite decreases with volume fraction of clay as follows:

Volume fraction clay	Modulus of elasticity (MPa)	Tensile strength (MPa)
0.3	830	24.0
0.6	2,070	3.4

Assuming both modulus and strength change linearly with volume fraction of clay, determine the allowable composition range that ensures a product with a modulus of at least 1,000 MPa and a strength of at least 10 MPa.

D 12.68. For the injection-molding process described in Problem 12.67, what specific composition would be preferred given that the cost per kg of polyethylene is 10 times that of clay?

CHAPTER 13
Electronic Materials

The modern microprocessor is a state-of-the-art application of semiconductor materials encased in a package of conventional structural materials. (Photograph courtesy of Intel Corporation.)

13.1 Charge Carriers and Conduction

13.2 Energy Levels and Energy Bands

13.3 Conductors

13.4 Insulators

13.5 Semiconductors

13.6 Composites

13.7 Electrical Classification of Materials

Materials selection is not limited to structural applications. We often choose materials because of their performance in electronic applications. This chapter shows that a classification of materials based on electrical conductivity rather than chemical bonding produces an additional category, viz. semiconductors. These materials have values of conductivity intermediate between the generally conductive metals and the generally insulating ceramics, glasses, and polymers. Superconductors are dramatic exceptions, in that certain metals and ceramics have no electrical resistivity and provide intriguing design possibilities. In this chapter, we focus on the wide range of semiconductor materials and the unique processing technologies that have evolved in the manufacturing of solid-state electronics. These technologies involve exceptional levels of structural perfection and chemical purity. The high degree of chemical purity allows the sophisticated control of impurity doping, an excellent example of the principles of solid-state diffusion from Chapter 5.

In Chapter 2, we found that an understanding of atomic bonding can lead to a useful classification system for engineering materials. In this chapter, we turn to a specific material property, *electrical conduction*, to reinforce our classification. This commonality should not be surprising in light of the electronic nature of bonding. Electrical conduction is the result of the motion of charge carriers (such as electrons) within the material. Once again, we find a manifestation of the concept that structure leads to properties. In Chapter 6, atomic and microscopic structures were found to lead to various mechanical properties. Electrical properties follow from electronic structure.

The ease or difficulty of electrical conduction in a material can be understood by returning to the concept of energy levels introduced in Chapter 2. In solid materials, discrete energy levels give way to energy bands. It is the relative spacing of these bands (on an energy scale) that determines the magnitude of conductivity. Metals, with large values of conductivity, are termed *conductors*. Ceramics, glasses, and polymers, with small values of conductivity, are termed *insulators*. *Semiconductors*, with intermediate values of conductivity, are best defined by the unique nature of their electrical conduction.

A link to the Materials Research Society Web site on the related Web site is an especially good portal to the applications of materials in the electronics industry.

In this chapter, we focus on electrical behavior in general and semiconduction in particular. A more detailed discussion of these topics is given in Chapters 13A and 13C on the related Web site. Also, Chapters 13B and 13D cover the closely related topics of optical behavior and magnetic materials, respectively.

13.1 | Charge Carriers and Conduction

The conduction of electricity in materials is by means of individual, atomic-scale species called **charge carriers**. The simplest example of a charge carrier is the *electron*, a particle with 0.16×10^{-18} C of negative charge (see Section 2.1). A more abstract concept is the **electron hole**, which is a missing electron in an electron cloud. The absence of the negatively charged electron gives the electron hole an effective positive charge of 0.16×10^{-18} C relative to its environment. Electron holes play a central role in the behavior of semiconductors and will be discussed in detail in Section 13.5. In ionic materials, anions can serve as negative charge carriers, and cations can serve as positive carriers. As seen in Section 2.2, the valence of each ion indicates positive or negative charge in multiples of 0.16×10^{-18} C.

A simple method for measurement of electrical conduction is shown in Figure 13.1. The magnitude of **current** flow, I, through the circuit with a given **resistance**, R, and **voltage**, V, is related by **Ohm's*** **law**,

$$V = IR, \tag{13.1}$$

where V is in units of volts,[†] I is in amperes[‡] (1 A = 1 C/s), and R is in ohms. The resistance value depends on the specific sample geometry; R increases with sample length, l, and decreases with sample area, A. As a result, a property more characteristic of a given material and independent of its geometry is **resistivity**, ρ, defined as

$$\rho = \frac{RA}{l}. \tag{13.2}$$

The units for resistivity are $\Omega \cdot$ m. An equally useful material property is the reciprocal of resistivity, **conductivity**, σ, where

$$\sigma = \frac{1}{\rho}, \tag{13.3}$$

with units of $\Omega^{-1} \cdot$ m^{-1}. Conductivity will be our most convenient parameter for establishing an electrical classification system for materials (Section 13.7).

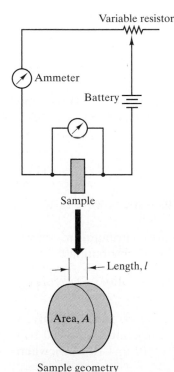

Variable resistor

Ammeter

Battery

Sample

Length, l

Area, A

Sample geometry

FIGURE 13.1 *Schematic of a circuit for measuring electrical conductivity. Sample dimensions relate to Equation 13.2.*

---

*Georg Simon Ohm (1787–1854), German physicist, who first published the statement of Equation 13.1. His definition of resistance led to the unit of resistance being named in his honor.

[†] Alessandro Giuseppe Antonio Anastasio Volta (1745–1827), Italian physicist, made major contributions to the development of understanding of electricity, including the first battery, or "voltage" source.

[‡] André Marie Ampère (1775–1836), French mathematician and physicist, was another major contributor to the field of *electrodynamics* (a term he introduced).

TABLE 13.1

Electrical Conductivities of Some Materials at Room Temperature		
Conducting range	**Material**	**Conductivity, $\sigma\,(\Omega^{-1} \cdot \mathrm{m}^{-1})$**
Conductors	Aluminum (annealed)	35.36×10^6
	Copper (annealed standard)	58.00×10^6
	Iron (99.99 + %)	10.30×10^6
	Steel (wire)	5.71–9.35×10^6
Semiconductors	Germanium (high purity)	2.0
	Silicon (high purity)	0.40×10^{-3}
	Lead sulfide (high purity)	38.4
Insulators	Aluminum oxide	10^{-10}–10^{-12}
	Borosilicate glass	10^{-13}
	Polyethylene	10^{-13}–10^{-15}
	Nylon 66	10^{-12}–10^{-13}

Source: Data from C. A. Harper, Ed., *Handbook of Materials and Processes for Electronics*, McGraw-Hill Book Company, New York, 1970; and J. K. Stanley, *Electrical and Magnetic Properties of Metals*, American Society for Metals, Metals Park, OH, 1963.

Conductivity is the product of the density of charge carriers, n, the charge carried by each, q, and the mobility of each carrier, μ:

$$\sigma = nq\mu. \tag{13.4}$$

The units for n are m^{-3}, for q are coulombs, and for μ are $\mathrm{m}^2/(\mathrm{V} \cdot \mathrm{s})$. The mobility is the average carrier velocity, or **drift velocity**, $\bar{v}$, divided by electrical field strength, E:

$$\mu = \frac{\bar{v}}{E}. \tag{13.5}$$

The drift velocity is in units of m/s, and the **electric field strength** ($E = V/l$) in units of V/m.

When both positive and negative charge carriers are contributing to conduction, Equation 13.4 must be expanded to account for both contributions:

$$\sigma = n_n q_n \mu_n + n_p q_p \mu_p. \tag{13.6}$$

The subscripts n and p refer to the negative and positive carriers, respectively. For electrons, electron holes, and monovalent ions, the magnitude of q is 0.16×10^{-18} C. For multivalent ions, the magnitude of q is $|Z_i| \times (0.16 \times 10^{-18}$ C$)$, where $|Z_i|$ is the magnitude of the valence (e.g., 2 for O^{2-}).

Table 13.1 lists values of conductivity for a wide variety of engineering materials. It is apparent that the magnitude of conductivity produces distinctive categories of materials consistent with the types outlined in Chapters 1 and 2. We shall discuss this electrical classification system in detail at the end of this chapter, but first we must look at the nature of electrical conduction in order to understand why conductivity varies by more than 20 orders of magnitude among common engineering materials.

EXAMPLE 13.1

A wire sample (1 mm in diameter by 1 m in length) of an aluminum alloy (containing 1.2% Mn) is placed in an electrical circuit such as that shown in Figure 13.1. A voltage drop of 432 mV is measured across the length of the wire as it carries a 10-A current. Calculate the conductivity of this alloy.

SOLUTION
From Equation 13.1,

$$R = \frac{V}{I}$$

$$= \frac{432 \times 10^{-3} \text{ V}}{10 \text{ A}} = 43.2 \times 10^{-3} \text{ } \Omega.$$

From Equation 13.2,

$$\rho = \frac{RA}{l}$$

$$= \frac{(43.2 \times 10^{-3} \text{ } \Omega)[\pi(0.5 \times 10^{-3} \text{ m})^2]}{1 \text{ m}}$$

$$= 33.9 \times 10^{-9} \text{ } \Omega \cdot \text{m}.$$

From Equation 13.3,

$$\sigma = \frac{1}{\rho}$$

$$= \frac{1}{33.9 \times 10^{-9} \text{ } \Omega \cdot \text{m}}$$

$$= 29.5 \times 10^{6} \text{ } \Omega^{-1} \cdot \text{m}^{-1}.$$

EXAMPLE 13.2

Assuming that the conductivity for copper in Table 13.1 is entirely due to free electrons [with a mobility of 3.5×10^{-3} $\text{m}^2/(\text{V} \cdot \text{s})$], calculate the density of free electrons in copper at room temperature.

SOLUTION
From Equation 13.4,

$$n = \frac{\sigma}{q\mu}$$

$$= \frac{58.00 \times 10^{6} \text{ } \Omega^{-1} \cdot \text{m}^{-1}}{0.16 \times 10^{-18} \text{ C} \times 3.5 \times 10^{-3} \text{ m}^2/(\text{V} \cdot \text{s})}$$

$$= 104 \times 10^{27} \text{ m}^{-3}.$$

EXAMPLE 13.3

Compare the density of free electrons in copper from Example 13.2 with the density of atoms.

SOLUTION
From Appendix 1,

$$\rho_{Cu} = 8.93 \text{ g} \cdot \text{cm}^{-3} \text{ with an atomic mass} = 63.55 \text{ amu}$$

and

$$\rho = 8.93 \frac{\text{g}}{\text{cm}^3} \times 10^6 \frac{\text{cm}^3}{\text{m}^3} \times \frac{1 \text{ g} \cdot \text{atom}}{63.55 \text{ g}} \times 0.6023 \times 10^{24} \frac{\text{atoms}}{\text{g} \cdot \text{atom}}$$

$$= 84.6 \times 10^{27} \text{ atoms/m}^3.$$

This solution compares with 104×10^{27} electrons/m^3 from Example 13.2; that is,

$$\frac{\text{free electrons}}{\text{atom}} = \frac{104 \times 10^{27} \text{ m}^{-3}}{84.6 \times 10^{27} \text{ m}^{-3}} = 1.23.$$

In other words, the conductivity of copper is high because each atom contributes roughly one free (conducting) electron. We shall see in Example 13.11 that, in semiconductors, the number of conducting electrons contributed per atom is considerably smaller.

EXAMPLE 13.4

Calculate the drift velocity of the free electrons in copper for an electric field strength of 0.5 V/m.

SOLUTION
From Equation 13.5,

$$\bar{v} = \mu E$$

$$= [3.5 \times 10^{-3} \text{ m}^2/(\text{V} \cdot \text{s})](0.5 \text{ V} \cdot \text{m}^{-1})$$

$$= 1.75 \times 10^{-3} \text{ m/s}.$$

PRACTICE PROBLEM 13.1

(a) The wire described in Example 13.1 shows a voltage drop of 432 mV. Calculate the voltage drop to be expected in a 0.5-mm-diameter ($\times$ 1-m-long) wire of the same alloy that also carries a current of 10 A. **(b)** Repeat part (a) for a 2-mm-diameter wire. **(Note that the solutions to all practice problems are provided on the related Web site.)**

PRACTICE PROBLEM 13.2

How many free electrons would there be in a spool of high-purity copper wire (1 mm diameter × 10 m long)? (See Example 13.2.)

PRACTICE PROBLEM 13.3

In Example 13.3, we compare the density of free electrons in copper with the density of atoms. How many copper atoms would be in the spool of wire described in Practice Problem 13.2?

PRACTICE PROBLEM 13.4

The drift velocity of the free electrons in copper is calculated in Example 13.4. How long would a typical free electron take to move along the entire length of the spool of wire described in Practice Problem 13.2 under the voltage gradient of 0.5 V/m?

13.2 | Energy Levels and Energy Bands

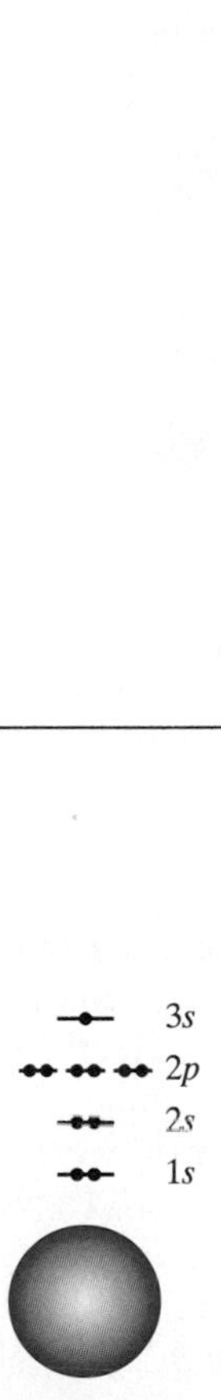

— $3s$

$2p$

$2s$

$1s$

Isolated Na atom

FIGURE 13.2 *Energy-level diagram for an isolated sodium atom.*

In Section 2.1 we saw how electron orbitals in a single atom are associated with discrete energy levels (Figure 2.3). Now, let us turn to a similar example. Figure 13.2 shows an energy-level diagram for a single sodium atom. As indicated in Appendix 1, the electronic configuration is $1s^2 2s^2 2p^6 3s^1$. The energy-level diagram indicates that there are actually three orbitals associated with the $2p$ energy level and that each of the $1s$, $2s$, and $2p$ orbitals is occupied by two electrons. This distribution of electrons among the various orbitals is a manifestation of the **Pauli* exclusion principle**, an important concept from quantum mechanics that indicates that no two electrons can occupy precisely the same state. Each horizontal line shown in Figure 13.2 represents a different orbital (i.e., a unique set of three quantum numbers). Each such orbital can be occupied by two electrons because they are in two different states; that is, they have opposite or antiparallel electron spins (representing different values for a fourth quantum number). In general, electron spins can be parallel or antiparallel. The outer orbital ($3s$) is half-filled by a single electron. Looking at Appendix 1, we note that the next element in the periodic table (Mg) fills the $3s$ orbital with two electrons (which would, by the Pauli exclusion principle, have opposite electron spins).

Consider next a hypothetical four-atom sodium molecule, Na$_4$ (Figure 13.3). The energy diagrams for the atom core electrons ($1s^2 2s^2 2p^6$) are essentially unchanged. However, the situation for the four outer orbital electrons is affected by the Pauli exclusion principle because the delocalized electrons are now being shared by all four atoms in the molecule. These electrons cannot all occupy a single orbital. The result is a "splitting" of the $3s$ energy level into four slightly different levels, which makes each level unique and satisfies the Pauli exclusion principle.

---

*Wolfgang Pauli (1900–1958), Austrian-American physicist, was a major contributor to the development of atomic physics. To a large extent, the understanding (which the exclusion principle provides) of outer-shell electron populations allows us to understand the order of the periodic table. These outer-shell electrons play a central role in the chemical behavior of the elements.

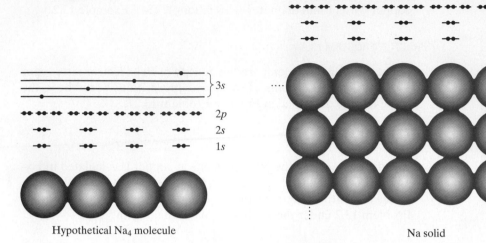

FIGURE 13.3 *Energy-level diagram for a hypothetical Na₄ molecule. The four shared, outer orbital electrons are "split" into four slightly different energy levels, as predicted by the Pauli exclusion principle.*

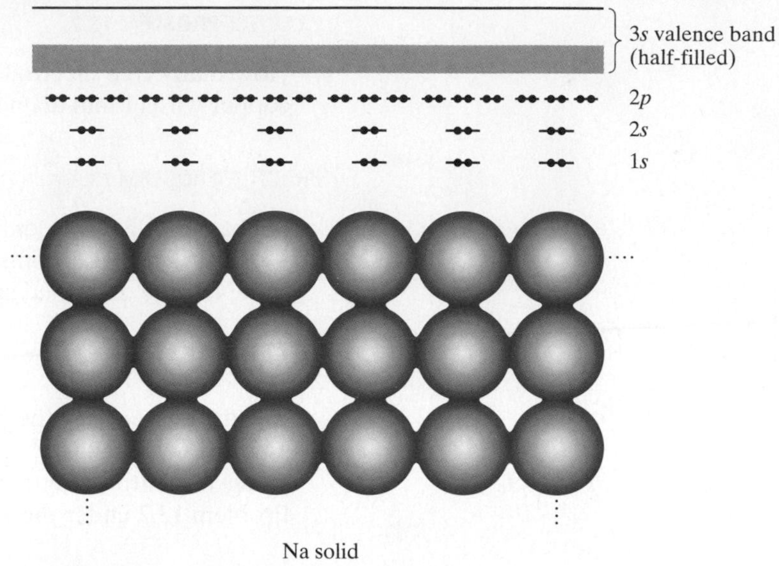

FIGURE 13.4 *Energy-level diagram for solid sodium. The discrete 3s energy level of Figure 13.2 has given way to a pseudocontinuous energy band (half-filled). Again, the splitting of the 3s energy level is predicted by the Pauli exclusion principle.*

It would be possible for the splitting to produce only two levels, each occupied by two electrons of opposite spin. In fact, electron pairing in a given orbital tends to be delayed until all levels of a given energy have a single electron, which is referred to as **Hund's*** **rule**. As another example, nitrogen (element 7) has three 2p electrons, each in a different orbital of equal energy. Pairing of two 2p electrons of opposite spin in a single orbital does not occur until element 8 (oxygen). The result of this splitting is a narrow *band* of energy levels corresponding to what was a single 3s level in the isolated atom. An important aspect of this electronic structure is that, as in the 3s level of the isolated atom, the 3s band of the Na₄ molecule is only half-filled. As a result, electron mobility between adjacent atoms is quite high.

A simple extension of the effect seen in the hypothetical four-atom molecule is shown in Figure 13.4, in which a large number of sodium atoms are joined by metallic bonding to produce a solid. In this metallic solid, the atom core electrons are again not directly involved in the bonding, and their energy diagrams remain essentially unchanged. However, the large number of atoms involved (e.g., on the order of Avogadro's number) produces an equally large number of energy-level splittings for the outer (3s) orbitals. The total range of energy values for the various 3s orbitals is not large. Rather, the spacing between adjacent 3s orbitals is extremely small. The result is a pseudocontinuous **energy band** corresponding to the 3s energy level of the isolated atom. As with the isolated Na atom and the hypothetical Na₄ molecule, the valence-electron energy band in the metallic solid is only half-filled, permitting high mobility of outer orbital electrons throughout

*F. Hund, *Z. Physik 42*, 93 (1927).

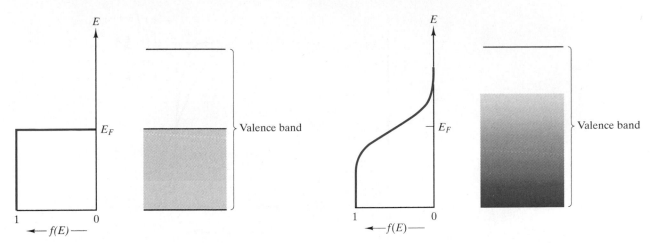

FIGURE 13.5 *The Fermi function, f(E), describes the relative filling of energy levels. At 0 K, all energy levels are completely filled up to the Fermi level, E_F, and are completely empty above E_F.*

FIGURE 13.6 *At T > 0 K, the Fermi function, f(E), indicates promotion of some electrons above E_F.*

the solid. Produced from valence electrons, the energy band of Figure 13.4 is also termed the **valence band**. An important conclusion is that metals are good electrical conductors because their valence band is only partially filled. This statement is valid, although the detailed nature of the partially filled valence band is different in some metals. For instance, in Mg (element 12), there are two 3*s* electrons that fill the energy band, which is only half-filled in Na (element 11). However, Mg has a higher empty band that overlaps the filled one. The net result is an outer valence band only partially filled.

A more detailed picture of the nature of electrical conduction in metals is obtained by considering how the nature of the energy band varies with temperature. Figure 13.4 implied that the energy levels in the valence band are completely full up to the midpoint of the band and completely empty above. In fact, this concept is true only at a temperature of absolute zero (0 K). Figure 13.5 illustrates this condition. The energy of the highest filled state in the energy band (at 0 K) is known as the **Fermi* level** (E_F). The extent to which a given energy level is filled is indicated by the **Fermi function**, $f(E)$. The Fermi function represents the probability that an energy level, E, is occupied by an electron and can have values between 0 and 1. At 0 K, $f(E)$ is equal to 1 up to E_F and is equal to 0 above E_F. This limiting case (0 K) is not conducive to electrical conduction. Since the energy levels below E_F are full, conduction requires electrons to increase their energy to some level just above E_F (i.e., to unoccupied levels). This energy promotion requires some external energy source. One means of providing this energy is from thermal energy obtained by heating the material to some temperature above 0 K. The resulting Fermi function, $f(E)$, is shown in Figure 13.6. For $T > 0$ K, some of the electrons just below E_F are promoted to unoccupied levels just above E_F.

---

*Enrico Fermi (1901–1954), Italian physicist, made numerous contributions to 20th-century science, including the first nuclear reactor in 1942. His work that improved understanding of the nature of electrons in solids had come nearly 20 years earlier.

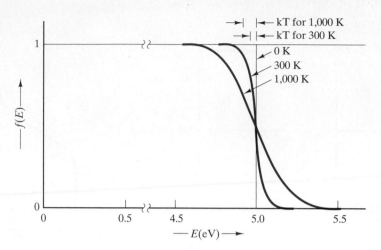

FIGURE 13.7 *Variation of the Fermi function, f(E), with the temperature for a typical metal (with $E_F = 5$ eV). Note that the energy range over which f(E) drops from 1 to 0 is equal to a few times kT.*

The relationship between the Fermi function, $f(E)$, and absolute temperature, T, is

$$f(E) = \frac{1}{e^{(E-E_F)/kT} + 1},$$ (13.7)

where k is the Boltzmann constant (13.8×10^{-24} J/K). In the limit of $T = 0$ K, Equation 13.7 correctly gives the step function of Figure 13.5. For $T > 0$ K, it indicates that $f(E)$ is essentially 1 far below E_F and essentially 0 far above. Near E_F, the value of $f(E)$ varies in a smooth fashion between these two extremes. At E_F, the value of $f(E)$ is precisely 0.5. As the temperature increases, the range over which $f(E)$ drops from 1 to 0 increases (Figure 13.7) and is on the order of the magnitude of kT. In summary, metals are good electrical conductors because thermal energy is sufficient to promote electrons above the Fermi level to otherwise unoccupied energy levels. At these levels ($E > E_F$), the accessibility of unoccupied levels in adjacent atoms yields high mobility of conduction electrons known as **free electrons** through the solid.

Our discussion of energy bands to this point has focused on metals and how they are good electrical conductors. Consider now the case of a nonmetallic solid, carbon in the diamond structure, which is a very poor electrical conductor. In Chapter 2, we saw that the valence electrons in this covalently bonded material are shared among adjacent atoms. The net result is that the valence band of carbon (diamond) is full. This valence band corresponds to the sp^3 hybrid energy level of an isolated carbon atom (Figure 2.3). To promote electrons to energy levels above the sp^3 level in an isolated carbon atom requires going above regions of forbidden energy. Similarly for the solid, promotion of an electron from the valence band to the **conduction band** requires going above an **energy band gap**, E_g (Figure 13.8). The concept of a Fermi level, E_F, still applies. However, E_F now falls in the center of the band gap. In Figure 13.8, the Fermi function, $f(E)$, corresponds to room temperature (298 K). One must bear in mind that the probabilities predicted by $f(E)$ can be realized only in the valence and conduction bands. Electrons are

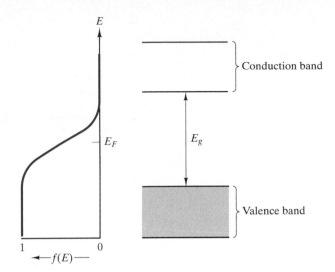

FIGURE 13.8 *Comparison of the Fermi function, f(E), with the energy band structure for an insulator. Virtually no electrons are promoted to the conduction band [f (E) = 0 there] because of the magnitude of the band gap (>2 eV).*

forbidden to have energy levels within the band gap. The important conclusion of Figure 13.8 is that $f(E)$ is essentially equal to 1 throughout the valence band and equal to 0 throughout the conduction band. The inability of thermal energy to promote a significant number of electrons to the conduction band gives diamond its characteristically poor electrical conductivity.

As a final example, consider silicon, element 14, residing just below carbon in the periodic table (Figure 2.2). In the same periodic table group, silicon behaves chemically in a way similar to carbon. In fact, silicon forms a covalently bonded solid with the same crystal structure as that of diamond, which is the namesake for the diamond cubic structure discussed in Section 3.5. The energy band structure of silicon (Figure 13.9) also looks very similar to that for diamond (Figure 13.8). The primary difference is that silicon has a smaller band gap ($E_g = 1.107\,\text{eV}$, compared to ~6 eV for diamond). The result is that, at room temperature (298 K), thermal energy promotes a small but significant number of electrons from the valence to the conduction band. Each electron promotion creates a pair of charge carriers referred to as an **electron-hole pair**. Consequently, electron holes are produced in the valence band equal in number to the conduction electrons. These electron holes are positive charge carriers, as mentioned in Section 13.1. With both positive and negative charge carriers present in moderate numbers, silicon demonstrates a moderate value of electrical conductivity intermediate between that for metals and insulators (Table 13.1). This simple semiconductor is discussed further in Section 13.5.

EXAMPLE 13.5

What is the probability of an electron being thermally promoted to the conduction band in diamond ($E_g = 5.6\,\text{eV}$) at room temperature (25°C)?

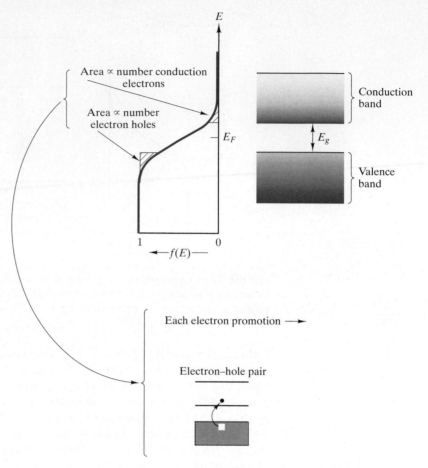

FIGURE 13.9 *Comparison of the Fermi function, $f(E)$, with the energy band structure for a semiconductor. A significant number of electrons is promoted to the conduction band because of a relatively small band gap (<2 eV). Each electron promotion creates a pair of charge carriers (i.e., an electron-hole pair).*

SOLUTION

From Figure 13.8, it is apparent that the bottom of the conduction band corresponds to

$$E - E_F = \frac{5.6}{2} \text{ eV} = 2.8 \text{ eV}.$$

From Equation 13.7 and using $T = 25°C = 298$ K,

$$f(E) = \frac{1}{e^{(E-E_F)/kT} + 1}$$

$$= \frac{1}{e^{(2.8 \text{ eV})/(86.2 \times 10^{-6} \text{ eV K}^{-1})(298 \text{ K})} + 1}$$

$$= 4.58 \times 10^{-48}.$$

EXAMPLE 13.6

What is the probability of an electron being thermally promoted to the conduction band in silicon ($E_g = 1.07$ eV) at room temperature (25°C)?

SOLUTION
As in Example 13.5,

$$E - E_F = \frac{1.107}{2} \text{ eV} = 0.5535 \text{ eV}$$

and

$$f(E) = \frac{1}{e^{(E-E_F)/kT} + 1}$$

$$= \frac{1}{e^{(0.5535 \text{ eV})/(86.2\times10^{-6} \text{ eV K}^{-1})(298 \text{ K})} + 1}$$

$$= 4.39 \times 10^{-10}.$$

While this number is small, it is 38 orders of magnitude greater than the value for diamond (Example 13.5) and is sufficient to create enough charge carriers (electron-hole pairs) to give silicon its semiconducting properties.

PRACTICE PROBLEM 13.5

What is the probability of an electron's being promoted to the conduction band in diamond at 50°C? (See Example 13.5.)

PRACTICE PROBLEM 13.6

What is the probability of an electron's being promoted to the conduction band in silicon at 50°C? (See Example 13.6.)

13.3 | Conductors

Conductors are materials with large values of conductivity. Table 13.1 indicates that the magnitude of conductivity for typical conductors is on the order of $10 \times 10^6 \ \Omega^{-1} \cdot \text{m}^{-1}$. The basis for this large value was discussed in the preceding section. Recalling Equation 13.6 as the general expression for conductivity, we can write the specific form for conductors as

$$\sigma = n_e q_e \mu_e, \tag{13.8}$$

where the subscript e refers to purely **electronic conduction**, which refers to σ specifically resulting from the movement of electrons. (**Electrical conduction** refers to a measurable value of σ that can arise from the movement of any type of charge carrier.) The dominant role of the band model of the preceding section is to indicate the importance of electron mobility, μ_e, in the conductivity of metallic conductors. This concept is well illustrated in the effect of two variables (temperature and composition) on conductivity in metals.

The effect of temperature on conductivity in metals is illustrated in Figure 13.10. In general, an increase in temperature above room temperature results in a drop in conductivity. This drop in conductivity is predominantly due to the drop in electron mobility, μ_e, with increasing temperature. The drop in electron mobility can, in turn, be attributed to the increasing thermal agitation of the crystalline structure of the metal as temperature increases. Because of the wavelike nature of electrons, these "wave packets" can move through crystalline structure most effectively when that structure is nearly perfect. Irregularity produced by thermal vibration diminishes electron mobility.

Equation 13.3 showed that resistivity and conductivity are inversely related. Therefore, the magnitude of resistivity for typical conductors is on the order of $0.1 \times 10^{-6}\ \Omega \cdot m$. Similarly, resistivity increases as temperature increases above room temperature. This relationship $[\rho(T)]$ is used more often than $\sigma(T)$ because the resistivity is found experimentally to increase quite linearly with temperature over this range; that is,

$$\rho = \rho_{\text{rt}}[1 + \alpha(T - T_{\text{rt}})], \tag{13.9}$$

where ρ_{rt} is the room temperature value of resistivity, α is the **temperature coefficient of resistivity**, T is the temperature, and T_{rt} is the room temperature. The data in Figure 13.10 are replotted in Figure 13.11 to illustrate Equation 13.9. Table 13.2 gives some representative values of ρ_{rt} and α for metallic conductors.

Inspection of Table 13.2 reveals that ρ_{rt} is a function of composition when forming solid solutions (e.g., $\rho_{\text{rt, pure Fe}} < \rho_{\text{rt, steel}}$). For small additions of impurity

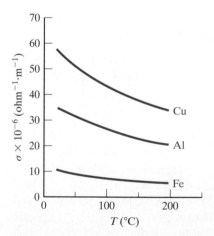

FIGURE 13.10 *Variation in electrical conductivity with temperature for some metals. (From J. K. Stanley,* Electrical and Magnetic Properties of Metals, *American Society for Metals, Metals Park, OH, 1963.)*

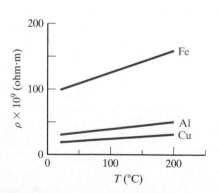

FIGURE 13.11 *Variation in electrical resistivity with temperature for the same metals shown in Figure 13.10. The linearity of these data defines the temperature coefficient of resistivity,* α.

TABLE 13.2

Resistivities and Temperature Coefficients of Resistivity for Some Metallic Conductors		
Material	**Resistivity at 20°C $\rho_{rt}(\Omega \cdot m)$**	**Temperature coefficient of resistivity at 20°C α (°C^{-1})**
Aluminum (annealed)	28.28×10^{-9}	0.0039
Copper (annealed standard)	17.24×10^{-9}	0.00393
Gold	24.4×10^{-9}	0.0034
Iron (99.99 +%)	97.1×10^{-9}	0.00651
Lead (99.73 +%)	206.48×10^{-9}	0.00336
Magnesium (99.80%)	44.6×10^{-9}	0.01784
Mercury	958×10^{-9}	0.00089
Nickel (99.95% + Co)	68.4×10^{-9}	0.0069
Nichrome (66% Ni + Cr and Fe)	$1,000 \times 10^{-9}$	0.0004
Platinum (99.99%)	106×10^{-9}	0.003923
Silver (99.78%)	15.9×10^{-9}	0.0041
Steel (wire)	$107-175 \times 10^{-9}$	0.006–0.0036
Tungsten	55.1×10^{-9}	0.0045
Zinc	59.16×10^{-9}	0.00419

Source: Data from J. K. Stanley, *Electrical and Magnetic Properties of Metals*, American Society for Metals, Metals Park, OH, 1963.

to a nearly pure metal, the increase in ρ is nearly linear with the amount of impurity addition (Figure 13.12). This relationship, which is reminiscent of Equation 13.9, can be expressed as

$$\rho = \rho_0(1 + \beta x), \qquad (13.10)$$

where ρ_0 is the resistivity of the pure metal, β is a constant for a given metal-impurity system (related to the slope of a plot such as that shown in Figure 13.12), and x is the amount of impurity addition. Of course, Equation 13.10 applies to a fixed temperature. Combined variations in temperature and composition would involve the effects of both α (from Equation 13.9) and β (from Equation 13.10). It is also necessary to recall that Equation 13.10 applies for small values of x only. For large values of x, ρ becomes a nonlinear function of x. A good example is shown in Figure 13.13 for the gold-copper alloy system. As for Figure 13.12, these data were obtained at a fixed temperature. For Figure 13.13 it is important to note that, as in Figure 13.12, pure metals (either gold or copper) have lower resistivities than alloys with impurity additions. For example, the resistivity of pure gold is less than that for gold with 10 at % copper. Similarly, the resistivity of pure copper is less than that for copper with 10 at % gold. The result of this trend is that the maximum resistivity for the gold–copper alloy system occurs at some intermediate composition (~45 at % gold, 55 at % copper). The reason that resistivity is increased by impurity additions is closely related to the reason that temperature increases resistivity. Impurity atoms diminish the degree of crystalline perfection of an otherwise pure metal.

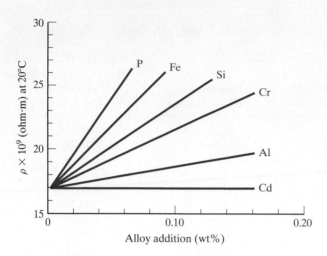

FIGURE 13.12 *Variation in electrical resistivity with composition for various copper alloys with small levels of elemental additions. Note that all data are at a fixed temperature (20°C). (From J. K. Stanley,* Electrical and Magnetic Properties of Metals, *American Society for Metals, Metals Park, OH, 1963.)*

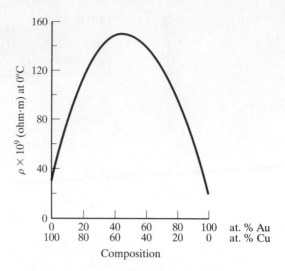

FIGURE 13.13 *Variation in electrical resistivity with large composition variations in the gold–copper alloy system. Resistivity increases with alloy additions for both pure metals. As a result, the maximum resistivity in the alloy system occurs at an intermediate composition (~45 at % gold, 55 at % copper). As with Figure 13.12, note that all data are at a fixed temperature (0°C). (From J. K. Stanley,* Electrical and Magnetic Properties of Metals, *American Society for Metals, Metals Park, OH, 1963.)*

A concept useful in visualizing the effect of crystalline imperfection on electrical conduction is the mean free path of an electron. As pointed out earlier in discussing the effect of temperature, the wavelike motion of an electron through an atomic structure is hindered by structural irregularity. The average distance that an electron wave can travel without deflection is termed its **mean free path**. Structural irregularities reduce the mean free path, which, in turn, reduces drift velocity, mobility, and finally, conductivity (see Equations 13.5 and 13.8). The nature of chemical imperfection was covered in detail in Section 4.1. For now, we only need to appreciate that any reduction in the periodicity of the metal's atomic structure hinders the movement of the periodic electron wave. For this reason, many of the structural imperfections discussed in Chapter 4 (e.g., point defects and dislocations) have been shown to cause increases in resistivity in metallic conductors.

THERMOCOUPLES

One important application of conductors is the measurement of temperature. A simple circuit, known as a **thermocouple**, which involves two metal wires for making such a measurement, is shown in Figure 13.14. The effectiveness of the thermocouple can ultimately be traced to the temperature sensitivity of the Fermi function (e.g., Figure 13.7). For a given metal wire (e.g., metal A in Figure 13.14)

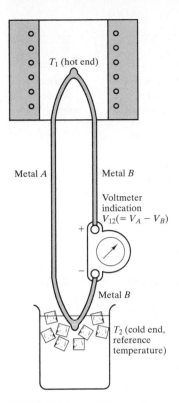

FIGURE 13.14 *Schematic illustration of a thermocouple. The measured voltage, V_{12}, is a function of the temperature difference, $T_1 - T_2$. The overall phenomenon is termed the Seebeck effect.*

connected between two different temperatures—T_1 (hot) and T_2 (cold)—more electrons are excited to higher energies at the hot end than at the cold end. As a result, there is a driving force for electron transport from the hot to the cold end. The cold end is then negatively charged and the hot end is positively charged with a voltage, V_A, between the ends of the wire. A useful feature of this phenomenon is that V_A depends only on the temperature difference $T_1 - T_2$, not on the temperature distribution along the wire. However, a useful voltage measurement requires a second wire (metal B in Figure 13.14) that contains a voltmeter. If metal B is of the same material as metal A, there will also be a voltage V_A induced in metal B, and the meter would read the net voltage ($= V_A - V_A = 0$ V). However, different metals will tend to develop different voltages between a given temperature difference ($T_1 - T_2$). In general, for a metal B different from metal A, the voltmeter in Figure 13.14 will indicate a net voltage, $V_{12} = V_A - V_B$. The magnitude of V_{12} will increase with increasing temperature difference, $T_1 - T_2$. The induced voltage, V_{12}, is called the **Seebeck* potential**, and the overall phenomenon illustrated by Figure 13.14 is called the **Seebeck effect**. The utility of the simple circuit for temperature measurement is apparent. By choosing a convenient reference temperature for T_2 (usually a fixed ambient temperature or an ice-water bath at $0°C$), the measured voltage, V_{12}, is a nearly linear function of T_1. The exact dependence of V_{12} on temperature is tabulated for several common thermocouple systems, such as those listed in Table 13.3. A plot of V_{12} with temperature for those common systems is given in Figure 13.15.

In this chapter, we shall find numerous examples of semiconductors competing with more traditional electronic materials. In the area of temperature measurement, semiconductors typically have a much more pronounced Seebeck effect than do metals. This concept is associated with the exponential (Arrhenius) nature of conductivity as a function of temperature in semiconductors. As a result, temperature-measuring semiconductors, or *thermistors*, are capable of measuring extremely small changes in temperature (as low as $10^{-6}°C$). However, due to a limited temperature operating range, thermistors have not displaced traditional thermocouples for general temperature-measuring applications.

SUPERCONDUCTORS

Figure 13.10 illustrated how the conductivity of metals rises gradually as temperature is decreased. This trend continues as temperature is decreased well below room temperature. But even at extremely low temperatures (e.g., a few degrees Kelvin), typical metals still exhibit a finite conductivity (i.e., a nonzero resistivity). A few materials are dramatic exceptions. Figure 13.16 illustrates such a case. At a critical temperature (T_c), the resistivity of mercury drops suddenly to zero, and mercury becomes a **superconductor**. Mercury was the first material known to exhibit this behavior. In 1911, H. Kamerlingh Onnes first reported the results illustrated in Figure 13.16 as a by-product of his research on the liquefaction

---

*Thomas Johann Seebeck (1770–1831), Russian-German physicist, in 1821 observed the famous effect that still bears his name. His treatment of closely related problems in thermoelectricity (interconversion of heat and electricity) was less successful, and others were to be associated with such problems (e.g., the Peltier and Thomson effects).

TABLE 13.3

Common Thermocouple Systems					
Type	Common name	Positive element[a]	Negative element[a]	Recommended service environment(s)	Maximum service temp. (°C)
B	Platinum-rhodium/ platinum-rhodium	70 Pt–30 Rh	94 Pt–6 Rh	Oxidizing Vacuum Inert	1,700
E	Chromel/constantan	90 Ni–9 Cr	44 Ni–55 Cu	Oxidizing	870
J	Iron/constantan	Fe	44 Ni–55 Cu	Oxidizing Reducing	760
K	Chromel/alumel	90 Ni–9 Cr	94 Ni–Al, Mn, Fe, Si, Co	Oxidizing	1,260
R	Platinum/platinum-rhodium	87 Pt–13 Rh	Pt	Oxidizing Inert	1,480
S	Platinum/platinum-rhodium	90 Pt–10 Rh	Pt	Oxidizing Inert	1,480
T	Copper/constantan	Cu	44 Ni–55 Cu	Oxidizing Reducing	370

[a] Alloy compositions expressed as weight percents.
Source: Data from *Metals Handbook*, 9th ed., Vol. 3, American Society for Metals, Metals Park, OH, 1980.

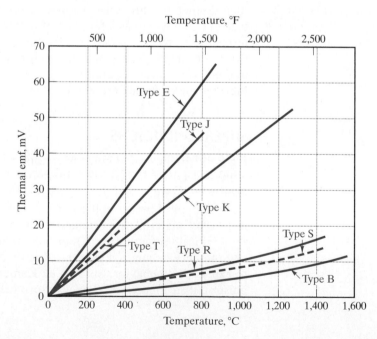

FIGURE 13.15 *Plot of thermocouple electromotive force (= V_{12} in Figure 13.14) as a function of temperature for some common thermocouple systems listed in Table 13.3. (From* Metals Handbook, *9th ed., Vol. 3, American Society for Metals, Metals Park, OH, 1980.)*

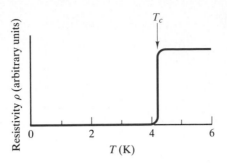

FIGURE 13.16 *Resistivity of mercury drops suddenly to zero at a critical temperature, T_c (= 4.12 K). Below T_c, mercury is a superconductor.*

and solidification of helium. Numerous other materials have since been found (e.g., niobium, vanadium, lead, and their alloys). Several empirical facts about superconductivity were known following the early studies. The effect was reversible and was generally exhibited by metals that were relatively poor conductors at room temperature. The drop in resistivity at T_c is sharp for pure metals, but may occur over a 1- to 2-K span for alloys. For a given superconductor, the transition temperature is reduced by increasing current density or magnetic field strength. Until the 1980s, attention was focused on metals and alloys (especially Nb systems), and T_c was below 25 K. In fact, the development of higher-T_c materials followed a nearly straight line on a time scale from 4.12 K in 1911 (for Hg) to 23.3 K in 1975 (for Nb_3Ge). As Figure 13.17 illustrates, a dramatic leap in T_c began in 1986 with the discovery that an $(La, Ba)_2CuO_4$ ceramic exhibited superconductivity at 35 K. In 1987, $YBa_2Cu_3O_7$ was found to have a T_c of 95 K, a major milestone in that the material is superconducting well above the temperature of liquid nitrogen (77 K), a relatively economical cryogenic level. By 1988, a Tl–Ba–Ca–Cu–O ceramic exhibited a T_c of 127 K. In spite of intense research activity involving a wide range of ceramic compounds, the 127 K T_c record was not broken for 5 years when, in 1993, the substitution of Hg for Tl produced a T_c of 133 K. Under extremely high pressures (e.g., 235,000 atm), the T_c of this material can be increased to as high as 150 K. The impracticality of this environmental pressure and the toxicity of Tl and Hg contribute to $YBa_2Cu_3O_7$ continuing to be the most fully studied of the high-T_c materials. Considerable work continues in the effort to gradually increase T_c, with some hope that another breakthrough could occur to accelerate the approach to an ultimate goal of a room-temperature superconductor.

The resistivity of a $YBa_2Cu_3O_7$ ceramic superconductor is shown in Figure 13.18. Extending the characteristic of metallic superconductors just described, we note that the drop in resistivity occurs over a wider temperature range (≈ 5 K) for this material with a relatively high T_c. Also, as poorly conductive metals exhibit superconductivity, the even more poorly conductive ceramic oxides are capable of exhibiting superconductivity to even higher temperatures.

The unit cell of $YBa_2Cu_3O_7$ is shown in Figure 13.19a. This material is frequently referred to as the **1–2–3 superconductor** due to the three metal ion subscripts. Although the structure of this superconductor appears relatively complex, it is closely related to the $CaTiO_3$ perovskite structure of Figure 13.19b. In simple perovskite, there is a ratio of two metal ions to three oxygen ions. The chemistry of the 1–2–3 superconductor has six metal ions to only seven oxygen ions, a deficiency

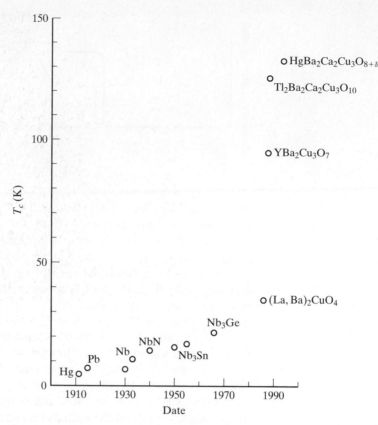

FIGURE 13.17 *The highest value of T_c increased steadily with time until the development of ceramic-oxide superconductors in 1986.*

of two oxygen ions accommodated by slight distortion of the perovskite arrangement. In fact, the unit cell in Figure 13.19a can be thought of as being equivalent to three distorted perovskite unit cells, with a Ba^{2+} ion centered in the top and bottom cells and a Y^{3+} centered in the middle cell. The boundaries between the perovskite-like subcells are distorted layers of copper and oxygen ions. A careful analysis of the charge balance between the cations and anions in the unit cell of Figure 13.19a indicates that, to preserve charge neutrality, one of the three copper ions must have the unusual valence of 3^+, while the other two have the common value of 2^+. The unit cell is orthorhombic. A chemically equivalent material with a tetragonal unit cell is not superconducting. Although the structure of Figure 13.19a is among the most complex considered in this text, it is still slightly idealized. The 1–2–3 superconductor, in fact, has a slight nonstoichiometry, $YBa_2Cu_3O_{7-x}$, with the value of $x \approx 0.1$.

Substantial progress has been made in the theoretical modeling of superconductivity. Ironically, lattice vibrations, which are the source of resistivity for normal conductors, are the basis of superconductivity in metals. At sufficiently low temperatures, an ordering effect occurs between lattice atoms and electrons. Specifically, the ordering effect is a synchronization between lattice atom vibrations and the wavelike motion of conductive electrons (associated in pairs of

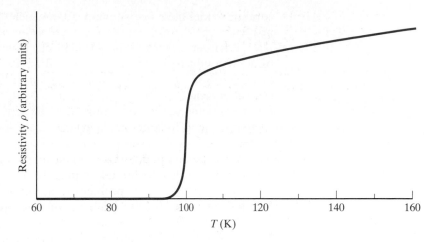

FIGURE 13.18 *The resistivity of $YBa_2Cu_3O_7$ as a function of temperature, indicating a $T_c \approx 95$ K.*

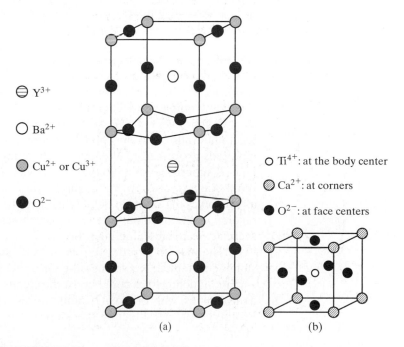

FIGURE 13.19 *(a) Unit cell of $YBa_2Cu_3O_7$. It is roughly equivalent to three distorted perovskite unit cells of the type shown for (b) $CaTiO_3$.*

opposite spin). This cooperative motion results in the complete loss of resistivity. The delicate nature of the lattice-electron ordering accounts for the traditionally low values of T_c in metals. Although superconductivity in high-T_c superconductors also involves paired electrons, the nature of the conduction mechanism is not as fully understood. Specifically, electron pairing does not seem to result from the same type of synchronization with lattice vibrations. What has become apparent is that the copper–oxygen planes in Figure 13.19a are the pathways for the supercurrent. In the 1–2–3 superconductor, that current is carried by electron holes. Other ceramic oxides have been developed in which the current is carried by electrons, not holes.

Whether the current is carried by electrons or by electron holes, the promise of superconductors produced by the dramatic increase in T_c has encountered an obstacle in the form of another important material parameter, namely the **critical current density**, which is defined as the current flow at which the material stops being superconducting. Metallic superconductors used in applications such as magnets in large-scale particle accelerators have critical current densities on the order of 10^{10} A/m^2. Such magnitudes have been produced in thin films of ceramic superconductors, but bulk samples give values about one-hundredth that size. Ironically, the limitation in current density becomes more severe for increasing values of T_c. The problem is caused by the penetration of a surrounding magnetic field into the material, creating an effective resistance due to the interaction between the current and mobile magnetic flux lines. This problem does not exist in metallic superconductors because the magnetic flux lines are not mobile at such low temperatures. In the important temperature range above 77 K, this effect becomes significant and is increasingly important with increasing temperature. The result appears to be little advantage to T_c values much greater than that found in the 1–2–3 material and, even there, the immobilization of magnetic flux lines may require a thin-film configuration or some special form of microstructural control.

If the limitation of critical density can be considered a materials science challenge, a materials engineering challenge exists in the need to fabricate these relatively complex and inherently brittle ceramic compounds into useable product shapes. The issues raised in Section 6.1 in discussing the nature of ceramics as structural materials come into play here also. As with the current density limitation, materials-processing challenges are directing efforts in the commercialization of high-T_c superconductors to thin-film device applications and in the production of small-diameter wire for cable and solenoid applications. Wire production generally involves the addition of metallic silver to 1–2–3 superconductor particles. The resulting composite has adequate mechanical performance without a significant sacrifice of superconducting properties.

The incentive to develop large-scale superconductors for power transmission is substantial. Replacing oil-cooled copper lines with liquid-nitrogen-cooled superconducting lines could boost the electrical transmission capacity by as much as five times, with obvious benefits to both the economy and the environment. Prototype superconducting tapes up to 100 m long are being produced for this effort.

In the meantime, one of the most promising applications of superconductors is the use of thin films as filters for cellular-telephone base stations. Compared with conventional copper-metal technology, the superconductor filters can enhance the range of the base stations, reduce channel interference, and decrease the number of dropped calls. Other applications of superconductors, both metallic and ceramic, are generally associated with their magnetic behavior and are discussed further in Chapter 13D on the related Web site.

Whether or not high-T_c superconductors will create a technological revolution on the scale provided by semiconductors, the breakthrough in developing the new family of high-T_c materials in the late 1980s still stands as one of the most exciting developments in materials science and engineering since the invention of the transistor.

EXAMPLE 13.7

Calculate the conductivity of gold at 200°C.

SOLUTION
From Equation 13.9 and Table 13.2,

$$\rho = \rho_{rt}[1 + \alpha(T - T_{rt})]$$
$$= (24.4 \times 10^{-9}\ \Omega \cdot m)[1 + 0.0034°C^{-1}(200 - 20)°C]$$
$$= 39.3 \times 10^{-9}\ \Omega \cdot m.$$

From Equation 13.3,

$$\sigma = \frac{1}{\rho}$$
$$= \frac{1}{39.3 \times 10^{-9}\ \Omega \cdot m}$$
$$= 25.4 \times 10^{6}\ \Omega^{-1} \cdot m^{-1}.$$

EXAMPLE 13.8

Estimate the resistivity of a copper–0.1 wt % silicon alloy at 100°C.

SOLUTION
Assuming that the effects of temperature and composition are independent and that the temperature coefficient of resistivity of pure copper is a good approximation to that for Cu–0.1 wt % Si, we can write

$$\rho_{100°C,\ Cu-0.1\ Si} = \rho_{20°C,\ Cu-0.1\ Si}[1 + \alpha(T - T_{rt})].$$

From Figure 13.12,

$$\rho_{20°C,\ Cu-0.1\ Si} \simeq 23.6 \times 10^{-9}\ \Omega \cdot m.$$

Then

$$\rho_{100°C,\ Cu-0.1\ Si} = (23.6 \times 10^{-9}\ \Omega \cdot m)[1 + 0.00393°C^{-1}(100 - 20)°C]$$
$$= 31.0 \times 10^{-9}\ \Omega \cdot m.$$

Note. The assumption that the temperature coefficient of resistivity for the alloy was the same as that for the pure metal is generally valid only for small alloy additions.

EXAMPLE 13.9

A chromel/constantan thermocouple is used to monitor the temperature of a heat-treatment furnace. The output relative to an ice-water bath is 60 mV.

(a) What is the temperature in the furnace?

(b) What would be the output relative to an ice-water bath for a chromel/alumel thermocouple?

SOLUTION

(a) Table 13.3 shows that the chromel/constantan thermocouple is "type E." Figure 13.15 shows that the type E thermocouple has an output of 60 mV at 800°C.

(b) Table 13.3 shows that the chromel/alumel thermocouple is "type K." Figure 13.15 shows that the type K thermocouple at 800°C has an output of 33 mV.

EXAMPLE 13.10

A $YBa_2Cu_3O_7$ superconductor is fabricated in a thin film strip with dimensions 1 μm thick $\times$ 1 mm wide $\times$ 10 mm long. At 77 K, super conductivity is lost when the current along the long dimension reaches a value of 17 A. What is the critical current density for this thin-film configuration?

SOLUTION

The current per cross-sectional area is

$$\text{critical current density} = \frac{17 \text{ A}}{(1 \times 10^{-6} \text{ m})(1 \times 10^{-3} \text{ m})}$$

$$= 1.7 \times 10^{10} \text{ A/m}^2.$$

PRACTICE PROBLEM 13.7

Calculate the conductivity at 200°C of **(a)** copper (annealed standard) and **(b)** tungsten. (See Example 13.7.)

PRACTICE PROBLEM 13.8

Estimate the resistivity of a copper–0.06 wt % phosphorus alloy at 200°C. (See Example 13.8.)

PRACTICE PROBLEM 13.9

In Example 13.9, we find the output from a type K thermocouple at 800°C. What would be the output from a Pt/90 Pt–10 Rh thermocouple?

PRACTICE PROBLEM 13.10

When the 1–2–3 superconductor in Example 13.10 is fabricated in a bulk specimen with dimensions 5 mm × 5 mm × 20 mm, the current in the long dimension at which superconductivity is lost is found to be 3.25×10^3 A. What is the critical current density for this configuration?

13.4 | Insulators

Insulators are materials with low conductivity. Table 13.1 gives magnitudes for conductivity in typical insulators from approximately 10^{-10} to 10^{-16} $\Omega^{-1} \cdot m^{-1}$. This drop in conductivity of roughly 20 orders of magnitude (compared with typical metals) is the result of energy band gaps greater than 2 eV (compared to zero for metals). It is important to note that these low-conductivity materials are an important part of the electronics industry. For example, roughly 80% of the industrial ceramics market worldwide is from this category, with the structural ceramics introduced in Chapter 11 representing only the remaining 20%. The dominant industrial use of electronic ceramics includes their applications based on the closely related magnetic behavior which is covered in Chapter 13D on the related Web site.

It is not a simple matter to rewrite Equation 13.6 to produce a specific conductivity equation for insulators comparable with Equation 13.8 for metals. Clearly, the density of electron carriers, n_e, is extremely small because of the large band gap. In many cases, the small degree of conductivity for insulators is not the result of thermal promotion of electrons across the band gap. Instead, it may be due to electrons associated with impurities in the material. It can also result from ionic transport (e.g., Na^+ in NaCl). Therefore, the specific form of Equation 13.6 depends on the specific charge carriers involved.

Figure 13.20 shows the nature of charge buildup in a typical insulator, or **dielectric**, application, a parallel-plate **capacitor**. On the atomic scale, the charge buildup corresponds to the alignment of electrical dipoles within the dielectric. This concept is explored in detail in conjunction with the discussion of ferroelectrics and piezoelectrics. A **charge density**, D (in units of C/m^2), is produced

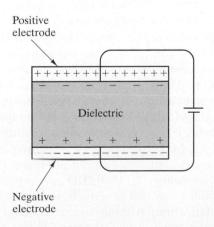

Positive electrode

Dielectric

Negative electrode

FIGURE 13.20 *A parallel-plate capacitor involves an insulator, or dielectric, between two metal electrodes. The charge density buildup at the capacitor surface is related to the dielectric constant of the material, as indicated by Equation 13.13.*

TABLE 13.4

Dielectric Constant and Dielectric Strength for Some Insulators		
Material	**Dielectric constant,[a] κ**	**Dielectric strength (kV/mm)**
Al_2O_3 (99.9%)	10.1	9.1[b]
Al_2O_3 (99.5%)	9.8	9.5[b]
BeO (99.5%)	6.7	10.2[b]
Cordierite	4.1–5.3	2.4–7.9[b]
Nylon 66-reinforced with 33% glass fibers (dry-as-molded)	3.7	20.5
Nylon 66-reinforced with 33% glass fibers (50% relative humidity)	7.8	17.3
Acetal (50% relative humidity)	3.7	19.7
Polyester	3.6	21.7

[a] At 10^3 Hz.

[b] Average root-mean-square (RMS) values at 60 Hz.

Source: Data from *Ceramic Source '86*, American Ceramic Society, Columbus, OH, 1985, and *Design Handbook for Du Pont Engineering Plastics.*

and is directly proportional to the electrical field strength, E (in units of V/m),

$$D = \varepsilon E, \tag{13.11}$$

where the proportionality constant, ε, is termed the **electric permittivity** of the dielectric and has units of C/(V · m). For the case of a vacuum between the plates in Figure 13.20, the charge density is

$$D = \varepsilon_0 E, \tag{13.12}$$

where ε_0 is the electric permittivity of a vacuum, which has a value of 8.854×10^{-12} C/(V · m). For the general dielectric, Equation 13.11 can be rewritten as

$$D = \varepsilon_0 \kappa E, \tag{13.13}$$

where κ is a dimensionless material constant called the relative permittivity, relative dielectric constant, or, more commonly, the **dielectric constant**. It represents the factor by which the capacitance of the system in Figure 13.20 is increased by inserting the dielectric in place of the vacuum. For a given dielectric, there is a limiting voltage gradient, termed the **dielectric strength**, at which an appreciable current flow (or breakdown) occurs and the dielectric fails. Table 13.4 gives representative values of dielectric constant and dielectric strength for various insulators.

An interesting family of insulators with unique and useful electrical properties is represented by the ceramic material barium titanate ($BaTiO_3$). The crystal structure is of the perovskite type, shown (for $CaTiO_3$) in Figure 13.19b. For $BaTiO_3$, the cubic structure shown in Figure 13.19b is found above 120°C. Upon cooling just below 120°C, $BaTiO_3$ undergoes a phase transformation to a tetragonal modification that exhibits **ferroelectric** behavior (i.e., it can undergo spontaneous polarization). Although ferroelectricity is an intriguing phenomenon, the most common applications of ferroelectrics stem from a closely related phenomenon, **piezoelectricity**. The prefix *piezo-* comes from the Greek word for pressure. Piezoelectric materials give an electrical response to mechanical pressure application. Conversely, electrical signals can make them pressure generators. This ability to convert electrical to mechanical energy and vice versa is a good example of a **transducer**, which, in general, is a device for converting one form of energy to another form.

The use of ultrasonics as a nondestructive testing technique was introduced in Section 8.4. Specifically, piezoelectric transducers are used as ultrasonic transmitters and/or receivers. In this common application, electrical signals (voltage oscillations) in the megahertz range produce or sense ultrasonic waves of that frequency. A full discussion of ferroelectric and piezoelectric materials is given in Chapter 13A on the related Web site.

13.5 | Semiconductors

Semiconductors are materials with conductivities intermediate between those of conductors and insulators. The magnitudes of conductivity in the semiconductors in Table 13.1 fall within the range 10^{-4} to 10^{+4} $\Omega^{-1} \cdot m^{-1}$. This intermediate range corresponds to band gaps of less than 2 eV. As shown in Figure 13.9, both conduction electrons and electron holes are charge carriers in a simple semiconductor. For the pure silicon example of Figure 13.9, the number of conduction electrons is equal to the number of electron holes. Pure, elemental semiconductors of this type are examples of **intrinsic semiconduction**.

For the case of intrinsic semiconductors, we can transform the general conductivity expression (Equation 13.6) into a specific form for intrinsic semiconductors,

$$\sigma = nq(\mu_e + \mu_h), \qquad \textbf{(13.14)}$$

where n is the density of conduction electrons (= density of electron holes), q is the magnitude of electron charge (= magnitude of hole charge = 0.16×10^{-18} C), μ_e is the mobility of a **conduction electron**, and μ_h is the mobility of an electron hole. Table 13.5 gives some representative values of μ_e and μ_h together with E_g,

TABLE 13.5

Properties of Some Common Semiconductors at Room Temperature (300 K)				
Material	Energy gap, E_g (eV)	Electron mobility, μ_e [m^2/(V·s)]	Hole mobility, μ_h [m^2/(V·s)]	Carrier density, n_e (= n_h) (m^{-3})
Si	1.107	0.140	0.038	14×10^{15}
Ge	0.66	0.364	0.190	23×10^{18}
CdS	2.59[a]	0.034	0.0018	—
GaAs	1.47	0.720	0.020	1.4×10^{12}
InSb	0.17	8.00	0.045	13.5×10^{21}

[a]This value is above our upper limit of 2 eV used to define a semiconductor. Such a limit is somewhat arbitrary. In addition, most commercial devices involve impurity levels that substantially change the nature of the band gap (see discussion of extrinsic semiconductors).

Source: Data from C. A. Harper, Ed., *Handbook of Materials and Processes for Electronics*, McGraw-Hill Book Company, New York, 1970.

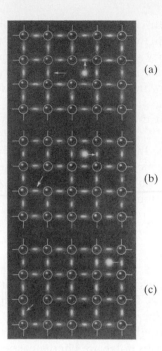

(a)

(b)

(c)

FIGURE 13.21 *Creation and motion of a conduction electron and an electron hole in a semiconductor. (a) An electron breaks away from the covalent bond, leaving a vacant bonding state, or a hole. The electron is now free to move in an electric field. In terms of the band model, the electron has gone from the valence band to the conduction band, leaving a hole in the valence band. The electron is shown moving upward, and the hole is shown moving to the left. (b) The conduction electron will now move to the right, and the hole will move down to the left. (c) The motions of (b) have been completed; the hole and electron continue to move outward. (From R. M. Rose, L. A. Shepard, and J. Wulff,* The Structures and Properties of Materials, *Vol. 4:* Electronic Properties, *John Wiley & Sons, Inc., New York, 1966.)*

the energy band gap and the carrier density at room temperature. Inspection of the mobility data indicates that μ_e is consistently higher than μ_h, sometimes dramatically so. The conduction of electron holes in the valence band is a relative concept. In fact, electron holes exist only in relation to the valence electrons; that is, an electron hole is a missing valence electron. The movement of an electron hole in a given direction is simply a representation that valence electrons have moved in the opposite direction (Figure 13.21). The cooperative motion of the valence electrons (represented by μ_h) is an inherently slower process than the motion of the conduction electron (represented by μ_e).

INTRINSIC, ELEMENTAL SEMICONDUCTORS

In Section 13.3, we found that the conductivity of metallic conductors dropped with increasing temperature. By contrast, the conductivity of semiconductors increases with increasing temperature (Figure 13.22). The reason for this opposite trend can be appreciated by inspection of Figure 13.9. The number of charge carriers depends on the overlap of the "tails" of the Fermi function curve with the valence and conduction bands. Figure 13.23 illustrates how increasing temperature extends the Fermi function, giving more overlap (i.e., more charge carriers). The specific temperature dependence of conductivity for semiconductors [i.e., $\sigma(T)$] follows from a mechanism of thermal activation, as implied by the illustrations in Figures 13.9 and 13.23. For this mechanism, the density of carriers increases exponentially with temperature; that is,

$$n \propto e^{-E_g/2kT}, \qquad (13.15)$$

where E_g is the band gap, k is the Boltzmann constant, and T is the absolute temperature. This equation is another occurrence of Arrhenius behavior (Section 5.1).

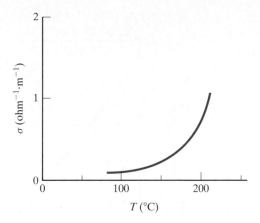

FIGURE 13.22 *Variation in electrical conductivity with temperature for semiconductor silicon. Contrast with the behavior shown for the metals in Figure 13.10. (This plot is based on the data in Table 13.5 using Equations 13.14 and 13.16.)*

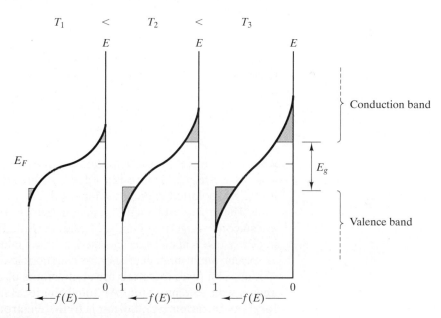

FIGURE 13.23 *Schematic illustration of how increasing temperature increases overlap of the Fermi function, f(E), with the conduction and valence bands giving increasing numbers of charge carriers. (Note also Figure 13.9.)*

Note that Equation 13.15 differs slightly from the general Arrhenius form given by Equation 5.1. There is a factor of 2 in the exponent of Equation 13.15 arising from the fact that each thermal promotion of an electron produces two charge carriers—an electron-hole pair.

Returning to Equation 13.14, we see that $\sigma(T)$ is determined by the temperature dependence of μ_e and μ_h as well as n. As with metallic conductors, μ_e and μ_h drop off slightly with increasing temperature. However, the exponential rise

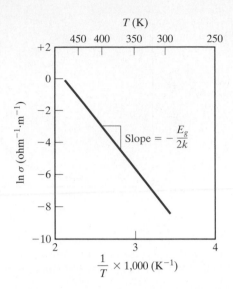

in n dominates the overall temperature dependence of σ, allowing us to write

$$\sigma = \sigma_0 e^{-E_g/2kT}, \tag{13.16}$$

where σ_0 is a preexponential constant of the type associated with Arrhenius equations (Section 5.1). By taking the logarithm of each side of Equation 13.16, we obtain

$$\ln \sigma = \ln \sigma_0 - \frac{E_g}{2k}\frac{1}{T}, \tag{13.17}$$

which indicates that a semilog plot of $\ln \sigma$ versus T^{-1} gives a straight line with a slope of $-E_g/2k$. Figure 13.24 demonstrates this linearity with the data of Figure 13.22 replotted in an Arrhenius plot.

The intrinsic, elemental semiconductors of group IV A of the periodic table are silicon (Si), germanium (Ge), and tin (Sn). This list is remarkably simple. No impurity levels need to be specified, as these materials are purposely prepared to an exceptionally high degree of purity. The role of impurities is treated in the next section. The three materials are all members of group IV A of the periodic table and have small values of E_g. Although this list is small, it has enormous importance for modern technology. Silicon is to the electronics industry what steel is to the automotive and construction industries. Pittsburgh is the "Steel City," and the Santa Clara Valley of California, once known for agriculture, is now known as "Silicon Valley" due to the major development of solid-state technology in the region.

Some elements in the neighborhood of group IV A (e.g., B from group III A and Te from group VI A) are also semiconductors. However, the predominant examples of commercial importance are Si and Ge from group IV A. Gray tin (Sn) transforms to white tin at 13°C. The transformation from the diamond cubic to a tetragonal structure near ambient temperature prevents gray tin from having any useful device application.

Again note that Table 13.5 gives values of band gap (E_g), electron mobility (μ_e), hole mobility (μ_h), and conduction electron density at room temperature (n)

for the two intrinsic, elemental semiconductors, Si and Ge. Because mobilities are not a strong function of composition, these values will also apply to the slightly impure, extrinsic semiconductors discussed next.

EXAMPLE 13.11

Calculate the fraction of Si atoms that provides a conduction electron at room temperature.

SOLUTION

As in Example 13.3, the atomic density can be calculated from data in Appendix 1:

$\rho_{Si} = 2.33 \text{ g} \cdot \text{cm}^{-3}$ with an atomic mass $= 28.09$ amu

$$\rho = 2.33 \frac{\text{g}}{\text{cm}^3} \times 10^6 \frac{\text{cm}^3}{\text{m}^3} \times \frac{1 \text{ g} \cdot \text{atom}}{28.09 \text{ g}} \times 0.6023 \times 10^{24} \frac{\text{atoms}}{\text{g} \cdot \text{atom}}$$

$$= 50.0 \times 10^{27} \text{ atoms/m}^3.$$

Table 13.5 indicates that

$$n_e = 14 \times 10^{15} \text{ m}^{-3}.$$

Then, the fraction of atoms providing conduction electrons is

$$\text{fraction} = \frac{14 \times 10^{15} \text{ m}^{-3}}{50 \times 10^{27} \text{ m}^{-3}} = 2.8 \times 10^{-13}.$$

Note. This result can be compared with the roughly 1:1 ratio of conducting electrons to atoms in copper (Example 13.3).

EXAMPLE 13.12

Calculate the conductivity of germanium at 200°C.

SOLUTION
From Equation 13.14,

$$\sigma = nq(\mu_e + \mu_h).$$

Using the data of Table 13.5, we have

$$\sigma_{300 \text{ K}} = (23 \times 10^{18} \text{ m}^{-3})(0.16 \times 10^{-18} \text{ C})(0.364 + 0.190) \text{ m}^2/(\text{V} \cdot \text{s})$$

$$= 2.04 \, \Omega^{-1} \cdot \text{m}^{-1}.$$

From Equation 13.16,

$$\sigma = \sigma_0 e^{-E_g/2kT}.$$

To obtain σ_0,

$$\sigma_0 = \sigma e^{+E_g/2kT}.$$

Again, using data from Table 13.5 we obtain

$$\sigma_0 = (2.04 \ \Omega^{-1} \cdot m^{-1})e^{+(0.66 \text{ eV})/2(86.2 \times 10^{-6} \text{ eV/K})(300 \text{ K})}$$

$$= 7.11 \times 10^5 \ \Omega^{-1} \cdot m^{-1}.$$

Then,

$$\sigma_{200^\circ C} = (7.11 \times 10^5 \ \Omega^{-1} \cdot m^{-1})e^{-(0.66 \text{ eV})/2(86.2 \times 10^{-6} \text{ eV/K})(473 \text{ K})}$$

$$= 217 \ \Omega^{-1} \cdot m^{-1}.$$

EXAMPLE 13.13

You are asked to characterize a new semiconductor. If its conductivity at 20°C is 250 $\Omega^{-1} \cdot m^{-1}$ and at 100°C is 1,100 $\Omega^{-1} \cdot m^{-1}$, what is its band gap, E_g?

SOLUTION
From Equation 13.17,

$$\ln \sigma_{T_1} = \ln \sigma_0 - \frac{E_g}{2k} \frac{1}{T_1} \quad \text{(a)}$$

and

$$\ln \sigma_{T_2} = \ln \sigma_0 - \frac{E_g}{2k} \frac{1}{T_2} \quad \text{(b)}.$$

Subtracting (b) from (a) yields

$$\ln \sigma_{T_1} - \ln \sigma_{T_2} = \ln \frac{\sigma_{T_1}}{\sigma_{T_2}}$$

$$= -\frac{E_g}{2k}\left(\frac{1}{T_1} - \frac{1}{T_2}\right).$$

Then,

$$-\frac{E_g}{2k} = \frac{\ln(\sigma_{T_1}/\sigma_{T_2})}{1/T_1 - 1/T_2},$$

or

$$E_g = \frac{(2k)\ln(\sigma_{T_2}/\sigma_{T_1})}{1/T_1 - 1/T_2}.$$

Taking $T_1 = 20^\circ C \ (= 293 \text{ K})$ and $T_2 = 100^\circ C \ (= 373 \text{ K})$ gives us

$$E_g = \frac{(2 \times 86.2 \times 10^{-6} \text{ eV/K})\ln(1,100/250)}{\frac{1}{373} \text{ K}^{-1} - \frac{1}{293}\text{K}^{-1}}$$

$$= 0.349 \text{ eV}.$$

PRACTICE PROBLEM 13.11

Using the data in Table 13.5, calculate **(a)** the total conductivity and **(b)** the resistivity of Si at room temperature. (See Example 13.11.)

PRACTICE PROBLEM 13.12

(a) Calculate the conductivity of germanium at 100°C, and **(b)** plot the conductivity over the range of 27 to 200°C as an Arrhenius-type plot similar to that shown in Figure 13.24. (See Example 13.12.)

PRACTICE PROBLEM 13.13

In characterizing a semiconductor in Example 13.13, we calculate its band gap. Using that result, calculate its conductivity at 50°C.

EXTRINSIC, ELEMENTAL SEMICONDUCTORS

Intrinsic semiconduction is a property of the pure material. **Extrinsic semiconduction** results from impurity additions known as **dopants**, and the process of adding these components is called *doping*. The term *impurity* has a different sense here compared with its use in previous chapters. For instance, many of the impurities in metal alloys and engineering ceramics were components "dragged along" from raw material sources. The impurities in semiconductors are added carefully after the intrinsic material has been prepared to a high degree of chemical purity. The conductivity of metallic conductors was shown to be sensitive to alloy composition in Section 13.3. Now we shall explore the composition effect for semiconductors. There are two distinct types of extrinsic semiconduction: (1) *n- type*, in which negative charge carriers dominate, and (2) *p-type*, in which positive charge carriers dominate. The philosophy behind producing each type is illustrated in Figure 13.25,

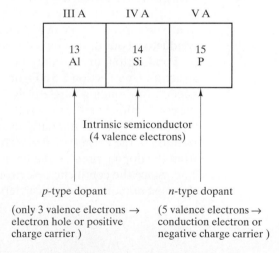

FIGURE 13.25 *Small section of the periodic table of elements. Silicon, in group IV A, is an intrinsic semiconductor. Adding a small amount of phosphorus from group V A provides extra electrons (not needed for bonding to Si atoms). As a result, phosphorus is an n-type dopant (i.e., an addition-producing negative charge carrier). Similarly, aluminum, from group III A, is a p-type dopant in that it has a deficiency of valence electrons leading to positive charge carriers (electron holes).*

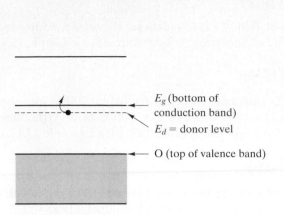

FIGURE 13.26 *Energy band structure of an n-type semiconductor. The extra electron from the group V A dopant produces a donor level (E_d) near the conduction band, which provides relatively easy production of conduction electrons. This figure can be contrasted with the energy band structure of an intrinsic semiconductor in Figure 13.9.*

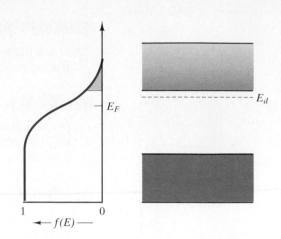

FIGURE 13.27 *Comparison of the Fermi function, f(E), with the energy band structure for an n-type semiconductor. The extra electrons shift the Fermi level (E_F) upward compared to Figure 13.9, in which it was in the middle of the band gap for an intrinsic semiconductor.*

which shows a small portion of the periodic table in the vicinity of silicon. The intrinsic semiconductor silicon has four valence (outer-shell) electrons. Phosphorus is an *n*-type dopant because it has five valence electrons. The one extra electron can easily become a conduction electron (i.e., a negative charge carrier). On the other hand, aluminum is a *p*-type dopant because of its three valence electrons. This deficiency of one electron (compared to silicon's four valence electrons) can easily produce an electron hole (a positive charge carrier).

The addition of a group V A atom, such as phosphorus, into solid solution in a crystal of group IV A silicon affects the energy band structure of the semiconductor. Four of the five valence electrons in the phosphorus atom are needed for bonding to four adjacent silicon atoms in the four-coordinated diamond-cubic structure (see Section 3.5). Figure 13.26 shows that the extra electron that is not needed for bonding is relatively unstable and produces a **donor level** (E_d) near the conduction band. As a result, the energy barrier to forming a conduction electron ($E_g - E_d$) is substantially less than in the intrinsic material (E_g). The relative position of the Fermi function is shown in Figure 13.27. Due to the extra electrons from the doping process, the Fermi level (E_F) is shifted upward.

Since the conduction electrons provided by the group V A atoms are by far the most numerous charge carriers, the conductivity equation takes the form

$$\sigma = nq\mu_e, \qquad \textbf{(13.18)}$$

where all terms were previously defined in Equation 13.14, with *n* being the number of electrons due to the dopant atoms. A schematic of the production of a conduction electron from a group V A dopant is shown in Figure 13.28. This schematic can be contrasted with the schematic of an intrinsic semiconductor shown in Figure 13.21.

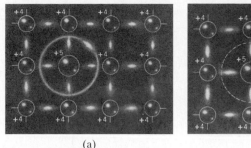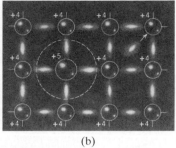

(a)   (b)

FIGURE 13.28 *Schematic of the production of a conduction electron in an n-type semiconductor. (a) The extra electron associated with the group V A atom can (b) easily break away, becoming a conduction electron and leaving behind an empty donor state associated with the impurity atom. This figure can be contrasted with the similar figure for intrinsic material in Figure 13.21. (From R. M. Rose, L. A. Shepard, and J. Wulff,* The Structure and Properties of Materials, *Vol. 4:* Electronic Properties, *John Wiley & Sons, Inc., New York, 1966.)*

Extrinsic semiconduction is another thermally activated process that follows the Arrhenius behavior. For **n-type semiconductors**, we can write

$$\sigma = \sigma_0 e^{-(E_g - E_d)/kT}, \tag{13.19}$$

where the various terms from Equation 13.16 again apply and where E_d is defined by Figure 13.25. One should note that there is no factor of 2 in the exponent of Equation 13.19 as there was for Equation 13.16. In extrinsic semiconduction, thermal activation produces a single charge carrier as opposed to the two carriers produced in intrinsic semiconduction. An Arrhenius plot of Equation 13.19 is shown in Figure 13.29.

The temperature range for n-type extrinsic semiconduction is limited. The conduction electrons provided by group V A atoms are much easier to produce thermally than are conduction electrons from the intrinsic process. However, the number of extrinsic conduction electrons can be no greater than the number of dopant atoms (i.e., one conduction electron per dopant atom). As a result, the Arrhenius plot of Figure 13.29 has an upper limit corresponding to the temperature at which all possible extrinsic electrons have been promoted to the conduction band. Figure 13.30 illustrates this concept. Figure 13.30 represents plots of both Equation 13.19 for extrinsic behavior and Equation 13.16 for intrinsic behavior. Note that the value of σ_0 for each region will be different. At low temperatures (large $1/T$ values), extrinsic behavior (Equation 13.19) dominates. The **exhaustion range** is a nearly horizontal plateau in which the number of charge carriers is fixed (= number of dopant atoms). As a practical matter, the conductivity does drop off slightly with increasing temperature (decreasing $1/T$) due to increasing thermal agitation. This change is comparable to the behavior of metals in which the number of conduction electrons is fixed, but mobility drops slightly with temperature (Section 13.3). As temperature continues to rise, the conductivity due to the intrinsic material (pure silicon) eventually is greater than that due to the extrinsic charge carriers (Figure 13.30). The exhaustion range is a useful concept for engineers who wish to minimize the need for temperature compensation in electrical circuits. In this range, conductivity is nearly constant with temperature.

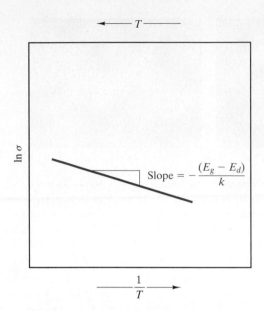

FIGURE 13.29 *Arrhenius plot of electrical conductivity for an n-type semiconductor. This plot can be contrasted with the similar plot for intrinsic material in Figure 13.24.*

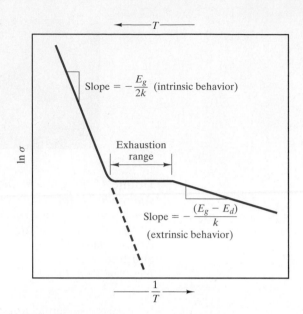

FIGURE 13.30 *Arrhenius plot of electrical conductivity for an n-type semiconductor over a wider temperature range than shown in Figure 13.29. At low temperatures (high 1/T), the material is extrinsic. At high temperatures (low 1/T), the material is intrinsic. In between is the exhaustion range, in which all "extra electrons" have been promoted to the conduction band.*

When a group III A atom such as aluminum goes into solid solution in silicon, its three valence electrons leave it one short of that needed for bonding to the four adjacent silicon atoms. Figure 13.31 shows that the result for the energy band structure of silicon is an **acceptor level** near the valence band. A silicon valence electron can be easily promoted to this acceptor level generating an electron hole (i.e., a positive charge carrier). As with n-type material, the energy barrier to forming a charge carrier (E_a) is substantially less than it is in the intrinsic material (E_g). The relative position of the Fermi function is shifted downward in p-type material (Figure 13.32). The appropriate conductivity equation is

$$\sigma = nq\mu_h, \tag{13.20}$$

where n is the density of electron holes. A schematic of the production of an electron hole from a group III A dopant is shown in Figure 13.33.

The Arrhenius equation for **p-type semiconductors** is

$$\sigma = \sigma_0 e^{-E_a/kT}, \tag{13.21}$$

where the terms from Equation 13.19 again apply and E_a is defined by Figure 13.31. As with Equation 13.19, there is no factor of 2 in the exponent due to the single (positive) charge carrier involved. Figure 13.34 shows the Arrhenius plot of ln σ versus $1/T$ for a p-type material. This plot is quite similar to that shown in

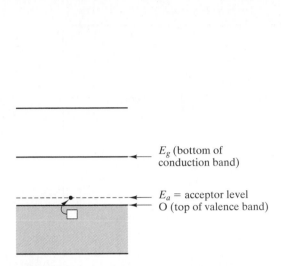

FIGURE 13.31 *Energy band structure of a p-type semi-conductor. The deficiency of valence electrons in the group III A dopant produces an acceptor level (E_a) near the valence band. Electron holes are produced as a result of thermal promotion over this relatively small energy barrier.*

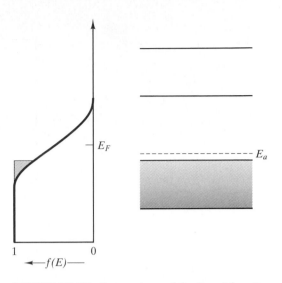

FIGURE 13.32 *Comparison of the Fermi function with the energy band structure for a p-type semiconductor. This electron deficiency shifts the Fermi level downward compared to that shown in Figure 13.9.*

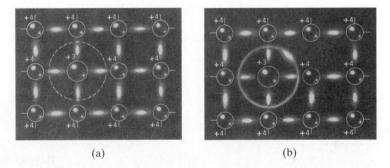

(a)   (b)

FIGURE 13.33 *Schematic of the production of an electron hole in a p-type semiconductor. (a) The deficiency in valence electrons for the group III A atom creates an empty state, or electron hole, orbiting about the acceptor atom. (b) The electron hole becomes a positive charge carrier as it leaves the acceptor atom behind with a filled acceptor state. (The motion of electron holes, of course, is due to the cooperative motion of electrons.) (From R. M. Rose, L. A. Shepard, and J. Wulff,* The Structure and Properties of Materials, *Vol. 4:* Electronic Properties, *John Wiley & Sons, Inc., New York, 1966.)*

Figure 13.30 for *n*-type semiconductors. The plateau in conductivity between the extrinsic and intrinsic regions is termed the **saturation range** for *p*-type behavior rather than exhaustion range. Saturation occurs when all acceptor levels (= number of group III A atoms) have become occupied with electrons.

Some of the common extrinsic, elemental semiconductor systems used in solid-state technology are listed in Table 13.6. Table 13.7 gives values of donor level relative to band gaps ($E_g - E_d$) and acceptor level (E_a) for various *n*-type and *p*-type donors, respectively.

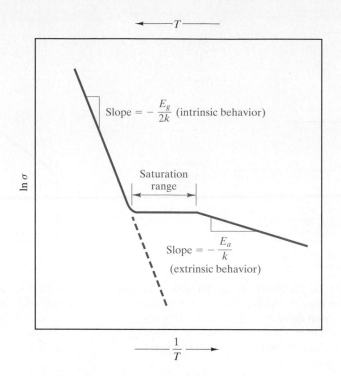

FIGURE 13.34 *Arrhenius plot of electrical conductivity for a p-type semiconductor over a wide temperature range. This plot is quite similar to the behavior shown in Figure 13.30. The region between intrinsic and extrinsic behavior is termed the saturation range, which corresponds to all acceptor levels being "saturated" or occupied with electrons.*

TABLE 13.6

Some Extrinsic, Elemental Semiconductors			
Element	**Dopant**	**Periodic table group of dopant**	**Maximum solid solubility of dopant (atoms/m^3)**
Si	B	III A	600×10^{24}
	Al	III A	20×10^{24}
	Ga	III A	40×10^{24}
	P	V A	$1{,}000 \times 10^{24}$
	As	V A	$2{,}000 \times 10^{24}$
	Sb	V A	70×10^{24}
Ge	Al	III A	400×10^{24}
	Ga	III A	500×10^{24}
	In	III A	4×10^{24}
	As	V A	80×10^{24}
	Sb	V A	10×10^{24}

Source: Data from W. R. Runyan and S. B. Watelski, in *Handbook of Materials and Processes for Electronics*, C. A. Harper, Ed., McGraw-Hill Book Company, New York, 1970.

TABLE 13.7

Impurity Energy Levels for Extrinsic Semiconductors			
Semiconductor	**Dopant**	$E_g - E_d$ **(eV)**	E_a **(eV)**
Si	P	0.044	—
	As	0.049	—
	Sb	0.039	—
	Bi	0.069	—
	B	—	0.045
	Al	—	0.057
	Ga	—	0.065
	In	—	0.160
	Tl	—	0.260
Ge	P	0.012	—
	As	0.013	—
	Sb	0.096	—
	B	—	0.010
	Al	—	0.010
	Ga	—	0.010
	In	—	0.011
	Tl	—	0.010
GaAs	Se	0.005	—
	Te	0.003	—
	Zn	—	0.024
	Cd	—	0.021

Source: Data from W. R. Runyan and S. B. Watelski, in *Handbook of Materials and Processes for Electronics*, C. A. Harper, Ed., McGraw-Hill Book Company, New York, 1970.

A note of comparison between semiconductors and metals is in order. The composition and temperature effects for semiconductors are opposite to those for metals. For metals, small impurity additions decreased conductivity [see Figure 13.12, which showed $\rho(= 1/\sigma)$ increasing with addition levels]. Similarly, increases in temperature decreased conductivity (see Figure 13.10). Both effects were due to reductions in electron mobility resulting from reductions in crystalline order. We have seen that for semiconductors, appropriate impurities and increasing temperature increase conductivity. Both effects are described by the energy band model and Arrhenius behavior.

In Section 4.5, the economic advantage of **amorphous** (noncrystalline) **semiconductors** was pointed out. The technology of these amorphous materials is somewhat behind that of their crystalline counterparts, and the scientific understanding of semiconduction in noncrystalline solids is less developed. However, commercial development of amorphous semiconductors has evolved into a wide market. These materials account for more than one-quarter of the photovoltaic (solar cell) market, largely for portable consumer products such as solar watches and laptop displays. One example is "amorphous silicon" that is often prepared by the decomposition of silane (SiH_4). This process is frequently incomplete and "amorphous silicon" is then, in reality, a silicon–hydrogen alloy. Other examples include a number of chalcogenides (S, Se, and Te) and amorphous selenium that has played a central role in the xerography process (as a photoconductive coating that permits the formation of a charged image).

EXAMPLE 13.14

An extrinsic silicon contains 100 ppb Al by weight. What is the atomic % Al?

SOLUTION

For 100 g of doped silicon, there will be

$$\frac{100}{10^9} \times 100 \text{ g Al} = 1 \times 10^{-5} \text{ g Al}.$$

Using the data of Appendix 1, we can calculate

$$\text{no. g} \cdot \text{atoms Al} = \frac{1 \times 10^{-5} \text{ g Al}}{26.98 \text{ g/g} \cdot \text{atom}} = 3.71 \times 10^{-7} \text{ g} \cdot \text{atom}$$

and

$$\text{no. g} \cdot \text{atoms Si} = \frac{(100 - 1 \times 10^{-5}) \text{ g Si}}{28.09 \text{ g/g} \cdot \text{atom}} = 3.56 \text{ g} \cdot \text{atoms},$$

which gives

$$\text{atomic \% Al} = \frac{3.71 \times 10^{-7} \text{ g} \cdot \text{atom}}{(3.56 + 3.7 \times 10^{-7}) \text{ g} \cdot \text{atom}} \times 100$$

$$= 10.4 \times 10^{-6} \text{ atomic \%}.$$

EXAMPLE 13.15

In a phosphorus-doped (*n*-type) silicon, the Fermi level (E_F) is shifted upward 0.1 eV. What is the probability of an electron's being thermally promoted to the conduction band in silicon ($E_g = 1.107$ eV) at room temperature (25°C)?

SOLUTION

From Figure 13.27 and Equation 13.7, it is apparent that

$$E - E_F = \frac{1.107}{2} \text{ eV} - 0.1 \text{ eV} = 0.4535 \text{ eV}$$

and then

$$f(E) = \frac{1}{e^{(E-E_F)/kT} + 1}$$

$$= \frac{1}{e^{(0.4535 \text{ eV})/(86.2 \times 10^{-6} \text{ eV} \cdot \text{K}^{-1})(298 \text{ K})} + 1}$$

$$= 2.20 \times 10^{-8}.$$

This number is small, but it is roughly two orders of magnitude higher than the value for intrinsic silicon as calculated in Example 13.6.

EXAMPLE 13.16

For a hypothetical semiconductor with n-type doping and $E_g = 1$ eV while $E_d = 0.9$ eV, the conductivity at room temperature ($25°C$) is $100 \ \Omega^{-1} \cdot m^{-1}$. Calculate the conductivity at $30°C$.

SOLUTION

Assuming that extrinsic behavior extends to $30°C$, we can apply Equation 13.19:

$$\sigma = \sigma_0 e^{-(E_g - E_d)/kT}.$$

At $25°C$,

$$\sigma_0 = \sigma e^{+(E_g - E_d)/kT}$$

$$= (100 \ \Omega^{-1} \cdot m^{-1}) e^{+(1.0 - 0.9) \ \text{eV}/(86.2 \times 10^{-6} \ \text{eV} \cdot K^{-1})(298 \ K)}$$

$$= 4.91 \times 10^3 \ \Omega^{-1} \cdot m^{-1}.$$

At $30°C$, then,

$$\sigma = (4.91 \times 10^3 \ \Omega^{-1} \cdot m^{-1}) e^{-(0.1 \ \text{eV})/(86.2 \times 10^{-6} \ \text{eV} \cdot K^{-1})(303 \ K)}$$

$$= 107 \ \Omega^{-1} \cdot m^{-1}.$$

EXAMPLE 13.17

For a phosphorus-doped germanium semiconductor, the upper temperature limit of the extrinsic behavior is $100°C$. The extrinsic conductivity at this point is $60 \ \Omega^{-1} \cdot m^{-1}$. Calculate the level of phosphorus doping in ppb by weight.

SOLUTION

In this case, all dopant atoms have provided a donor electron. As a result, the density of donor electrons equals the density of impurity phosphorus. The density of donor electrons is given by rearrangement of Equation 13.18:

$$n = \frac{\sigma}{q\mu_e}.$$

Using data from Table 13.9, we obtain

$$n = \frac{60 \ \Omega^{-1} \cdot m^{-1}}{(0.16 \times 10^{-18} \ C)[0.364 \ m^2/(V \cdot s)]} = 1.03 \times 10^{21} \ m^{-3}.$$

Important Note. As pointed out in the text, the carrier mobilities of Tables 13.5 and 13.9 apply to extrinsic as well as intrinsic materials. For example, conduction electron mobility in germanium does not change significantly with impurity addition as long as the addition levels are not great.

Using data from Appendix 1 gives us

$$[P] = 1.03 \times 10^{21} \frac{\text{atoms P}}{\text{m}^3} \times \frac{30.97 \text{ g P}}{0.6023 \times 10^{24} \text{ atoms P}}$$

$$\times \frac{1 \text{ cm}^3 \text{ Ge}}{5.32 \text{ g Ge}} \times \frac{1 \text{ m}^3}{10^6 \text{ cm}^3}$$

$$= 9.96 \times 10^{-9} \frac{\text{g P}}{\text{g Ge}} = \frac{9.96 \text{ g P}}{10^9 \text{ g Ge}} = \frac{9.96 \text{ g P}}{\text{billion g Ge}}$$

$$= 9.96 \text{ ppb P}.$$

EXAMPLE 13.18

For the semiconductor in Example 13.17:

(a) Calculate the upper temperature for the exhaustion range.
(b) Determine the extrinsic conductivity at 300 K.

SOLUTION

(a) The upper temperature for the exhaustion range (see Figure 13.30) corresponds to the point where intrinsic conductivity equals the maximum extrinsic conductivity. Using Equations 13.14 and 13.16 together with data from Table 13.9, we obtain

$$\sigma_{300 \text{ K}} = (23 \times 10^{18} \text{ m}^{-3})(0.16 \times 10^{-18} \text{ C})(0.364 + 0.190) \text{ m}^2/(\text{V} \cdot \text{s})$$

$$= 2.04 \ \Omega^{-1} \cdot \text{m}^{-1}$$

and

$$\sigma = \sigma_0 e^{-E_g/2kT} \text{ or } \sigma_0 = \sigma e^{+E_g/2kT},$$

giving

$$\sigma_0 = (2.04 \ \Omega^{-1} \cdot \text{m}^{-1}) e^{+(0.66 \text{ eV})/2(86.2 \times 10^{-6} \text{ eV/K})(300 \text{ K})}$$

$$= 7.11 \times 10^5 \ \Omega^{-1} \cdot \text{m}^{-1}.$$

Then, using the value from Example 13.17,

$$60 \ \Omega^{-1} \cdot \text{m}^{-1} = (7.11 \times 10^5 \ \Omega^{-1} \cdot m^{-1}) e^{-(0.66 \text{ eV})/2(86.2 \times 10^{-6} \text{ eV/K})T},$$

giving

$$T = 408 \text{ K} = 135°\text{C}.$$

(b) Calculating extrinsic conductivity for this n-type semiconductor requires Equation 13.19 to be used:

$$\sigma = \sigma_0 e^{-(E_g - E_d)/kT}.$$

In Example 13.17, it is given that $\sigma = 60\ \Omega^{-1} \cdot m^{-1}$ at 100°C. In Table 13.7, we see that $E_g - E_d$ is 0.012 eV for P-doped Ge. As a result,

$$\sigma_0 = \sigma e^{+(E_g - E_d)/kT}$$

$$= (60\ \Omega^{-1} \cdot m^{-1})e^{+(0.012\ eV)/(86.2 \times 10^{-6}\ eV/K)(373\ K)}$$

$$= 87.1\ \Omega^{-1} \cdot m^{-1}.$$

At 300 K,

$$\sigma = (87.1\ \Omega^{-1} \cdot m^{-1})e^{-(0.012\ eV)/(86.2 \times 10^{-6}\ eV/K)(300\ K)}$$

$$= 54.8\ \Omega^{-1} \cdot m^{-1}.$$

EXAMPLE 13.19

Plot the conductivity of the phosphorus-doped germanium of Examples 13.17 and 13.18 in a manner similar to the plot in Figure 13.30.

SOLUTION
The key data are extrinsic and intrinsic.

Extrinsic data:

$$\sigma_{100°C} = 60\ \Omega^{-1} \cdot m^{-1} \quad or \quad \ln \sigma = 4.09\ \Omega^{-1} \cdot m^{-1}$$

$$at\ T = 100°C = 373\ K \quad or \quad 1/T = 2.68 \times 10^{-3}\ K^{-1}$$

and

$$\sigma_{300\ K} = 54.8\ \Omega^{-1} \cdot m^{-1} \quad or \quad \ln \sigma = 4.00\ \Omega^{-1} \cdot m^{-1}$$

$$at\ T = 300\ K \quad or \quad 1/T = 3.33 \times 10^{-3}\ K^{-1}.$$

Intrinsic data:

$$\sigma_{408\ K} = 60\ \Omega^{-1} \cdot m^{-1} \quad or \quad \ln \sigma = 4.09\ \Omega^{-1} \cdot m^{-1}$$

$$at\ T = 408\ K \quad or \quad 1/T = 2.45 \times 10^{-3}\ K^{-1}$$

and

$$\sigma_{300\ K} = 2.04\ \Omega^{-1} \cdot m^{-1} \quad or \quad \ln \sigma = 0.713\ \Omega^{-1} \cdot m^{-1}$$

$$at\ T = 300\ K \quad or \quad 1/T = 3.33 \times 10^{-3}\ K^{-1}.$$

TABLE 13.9

Electrical Properties for Some Intrinsic, Compound Semiconductors at Room Temperature (300 K)					
Group	Semiconductor	E_g (eV)	μ_e [m²/(V·s)]	μ_h [m²/(V·s)]	n_e (= n_h) (m⁻³)
III–V	AlSb	1.60	0.090	0.040	—
	GaP	2.25	0.030	0.015	—
	GaAs	1.47	0.720	0.020	1.4×10^{12}
	GaSb	0.68	0.500	0.100	—
	InP	1.27	0.460	0.010	—
	InAs	0.36	3.300	0.045	—
	InSb	0.17	8.000	0.045	13.5×10^{21}
II–VI	ZnSe	2.67	0.053	0.002	—
	ZnTe	2.26	0.053	0.090	—
	CdS	2.59	0.034	0.002	—
	CdTe	1.50	0.070	0.007	—
	HgTe	0.025	2.200	0.016	—

Source: Data from W. R. Runyan and S. B. Watelski, in *Handbook of Materials and Processes for Electronics*, C. A. Harper, ed., McGraw-Hill Book Company, New York, 1970.

SOLUTION

Considering 100 g of doped GaAs and following the method of Example 13.14, we find

$$\frac{100}{10^9} \times 100 \text{ g Se} = 1 \times 10^{-5} \text{ g Se.}$$

Using data from Appendix 1, we obtain

$$\text{no. g} \cdot \text{atom Se} = \frac{1 \times 10^{-5} \text{ g Se}}{78.96 \text{ g/g} \cdot \text{atom}} = 1.27 \times 10^{-7} \text{ g} \cdot \text{atom}$$

and

$$\text{no. moles GaAs} = \frac{(100 - 1 \times 10^{-5}) \text{ g GaAs}}{(69.72 + 74.92) \text{ g/mol}} = 0.691 \text{ mol.}$$

Finally,

$$\text{mol\%Se} = \frac{1.27 \times 10^{-7} \text{ g} \cdot \text{atom}}{(0.691 + 1.27 \times 10^{-7}) \text{ mol}} \times 100$$

$$= 18.4 \times 10^{-6} \text{ mol \%.}$$

EXAMPLE 13.21

Calculate the intrinsic conductivity of GaAs at 50°C.

SOLUTION

From Equation 13.14,

$$\sigma = nq(\mu_e + \mu_h).$$

Using the data of Table 13.9, we obtain

$$\sigma_{300\ K} = (1.4 \times 10^{12}\ m^{-3})(0.16 \times 10^{-18}\ C)(0.720 + 0.020)\ m^2/(V \cdot s)$$
$$= 1.66 \times 10^{-7}\ \Omega^{-1} \cdot m^{-1}.$$

From Equation 13.16,

$$\sigma = \sigma_0 e^{-E_g/2kT},$$

or

$$\sigma_0 = \sigma e^{+E_g/2kT}.$$

Again using data from Table 13.9, we obtain

$$\sigma_0 = (1.66 \times 10^{-7}\ \Omega^{-1} \cdot m^{-1})e^{+(1.47\ eV)/2(86.2 \times 10^{-6}\ eV/K)(300\ K)}$$
$$= 3.66 \times 10^{5}\ \Omega^{-1} \cdot m^{-1}.$$

Then,

$$\sigma_{50°C} = (3.66 \times 10^{5}\ \Omega^{-1} \cdot m^{-1})e^{-(1.47\ eV)/2(86.2 \times 10^{-6}\ eV/K)(323\ K)}$$
$$= 1.26 \times 10^{-6}\ \Omega^{-1} \cdot m^{-1}.$$

EXAMPLE 13.22

In intrinsic semiconductor CdTe, what fraction of the current is carried by electrons, and what fraction is carried by holes?

SOLUTION
By using Equation 13.14,

$$\sigma = nq(\mu_e + \mu_h),$$

it is apparent that

$$\text{fraction due to electrons} = \frac{\mu_e}{\mu_e + \mu_h}$$

and

$$\text{fraction due to holes} = \frac{\mu_h}{\mu_e + \mu_h}.$$

Using the data of Table 13.9 yields

$$\text{fraction due to electrons} = \frac{0.070}{0.070 + 0.007} = 0.909$$

and

$$\text{fraction due to holes} = \frac{0.007}{0.070 + 0.007} = 0.091.$$

PRACTICE PROBLEM 13.18

Example 13.20 describes a GaAs semiconductor with 100-ppb Se doping. What is the atomic density of Se atoms in this extrinsic semiconductor? (The density of GaAs is 5.32 Mg/m^3.)

PRACTICE PROBLEM 13.19

Calculate the intrinsic conductivity of InSb at 50°C. (See Example 13.21.)

PRACTICE PROBLEM 13.20

For intrinsic InSb, calculate the fraction of the current carried by electrons and the fraction carried by electron holes. (See Example 13.22.)

PROCESSING OF SEMICONDUCTORS

The most striking feature of semiconductor processing is the ability to produce materials of unparalleled structural and chemical perfection. Feature boxes in Chapters 3 and 9 illustrate how this perfection is achieved. There are a number of crystal-growing techniques used for producing high-quality single crystals of semiconducting materials by growth from the melt. The chemical perfection is due to the process of **zone refining**, as illustrated in the feature box in Chapter 9. The phase diagram shown there illustrates that the impurity content in the liquid is substantially greater than that in the solid. These data allow us to define a **segregation coefficient, K**, as

$$K = \frac{C_s}{C_l},\qquad\qquad (13.22)$$

where C_s and C_l are the impurity concentrations in solid and liquid, respectively. Of course, K is much less than 1 for the case shown in the feature box of Chapter 9, and near the edge of the phase diagram, the solidus and liquidus lines are fairly straight, giving a constant value of K over a range of temperatures.

Structural defects, such as dislocations, adversely affect the performance of the silicon-based devices discussed in the next section. Such defects are a common byproduct of oxygen solubility in the silicon. A process known as **gettering** is used to "getter" (or *capture*) the oxygen, removing it from the region of the silicon where the device circuitry is developed. Ironically, the dislocation-producing oxygen is often removed by introducing dislocations on the "back" side of wafers, where the "front" side is defined as that where the circuitry is produced. Mechanical damage (e.g., by abrasion or laser impact) produces dislocations that serve as gettering sites at which SiO_2 precipitates are formed. This approach is known as **extrinsic gettering**. A more subtle approach is to heat treat the silicon wafer so that SiO_2 precipitates form within the wafer, but sufficiently far below the front face to prevent interference with circuit development. This latter approach is known as **intrinsic gettering** and may involve as many as three separate annealing steps between 600 and 1,250°C extending over a period of several hours. Both extrinsic and intrinsic gettering are commonly used in semiconductor processing.

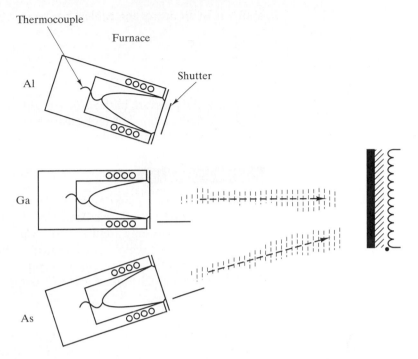

Thermocouple

Furnace

Al

Shutter

Ga

As

FIGURE 13.35 *Schematic illustration of the molecular-beam epitaxy technique. Resistance-heated source furnaces (also called effusion or Knudsen cells) provide the atomic or molecular beams (approximately 10-mm radius). Shutters control the deposition of each beam onto the heated substrate. (From J. W. Mayer and S. S. Lau,* Electronic Materials Science: For Integrated Circuits in Si and GaAs, *Macmillan Publishing Company, New York, 1990.)*

Following the production of the crystals, they are sliced into thin *wafers*, usually by a diamond-impregnated blade or wire. After grinding and polishing, the wafers are ready for the complex sequence of steps necessary to build a microcircuit. This processing sequence will be discussed in conjunction with our discussion of semiconductor devices below. In addition, many modern electronic devices are based on the buildup of thin-film layers of one semiconductor on another while maintaining some particular crystallographic relationship between the layer and the substrate. This **vapor deposition** technique is called **epitaxy**. **Homoepitaxy** involves the deposition of a thin film of essentially the same material as the substrate (e.g., Si on Si). **Heteroepitaxy** involves two materials of significantly different compositions (e.g., Al_xGa_{1-x} As on GaAs). Advantages of epitaxial growth include careful compositional control and reduced concentrations of unwanted defects and impurities. Figure 13.35 illustrates the process of **molecular-beam epitaxy** (MBE), a highly controlled ultrahigh-vacuum deposition process. The epitaxial layers are grown by impinging heated beams of the appropriate atoms or molecules on a heated substrate. The **effusion cells**, or **Knudsen*** cells, provide

---

*Martin Hans Christian Knudsen (1871–1949), Danish physicist. His brilliant career at the University of Copenhagen centered on many pioneering studies of the nature of gases at low pressure. He also developed a parallel interest in hydrography, establishing methods to define the various properties of seawater.

a flux, F, of atoms (or molecules) per second given by

$$F = \frac{pA}{\sqrt{2\pi mkT}}, \tag{13.23}$$

where p is the pressure in the effusion cell, A is the area of the aperture, m is the mass of a single atomic or molecular species, k is the Boltzmann's constant, and T is the absolute temperature.

EXAMPLE 13.23

We can use the Al–Si phase diagram (Figure 9.13) to illustrate the principle of zone refining. Assuming that we have a bar of silicon with aluminum as its only impurity, **(a)** calculate the segregation coefficient, K, in the Si-rich region and **(b)** calculate the purity of a 99 wt % Si bar after a single pass of the molten zone. (Note that the solidus can be taken as a straight line between a composition of 99.985 wt % Si at 1,190°C and 100 wt % Si at 1,414°C.)

SOLUTION

(a) Close inspection of Figure 9.13 indicates that the liquidus curve crosses the 90% Si composition line at a temperature of 1,360°C.

The solidus line can be expressed in the form

$$y = mx + b,$$

with y being the temperature and x being the silicon composition (in wt %). For the conditions stated,

$$1,190 = m(99.985) + b$$

and

$$1,414 = m(100) + b.$$

Solving the equations gives

$$m = 1.493 \times 10^4 \quad \text{and} \quad b = -1.492 \times 10^6.$$

At 1,360°C (where the liquidus composition is 90% Si), the solid composition is given by

$$1,360 = 1.493 \times 10^4 x - 1.492 \times 10^6,$$

or

$$x = \frac{1,360 + 1.492 \times 10^6}{1.493 \times 10^4}$$
$$= 99.99638.$$

The segregation coefficient is calculated in terms of impurity levels; that is,

$$c_s = 100 - 99.99638 = 0.00362 \text{ wt \% Al}$$

and

$$c_l = 100 - 90 = 10 \text{ wt \% Al},$$

yielding

$$K = \frac{c_s}{c_l} = \frac{0.00362}{10} = 3.62 \times 10^{-4}.$$

(b) For the liquidus line, a similar straight-line expression takes on the values

$$1{,}360 = m(90) + b$$

and

$$1{,}414 = m(100) + b,$$

yielding

$$m = 5.40 \quad \text{and} \quad b = 874.$$

A 99 wt % Si bar will have a liquidus temperature

$$T = 5.40(99) + 874 = 1{,}408.6°\text{C}.$$

The corresponding solidus composition is given by

$$1{,}408.6 = 1.493 \times 10^4 x - 1.492 \times 10^6,$$

or

$$x = \frac{1{,}408.6 + 1.492 \times 10^6}{4{,}924} = 99.999638 \text{ wt \% Si}.$$

An alternate composition expression is

$$\frac{(100 - 99.999638) \text{ \% Al}}{100\%} = 3.62 \times 10^{-6} \text{Al},$$

or 3.62 parts per million Al.

Note. These calculations are susceptible to round-off errors. Values for m and b in the solidus-line equation must be carried to several places.

PRACTICE PROBLEM 13.21

The purity of a 99 wt % Si bar after one zone-refining pass is found in Example 13.23. What would be the purity after two passes?

SEMICONDUCTOR DEVICES

Little space in this book is devoted to details of the final applications of engineering materials. Our focus is on the nature of the material itself. For structural and optical applications, detailed descriptions are often unnecessary. The structural steel and glass windows of modern buildings are familiar to us all, but the miniaturized applications of semiconductor materials are generally less so. In this section, we look briefly at some solid-state **devices**.

Miniaturized electrical circuits are the result of the creative combination of p-type and n-type semiconducting materials. An especially simple example is the **rectifier**, or **diode**, shown in Figure 13.36. This diode contains single **p–n junction** (i.e., a boundary between adjacent regions of p-type and n-type materials). This junction can be produced by physically joining two pieces of material, one p-type and one n-type. Later, we shall see more subtle ways of forming such junctions by diffusing different dopants (p-type and n-type) into adjacent regions of an (initially) intrinsic material. When voltage is applied to the device, as shown in Figure 13.36b, the charge carriers are driven away from the junction (the positive holes toward the negative electrode, and the negative electrons toward the positive electrode). This **reverse bias** quickly leads to polarization of the rectifier. Majority charge carriers in each region are driven to the adjacent electrodes,

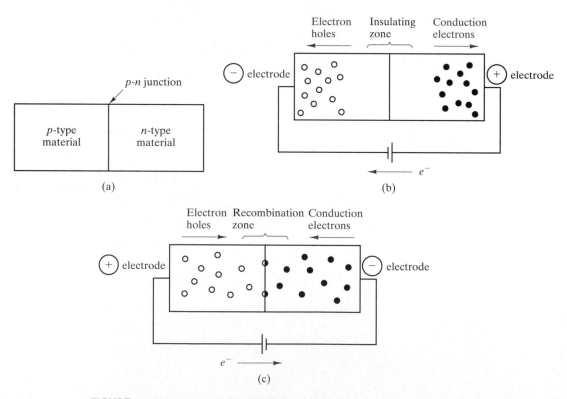

FIGURE 13.36 *(a) A solid-state rectifier, or diode, contains a single p–n junction. (b) In reverse bias, polarization occurs and little current flows. (c) In forward bias, majority carriers in each region flow toward the junction, where they are continuously recombined.*

and only a minimal current (due to intrinsic charge carriers) can flow. Reversing the voltage produces a **forward bias**, as shown in Figure 13.36c. In this case, the majority charge carriers in each region flow toward the junction where they are continuously recombined (each electron filling an electron hole). This process allows a continuous flow of current in the overall circuit. This continuous flow is aided at the electrodes. Electron flow in the external circuit provides a fresh supply of electron holes (i.e., removal of electrons) at the positive electrode and a fresh supply of electrons at the negative electrode. Figure 13.37a shows the current flow as a function of voltage in an ideal rectifier, while Figure 13.37b shows current flow in an actual device. The ideal rectifier allows zero current to pass in reverse bias and has zero resistivity in forward bias. The actual device has some small current in reverse bias (from minority carriers) and a small resistivity in forward bias. This simple solid-state device replaced the relatively bulky vacuum-tube rectifier (Figure 13.38). The solid-state devices to replace the various vacuum tubes allowed a substantial miniaturization of electrical circuitry in the 1950s.

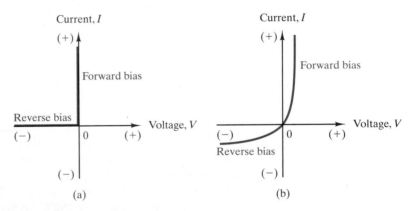

FIGURE 13.37 *Current flow as a function of voltage in (a) an ideal rectifier and (b) an actual device such as that shown in Figure 13.36.*

FIGURE 13.38 *Comparison of a vacuum-tube rectifier with a solid-state counterpart. Such components allowed substantial miniaturization in the early days of solid-state technology. (Courtesy of R. S. Wortman.)*

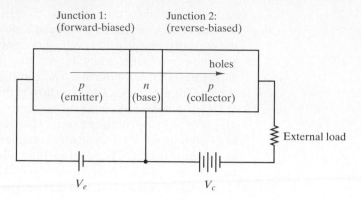

FIGURE 13.39 *Schematic of a transistor (a p–n–p sandwich). The overshoot of electron holes across the base (n-type region) is an exponential function of the emitter voltage, V_e. Because the collector current (I_c) is similarly an exponential function of V_e, this device serves as an amplifier. An n–p–n transistor functions similarly except that electrons rather than holes are the overall current source.*

Perhaps the most heralded component in this solid-state revolution was the **transistor** shown in Figure 13.39. This device consists of a pair of nearby *p–n* junctions. Note that the three regions of the transistor are termed **emitter**, **base**, and **collector**. Junction 1 (between the emitter and base) is forward biased. As such, it looks identical to the rectifier in Figure 13.36c. However, the function of the transistor requires behavior not considered in our description of the rectifier. Specifically, the recombination of electrons and holes shown in Figure 13.36 does not occur immediately. In fact, many of the charge carriers move well beyond the junction. If the base (*n*-type) region is narrow enough, a large number of the holes (excess charge carriers) pass across junction 2. A typical base region is less than 1 μm wide. Once in the collector, the holes again move freely (as majority charge carriers). The extent of "overshoot" of holes beyond junction 1 is an exponential function of the emitter voltage, V_e. As a result, the current in the collector, I_c, is an exponential function of V_e,

$$I_c = I_0 e^{V_e/B}, \tag{13.24}$$

where I_0 and B are constants. The transistor is an **amplifier**, since slight increases in emitter voltage can produce dramatic increases in collector current. The *p–n–p* "sandwich" in Figure 13.39 is not a unique transistor design. An *n–p–n* system functions in a similar way, with electrons, rather than holes, being the overall current source. In integrated circuit technology, the configuration in Figure 13.39 is also called a **bipolar junction transistor** (BJT).

A contemporary variation of transistor design is illustrated in Figure 13.40. The **field-effect transistor** (FET) incorporates a "channel" between a **source** and a **drain** (corresponding to the emitter and collector, respectively, in Figure 13.39). The *p*-channel (under an insulating layer of vitreous silica) becomes conductive upon application of a negative voltage to the **gate** (corresponding to the base in Figure 13.39). The channel's field, which results from the negative gate voltage, produces an attraction for holes from the substrate. (In effect, the *n*-type material just under the silica layer is distorted by the field to *p*-type character.) The result

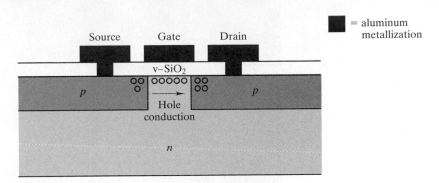

FIGURE 13.40 *Schematic of a field-effect transistor (FET). A negative voltage applied to the gate produces a field under the vitreous silica layer and a resulting p-type conductive channel between the source and the drain. The width of the gate is less than 50 nm in contemporary integrated circuits.*

is the free flow of holes from the p-type source to the p-type drain. The removal of the voltage on the gate effectively stops the overall current.

An n-channel FET is comparable to that shown in Figure 13.40, but with the p- and n-type regions reversed and electrons, rather than electron holes, serving as charge carriers. The operating frequency of high-speed electronic devices is limited by the time required for an electron to move from the source to the drain across such an n-channel. A primary effort in silicon-based **integrated-circuit** (IC) technology is to reduce the gate length, with a value of <1 μm being typical and ≈ 50 nm being the current minimum. An alternative is to go to a semiconductor with a higher electron mobility, such as GaAs. (Note the relative values of μ_e for Si and GaAs in Table 13.5.) The use of GaAs must be balanced against its higher cost and more difficult processing technology.

Modern technology has moved with great speed, but nowhere has progress moved at quite such a dizzy pace as in solid-state technology. Solid-state circuit elements such as diodes and transistors allowed substantial miniaturization when they replaced vacuum tubes. Further increases in miniaturization have occurred by eliminating separate solid-state elements. A sophisticated electrical **microcircuit** such as that shown in Figure 1.17 can be produced by application of precise patterns of diffusible n-type and p-type dopants to give numerous elements within a single **chip** of single-crystal silicon. Figure 13.41 shows an array of many such chips that have been produced on a single silicon **wafer**, a thin slice from a cylindrical single crystal of high-purity silicon. A typical wafer is 150 mm (6 in.), 200 mm (8 in.), or 300 mm (12 in.) in diameter and 250 μm thick, with chips being 5 to 10 mm on an edge. Individual circuit element patterns are produced by **lithography**, originally a print-making technique involving patterns of ink on a porous stone (giving the prefix *litho* from the Greek word *lithos*, meaning "stone"). The sequence of steps used to produce a vitreous SiO_2 pattern on silicon is shown in Figure 13.42. The original uniform SiO_2 layer is produced by thermal oxidation of the Si between 900 and 1,100°C. The key to the IC lithography process is the use of a polymeric **photoresist**. In Figure 13.42, a "positive" photoresist is used in which the material is depolymerized by exposure to ultraviolet radiation. A solvent is used to remove the *exposed* photoresist. A "negative" photoresist is used in the metallization process of Figure 13.43, in which the ultraviolet radiation leads to cross-linking of the polymer, allowing the solvent to remove the *unexposed* material.

FIGURE 13.41 *A silicon wafer (300 mm diameter) containing numerous chips of the type illustrated in Figure 1.17. (Photograph courtesy of Intel Corporation.)*

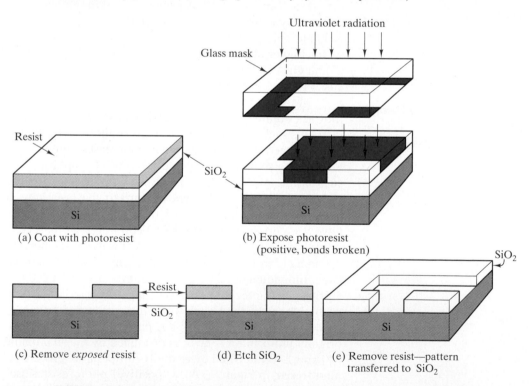

Ultraviolet radiation

Glass mask

SiO_2

Resist

Si

(a) Coat with photoresist

(b) Expose photoresist
(positive, bonds broken)

Resist

SiO_2

SiO_2

Si

Si

Si

(c) Remove *exposed* resist

(d) Etch SiO_2

(e) Remove resist—pattern
transferred to SiO_2

FIGURE 13.42 *Schematic illustration of the lithography process steps for producing vitreous SiO_2 patterns on a silicon wafer. (From J. W. Mayer and S. S. Lau,* Electronic Materials Science: For Integrated Circuits in Si and GaAs, *Macmillan Publishing Company, New York, 1990.)*

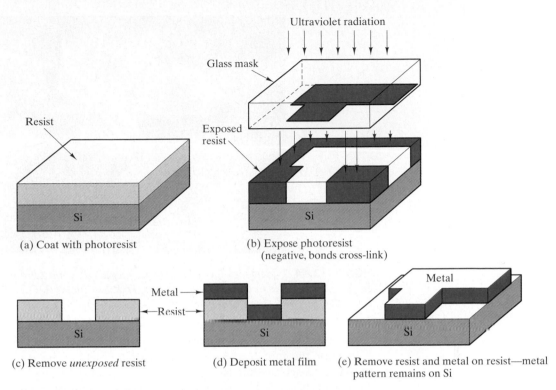

FIGURE 13.43 *Schematic illustration of the lithography process steps for producing metal patterns on a silicon wafer. (From J. W. Mayer and S. S. Lau,* Electronic Materials Science: For Integrated Circuits in Si and GaAs, *Macmillan Publishing Company, New York, 1990.)*

During the later stages of circuit fabrication, oxide and nitride layers are often deposited on silicon to serve as insulating films between metal lines or as insulating, protective covers. These layers are generally deposited by **chemical vapor deposition** (CVD). SiO_2 films can be produced between 250 and 450°C by the reaction of silane and oxygen:

$$SiH_4 + O_2 \rightarrow SiO_2 + 2\,H_2. \qquad (13.25)$$

Silicon nitride films involve the reaction of silane and ammonia:

$$3\,SiH_4 + 4\,NH_3 \rightarrow Si_3N_4 + 12\,H_2. \qquad (13.26)$$

Ultimately, the submicron-scale IC patterns must be connected to the macroscopic electronic "package" by relatively large-scale metal wires with diameters of 25 to 75 μm (Figure 13.44).

An active area of development at the current time is the production of **quantum wells**, which are thin layers of semiconducting material in which the wavelike electrons are confined within the layer thickness, a dimension as small as 2 nm. Advanced processing techniques are being used to develop such confined regions in two dimensions (**quantum wires**) or in three dimensions (**quantum dots**, again as small as 2 nm on a side). These small dimensions permit electron transit times of less than a picosecond and correspondingly high device operating speeds.

FIGURE 13.44 *Typical metal wire bond to an integrated circuit. (From C. Woychik and R. Senger, in* Principles of Electronic Packaging, *D. P. Seraphim, R. C. Lasky, and C.-Y. Li, Eds., McGraw-Hill Book Company, New York, 1989.)*

In summary, semiconductor devices have revolutionized modern life by providing for the miniaturization of electronic circuits. The replacement of traditional, large-scale elements such as diodes and transistors with separate solid-state counterparts started a revolution. The development of integrated microcircuits accelerated this revolution. Miniaturization is continuing by reducing the size of microcircuit elements. At the outset, miniaturization of electronics was driven by the aerospace industry, for which computers and circuitry needed to be small and low in power consumption. The steady and dramatic reduction in cost that has accompanied these developments has led to the increasing crossover of applications into industrial production and control, as well as consumer electronics.

The dramatic changes in our technology are nowhere more evident than in the computer field. In addition to the amplification of electrical signals discussed previously, transistors and diodes can also serve as switching devices. This application is the basis of the computational and information storage functions of computers. The elements within the microcircuit represent the two states ("off" or "on") of the binary arithmetic of the digital circuit. In this application, miniaturization has led to the trend away from mainframes to personal computers and, further, to portable (laptop) computers. Simultaneous with this trend has been a steady reduction in cost and an increase in computational power.

Figure 13.45 illustrates the dramatic progress in the miniaturization of computer chips. The number of transistors produced in the microcircuitry of a single chip has, over roughly 3 decades, gone from a few thousand to hundreds of millions. These numbers have generally doubled every 2 years. This steady pace of miniaturization has become widely known as **Moore's law** after Gordon Moore, cofounder of Intel Corporation, who predicted this capability in the early days of IC technology. Current lithography techniques can produce features on the order of 50 nm in width. This achievement is the result of intense research on the use of short wavelength ultraviolet and x-ray lithography. UV and x-ray wavelengths are much shorter than those of visible light and can potentially allow production of lithographic features down to 10 nm or less. In this way, microcircuit technology could continue to follow Moore's law into a level of one billion transistors per chip.

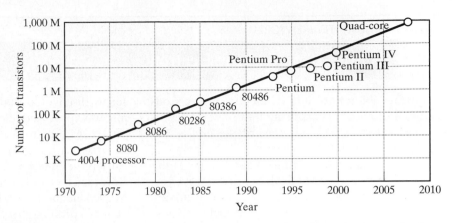

Moore's law is the quintessential example of Powers of Ten *in modern engineering. There has been a five orders of magnitude increase in the number of transistors on a millimeter scale computer chip during the relatively short history of the microprocessor. Current chip densities would suggest that* nanoprocessor *might be a better term.*

FIGURE 13.45 *The rapid and steady growth in the number of transistors contained on a single microcircuit chip has generally followed Moore's law, which states that the number doubles roughly every two years. (After data from Intel Corporation.)*

EXAMPLE 13.24

A given transistor has a collector current of 5 mA when the emitter voltage is 5 mV. Increasing the emitter voltage to 25 mV (a factor of 5) increases the collector current to 50 mA (a factor of 10). Calculate the collector current produced by further increasing the emitter voltage to 50 mV.

SOLUTION
Using Equation 13.24,

$$I_c = I_0 e^{V_e/B},$$

we are given

$$I_c = 5 \text{ mA} \quad \text{when } V_e = 5 \text{ mV}$$

and

$$I_c = 50 \text{ mA} \quad \text{when } V_e = 25 \text{ mV}.$$

Then,

$$\frac{50 \text{ mA}}{5 \text{ mA}} = e^{25 \text{ mV}/B - 5 \text{ mV}/B},$$

giving

$$B = 8.69 \text{ mV}$$

and

$$I_0 = 5 \text{ mA} e^{-(5 \text{ mV})/(8.69 \text{ mV})}$$

$$= 2.81 \text{ mA}.$$

Therefore,

$$I_{c,50 \text{ mV}} = (2.81 \text{ mA}) e^{50 \text{ mV}/8.69 \text{ mV}}$$

$$= 886 \text{ mA}.$$

THE MATERIAL WORLD

A Brief History of the Electron

Given the ubiquitous role of electronics in modern life, it is perhaps surprising that the electron itself was "discovered" only toward the end of the 19th century. In 1897, Professor J. J. Thomson of Cambridge University in England showed that the cathode rays in a device equivalent to a primitive version of the television tube were streams of negatively charged particles. He called these particles corpuscles. By varying the magnitudes of electric and magnetic fields through which the corpuscles traveled, he was able to measure the ratio of mass m to charge q. Thomson furthermore made the bold claim that his corpuscles were a basic constituent of all matter and that they were more than 1,000 times lighter than the lightest known atom, hydrogen. He was right on both counts. (The electron's mass is 1/1,836.15 that of the hydrogen atom.) As a result of his m/q measurement combined with his bold and accurate claims, Thomson is remembered as the discoverer of the electron.

Fifty years later, John Bardeen, Walter Brattain, and William Shockley at Bell Labs in New Jersey discovered that a small single crystal of germanium provided electronic signal amplification. This first transistor ushered in the era of solid-state electronics. In 1955, Dr. Shockley left Bell Labs and started the Shockley Semiconductor Laboratory of Beckman Instruments in Mountain View, California. In 1963, he moved on to Stanford University, where he was a professor of Engineering Science for many years. Shockley's experience at the Shockley Semiconductor Laboratory soured in 1957 when a group of young engineers he dubbed the "traitorous eight" left to form their own company, Fairchild Semiconductor, with the backing of the Fairchild Camera and Instrument Company. Among these young rebels were Gordon Moore and Robert Noyce.

By 1968, Moore and Noyce had become disenchanted with Fairchild. They were not alone. Many of the engineers were leaving with the feeling that technology was being supplanted by the politics of the workplace. Moore and Noyce left Fairchild to form a new company called Intel. They were joined in this venture by another former Fairchild employee

THE

LONDON, EDINBURGH, AND DUBLIN

PHILOSOPHICAL MAGAZINE

*AND

JOURNAL OF SCIENCE.

[FIFTH SERIES.]

OCTOBER 1897.

XL. *Cathode Rays.* By J. J. THOMSON, *M.A., F.R.S., Cavendish Professor of Experimental Physics, Cambridge*.*

THE experiments† discussed in this paper were undertaken in the hope of gaining some information as to the nature of the Cathode Rays. The most diverse opinions are held as to these rays; according to the almost unanimous opinion of German physicists they are due to some process in the æther to which—inasmuch as in a uniform magnetic field their course is circular and not rectilinear—no phenomenon hitherto observed is analogous; another view of these rays is that, so far from being wholly ætherial, they are in fact

(Courtesy of the Philosophical Magazine.*)*

Andrew Grove, Robert Noyce, and Gordon Moore in 1975. (Courtesy of Intel Corporation.)

(Continued)

named Andrew Grove. At this time, there were 30,000 computers in the world. Most were mainframes big enough to fill a room; the rest were minicomputers roughly the size of a refrigerator. Computer codes were entered mechanically using punch cards. The young Intel company had reasonably good success making computer memory, but in 1971 it took a bold step in developing for a client, Busicom of Japan, a radically new product called a microprocessor. The 4004 product took 9 months to develop and contained 2,300 transistors on a single chip of silicon. Modest by today's standards, the 4004 had as much computing power as the pioneering ENIAC computer invented in 1946, which weighed 27 Mg and contained 18,000 vacuum tubes. Realizing it had a product with substantial potential, Intel bought the design and marketing rights to the 4004 microprocessor back from Busicom for $60,000. Shortly thereafter, Busicom went bankrupt. The rest, as they say, is history.

PRACTICE PROBLEM 13.22

In Example 13.24, we calculate, for a given transistor, the collector current produced by increasing the emitter voltage to 50 mV. Make a continuous plot of collector current versus emitter voltage for this device over the range of 5 to 50 mV.

13.6 | Composites

There is no particular magnitude of conductivity characteristic of composites. As discussed in Chapter 12, composites are defined in terms of combinations of the four fundamental material types. A composite of two or more metals will be a conductor. A composite of two or more insulators will be an insulator. However, a composite containing both a metal and an insulator could have a conductivity characteristic of either extreme or some intermediate value, depending on the geometrical distribution of the conducting and nonconducting phases. We found in Section 12.2 that many properties of composites, including electrical conductivity, are geometry sensitive (e.g., Equation 12.15).

13.7 | Electrical Classification of Materials

We are now ready to summarize the classification system implied by the data of Table 13.1. Figure 13.46 shows these data along a log scale. The four fundamental materials categories defined by atomic bonding in Chapter 2 are now sorted based on their relative ability to conduct electricity. Metals are good conductors. Semiconductors are best defined by their intermediate values of σ caused by a small, but measurable, energy barrier to electronic conduction (the band gap). Ceramics, glasses, and polymers are insulators characterized by a large barrier to electronic conduction. We should note, however, that certain materials such as ZnO can be a semiconductor in an electrical classification or a ceramic in a bonding (Chapter 2) classification. Also, in Section 13.3, we noted that certain oxides have been found to be superconducting. However, on the whole, ceramics, as discussed in Chapter 11, are usually insulators. Composites can be found anywhere along the conductivity scale depending on the nature of their components and the geometrical distribution of those components.

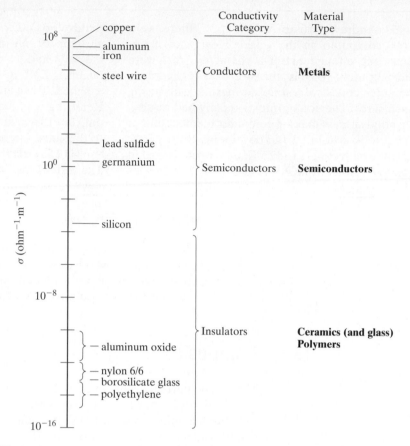

FIGURE 13.46 *Plot of the electrical conductivity data from Table 13.1. The conductivity ranges correspond to the four fundamental types of engineering materials.*

Summary

Electrical conduction, like atomic bonding, provides a basis for classifying engineering materials. The magnitude of electrical conductivity depends on both the number of charge carriers available and the relative mobility of those carriers. Various charged species can serve as carriers, but our primary interest is in the electron. In a solid, energy bands exist that correspond to discrete energy levels in isolated atoms. Metals are termed conductors because of their high values of electrical conductivity, which is the result of an unfilled valence band. Thermal energy, even at room temperature, is sufficient to promote a large number of electrons above the Fermi level into the upper half of the valence band. Temperature increase or impurity addition causes the conductivity of metals to decrease (and

the resistivity to increase). Any such decrease in the perfection of crystal structure decreases the ability of electron "waves" to pass through the metal. Important examples of conductors include the thermocouples and the superconductors.

Ceramics and glasses and polymers are termed insulators because their electrical conductivity is typically 20 orders of magnitude lower than that for metallic conductors. This difference occurs because there is a large energy gap (greater than 2 eV) between their filled valence bands and their conduction bands so that thermal energy is insufficient to promote a significant number of electrons above the Fermi level into a conduction band. Important examples of insulators are the ferroelectrics and piezoelectrics. (A dramatic exception is the ability of certain oxide ceramics to exhibit superconductivity at relatively high temperatures.)

Semiconductors with intermediate values of conductivity are best defined by the nature of this conductivity. Their energy gap is sufficiently small (generally less than 2 eV) so that a small but significant number of electrons are promoted above the Fermi level into the conduction band at room temperature. The charge carriers, in this case, are both the conduction electrons and the electron holes created in the valence band by the electron promotion. We consider the effect of impurities in extrinsic, elemental semiconductors. Doping a group IV A material, such as Si, with a group V A impurity, such as P, produces an *n*-type semiconductor in which negative charge carriers (conduction electrons) dominate. The "extra" electron from the group V A addition produces a donor level in the energy band structure of the semiconductor. As with intrinsic semiconductors, extrinsic semiconduction exhibits Arrhenius behavior. In *n*-type material, the temperature span between the regions of extrinsic and intrinsic behavior is called the exhaustion range. A *p*-type semiconductor is produced by doping a group IV A material with a group III A impurity, such as Al. The group III A element has a "missing" electron, producing an acceptor level in the band structure and leading to formation of positive charge carriers (electron holes). The region between extrinsic and intrinsic behavior for *p*-type semiconductors is called the saturation range.

The major electrical properties needed to specify an intrinsic semiconductor are band gap, electron mobility, hole mobility, and conduction electron density (= electron-hole density) at room temperature. For extrinsic semiconductors, one needs to specify either the donor level (for *n*-type material) or the acceptor level (for *p*-type material).

Compound semiconductors usually have an MX composition with an average of four valence electrons per atom. The III–V and II–VI compounds are the common examples. Amorphous semiconductors are the noncrystalline materials with semiconducting behavior. Elemental and compound materials are both found in this category. Chalcogenides are important members of this group.

Semiconductor processing is unique in the production of commercial materials in that exceptionally high-quality crystalline structure is required, together with chemical impurities in the ppb level. Structural perfection is approached with various crystal-growing techniques. Chemical perfection is approached by the process of zone refining. Vapor deposition techniques, such as molecular-beam epitaxy, are used to produce thin films for advanced electronic devices.

To appreciate the applications of semiconductors, we review a few of the solid-state devices that have been developed in the past few decades. The solid-state rectifier, or diode, contains a single *p–n* junction. Current flows readily when this junction is forward biased but is almost completely choked off when reverse biased. The transistor is a device consisting of a pair of nearby *p–n* junctions. The net result is a solid-state amplifier. Replacing vacuum tubes with solid-state elements produced substantial miniaturization of electrical circuits. Further miniaturization has resulted from the production of microcircuits consisting of precise patterns of *n*-type and *p*-type regions on a single-crystal chip. An increasingly finer degree of miniaturization is an ongoing goal of this IC technology.

Composites can have conductivity values anywhere from the conductor to the insulator range depending on the components and the geometrical distribution of those components.

Key Terms

acceptor level (460)
amorphous semiconductor (463)
amplifier (478)
base (478)
bipolar junction transistor (BJT) (478)
capacitor (449)
charge carrier (427)

charge density (449)
chemical vapor deposition (CVD) (481)
chip (479)
collector (478)
compound semiconductor (469)
conduction band (434)
conduction electron (451)

conductivity (427)
conductor (437)
critical current density (446)
current (427)
device (476)
dielectric (449)
dielectric constant (450)
dielectric strength (450)

diode (476)
donor level (458)
dopant (457)
drain (478)
drift velocity (428)
effusion cell (473)
electrical conduction (437)
electric field strength (428)
electric permittivity (450)
electron hole (427)
electron-hole pair (435)
electronic conduction (437)
emitter (478)
energy band (432)
energy band gap (434)
epitaxy (473)
exhaustion range (459)
extrinsic gettering (472)
extrinsic semiconduction (457)
Fermi function (433)
Fermi level (433)
ferroelectric (450)
field-effect transistor (FET) (478)
forward bias (477)
free electron (434)

gate (478)
gettering (472)
heteroepitaxy (473)
homoepitaxy (473)
Hund's rule (432)
insulator (449)
integrated circuit (IC) (479)
intrinsic gettering (472)
intrinsic semiconduction (451)
Knudsen cell (473)
lithography (479)
mean free path (440)
microcircuit (479)
molecular-beam epitaxy (MBE) (473)
Moore's law (482)
n-type semiconductor (459)
Ohm's law (427)
1–2–3 superconductor (443)
Pauli exclusion principle (431)
p–n junction (476)
p-type semiconductor (460)
photoresist (479)
piezoelectricity (451)
quantum dot (481)
quantum well (481)

quantum wire (481)
rectifier (476)
resistance (427)
resistivity (427)
reverse bias (476)
saturation range (461)
Seebeck effect (441)
Seebeck potential (441)
segregation coefficient (472)
semiconductor (451)
source (478)
superconductor (441)
temperature coefficient of
    resistivity (438)
thermocouple (440)
III–V compound (469)
transducer (451)
transistor (478)
II–VI compound (469)
valence band (433)
vapor deposition (473)
voltage (427)
wafer (479)
zone refining (472)

References

Harper, C. A., and **R. N. Sampson**, *Electronic Materials and Processes Handbook*, 3rd ed., McGraw-Hill, New York, 2004.

Kittel, C., *Introduction to Solid State Physics*, 8th ed., John Wiley & Sons, Inc., New York, 2005. Although this text is at a more advanced level, it is a classic source for information on the properties of solids.

Mayer, J. W., and **S. S. Lau**, *Electronic Materials Science: For Integrated Circuits in Si and GaAs*, Macmillan Publishing Company, New York, 1990.

Tu, K. N., J. W. Mayer, and **L. C. Feldman**, *Electronic Thin Film Science*, Macmillan Publishing Company, New York, 1992.

Problems

13.1. • Charge Carriers and Conduction

13.1. **(a)** Assume that the circuit in Figure 13.1 contains, as a sample, a cylindrical steel bar 1 cm diameter × 10 cm long with a conductivity of $7.00 \times 10^6\ \Omega^{-1} \cdot m^{-1}$. What would be the current in this bar due to a voltage of 10 mV? **(b)** Repeat part (a) for a bar of high-purity silicon of the same dimensions. (See Table 13.1.) **(c)** Repeat part (a) for a bar of borosilicate glass of the same dimensions. (Again, see Table 13.1.)

13.2. A lightbulb operates with a line voltage of 110 V. If the filament resistance is 200 Ω, calculate the number of electrons per second traveling through the filament.

13.3. A semiconductor wafer is 0.5 mm thick. A potential of 100 mV is applied across this thickness. **(a)** What is the electron drift velocity if their mobility is 0.2 m²/(V·s)? **(b)** How much time is required for an electron to move across this thickness?

D 13.4. A 1-mm-diameter wire is required to carry a current of 10 A, but the wire must not have a power dissipation (I^2R) greater than 10 W/m of wire. Of the materials listed in Table 13.1, which are suitable for this wire application?

13.5. A strip of aluminum metallization on a solid-state device is 1 mm long with a thickness of 1 μm and a width of 5 μm. What is the resistance of this strip?

13.6. For a current of 10 mA along the aluminum strip in Problem 13.5, calculate **(a)** the voltage along the length of the strip and **(b)** the power dissipated (I^2R).

D 13.7. A structural design involves a steel wire 2 mm in diameter that will carry an electrical current. If the resistance of the wire must be less than 25 Ω, calculate the maximum length of the wire, given the data in Table 13.1.

D 13.8. For the design discussed in Problem 13.7, calculate the allowable wire length if a 3-mm diameter is permitted.

13.2. • **Energy Levels and Energy Bands**

13.9. At what temperature will the 5.60-eV energy level for electrons in silver be 25% filled? (The Fermi level for silver is 5.48 eV.)

13.10. Generate a plot comparable to that shown in Figure 13.7 at a temperature of 1,000 K for copper, which has a Fermi level of 7.04 eV.

13.11. What is the probability of an electron's being promoted to the conduction band in indium antimonide, InSb, at **(a)** 25°C and **(b)** 50°C? (The band gap of InSb is 0.17 eV.)

13.12. At what temperature will diamond have the same probability for an electron's being promoted to the conduction band as silicon has at 25°C? (The answer to this question indicates the temperature range in which diamond can be properly thought of as a semiconductor rather than as an insulator.)

13.13. Gallium forms semiconducting compounds with various group VA elements. The band gap systematically drops with increasing atomic number of the VA elements. For example, the band gaps for the III–V semiconductors GaP, GaAs, and GaSb are 2.25 eV, 1.47 eV, and 0.68 eV, respectively. Calculate the probability of an electron's being promoted to the conduction band in each of these semiconductors at 25°C.

13.14. The trend discussed in Problem 13.13 is a general one. Calculate the probability of an electron's being promoted to the conduction band at 25°C in the II–VI semiconductors CdS and CdTe, which have band gaps of 2.59 eV and 1.50 eV, respectively.

13.3. • **Conductors**

13.15. A strip of copper metallization on a solid-state device is 1 mm long with a thickness of 1 μm and a width of 5 μm. If a voltage of 0.1 V is applied along the long dimension, what is the resulting current?

13.16. A metal wire 1 mm in diameter × 10 m long carries a current of 0.1 A. If the metal is pure copper at 30°C, what is the voltage drop along this wire?

13.17. Repeat Problem 13.16, assuming that the wire is a Cu–0.1 wt % Al alloy at 30°C.

13.18. A type K thermocouple is operated with a reference temperature of 100°C (established by the boiling of distilled water). What is the temperature in a crucible for which a thermocouple voltage of 30 mV is obtained?

13.19. Repeat Problem 13.18 for the case of a chromel/constantan thermocouple.

13.20. A furnace for oxidizing silicon is operated at 1,000°C. What would be the output (relative to an ice-water bath) for **(a)** a type S, **(b)** a type K, and **(c)** a type J thermocouple?

13.21. An important application of metal conductors in the field of materials processing is in the form of metal wire for resistance-heated furnace elements. Some of the alloys used as thermocouples also serve as furnace elements. For example, consider the use of a 1-mm-diameter chromel wire to produce a 1-kW furnace coil in a laboratory furnace operated at 110 V. What length of wire is required for this furnace design? (*Note:* The power of the resistance-heated wire is equal to I^2R, and the resistivity of the chromel wire is 1.08×10^6 $\Omega \cdot$ m.)

13.22. Given the information in Problem 13.21, calculate the power requirement of a furnace constructed of a chromel wire 5 m long and 1 mm diameter operating at 110 V.

13.23. What would be the power requirement for the furnace in Problem 13.22 if it is operated at 208 V?

13.24. A tungsten lightbulb filament is 10 mm long and 100 μm in diameter. What is the current in the filament when operating at 1,000°C with a line voltage of 110 V?

13.25. What is the power dissipation (I^2R) in the filament of Problem 13.24?

13.26. For a bulk 1–2–3 superconductor with a critical current density of 1×10^8 A/m^2, what is the maximum supercurrent that could be carried in a 1-mm-diameter wire of this material?

13.27. If progress in increasing T_c for superconductors had continued at the linear rate followed through 1975, by what year would a T_c of 95 K be achieved?

13.28. Verify the comment regarding the presence of one Cu^{3+} valence in the $YBa_2Cu_3O_7$ unit cell in the discussion of superconductors in Section 13.3.

13.29. Verify the chemical formula for $YBa_2Cu_3O_7$ using the unit-cell geometry of Figure 13.19a.

*__13.30.__ Describe the similarities and differences between the perovskite unit cell of Figure 13.19b and **(a)** the upper and lower thirds and **(b)** the middle third of the $YBa_2Cu_3O_7$ unit cell of Figure 13.19a.

13.4. • Insulators

13.31. Calculate the charge density on a 2 mm-thick capacitor made of 99.5% Al_2O_3 under an applied voltage of 1 kV.

13.32. Repeat Problem 13.31 for the same material at its breakdown voltage gradient (= dielectric strength).

13.33. Calculate the charge density on a capacitor made of cordierite at its breakdown dielectric strength of 3 kV/mm. The dielectric constant is 4.5.

13.34. By improved processing, a new cordierite capacitor can be made with properties superior to those of the one described in Problem 13.33. If the dielectric constant is increased to 5.0, calculate the charge density on a capacitor operated at a voltage gradient of 3 kV/mm (which is now below the breakdown strength).

13.5. • Semiconductors

13.35. Calculate the fraction of Ge atoms that provides a conduction electron at room temperature.

13.36. What fraction of the conductivity of intrinsic silicon at room temperature is due to **(a)** electrons and **(b)** electron holes?

13.37. What fraction of the conductivity at room temperature for **(a)** germanium and **(b)** CdS is contributed by (i) electrons and (ii) electron holes?

13.38. Using the data in Table 13.5, calculate the room-temperature conductivity of intrinsic gallium arsenide.

13.39. Using the data in Table 13.5, calculate the room-temperature conductivity of intrinsic InSb.

13.40. What fraction of the conductivity calculated in Problem 13.39 is contributed by **(a)** electrons and **(b)** electron holes?

13.41. Using data from Table 13.5, make a plot similar to Figure 13.24 showing both intrinsic silicon and intrinsic germanium over the temperature range of 27 to 200°C.

13.42. Superimpose a plot of the intrinsic conductivity of GaAs on the result of Problem 13.41.

13.43. There is a slight temperature dependence for the band gap of a semiconductor. For silicon, this dependence can be expressed as

$$E_g(T) = 1.152\,\text{eV} - \frac{AT^2}{T + B},$$

where $A = 4.73 \times 10^{-4}$ eV/K, $B = 636$ K, and T is in Kelvin. What is the percentage error in taking the band gap at 200°C to be the same as that at room temperature?

13.44. Repeat Problem 13.43 for GaAs, in which

$$E_g(T) = 1.567\,\text{eV} - \frac{AT^2}{T + B},$$

where $A = 5.405 \times 10^{-4}$ eV/K and $B = 204$ K.

13.45. An n-type semiconductor consists of 100 ppb of P doping, by weight, in silicon. What is **(a)** the mole percentage P and **(b)** the atomic density of P atoms? Compare your answer in part (b) with the maximum solid solubility level given in Table 13.6.

13.46. An As-doped silicon has a conductivity of 2.00×10^{-2} $\Omega^{-1} \cdot \text{m}^{-1}$ at room temperature. **(a)** What is the predominant charge carrier in this material? **(b)** What is the density of these charge carriers? **(c)** What is the drift velocity of these carriers under an electrical field strength of 200 V/m? (The μ_e and μ_h values given in Table 13.5 also apply for an extrinsic material with low impurity levels.)

13.47. Repeat Problem 13.46 for the case of a Ga-doped silicon with a conductivity of $2.00 \times 10^{-2} \Omega^{-1} \cdot m^{-1}$ at room temperature.

13.48. Calculate the conductivity for the saturation range of silicon doped with 10-ppb boron.

13.49. Calculate the conductivity for the saturation range of silicon doped with 20-ppb boron. (Note Problem 13.48.)

13.50. Calculate the conductivity for the exhaustion range of silicon doped with 10-ppb antimony.

13.51. Calculate the upper temperature limit of the saturation range for silicon doped with 10-ppb boron. (Note Problem 13.48.)

13.52. Calculate the upper temperature limit of the exhaustion range for silicon doped with 10-ppb antimony. (Note Problem 13.50.)

13.53. If the lower temperature limit of the saturation range for silicon doped with 10-ppb boron is 110°C, calculate the extrinsic conductivity at 300 K. (Note Problems 13.48 and 13.51.)

13.54. Plot the conductivity of the B-doped Si of Problem 13.53 in a manner similar to the plot in Figure 13.34.

13.55. If the lower temperature limit of the exhaustion range for silicon doped with 10-ppb antimony is 80°C, calculate the extrinsic conductivity at 300 K. (Note Problems 13.50 and 13.52.)

13.56. Plot the conductivity of the Sb-doped Si of Problem 13.55 in a manner similar to the plot in Figure 13.30.

13.57. In designing a solid-state device using B-doped Si, it is important that the conductivity not increase more than 10% (relative to the value at room temperature) during the operating lifetime. For this factor alone, what is the maximum operating temperature to be specified for this design?

13.58. In designing a solid-state device using As-doped Si, it is important that the conductivity not increase more than 10% (relative to the value at room temperature)

during the operating lifetime. For this factor alone, what is the maximum operating temperature to be specified for this design?

∗13.59. **(a)** It was pointed out in Section 13.3 that the temperature sensitivity of conductivity in semiconductors makes them superior to traditional thermocouples for certain high-precision temperature measurements. Such devices are referred to as thermistors. As a simple example, consider a wire 0.5 mm in diameter × 10 mm long made of intrinsic silicon. If the resistance of the wire can be measured to within 10^{-3} Ω, calculate the temperature sensitivity of this device at 300 K. (*Hint:* The very small differences here may make you want to develop an expression for $d\sigma/dT$.) **(b)** Repeat the calculation for an intrinsic germanium wire of the same dimensions. **(c)** For comparison with the temperature sensitivity of a metallic conductor, repeat the calculation for a copper (annealed standard) wire of the same dimensions. (The necessary data for this case can be found in Table 13.2.)

13.60. The band gap of intrinsic InSb is 0.17 eV. What temperature increase (relative to room temperature = 25°C) is necessary to increase its conductivity by **(a)** 10%, **(b)** 50%, and **(c)** 100%?

13.61. Illustrate the results of Problem 13.60 on an Arrhenius-type plot.

13.62. The band gap of intrinsic ZnSe is 2.67 eV. What temperature increase (relative to room temperature = 25°C) is necessary to increase its conductivity by **(a)** 10%, **(b)** 50%, and **(c)** 100%?

13.63. Illustrate the results of Problem 13.62 on an Arrhenius-type plot.

13.64. Starting from an ambient temperature of 300 K, what temperature increase is necessary to double the conductivity of intrinsic InSb?

13.65. Starting from an ambient temperature of 300 K, what temperature increase is necessary to double the conductivity of intrinsic GaAs?

13.66. Starting from an ambient temperature of 300 K, what temperature increase is necessary to double the conductivity of intrinsic CdS?

13.67. What temperature increase (relative to room temperature) is necessary to increase the conductivity of intrinsic GaAs by 1%?

13.68. What temperature increase (relative to room temperature) is necessary to increase by 1% the conductivity of **(a)** Se-doped GaAs and **(b)** Cd-doped GaAs?

13.69. In intrinsic semiconductor GaAs, what fraction of the current is carried by electrons, and what fraction is carried by holes?

13.70. What fraction of the current is carried by electrons and what fraction is carried by holes in **(a)** Se-doped GaAs and **(b)** Cd-doped GaAs, in the extrinsic behavior range?

13.71. When the aluminum impurity level in a silicon bar has reached 1 ppb, what would have been the purity of the liquid on the previous pass?

13.72. Suppose you have a bar of 99 wt % Sn, with the impurity being Pb. Determine the impurity level after one zone refining pass. (Recall the phase diagram for the Pb–Sn system in Figure 9.16.)

13.73. For the bar in Problem 13.72, what would be the impurity level after **(a)** two passes or **(b)** three passes?

13.74. **(a)** Calculate the flux of gallium atoms out of an MBE effusion cell at a pressure of 2.9×10^{-6} atm and a temperature of $970°C$ with an aperture area of 500 mm^2. **(b)** If the atomic flux in part (a) is projected onto an area of 45,000 mm^2 on the substrate side of the growth chamber, how much time is required to build up a monolayer of gallium atoms? (Assume, for simplicity, a square grid of adjacent Ga atoms.)

13.75. The high-frequency operation of solid-state devices can be limited by the transit time of an electron across the gate between the source and drain of an FET. For a device to operate at 1 gigahertz (10^9 s^{-1}), a transit time of 10^{-9} s is required. **(a)** What electron velocity is required to achieve this transit time across a 1-μm gate? **(b)** What electric field strength is required to

achieve this electron velocity in silicon? **(c)** For the same gate width and electric field strength, what operating frequency would be achieved with GaAs, a semiconductor with a higher electron mobility?

13.76. Make a schematic illustration of an n–p–n transistor analogous to the p–n–p case shown in Figure 13.39.

13.77. Make a schematic illustration of an n-channel FET analogous to the p-channel FET shown in Figure 13.40.

13.78. Figure 13.44 illustrates the relatively large metal connection needed to communicate with an IC. A limit to the increasing scale of integration in ICs is the density of interconnection. A useful empirical equation to estimate the number of signal input/output (I/O) pins in a device package is $P = KG^\alpha$, where K and α are empirical constants and G is the number of gates of the IC. (One gate is equal to approximately four transistors.) For K and α values of 7 and 0.2, respectively, calculate the number of pins for devices with **(a)** 1,000, **(b)** 10,000, and **(c)** 100,000 gates.

13.79. As pointed out in this chapter, the progress in improving the operating frequency of high-speed electronic devices is directly coupled with the reduction in the gate length in ICs. With this in mind, repeat Problem 13.75 for a 2 gigahertz silicon device using a 0.5-μm gate.

13.80. Repeat Problem 13.79 for a 10 gigahertz silicon device using a state-of-the-art 0.1-μm gate.

13.6. • Composites

13.81. Calculate the conductivity at $20°C$ parallel to the W filaments in the Cu-matrix composite in Table 12.7.

13.82. Repeat Problem 13.81 assuming a similar composite but with 25 vol % W filaments.

CHAPTER 14
Materials in Engineering Design

As seen in this chapter, the enjoyment of the sport of windsurfing is enhanced by sophisticated materials selection.

14.1 Material Properties—Engineering Design Parameters

14.2 Selection of Structural Materials—Case Studies

14.3 Selection of Electronic Materials—Case Studies

14.4 Materials and Our Environment

W e conclude our exploration of the field of *materials science and engineering* by focusing on *materials engineering*. This chapter explores the theme of "materials in engineering design." The topic of *materials selection* indicates how the material properties introduced throughout the book are parameters in the overall engineering design process. So-called Ashby charts, in which pairs of material properties are plotted against each other, are especially useful for visualizing the relative performance of different classes of materials. Illustrations of materials selection are provided by case studies for both structural and electronic materials. A discussion of the engineering design process is not complete without a brief introduction to the nature of environmental degradation of materials and acknowledging the various environmental aspects of the design process along with the need to recycle materials where possible.

14.1 | Material Properties—Engineering Design Parameters

In looking at fundamentals in the first ten chapters of this book and material categories in Chapters 11 through 13, we have defined dozens of the basic properties of engineering materials. For materials *science*, the nature of these properties is an end in itself. They serve as the basis of our understanding of the solid state. For materials *engineering*, the properties assume a new role. They are the **design parameters**, which are the basis for selecting a given material for a given application. A graphic example of this engineering perspective is shown in Figure 14.1, in which some of the basic mechanical properties defined in Chapter 6 appear as parameters in an engineering handbook. A broader view of the role of material properties as a link between materials science and materials engineering is given in Figure 14.2. Also, we can recall Figure 1.19 that illustrates the integral relationship among materials (and their properties), the processing of the materials, and the effective use of the materials in engineering design.

A convenient format for facilitating materials selection is to have tabular data listed in order of increasing property values. For example, Table 14.1 gives

493

Specifications
UNS number. A92036

Chemical Composition
Composition limits. 0.50 max Si; 0.50 max Fe; 2.2 max Cu; 0.10 to 0.40 Mn; 0.30 to 0.6 Mg; 0.10 max Cr; 0.25 max Zn; 0.15 max Ti; 0.05 max others (each); 0.15 max others (total); bal Al

Applications
Typical uses. Sheet for auto-body panels

Mechanical Properties

Tensile properties. Typical, for 0.64 to 3.18 mm (0.025 to 0.125 in.) flat sheet, T4 temper; tensile strength, 340 MPa (49 ksi); yield strength, 195 MPa (28 ksi); elongation, 24% in 50 mm or 2 in. Minimum, for 0.64 to 3.18 mm flat sheet, T4 temper; tensile strength, 290 MPa (42 ksi); yield strength, 160 MPa (23 ksi); elongation, 20% in 50 mm or 2 in.

Hardness. Typical, T4 temper: 80 HR15T strain-hardening exponent, 0.23

Elastic modulus. Tension, 70.3 GPa (10.2×10^6 ksi); compression, 71.7 GPa (10.4×10^6 ksi)

Fatigue strength. Typical, T4 temper: 124 MPa (18 ksi) at 10^7 cycles for flat sheet tested in reversed flexure

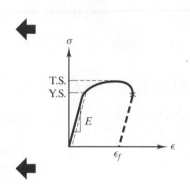

FIGURE 14.1 *The basic mechanical properties obtained from the tensile test introduced in Chapter 6 lead to a list of engineering design parameters for a given alloy. (The parameters are reproduced from a list in* ASM Handbook, *Vol. 2, ASM International, Materials Park, OH, 1990.)*

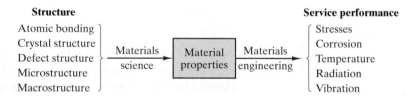

FIGURE 14.2 *Schematic illustration of the central role played by properties in the selection of materials. Properties are a link between the fundamental issues of materials science and the practical challenges of materials engineering. (From G. E. Dieter, in* ASM Handbook, *Vol. 20: Materials Selection and Design, ASM International, Materials Park, OH, 1997, p. 245.)*

the tensile strength of various tool steels listed in order of increasing strength. Similarly, Table 14.2 gives the ductility as % elongation for the same set of tool steels listed in order of increasing ductility.

A global view of the relative behavior of different classes of engineering materials is given by so-called **Ashby*** **charts** such as that shown in Figure 14.3, in

*Michael F. Ashby (1935–), English materials scientist and engineer. Ashby has been one of our most prolific materials scientists working on the most fundamental scientific problems and applying these principles to potential problems of materials selection, as exemplified by the Ashby charts.

TABLE 14.1

Selecting Tensile Strength of Tool Steels		
Type	**Condition**	**Tensile strength (MPa)**
S7	Annealed	640
L6	Annealed	655
S1	Annealed	690
L2	Annealed	710
S5	Annealed	725
L2	Oil quenched from 855°C and single tempered at 650°C	930
L6	Oil quenched from 845°C and single tempered at 650°C	965
S5	Oil quenched from 870°C and single tempered at 650°C	1,035
S7	Fan cooled from 940°C and single tempered at 650°C	1,240
L2	Oil quenched from 855°C and single tempered at 540°C	1,275
L6	Oil quenched from 845°C and single tempered at 540°C	1,345
S5	Oil quenched from 870°C and single tempered at 540°C	1,520
L2	Oil quenched from 855°C and single tempered at 425°C	1,550
L6	Oil quenched from 845°C and single tempered at 425°C	1,585
L2	Oil quenched from 855°C and single tempered at 315°C	1,790
S7	Fan cooled from 940°C and single tempered at 540°C	1,820
S5	Oil quenched from 870°C and single tempered at 425°C	1,895
S7	Fan cooled from 940°C and single tempered at 425°C	1,895
S7	Fan cooled from 940°C and single tempered at 315°C	1,965
L2	Oil quenched from 855°C and single tempered at 205°C	2,000
L6	Oil quenched from 845°C and single tempered at 315°C	2,000
S7	Fan cooled from 940°C and single tempered at 205°C	2,170
S5	Oil quenched from 870°C and single tempered at 315°C	2,240
S5	Oil quenched from 870°C and single tempered at 205°C	2,345

which pairs of materials properties are plotted against each other. In Figure 14.3, elastic modulus, E, is plotted against density, ρ, on logarithmic scales. Clearly, the various categories of structural materials tend to group together with, for example, the modulus-density combination for metal alloys generally being distinct from ceramics and glasses, polymers, and composites.

The use of Ashby charts in materials selection follows a general four-step design philosophy:

1. Translate design requirements into a material specification,

2. Screen out materials that fail constraints,

3. Rank by ability to meet objectives (using appropriate material indices), and

4. Search for supporting information for promising candidates.

The first case study in the next section (materials for windsurfer masts) illustrates this process in some detail. Figure 14.4 shows the effectiveness of the Ashby chart in a relatively simple and practical exercise, optimizing the trade-off between the cost and mass of a bicycle. One finds that the many designs summarized by the chart are bounded by an envelope, or *trade-off surface*, that represents optimal choices. Of course, there remains substantial subjectivity as one balances one's desire for a light bicycle against the cost one is willing to pay.

TABLE 14.2

Selecting Elongation of Tool Steels		
Type	**Condition**	**Elongation (%)**
L6	Oil quenched from 845°C and single tempered at 315°C	4
L2	Oil quenched from 855°C and single tempered at 205°C	5
S5	Oil quenched from 870°C and single tempered at 205°C	5
S5	Oil quenched from 870°C and single tempered at 315°C	7
S7	Fan cooled from 940°C and single tempered at 205°C	7
L6	Oil quenched from 845°C and single tempered at 425°C	8
S5	Oil quenched from 870°C and single tempered at 425°C	9
S7	Fan cooled from 940°C and single tempered at 315°C	9
L2	Oil quenched from 855°C and single tempered at 315°C	10
S5	Oil quenched from 870°C and single tempered at 540°C	10
S7	Fan cooled from 940°C and single tempered at 425°C	10
S7	Fan cooled from 940°C and single tempered at 540°C	10
L2	Oil quenched from 855°C and single tempered at 425°C	12
L6	Oil quenched from 845°C and single tempered at 540°C	12
S7	Fan cooled from 940°C and single tempered at 650°C	14
L2	Oil quenched from 855°C and single tempered at 540°C	15
S5	Oil quenched from 870°C and single tempered at 650°C	15
L6	Oil quenched from 845°C and single tempered at 650°C	20
S1	Annealed	24
L2	Annealed	25
L2	Oil quenched from 855°C and single tempered at 650°C	25
L6	Annealed	25
S5	Annealed	25
S7	Annealed	25

Source: (Tables 14.1 and 14.2) Data from *ASM Metals Reference Book*, 2nd ed., American Society for Metals, Metals Park, OH, 1984, as reconfigured in J. F. Shackelford, W. Alexander, and J. S. Park, *CRC Practical Handbook of Materials Selection*, CRC Press, Boca Raton, FL, 1995.

D | **EXAMPLE 14.1**

In selecting a tool steel for a machining operation, the design specification calls for a material with a tensile strength of $\geq 1{,}500$ MPa and a % elongation of $\geq 10\%$. Which specific alloys would meet this specification?

SOLUTION

By inspection of Tables 14.1 and 14.2, we see that the following alloy/heat treatment combinations would meet these specifications:

Type	Condition
S5	Oil quenched from 870°C and single tempered at 540°C
L2	Oil quenched from 855°C and single tempered at 425°C
L2	Oil quenched from 855°C and single tempered at 315°C
S7	Fan cooled from 940°C and single tempered at 540°C
S7	Fan cooled from 940°C and single tempered at 425°C

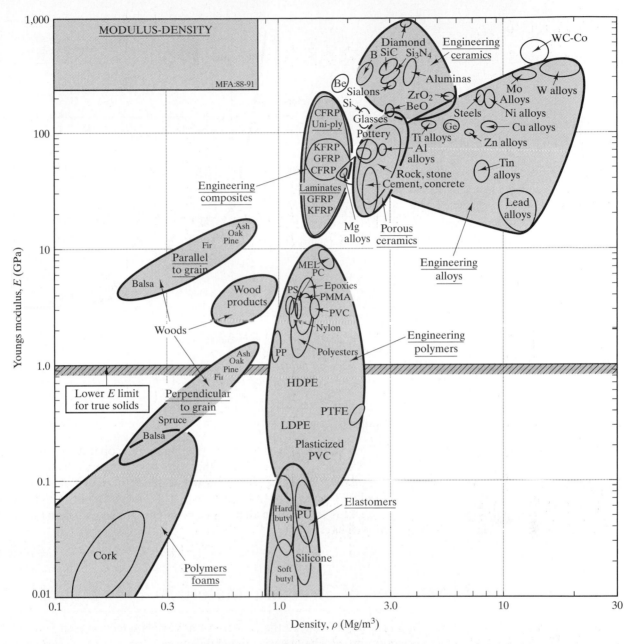

FIGURE 14.3 *A materials property chart with a global view of relative materials performance. In this case, plots of elastic modulus and density data (on logarithmic scales) for various materials indicate that members of the different categories of structural materials tend to group together. (After M. F. Ashby,* Materials Selection in Engineering Design, *Pergamon Press, Inc., Elmsford, NY, 1992.)*

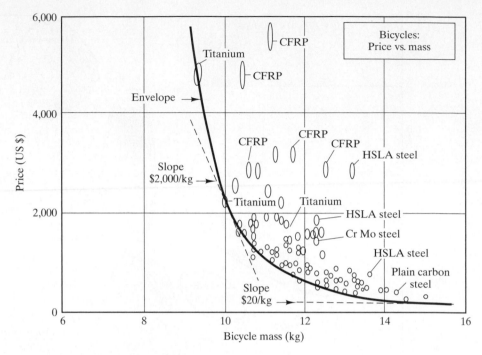

FIGURE 14.4 *A trade-off plot between cost and mass of bicycles. The optimal choices fall along the envelope, or tradeoff surface. (From M. F. Ashby, Granta Design Limited.)*

D PRACTICE PROBLEM 14.1

In reordering tool steel for the machining operation discussed in Example 14.1, you notice that the design specification has been updated. Which of the alloys would meet the new criteria of a tensile strength of $\geq$ 1,800 MPa and a % elongation of $\geq$ 10%? **(Note that the solutions to all practice problems are provided on the related Web site.)**

14.2 | Selection of Structural Materials—Case Studies

In Chapter 1 we introduced the concept of materials selection (see Figure 1.19). The selection process begins with the choice of a category of structural material (metal, ceramic, glass, polymer, or composite). With the wide range of commercial materials and properties (parameters) introduced in the previous thirteen chapters, we now have a clearer idea of our range of selection. The foundation for the fine-tuning of structural design parameters was laid in Chapters 9 and 10. In general, we look for an optimal balance of strength and ductility for a given application. Some of the common examples of composite materials in Chapter 12 provided this balance by the combination of a strong (but brittle) dispersed phase with a ductile (but weak) matrix. Many of the polymers of Chapter 12 provide adequate mechanical properties in combination with both formability and modest cost. Materials selection, however, can be more than an objective consideration of design parameters. The subjective factor of consumer appeal can play an equal part. In 1939, nylon 66 stockings were introduced, with 64 million pairs being sold in the first year alone.

When ductility is not essential, traditional, brittle ceramics (Chapter 11) can be used for their other attributes, such as high-temperature resistance or chemical durability. Many glasses (Chapter 11) and polymers (Chapter 12) are selected for their optical properties, such as transparency and color. Also, these traditional considerations are being modified by the development of new materials, such as high-fracture-toughness ceramics.

No material selection is complete until the issues of failure prevention and environmental degradation are taken into account (see Chapter 8 and Section 14.4). Consideration of the design configuration is important (e.g., whether stress-concentrating geometries or fatigue-producing cyclic loads are specified.) Chemical reaction, electrochemical corrosion, radiation damage, and wear can eliminate an otherwise attractive material.

The design selection process can be illustrated by some specific case studies. We shall look in some detail at the materials selection for windsurfer masts and more briefly at a variety of other case studies.

MATERIALS FOR WINDSURFER MASTS

Windsurfing is a sport that has become widely popular due to a successful merger of mechanical and materials engineering in the overall design process. Figure 14.5 illustrates the basic components of a windsurfer. Although the idea of a "freesail"

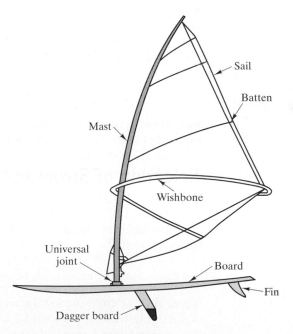

FIGURE 14.5 *Components of a windsurfer design. The stiffness of the mast controls the sail shape, and the pivoting of the mast about the universal joint controls the response of the craft. (After M. F. Ashby, "Performance Indices," in* ASM Handbook, *Vol. 20:* Materials Selection and Design, *ASM International, Materials Park, OH, 1997, pp. 281–290.)*

TABLE 14.3

Design Criteria for Winsurfer Masts[a]		
Criterion	**Constraint**	**Basis of constraint**
Stiffness	Specify	Sailing characteristics
Outer diameter	Limit	Reduce influence on airflow
Mass	Minimize	Stability

[a]After M. F. Ashby, "Performance Indices," in *ASM Handbook*, Vol. 20: *Materials Selection and Design*, ASM International, Materials Park, OH, 1997, pp. 281–290.

was put forward in the mid-1960s, a practical device awaited the development of a universal joint that allowed the mast to swivel freely about the surfboard. The mast is central to the design, and the selection of its specific shape and material of construction follows from a careful consideration of design criteria. The mast influences the dynamics of the sail and must flex under wind pressure. The resulting need to specify stiffness must be done in conjunction with a limit on outer diameter in order to reduce its influence on air flow. For stability, the mass of the mast must be minimized. The design criteria for windsurfer masts are summarized in Table 14.3.

The length, L, of a typical windsurfer mast is 4.6 m. The mast stiffness, S, is defined by

$$S = \frac{P}{\delta},$$ 　(14.1)

where P is the load created by hanging a weight at the midpoint of the mast supported horizontally at its ends and δ is the resulting deflection. As a practical matter, the stiffness of masts is characterized by the International Mast Check System (IMCS) number defined as

$$\text{IMCS number} = \frac{L}{\delta},$$ 　(14.2)

with the deflection, δ, produced by a standard weight of 30 kg. The IMCS number ranges between 20 (for a soft mast) and 32 (for a hard one).

The limit on the outer radius of the mast, $r_{\max}$, can be stated in a formal way. The outer radius of the mast, r, must be

$$r \leq r_{\max}.$$ 　(14.3)

For lightness, the mast is hollow, and the mass, m, which is minimized in the design, is given by

$$m = AL\rho,$$ 　(14.4)

where A is the cross-sectional area of the hollow mast, ρ is the density, and L was defined relative to Equation 14.2. A typical mast weighs between 1.8 and 3.0 kg.

Mast design is a specific application of a thin-walled tube, for which we can define a shape factor, ϕ, as

$$\phi = \frac{r}{t},$$ 　(14.5)

TABLE **14.4**

Design Results for Windsurfer Mast Materials[a]

Material	Shape factor $\phi(=r/t)$	Performance index M (GPa$^{1/2}$/[Mg/m^3])	Mass[b] m (kg)
Carbon fiber-reinforced polymer (CFRP)	14.3	22.9	2.0
Wood (spruce)	1.7	9.0	5.0
Aluminum	8.0	8.5	5.3
Glass fiber-reinforced polymer (GFRP)	4.3	6.2	7.3

[a]After M. F. Ashby, "Performance Indices," in *ASM Handbook*, Vol. 20: *Materials Selection and Design*, ASM International, Materials Park, OH, 1997, pp. 281–290.
[b]For r_{max} = 20 mm and International Mast Check System (IMCS) number = 26.

where t is the wall thickness and r was defined relative to Equation 14.3. An analysis of the mechanics of the thin-walled tube shows that the mass is related to the shape factor and the material properties elastic modulus, E, and density by

$$m = B\left(\frac{\rho}{[\phi E]^{1/2}}\right),\qquad (14.6)$$

where B is a constant specific to design loading conditions and includes the bending stiffness. Equation 14.6 indicates that the mass can be minimized by maximizing a performance index, M, defined as

$$M = \frac{(\phi E)^{1/2}}{\rho}.\qquad (14.7)$$

Table 14.4 summarizes the values of the shape factor, ϕ, performance index, M, and minimum mass, m, for a range of candidate mast materials for a midrange IMCS number of 26 and a maximum radius of r_{max} = 20 mm. The optimal materials tend to be those with the largest values of M. Furthermore, we can modify Equation 14.7 so that

$$M = \frac{(1/\phi)(\phi E)^{1/2}}{(1/\phi)\rho} = \frac{(E/\phi)^{1/2}}{(\rho/\phi)}.\qquad (14.8)$$

Then, the E/ϕ term can be given by a logarithmic expression by noting that

$$M^2(\rho/\phi)^2 = E/\phi,\qquad (14.9)$$

or

$$\ln(E/\phi) = 2\ln(\rho/\phi) + 2\ln M.\qquad (14.10)$$

The practical consequence of Equation 14.10 is that the behavior of a specific design geometry (the thin-walled tube) can be superimposed on the materials property chart of Figure 14.3 normalized by the shape factor, ϕ, as shown in Figure 14.6. For example, the carbon fiber-reinforced polymer (CFRP) location for the bulk material in Figure 14.3 corresponds to a shape factor of $\phi = 1$. In Figure 14.6, that position is connected to the tubing position corresponding to E/ϕ and ρ/ϕ, where $\phi = 14.3$, as given in Table 14.4.

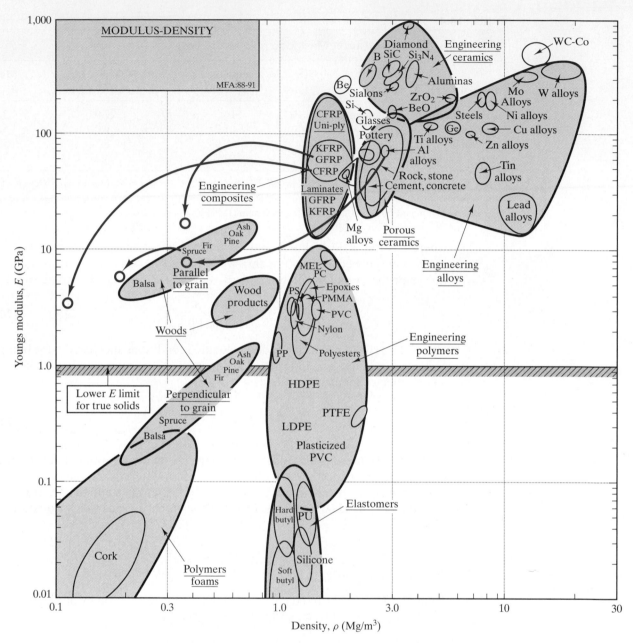

FIGURE 14.6 *The behavior of the windsurfer mast materials in Table 14.4 are superimposed on the* ln *E versus* ln *ρ chart of Figure 14.3 normalized by the shape factor, ϕ, of a thin-walled tube. For example, the CFRP mast with a shape factor of ϕ = 14.3 is shown at a position of (E/14.3, ρ/14.3) relative to the (E, ρ) position of the bulk material for which ϕ = 1. (After M. F. Ashby, "Performance Indices," in* ASM Handbook, *Vol. 20:* Materials Selection and Design, *ASM International, Materials Park, OH, 1997, pp. 281–290.)*

Finally, we should acknowledge that windsurfer design blurs the boundary between art and science. For example, the mast is routinely "tuned" by varying the stiffness along its length in order to match the weight of the surfer as well as the type of windsurfing (slalom, race, or wave).

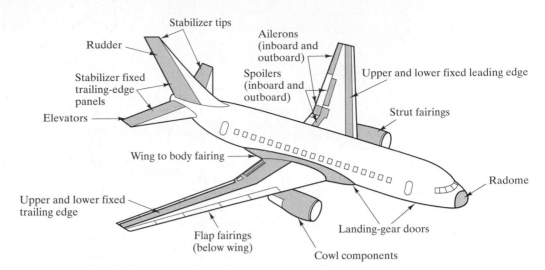

FIGURE 14.7 *Schematic illustration of the composite structural applications for the exterior surface of a Boeing 767 aircraft. (After data from the Boeing Airplane Company.)*

METAL SUBSTITUTION WITH COMPOSITES

A key example of the driving force for replacing metals with lower-density composites is in the commercial aircraft industry. Manufacturers had developed parts of fiberglass for improved dynamics and cost savings by the early 1970s. In the mid-1970s, the "oil crisis" led to a rapid rise in fuel costs, from 18% of direct operating costs to 60% within a few years. (One kilogram of "dead weight" on a commercial jet aircraft can consume 830 liters of fuel per year.) An early response to the need for materials substitution for fuel savings was the use of more than 1,100 kg of Kevlar-reinforced composites in the Lockheed L–1011–500 long-range aircraft. The result was a net 366-kg weight savings on the secondary exterior structure. Similar substitutions were made later on all L–1011 models. An excellent example of this effort is the design of the Boeing 767. Figure 14.7 illustrates this case. A significant fraction of the exterior surface consists of advanced composites, primarily with Kevlar and graphite reinforcements. The resulting weight savings using advanced composites is 570 kg.

MATERIALS FOR HIP–JOINT REPLACEMENT

The related Web site contains an article "Medical Materials" that appeared in the journal Advanced Materials and Processes, *published by ASM International. Used by permission.*

Some of the most dramatic developments in the applications of advanced materials have come in the field of medicine. One of the most successful has been the artificial-hip **prosthesis** (a device for replacing a missing body part). Figure 14.8 illustrates the surgical procedure involved in replacing a damaged or diseased hip joint and replacing it with a prosthesis. Figure 14.9 shows a typical example, a cobalt–chrome alloy (e.g., 50 wt % Co, 20 wt % Cr, 15 wt % W, 10 wt % Ni, and 5 wt % other elements) constituting the main stem and head, with an ultrahigh-molecular-weight polyethylene (with a molecular weight of 1 to 4×10^6 amu) cup completing the ball and socket system. The term total hip replacement (THR) refers to the simultaneous replacement of both the ball and socket with engineered materials. The orthopedic surgeon removes the degenerative hip joint and drills

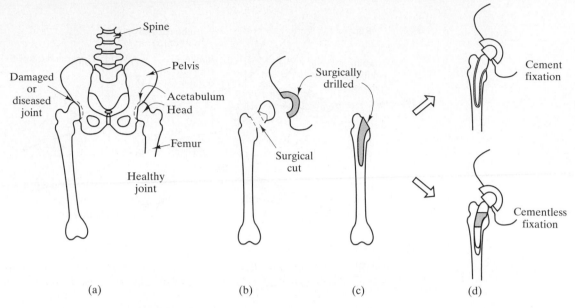

(a)   (b)   (c)   (d)

FIGURE 14.8 *Schematic of the total hip replacement (THR) surgery. In general, the femoral implant stem is anchored to the bone by either a thin layer (a few mm thick) of polymethylmethacrylate (PMMA) cement or a cementless system involving a snug fit of the stem in the femoral shaft. In typical cementless fixation, the upper one-third of the stem is covered with a porous coating of sintered metal-alloy beads. Bone growth into the porous surface provides a mechanical anchoring.*

FIGURE 14.9 *A cobalt-chrome stem and ball, with a polyethylene cup, form a ball and socket system for an artificial hip joint. (Courtesy of DePuy, a Division of Boehringer Mannheim Corporation.)*

out a cavity in the femoral bone to accommodate the stem. The stem is either anchored to the skeletal system with a polymethylmethacrylate (PMMA) cement or by bone ingrowth into a porous surface coating ("cementless fixation"). The cup is generally attached to a metal backing, which, in turn, is attached to the hip by metallic bone screws. The titanium alloy Ti–6Al–4V (Table 6.1) is generally preferred for the stem in cementless applications. The elastic modulus of Ti–6Al–4V (1.10×10^5 MPa) is closer to that of bone (1.4×10^4 MPA) than is cobalt–chrome (2.42×10^5 MPa) and, therefore, creates less stress on the bone due to modulus mismatch. (Cobalt–chrome is preferred for cemented implants for a similar reason. The lower elastic modulus of Ti–6Al–4V leads to excessive load on the interfacial cement.)

The metal/polymer interface provides a low-friction contact surface, and each material (the metal and the polymer) has good resistance to degradation by highly corrosive body fluids. Early artificial-hip designs with cement fixation had a typical lifetime of 5 years (limited primarily by mechanical loosening), adequate for elderly patients, but requiring painful replacement surgeries for younger people. The cementless fixation can extend the implant lifetime by a factor of three. More than 200,000 THR surgeries are performed in the United States each year, with a similar number in Europe.

The metal alloys and polymers involved in the THR are examples of **biomaterials**, which can be defined as engineered materials created for applications in biology and medicine. Biomaterials can be contrasted with the naturally occurring bone, which would be an example of a **biological material**. Both of these new terms are reminiscent of the biomimetic materials introduced in Chapter 11. Some of those engineered materials being produced by low-temperature, liquid-phase processing routes are prime candidates to be the next generation of biomaterials.

For roughly 4 decades, ceramics and glasses have been the focus of substantial research on their potential applications as biomaterials. The compressive load on the THR ball makes a high-density structural ceramic, such as Al_2O_3, a good candidate for that application. An attractive feature of this substitution would be the typically low surface wear of structural ceramics. In a similar way, the THR cup can be fabricated from Al_2O_3. As with many potential applications for ceramics in engineering designs, the inherent brittleness and low fracture toughness of these materials have limited their use in the THR.

In the past decade, however, a significant biomedical application has been found for one ceramic system. A ceramic substitute has been found for the traditional porous surfaces of cementless designs, namely hydroxyapatite coatings. It is ironic that such an obvious candidate as hydroxyapatite has only recently come into fashion as a biomaterial. Hydroxyapatite, $Ca_{10}(PO_4)_6(OH)_2$, is the primary mineral content of bone, representing 43% by weight. It has the distinct physiochemical advantages of being stable, inert, and biocompatible. The most successful application of hydroxyapatite in biomedicine has been in the form of a thin coating on a prosthetic implant, as shown in Figure 14.10. These coatings have been plasma-sprayed on both Co–Cr and Ti–6 Al–4V alloys. Optimal performance has come from coating thicknesses on the order of 25–30 μm. Interfacial strengths between the implant and bone are as much as five to seven times as great as with the uncoated specimens. The enhanced interfacial development corresponds to the mineralization of bone directly onto the hydroxyapatite surface with no signs of

FIGURE 14.10 *The Omnifit® HA Hip Stem consists of hydroxyapatite coating on a hip-replacement prosthesis for the purpose of improved adhesion between the prosthesis and bone. Hydroxyapatite is the predominant mineral phase in natural bone. (Courtesy of Osteonics, Allendale, NJ.)*

intermediate, fibrous tissue layers. Unlike the porous, metallic coatings that they replace, this ceramic substitute does not have to be porous. The bone adhesion can occur directly on a smooth hydroxyapatite surface. The substantial success of this coating system has led to its widespread use in THR surgery.

EXAMPLE 14.2

Estimate the annual fuel savings due to the weight reduction provided by composites in a fleet of 50 767 aircraft owned by a commercial airline.

SOLUTION

Using the information from the discussion of the case study in Section 14.2, we have

$$\text{fuel savings} = (\text{wt savings/aircraft}) \times \frac{(\text{fuel/year})}{(\text{wt savings})} \times 50 \text{ aircraft}$$

$$= (570 \text{ kg}) \times (830 \text{ l/yr})/\text{kg} \times 50$$

$$= 23.7 \times 10^6 \text{ l.}$$

PRACTICE PROBLEM 14.2

An annual fuel savings is calculated in Example 14.2. For this commercial airline, estimate the fuel savings that would have been provided by a fleet of 50 L–1011 aircraft. (See the case study in Section 14.2.)

14.3 | Selection of Electronic Materials—Case Studies

In Chapter 13, we saw that solid-state materials are naturally sorted into one of three categories: conductor, insulator, or semiconductor. Conductor selection is frequently determined by formability and cost as much as by specific conductivity values. The selection of an insulator can also be dominated by these same factors. A ceramic substrate may be limited by its ability to bond to a metallic conductor, or a polymeric insulation for conductive wire may be chosen due to its low cost. In some cases, new concepts in engineering design can make possible radically new materials choices (e.g., the optical glass fibers discussed in Chapter 1).

In Chapter 1, we found that design and materials selection are combined in an especially synergistic way in the semiconductor industry. Materials engineers and electronics engineers must work together effectively to ensure that the complex pattern of n- and p-type semiconducting regions in a chip provides a microcircuit of optimal utility.

As with the structural materials, we can obtain a fuller appreciation of the process of selecting electronic materials by looking at representative case studies. We shall look in some detail at the use of amorphous metals for electric-power distribution and more briefly at light-emitting diodes.

FIGURE 14.11 *Transformer core winding using an amorphous ferrous alloy wire.* *(Courtesy of Allied-Signal, Inc.)*

AMORPHOUS METAL FOR ELECTRIC–POWER DISTRIBUTION

The development of amorphous metals in recent decades has provided an attractive new choice for transformer cores. A key to the competitiveness of amorphous metals is the absence of grain boundaries, allowing for easier domain-wall motion. High resistivity (which damps eddy currents) and the absence of crystal anisotropy also contribute to magnetic domain-wall mobility. Ferrous "glasses" are among the most easily magnetized of all ferromagnetic materials. Figure 14.11 illustrates a transformer core application using an amorphous ferrous alloy. These ferrous alloys have especially low core losses. As manufacturing costs for the amorphous ribbons and wires were reduced, the energy conservation due to low core losses led to commercial applications. Fe–B–Si cores were put in commercial service as early as 1982, providing electrical power to homes and industrial facilities. The replacement of grain-oriented silicon steel with amorphous metals in transformer cores reduced core losses by 75%.

Given the large number of units in service and the continuous magnetization and demagnetization of the cores at line frequencies, transformers account for the largest portion of the energy losses in electric-power distribution systems. It is estimated that more than 50×10^9 kW·h of electrical energy is dissipated annually in the United States in the form of core losses. At an average generating cost of $0.035/kW · h, these core losses are valued at more than $1.5 billion.

Since 1982, well over one million amorphous metal distribution transformers have been installed worldwide, providing substantial improvement in the efficiency of electric-power distribution. The substitution of Fe-based amorphous metals for grain-oriented silicon steel is not, however, without design challenges. The amorphous metal tends to be thinner, harder, and more fragile than the silicon steel. Amorphous metal transformer installations must be compatible with existing electric-power distribution systems and must survive 30 years of continuous service. These fundamental requirements have affected the transformer manufacturing process and have mandated laboratory and field testing to ensure that potential energy savings could be provided reliably at a practical cost.

An early observation of alloy compositions that could, at reasonable quench rates, form amorphous structures was that they tended to involve a metallic element alloyed with a metalloid. Furthermore, such systems tend to form a "deep eutectic" at a composition of approximately 20 atomic % metalloid, where "deep eutectic" is defined as a low eutectic temperature compared to the melting points of the pure metal and metalloid. The classic example was an $Au_{80}Si_{20}$ alloy, which has a eutectic temperature of $363°C$, in contrast to melting points of Au and Si of $1,064°C$ and $1,414°C$, respectively. In effect, at a relatively low temperature these eutectic compositions provide liquid metal, which can more easily be "frozen" in the amorphous state. The association of amorphous metals with rapid solidification and eutectic compositions continues to be the basis of alloy design. Note that commercial amorphous metal transformer cores are generally 80 atomic % Fe, with the balance largely metalloids such as B, P, and C.

An amorphous $Fe_{80}P_{13}C_7$ alloy was produced in the laboratory in the late 1960s and the first commercial product, $Fe_{40}Ni_{40}P_{14}B_6$, known as METGLAS* 2826, was available in the early 1970s. Although the properties of this first product were optimized by annealing, relatively low saturation induction and critical temperature (above which it is nonmagnetic) limited its use to low-power, high-frequency applications. The development of amorphous alloys for electric power distribution have since focused primarily on Fe-based alloys. Table 14.5 represents distinct families of amorphous ferrous alloys in comparison to grain-oriented silicon steel. $Fe_{80}P_{13}C_7$ alloy offers the lowest raw-materials cost by the use of P and C. $Fe_{80}B_{20}$ and $Fe_{86}B_8C_6$ alloys offer higher saturation induction by replacing P and C with B and using a higher Fe content. $Fe_{80}B_{11}Si_9$ alloy offers the best thermal stability (highest critical temperature). Thermal stability has proven to be a primary design consideration in the commercialization of amorphous alloys for electric-power distribution. The primary concern in this regard is the crystallization of the alloy either during processing or in service leading to diminished performance. As a result, $Fe_{80}B_{11}Si_9$ is the most commonly used amorphous alloy in power applications. Although its saturation induction is only 80% of that for the grain-oriented silicon steel, the amorphous alloy generates only 30% of the core loss.

The processing of amorphous alloys has also provided numerous engineering challenges. To achieve the necessary $10^{5}°C/s$ quench rates to form these alloys by rapid solidification, at least one dimension must be small. The original laboratory demonstration of feasibility involved the splat-quenching of liquid droplets against a copper plate. Subsequent development led to the use of the chill block melt spinning technique, available since the 1870s for producing solder wire, in which a continuous stream of molten alloy is projected against the outer surface of a rotating drum. Refinements of this method led to the production of the initial commercial product ($Fe_{40}Ni_{40}P_{14}B_6$ alloy) in a continuous ribbon 50 μm thick and

---

*METGLAS is a registered trademark.

TABLE 14.5

Characteristics of Traditional Electrical Steel and Fe–Based Amorphous Metals[a]

Material	Saturation induction B_s (Wb/m^2)	Critical temperature T_c (K)	Coercive force H_c (A/m)	Core loss @ 60 Hz, 1.4 Wb/m^2 CL (W/kg)
Grain-oriented Fe–3.2% Si	2.01	1,019	24	0.7
Amorphous $Fe_{80}P_{13}C_7$	1.40	587	5	—
Amorphous $Fe_{80}P_{20}$	1.60	647	3	0.3
Amorphous $Fe_{86}P_8C_6$	1.75	<600	4	0.4 (est.)
Amorphous $Fe_{80}P_{11}Si_9$	1.59	665	2	0.2

[a]After N. DeCristofaro, "Amorphous Metals in Electric-Power Distribution Applications," *MRS Bulletin 23*, 50 (1998).

FIGURE 14.12 *Pole-mounted amorphous metal distribution transformer. (Courtesy of Metglas.)*

1.7 mm wide. Furthermore, refinements of this processing technology could only produce ribbons up to 5 mm in width. The production of transformer cores was significantly enhanced by the availability of wide sheets of alloy rather than narrow ribbon. Such sheets became possible with the development of planar flow casting, in which the molten alloy is forced through a slotted nozzle in close proximity ($\approx$0.5 mm) to the surface of a moving substrate. The melt puddle is then constrained between the nozzle and substrate, producing a stable, rectangular cross section. The planar-flow casting technique has produced amorphous metal sheets up to 300 mm in width, with 210-mm widths commercially available.

The last quarter of the 20th century saw major changes in the global electric-power industry. After the 1973 oil embargo, electric utilities developed an understandable interest in higher-efficiency transformers, even as energy supplies and prices stabilized in the 1980s. In the United States, the Electric Power Research Institute (EPRI) focused attention on distribution transformers, typically mounted on utility poles or concrete pads (Figure 14.12). Distribution transformers step

TABLE 14.6

Environmental Impact of Amorphous Metal Transformers[a]					
	Country or Region				
Benefit	**U.S.**	**Europe**	**Japan**	**China**	**India**
Energy savings (10^9 kW·h)	40	25	11	9	2
Oil (10^6 barrels)	70	45	20	15	4
CO_2 (10^9 kg)	32	18	9	11	3
NO_x (10^6 kg)	100	63	27	82	20
SO_2 (10^6 kg)	240	150	68	190	47

[a]After N. DeCristofaro, "Amorphous Metals in Electric-Power Distribution Applications," *MRS Bulletin 23*, 50 (1998).

down electrical voltages from the 5–14 kV range used for local transmission to the 120–240 V range used in homes and businesses. Given the thinner geometry and more brittle mechanical behavior of amorphous alloys, the specific design of the transformer core has been modified somewhat in comparison to the traditional grain-oriented silicon steel. To simulate the geometry of silicon steel sheet, thin amorphous sheets are "prespooled" in multiple layered packages. The resulting thick package makes handling and installation of the core substantially more practical.

Clearly, the discussion of transformer cores reminds us that "electronic" materials are also "electromagnetic" in nature. Some of the magnetic terms used in our discussion that might not be familiar can be found in Chapter 13D on the related Web site.

Global political and economic forces, including continuing concerns over energy resources, are motivating electric-power companies to reduce costs and improve service. These changes must be balanced, however, against substantial environmental concerns. The Third Conference of the Parties of the United Nations Framework Convention on Climate Change, held in Kyoto, Japan, in December 1997, adopted an agreement known as the Kyoto Protocol to cut greenhouse gas emissions by 5% from 1990 levels. Table 14.6 indicates the potential benefits, in this regard, associated with amorphous metal distribution transformers. Note that in the United States alone, an energy equivalent of 70 million barrels of oil could be saved along with an attendant reduction in CO_2, NO_x, and SO_2 gases.

Amorphous metal transformers are often more expensive than silicon steel models. The amorphous metal units can, however, be more cost effective in many electric-power systems. Utility engineers commonly use a "loss-evaluation" method, which includes economic factors such as transformer loading patterns, energy cost, inflation, and interest rates. Their goal is to combine the initial transformer cost with the cost of operation, thus creating a total owning cost (TOC) defined by

$$TOC = BP + (F_{CL} \times CL) + (F_{LL} \times LL), \qquad (14.11)$$

where BP is the bid price, F_{CL} is the core loss factor, CL is the core loss, F_{LL} is the load loss factor, and LL is the load loss (defined as the energy loss by the system other than the transformer core). Table 14.7 shows that for this case the

TABLE 14.7

Economic Comparison of Traditional Electrical Steel and Fe-Based Amorphous Metal Transformers[a]		
Distribution transformer @ 60 Hz, 500 kW (15 kV/480–277 V)	**Amorphous metal core**	**Grain-oriented silicon steel core**
1. Core loss (W)	230	610
2. Core loss factor ($/W)	$5.50	$5.50
3. Load loss (W)	3,192	3,153
4. Load loss factor ($/W)	$1.50	$1.50
5. Efficiency (%)	99.6	99.4
6. Bid price	$11,500	$10,000
7. Core loss value (1×2)	$1,265	$3,355
8. Load loss value (3×4)	$4,788	$4,730
9. Total owing cost ($6 + 7 + 8$)	$17,558	$18,085

[a] After N. DeCristofaro, "Amorphous Metals in Electric-Power Distribution Applications," *MRS Bulletin 23*, 50 (1998).

amorphous metal transformer costs 15% more than its traditional counterpart but provides an overall 3% reduction in TOC.

LIGHT-EMITTING DIODE

Luminescence can be defined as light emission accompanying various forms of energy absorption. Electroluminescence is the term for electron-induced light emission. An important form of electroluminescence results when a forward-biased potential is applied across the *p–n* junction introduced in Section 13.5. Within a recombination region near the junction, electrons and holes can annihilate each other and emit photons of visible light. The emitted wavelength is given by

$$\lambda = hc/E_g, \tag{14.12}$$

where h is Planck's* constant ($= 0.6626 \times 10^{-33}$ J $\cdot$ s) and c is the speed of light ($= 0.2998 \times 10^9$ m/s), and E_g is the band gap in Joules.

The E_g and corresponding wavelength emitted are functions of the semiconductor composition, with a variety of common examples given in Table 14.8. As a practical matter, the electron transition is from a small range of energies at the lower end of the conduction band to a small range of energies at the upper end of the valence band (Figure 14.13). The resulting range of values for E_g can produce a **spectral width** of light on the order of a few nm. Figure 14.14 shows both surface- and edge-emitting configurations for these **light-emitting diodes** (LEDs). The construction of the ubiquitous digital display based on LEDs is illustrated in Figure 14.15. Many lasers provide the carrier waves for most optical fiber networks and, in fact, are important examples of LEDs.

---

*Max Karl Ernst Ludwig Planck (1858–1947), German physicist. His lifetime spanned the transition between the 19th and 20th centuries, which is symbolic of his contribution in bridging classical (19th-century) and modern (20th-century) physics. He introduced the relationship between photon energy and wavelength and the term *quantum* in 1900 while developing a successful model of the energy spectrum from a "blackbody radiator." The prestigious Max Planck Institutes in Germany bear his name.

TABLE 14.8

Common Light–Emitting Diode Compounds and Wavelengths[a]		
Compound	**Wavelength (nm)**	**Color**
GaP	565	Green
GaAsP	590	Yellow
GaAsP	632	Orange
GaAsP	649	Red
GaAlAs	850	Near IR
GaAs	940	Near IR
InGaAs	1,060	Near IR
InGaAsP	1,300	Near IR
InGaAsP	1,550	Near IR

[a] From R. C. Dorf, *Engineering Handbook*, CRC Press, Boca Raton, FL, 1993, p. 751.

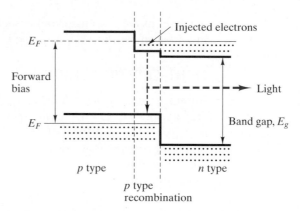

FIGURE 14.13 *Schematic illustration of the energy band structure for a light-emitting diode (LED).* *(After R. C. Dorf,* Electrical Engineering Handbook, *CRC Press, Boca Raton, FL, 1993, p. 750.)*

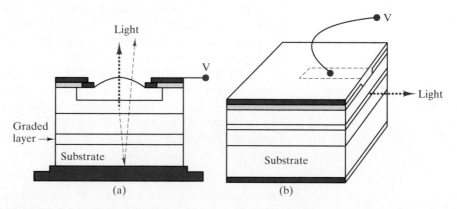

FIGURE 14.14 *Schematic illustration of (a) surface-emitting and (b) edge-emitting light-emitting diodes (LEDs).* *(After R. C. Dorf,* Electrical Engineering Handbook, *CRC Press, Boca Raton, FL, 1993, p. 750.)*

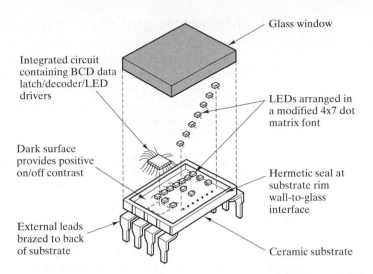

Integrated circuit containing BCD data latch/decoder/LED drivers

Glass window

LEDs arranged in a modified 4x7 dot matrix font

Dark surface provides positive on/off contrast

Hermetic seal at substrate rim wall-to-glass interface

External leads brazed to back of substrate

Ceramic substrate

FIGURE 14.15 *Schematic illustration of a digital display employing an array of light-emitting diodes (LED). (From S. Gage et al., Optoelectronics/Fiber-Optics Applications Manual, 2nd ed., Hewlett-Packard/McGraw-Hill, New York, 1981.)*

A greater appreciation of the optical behavior described in this case study can be gained from Chapter 13B on the related Web site.

EXAMPLE 14.3

Relative to a GaAs LED, calculate the photon wavelength that corresponds to the GaAs band gap given in Table 13.5.

SOLUTION

Using Equation 14.12 and the GaAs band gap from Table 13.5, we have

$$\lambda = hc/E_g$$
$$= ([0.663 \times 10^{-33} \text{ J} \cdot \text{s}][0.300 \times 10^9 \text{ m/s}]/[1.47 \text{ eV}])$$
$$\times (6.242 \times 10^{18} \text{ eV/J})$$
$$= 844 \times 10^{-9} \text{ m} = 844 \text{ nm}.$$

Note. This wavelength is in the infrared region of the electromagnetic spectrum. Variations in the temperature and chemical composition of the semiconductor can be used to increase the band gap, thereby reducing the wavelength into the visible light region.

PRACTICE PROBLEM 14.3

As noted in Example 14.3, the band gaps of semiconductors are a function of composition. By adding some GaP to the GaAs, the band gap can be increased to 1.78 eV. Calculate the photon wavelength that corresponds to this larger band gap.

14.4 | Materials and Our Environment

The application of engineering materials can be limited by reaction with their environment. As engineers, we are increasingly responsible for ensuring that, in turn, engineering materials do not adversely affect the environment in which all of us must live. Recycling is one effective way to limit the impact of materials on the environment.

ENVIRONMENTAL DEGRADATION OF MATERIALS

A link to the National Association of Corrosion Engineers (NACE) Web site on the related Web site is an especially good portal to the subject of corrosion.

In general, metals and alloys form stable oxide compounds under exposure to air at elevated temperatures. A few notable exceptions such as gold are highly prized. The stability of metal oxides is demonstrated by their relatively high melting points compared to the pure metal. For example, Al melts at 660°C, whereas Al_2O_3 melts at 2,054°C. Even at room temperature, thin surface layers of oxide can form on some metals. For some metals, the reactivity with atmospheric oxygen, or **oxidation**, can be a primary limitation to their engineering application. For others, surface oxide films can protect the metal from more serious environmental attack.

There are four mechanisms commonly identified with metal oxidation, as illustrated in Figure 14.16. The oxidation of a given metal or alloy can usually be characterized by one of these four diffusional processes, including (a) an "unprotective," porous oxide film through which molecular oxygen (O_2) can continuously pass and react at the metal-oxide interface; (b) a nonporous film through which cations diffuse in order to react with oxygen at the outer (air-oxide) interface; (c) a nonporous film through which O^{2-} ions diffuse in order to react with the metal at the metal-oxide interface; and (d) a nonporous film in which both cations and O^{2-} nions diffuse at roughly the same rate, causing the oxidation reaction to occur within the oxide film rather than at an interface. At this point, you may wish to review the discussion of ionic diffusion in Section 5.3. You should also note

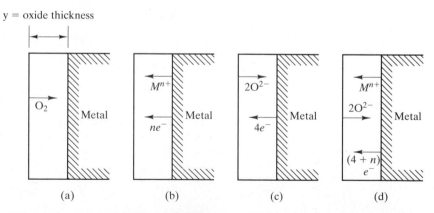

FIGURE 14.16 *Four possible metal oxidation mechanisms. (a) "Unprotective" film is sufficiently porous to allow continuous access of molecular O_2 to the metal surface. Mechanisms (b)–(d) represent nonporous films that are "protective" against O_2 permeation. In (b), cations diffuse through the film reacting with oxygen at the outer surface. In (c), O^{2-} ions diffuse to the metal surface. In (d), both cations and anions diffuse at nearly equal rates, leading to the oxidation reaction occurring within the oxide film.*

TABLE 14.9

Pilling–Bedworth Ratios
for Various Metal Oxides

Protective oxides	Nonprotective oxides
Be—1.59	Li—0.57
Cu—1.68	Na—0.57
Al—1.28	K—0.45
Si—2.27	Ag—1.59
Cr—1.99	Cd—1.21
Mn—1.79	Ti—1.95
Fe—1.77	Mo—3.40
Co—1.99	Hf—2.61
Ni—1.52	Sb—2.35
Pd—1.60	W—3.40
Pb—1.40	Ta—2.33
Ce—1.16	U—3.05
	V—3.18

Source: B. Chalmers, *Physical Metallurgy,* John Wiley & Sons, Inc., New York, 1959.

that charge neutrality requires cooperative migration of electrons with the ions in mechanisms (b)–(d). It may not be obvious that the diffusing ions in mechanisms (b)–(d) are the result of two distinct mechanisms: $M \rightarrow M^{n+} + ne^-$ at the metal-oxide interface and $O_2 + 4e^- \rightarrow 2O^{2-}$ at the air-oxide interface.

The tendency of a metal to form a protective oxide coating is indicated by an especially simple parameter known as the **Pilling–Bedworth[*] ratio**, R, given as

$$R = \frac{Md}{am\,D},\tag{14.13}$$

where M is the molecular weight of the oxide (with formula M_aO_b and density D) and m is the atomic weight of the metal (with density d). Careful analysis of Equation 14.13 reveals that R is simply the ratio of the oxide volume produced to the metal volume consumed. For $R < 1$, the oxide volume tends to be insufficient to cover the metal substrate. The resulting oxide coating tends to be porous and unprotective. For R equal to or slightly greater than 1, the oxide tends to be protective. For R greater than 2, there are likely to be large compressive stresses in the oxide that cause the coating to buckle and flake off, a process known as **spalling**. The general utility of the Pilling–Bedworth ratio to predict the protective nature of an oxide coating is illustrated in Table 14.9. The protective oxides generally have R values between 1 and 2. The nonprotective oxides generally have R values less than 1 or greater than 2. There are exceptions, such as Ag and Cd. A number of factors, in addition to R, must be favorable to produce a protective coating. Similar coefficients of thermal expansion and good adherence are such additional factors.

Two of our most familiar protective oxide coatings in everyday applications are those on anodized aluminum and stainless steel. **Anodized aluminum** represents a broad family of aluminum alloys with Al_2O_3 as the protective oxide. However, the oxide coating is produced in an acid bath rather than by routine atmospheric oxidation. The stainless steels were introduced in Section 11.1. The critical alloy addition is chromium, and the protective coating is an iron–chromium oxide. We shall refer to these coatings later as various forms of corrosion protection are reviewed.

Before leaving the subject of oxidation, we must note that oxygen is not the only chemically reactive component of environments to which engineering materials are subjected. Under certain conditions, atmospheric nitrogen can react to form nitride layers. A more common problem is the reaction of sulfur from hydrogen sulfide and other sulfur-bearing gases from various industrial processes. In jet engines, even nickel-based superalloys show rapid reaction with sulfur-containing combustion products. Cobalt-based superalloys are alternatives, although cobalt sources are limited. An especially insidious example of atmospheric attack is **hydrogen embrittlement**, in which hydrogen gas, also commonly found in a variety of industrial processes, permeates into a metal such as titanium, creating substantial internal pressure and even reacting to form brittle hydride compounds. The result, in either case, is a general loss of ductility.

Corrosion is the dissolution of a metal into an aqueous environment. The metal atoms dissolve as ions. A simple model of one form, **galvanic corrosion,**

[*]N. B. Pilling and R. E. Bedworth, *J. Inst. Met.* 29, 529 (1923).

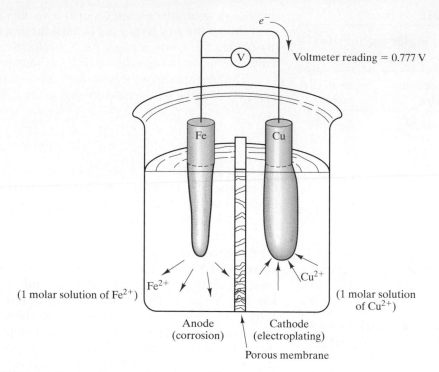

FIGURE 14.17 *A galvanic cell is produced by two dissimilar metals. The more "anodic" metal corrodes.*

is given in Figure 14.17. In this **electrochemical cell**, chemical change (e.g., the corrosion of the anodic iron) is accompanied by an electrical current. The metal bar on the left side of the electrochemical cell is an **anode**, which is a metal that supplies electrons to the external circuit and dissolves, or corrodes. The **anodic reaction** can be given as

$$Fe^0 \rightarrow Fe^{2+} + 2e^-. \tag{14.14}$$

This reaction is driven by the tendency to equilibrate the ionic concentration in both sides of the overall cell. The porous membrane allows the transport of Fe^{2+} ions between the two halves of the cell (thereby completing the total electrical circuit). The overall electrochemical cell is generated by two different metals even though each is surrounded by an equal concentration of its ions and aqueous solution. In this **galvanic** * **cell**, the iron bar, surrounded by a 1-molar solution of Fe^{2+}, is the anode and is corroded. (Recall that a 1-molar solution contains 1 gram atomic

---

*Luigi Galvani (1737–1798), Italian anatomist. In Section 5.3 we saw that a major contribution to materials science came from the medical sciences (the diffusional laws of Adolf Fick). In a similar way, we owe much of our basic understanding of electricity to Galvani, a professor of anatomy at the University of Bologna. He used the twitching of frog leg muscles to monitor electrical current. To duplicate Benjamin Franklin's lightning experiment, Galvani laid the frog leg muscles on brass hooks near an iron lattice work. During a thunderstorm, the muscles did indeed twitch, demonstrating again the electrical nature of lightning. But Galvani noticed that they also twitched whenever the muscles simultaneously touched the brass and iron. He thus identified the *galvanic cell*. When an instrument was developed to measure electrical current in 1820, Ampère (see Section 13.1) suggested that it be known as a galvanometer.

weight of ions in 1 liter of solution.) The copper bar, surrounded by a 1 molar solution of Cu^{2+}, is the **cathode**, and Cu^0 plates out on the bar. The **cathodic reaction** is

$$Cu^{2+} + 2e^- \rightarrow Cu^0. \tag{14.15}$$

The driving force for the overall cell of Figure 14.17 is the relative tendency for each metal to ionize. The net flow of electrons from the iron bar to the copper bar is a result of the stronger tendency of iron to ionize. A voltage of 0.777 V is associated with the overall electrochemical process. Each side of the electrochemical cell is appropriately termed a **half cell**, and Equations 14.14 and 14.15 are each **half-cell reactions**.

Because of the common occurrence of galvanic cells, a systematic collection of voltages associated with half-cell reactions has been made. This **electromotive force** (emf) **series** is given in Table 14.10. Of course, half-cells exist only as pairs. All emf values in Table 14.10 are defined relative to a reference electrode that, by convention, is taken as the ionization of H_2 gas over a platinum surface. The metals toward the bottom of the emf series are said to be more *active* (i.e., anodic). Those toward the top are more *noble* (i.e., cathodic). The total voltage measured in Figure 14.17 is the difference of two half-cell potentials $[+0.337\ V - (-0.440\ V) = 0.777\ V]$.

While a useful guide to galvanic corrosion tendencies, Table 14.10 is somewhat idealistic. Engineering designs seldom involve pure metals in standard

TABLE 14.10

Electromotive Force Series		
	Metal–metal ion equilibrium (unit activity)	**Electrode potential versus normal hydrogen electrode at 25°C (V)**
	$Au–Au^{3+}$	+1.498
↑	$Pt–Pt^{2+}$	+1.200
Noble or	$Pd–Pd^{2+}$	+0.987
cathodic	$Ag–Ag^+$	+0.799
	$Hg–Hg_2^{2+}$	+0.788
	$Cu–Cu^{2+}$	+0.337
	$H_2–H^+$	0.000
	$Pb–Pb^{2+}$	−0.126
	$Sn–Sn^{2+}$	−0.136
	$Ni–Ni^{2+}$	−0.250
	$Co–Co^{2+}$	−0.277
	$Cd–Cd^{2+}$	−0.403
	$Fe–Fe^{2+}$	−0.440
	$Cr–Cr^{3+}$	−0.744
	$Zn–Zn^{2+}$	−0.763
Active or	$Al–Al^{3+}$	−1.662
anodic	$Mg–Mg^{2+}$	−2.363
↓	$Na–Na^+$	−2.714
	$K–K^+$	−2.925

Source: After A. J. de Bethune and N. A. S. Loud, as summarized in M. G. Fontana and N. D. Greene, *Corrosion Engineering*, 2nd ed., John Wiley & Sons, Inc., New York, 1978.

TABLE 14.11

Galvanic Series in Seawater[a]

	Platinum
↑	Gold
Noble or cathodic	Graphite
	Titanium
	Silver

⌈ Chlorimet 3 (62 Ni, 18 Cr, 18 Mo)
⌊ Hastelloy C (62 Ni, 17 Cr, 15 Mo)

⌈ 18–8 Mo stainless steel (passive)
| 18–8 stainless steel (passive)
⌊ Chromium stainless steel 11–30% Cr (passive)

⌈ Inconel (passive) (80 Ni, 13 Cr, 7 Fe)
⌊ Nickel (passive)

Silver solder

⌈ Monel (70 Ni, 30 Cu)
| Cupronickels (60–90 Cu, 40–10 Ni)
| Bronzes (Cu–Sn)
| Copper
⌊ Brasses (Cu–Zn)

⌈ Chlorimet 2 (66 Ni, 32 Mo, 1 Fe)
⌊ Hastelloy B (60 Ni, 30 Mo, 6 Fe, 1 Mn)

⌈ Inconel (active)
⌊ Nickel (active)

Tin
Lead
Lead–tin solders

⌈ 18–8 Mo stainless steel (active)
⌊ 18–8 stainless steel (active)

Ni-resist (high-nickel cast iron)
Chromium stainless steel, 13% Cr (active)

⌈ Cast iron
⌊ Steel or iron

	2024 aluminum (4.5 Cu, 1.5 Mg, 0.6 Mn)
Active or anodic	Cadmium
↓	Commercially pure aluminum (1100)
	Zinc
	Magnesium and magnesium alloys

[a](Active) and (passive) designations indicate whether or not a passive, oxide film has been developed on the alloy surface.

Source: From tests conducted by International Nickel Company and summarized in M. G. Fontana and N. D. Greene, *Corrosion Engineering,* 2nd ed., John Wiley & Sons, Inc., New York, 1978.

concentration solutions. Instead, they involve commercial alloys in various aqueous environments. For such cases, a table of standard voltages is less useful than a simple, qualitative ranking of alloys by their relative tendencies to be active or noble. As an example, the **galvanic series** in seawater is given in Table 14.11. This listing is a useful guide for the design engineer in predicting the relative behavior of

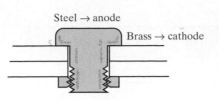

FIGURE 14.18 *A steel bolt in a brass plate creates a galvanic cell analogous to the model system in Figure 14.17.*

These illustrations show how environmental challenges to engineering designs can occur at both the visible millimeter to meter scale (Figure 14.18) and at the invisible microstructural scale (Figure 14.19).

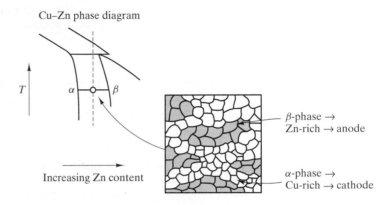

FIGURE 14.19 *A galvanic cell can be produced on the microscopic scale. Here, β-brass (bcc structure) is zinc rich and anodic relative to α-brass (fcc structure), which is copper rich.*

adjacent materials in marine applications. Close inspection of this galvanic series indicates that alloy composition can radically affect the tendency toward corrosion. For example, (plain carbon) steel is near the active end of the series, whereas some passive stainless steels are among the most noble alloys. Steel and brass are fairly widely separated, and Figure 14.18 shows a classic example of galvanic corrosion in which a steel bolt is unwisely selected for securing a brass plate in a marine environment. Figure 14.19 reminds us that two-phase microstructures can provide a small-scale galvanic cell, leading to corrosion even in the absence of a separate electrode on the macroscopic scale.

A more comprehensive set of examples of corrosion cells is given in Chapter 14A on the related Web site. For now, we need to consider the general strategy for preventing unwanted corrosion in engineering designs. We need not be surprised that corrosion is now an annual multi-billion-dollar expense to modern society. Even thin layers of condensed atmospheric moisture are sufficient aqueous environments for metallic alloys to lead to appreciable corrosion. A major challenge to all engineers employing metals in their designs is to prevent corrosive attack. When complete prevention is impossible, losses should be minimized. Consistent with the wide variety of corrosion problems, a wide range of preventative measures is available.

Our primary means of corrosion prevention is **materials selection**. Boating enthusiasts quickly learn to avoid steel bolts for brass hardware. A careful application of the principles of this chapter allows the materials engineer to find those alloys least susceptible to given corrosive environments. In a similar fashion, **design selection** can minimize damage. Threaded joints and similar high-stress regions are to be avoided when possible. When galvanic couples are required, a small-area anode next to a large-area cathode should be avoided. The resulting large current density at the anode accelerates corrosion.

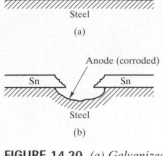

FIGURE 14.20 *(a) Galvanized steel consists of a zinc coating on a steel substrate. Since zinc is anodic to iron, a break in the coating does not lead to corrosion of the substrate. (b) In contrast, a more noble coating such as "tin plate" is protective only as long as the coating is free of breaks. At a break, the anodic substrate is preferentially attacked.*

TABLE 14.12

Protective Coatings for Corrosion Prevention	
Category	**Examples**
Metallic	Chrome plating
	Galvanized steel
Ceramic	Stainless steel
	Porcelain enamel
Polymeric	Paint

When an alloy must be used in an aqueous environment in which corrosion could occur, additional techniques are available to prevent degradation. **Protective coatings** provide a barrier between the metal and its environment. Table 14.12 lists various examples. These coatings are divided into three categories, corresponding to the fundamental structural materials: metals, ceramics, and polymers. Chrome plating has traditionally been used on decorative trim for automobiles. **Galvanized steel** operates on a somewhat different principle. As seen in Figure 14.20, protection is provided by a zinc coating. Because zinc is anodic relative to the steel, any break in the coating does not lead to corrosion of the steel, which is cathodic and preserved. Zinc can be contrasted with more noble coatings (e.g., tin on steel in Figure 14.20b), in which a break leads to accelerated corrosion of the substrate. As discussed earlier, stable oxide coatings on a metal can be protective. The (Fe, Cr) oxide coating on stainless steel is a classic example. However, Figure 14.21 illustrates a limitation for this material. Excessive heating (e.g., welding) can cause precipitation of chromium carbide at grain boundaries. The result is chromium

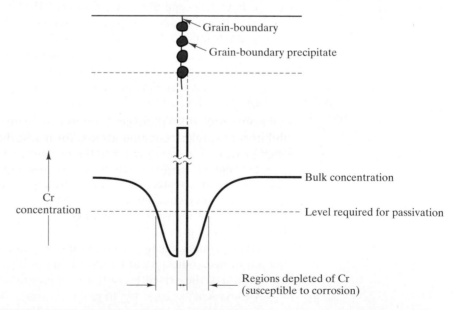

FIGURE 14.21 *Heating a stainless steel can cause precipitation of chromium carbide particles, leaving adjacent regions of the microstructure depleted in chromium and thereby susceptible to corrosion. This effect is the basis of the common warning to avoid welding of stainless-steel components.*

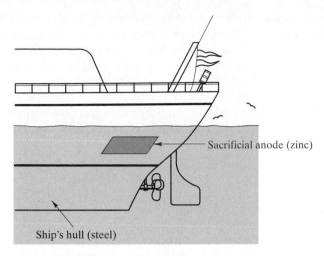

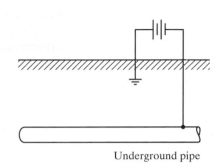

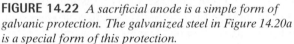

FIGURE 14.22 *A sacrificial anode is a simple form of galvanic protection. The galvanized steel in Figure 14.20a is a special form of this protection.*

FIGURE 14.23 *An impressed voltage is a form of galvanic protection that counters the corroding potential.*

depletion adjacent to the precipitates and susceptibility to corrosive attack in that area. An alternative to an oxide reaction layer is a deposited ceramic coating. **Porcelain enamels** are silicate glass coatings with thermal-expansion coefficients reasonably close to those of their metal substrates. Polymeric coatings can provide similar protection, usually at a lower cost. Paint is our most common example. (We should distinguish enamel paints, which are organic polymeric coatings, from the porcelain enamels, which are silicate glasses.)

A galvanized steel coating (see Table 14.12) is a specialized example of a **sacrificial anode**. A general, noncoating example, given in Figure 14.22, is one type of **galvanic protection**. Another is the use of an **impressed voltage**, in which an external voltage is used to oppose the one due to the electrochemical reaction. The impressed voltage stops the flow of electrons needed for the corrosion reaction to proceed. Figure 14.23 illustrates a common example of this technique.

A final approach to corrosion prevention is the use of an **inhibitor**, defined as a substance used in small concentrations that decreases the rate of corrosion. Most inhibitors are organic compounds that form adsorbed layers on the metal surface, which provides a system similar to the protective coatings discussed earlier.

The high electrical resistivity of ceramics and polymers removes them from consideration of corrosive mechanisms. The use of ceramic and polymeric protective coatings on metals leads to the general view of these nonmetallic materials as "inert." In fact, any material will undergo chemical reaction under suitable circumstances. As a practical matter, ceramics and polymers are relatively resistant to the environmental reactions associated with typical metals. Although electrochemical mechanisms are not significant, some direct chemical reactions can limit utility. A good example was the reaction of H_2O with silicates, leading to the phenomenon of static fatigue (see Section 8.3). Also, ceramic refractories are selected, as much as possible, to resist chemical reaction with molten metals that they contain in metal-casting processes.

We have concentrated on chemical reactions between materials and their environments. Increasingly, materials are also subjected to radiation fields. Nuclear power generation, radiation therapy, and communication satellites are a few of the applications in which materials must withstand severe radiation environments. Damaging forms of **radiation** include the high energy (short wavelength) end of the electromagnetic radiation spectrum (see Figure 3.30) and subatomic particles such as neutrons and electrons. The response of different materials to a given type of radiation varies considerably. Similarly, a given material can be affected quite differently by different types of radiation. Radiation can also strongly affect the performance of electrical and magnetic materials. Radiation damage to semiconductors used in communication satellites can be a major limitation for their design applications.

As with radiation damage, **wear** is generally a physical (rather than chemical) form of material degradation. Specifically, wear can be defined as the removal of surface material as a result of mechanical action. The amount of wear need not be great to be relatively devastating. (A 1,500-kg automobile can be "worn out" as a result of the loss of only a few grams of material from surfaces in sliding contact.) The systematic study of wear has revealed several key aspects of this phenomenon. Four main forms of wear have been identified: (1) **Adhesive wear** occurs when two smooth surfaces slide over each other and fragments are pulled off one surface and adhere to the other. The adjective for this category comes from the strong bonding or "adhesive" forces between adjacent atoms across the intimate contact surface. (2) **Abrasive wear** occurs when a rough, hard surface slides on a softer surface. The result is a series of grooves in the soft material and the resulting formation of wear particles. (3) **Surface-fatigue wear** occurs during repeated sliding or rolling over a track. Surface or subsurface crack formation leads to breakup of the surface. (4) **Corrosive wear** takes place with sliding in a corrosive environment and, of course, adds chemical degradation to the physical effects of wear. The sliding action can break down passivation layers and, thereby, maintain a high corrosion rate. Mechanical wear in combination with aqueous corrosion provides additional forms of environmental degradation. Examples include **stress-corrosion cracking** and **corrosion fatigue**, first introduced in Chapter 8.

Nonmetallic structural materials are frequently selected for their superior wear resistance. High-hardness ceramics generally provide excellent resistance to wear, for example, aluminum oxide, partially stabilized zirconia, and tungsten carbide (as a coating). Ceramic-on-ceramic hip prostheses, as noted in Section 14.2, are good examples. Polymers and polymer-matrix composites are increasingly replacing metals in bearings, cams, gears, and other sliding components. PTFE is an example of a self-lubricating polymer that is widely used for its wear resistance. Fiber reinforcement of PTFE improves other mechanical properties without sacrificing the wear performance.

EXAMPLE 14.4

Given that the density of Cu_2O is 6.00 Mg/m^3, calculate the Pilling–Bedworth ratio for copper.

SOLUTION
The expression for the Pilling–Bedworth ratio is given by Equation 14.13:

$$R = \frac{Md}{am\,D}.$$

Additional data, other than the density of Cu_2O, are available in Appendix 1:

$$R = \frac{[2(63.55) + 16.00](8.93)}{(2)(63.55)(6.00)}$$
$$= 1.68,$$

which is the value in Table 14.9.

EXAMPLE 14.5

Suppose that you set up a laboratory demonstration of a galvanic cell similar to that shown in Figure 14.17 but are forced to use zinc and iron for the electrodes. **(a)** Which electrode will be corroded? **(b)** If the electrodes are immersed in 1-molar solutions of their respective ions, what will be the measured voltage between the electrodes?

SOLUTION

(a) Inspection of Table 14.10 indicates that zinc is anodic relative to iron. Therefore, zinc will be corroded.

(b) Again, using Table 14.10 the voltage will be

$$\text{voltage} = (-0.440 \text{ V}) - (-0.763 \text{ V})$$
$$= 0.323 \text{ V}.$$

PRACTICE PROBLEM 14.4

(a) The Pilling–Bedworth ratio for copper is calculated in Example 14.4. In this case, we assume that cuprous oxide, Cu_2O, is formed. Calculate the Pilling–Bedworth ratio for the alternate possibility that cupric oxide, CuO, is formed. (The density of CuO is 6.40 Mg/m^3.) **(b)** Do you expect CuO to be a protective coating? Briefly explain.

PRACTICE PROBLEM 14.5

In Example 14.5, we analyze a simple galvanic cell composed of zinc and iron electrodes. Make a similar analysis of a galvanic cell composed of copper and zinc electrodes immersed in 1-molar ionic solutions.

ENVIRONMENTAL ASPECTS OF DESIGN

In Section 14.3, we looked in some detail at a case study in which conventional silicon steel was replaced by amorphous metals in transformer cores. The driving forces for this materials substitution were seen to be economic and environmental. The reduced core loss for amorphous metals led to substantial reductions in both energy costs and pollution emissions.

TABLE 14.13

Major Environmental Legislation in the United States[a]	
Clean Air Act (CAA), 1970	Controlled chemicals list, which allowed EPA to assess risk and act to prevent harm. EPA is allowed to enforce the law without proof of harm. In 1990, the Maximum Achievable Control Technology (MACT) was established in place of previous public health standards. Emission levels are set by EPA or the states (reauthorized 1996, amendments 1977, 1990, 1995).
Clean Water Act (CWA), 1972	Gave the EPA authority to control industrial discharge to water by imposing discharge requirements to industry, placing special controls on toxic discharge and requiring a variety of safety and construction measures to reduce spills to waterways.
Toxic Substances Control Act (TSCA), 1976	All new (toxic) substances or new uses of substances entering the marketplace are evaluated for health and environmental effects.
Resource Conservation and Recovery Act (RCRA), 1976	Clarified waste disposal issues and established cradle-to-grave control of hazardous waste.
Comprehensive Environmental Response, Compensation, and Liability (CERCLA), 1980	Reauthorized and amended RCRA, clarified responsibility and liability of parties involved in hazardous material management, established Superfund to pay for remediation (by 1984, the EPA identified 378,000 sites requiring corrective action).
Hazardous and Solid Waste Amendments (HSWA), 1984	Step up national efforts to improve hazardous waste management.
Emergency Planning and Community Right to Know Act (EPCRA), 1986, also known as Superfund Amendments and Reauthorization Act (SARA)	Section 313 of Title III of this act includes the Toxics Release Inventory (TRI), which makes hazardous waste and toxins a matter of public record. This act sets fines for violations.
Pollution Prevention Act (PPA), 1990	Establishes hierarchy of Pollution Prevention (PP or P^2) and places source reduction at the head of the list for pollution prevention. Firms that must report under TRI must also report level of pollution-prevention accomplishments.
Examples of Nonlegislative EPA Measures/Actions	
Environmental auditing, 1986	The EPA attempts to formalize procedures for environmental auditing by developing a generic protocol for environmental auditing.
Industrial Toxics Project (33/50 Program), 1991	Companies with large releases of the 17 chemicals of the TRI reported with largest volume releases were asked to target 33% reduction by 1992 and 50% reduction by 1995. This is a voluntary measure. There are hopes to foster a pollution-prevention ethic in business.

[a] After S. T. Fleischmann, "Environmental Aspects of Design," in *ASM Handbook, Vol. 20: Materials Selection and Design*, ASM International, Materials Park, OH, 1997, pp. 131–138.

Public policy is increasingly dictating reduced pollution emissions, regardless of whether this reduction correlates with economic benefits to industry. In the United States, the Environmental Protection Agency (EPA) has numerous regulations regarding environmental performance. Major environmental legislation of the past 3 decades is summarized in Table 14.13. Enforcement of these regulations by the EPA falls under the criminal code of law. Similar regulations also exist at the state level. Incentives for environmental regulation compliance are sufficiently great that systematic strategies have evolved for ensuring that the engineering design process incorporates regulatory compliance. Specifically, **design for the environment** (DFE) includes carrying out a **life-cycle assessment** (LCA), a "cradle-to-grave" evaluation of environmental and energy impacts of a given product design. An **environmental impact assessment** (EIA) is related to the LCA and is illustrated in Figure 14.24. The EIA is intended to structure a study of the impact of emissions so that potential major problems are identified. Clearly, environmental regulations have spawned a substantial new vocabulary. Some have labeled this overall exercise of environmentally sensitive design as simply **green engineering**.

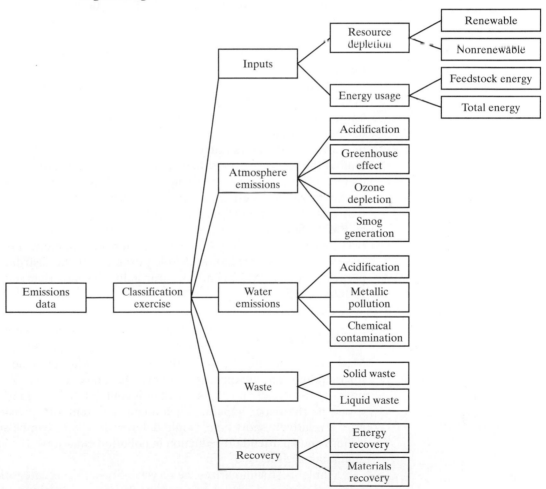

FIGURE 14.24 *Schematic illustration of an environmental impact assessment (EIA) of emissions data. [After L. Holloway et al.,* Materials and Design *15, 259 (1994).]*

TABLE 14.14

Chemicals Included under the National Emission Standard for Hazardous Air Pollutants (NESHAPS)[a]	
Asbestos	Rn–222
Benzene	Radionuclides
Beryllium	Copper
Coke oven emissions	Nickel
Mercury	Phenol
Vinyl chloride	Zinc and zinc oxide
Inorganic arsenic	

[a] After S. T. Fleischmann, "Environmental Aspects of Design," in *ASM Handbook*, Vol. 20: *Materials Selection and Design*, ASM International, Materials Park, OH, 1997, pp. 131–138.

Representative of specific challenges to materials engineers is the National Emission Standard for Hazardous Air Pollutants (NESHAPS), which is section 112 under section 40 of the Clean Air Act (see Table 14.13). NESHAPS regulates the emissions of the chemicals listed in Table 14.14, various combinations of which can be by-products of many of the common materials-processing technologies outlined earlier in this book.

Airborne particles known as atmospheric **aerosols** are an increasing focus of environmental regulators. The deadly potential of such pollution has been well known for some time. Especially infamous was a weeklong episode in London during 1952 in which thousands of children and elderly people died as a result of high levels of soot and sulfur dioxide. Since the Clean Air Act of 1970, the EPA has limited the levels of atmospheric particles. Originally, the rules covered particle diameters up to 50 μm. Health studies showed, however, that such coarse particles tend to be safely expelled from the body's upper airways. Since 1987, the EPA has only restricted particulate matter less than 10 μm in diameter, referred to simply as **PM10**. Several health studies in the 1990s have suggested that the greatest risk comes from an even finer scale of aerosols; namely, those with diameters less than 2.5 μm (**PM2.5**). A detailed example of the distribution of fine metal particles within a PM2.5 sample from an industrialized urban setting is summarized in Table 14.15.

RECYCLING

Materials that are highly inert in their environmental surroundings are candidates for **recycling**. The ubiquitous aluminum beverage can is a prime example. The opposite case would be a material that is completely biodegradable. An example would be the paper napkin, which might be a temporary visual pollutant, but, after a relatively short time, would deteriorate. There can be substantial energy savings and an attendant reduction in pollution emissions when materials such as aluminum are recycled.

Table 14.16 summarizes the recyclability of the six categories of engineering materials discussed in this book. Among the metals, aluminum alloys are excellent examples of recycling. The energy requirement for aluminum recycling is small in comparison to that for initial production. Many commercial aluminum alloys have

TABLE 14.15

Distribution of Metallic ≤ 2.5 μm Particulate Matter (PM2.5) from an Urban Setting[a]

	Concentration (ng/m^3) for size range:				
	0.069–0.24 μm	0.24–0.34 μm	0.34–0.56 μm	0.56–1.15 μm	1.15–2.5 μm
Vanadium	2.5	6.10	10.5	12.2	8.60
Nickel	1.3	4.40	7.7	4.5	0.50
Zinc	17.6	46.30	140.4	189.4	39
Selenium	<0.3	0.32	3.0	1.4	0.65
Lead	71.4	47.60	59.9	69.9	25.40

[a] Averaged data collected over a 6-day period in November 1987 in Long Beach, California, by T. A. Cahill et al., University of California, Davis.

TABLE 14.16

Recycling Characteristics of the Five Materials Categories

Category	Characteristics
Metals	Many commercial alloys are recyclable. Aluminum cans are prime examples. Difficulty associated with dissimilar material design configurations.
Ceramics and glasses	Crystalline ceramics generally not recycled. Glass containers widely recycled. Greater difficulty associated with combinations of dissimilar materials.
Polymers	Recycling code system in place for thermoplastic polymers. Thermosetting polymers generally not recycled. Elastomers, other than thermoplastic variety, are generally difficult to recycle.
Composites	The inherent fine-scale combination of dissimilar materials makes recycling impractical.
Semiconductors	Wafer reclaiming (stripping away surface circuitry) is widely practiced. Associated recycling of other solid and chemical waste is done at fabrication laboratories.

been designed to accommodate impurity contamination. Many other alloys, both ferrous and nonferrous, are recyclable. Difficult-to-recycle metals are often the result of design configurations more than the nature of an isolated alloy. Examples include the galvanized (zinc-coated) steel introduced earlier in this section and rivets fastened to nonmetallic components. In general, dissimilar material combinations are a challenge for recycling.

Glass containers are as well known for recycling as are aluminum cans. Glass-container manufacturing routinely incorporates a combination of recycled glass (termed "cullet") and raw materials (sand and carbonates of sodium and calcium). This use of recycled glass provides increased production rates and reduced pollution emissions. On the other hand, care is required to sort the recycled glass by

THE MATERIAL WORLD

Lead-Free Solder

The widespread use of solder to join components in contemporary electronics has created a major environmental concern. Most of these solder joins contain lead, which has resulted in the routine distribution of this toxic element in homes and offices and has led to concentrated levels of lead at waste-disposal sites. In response to this concern, the European Union and Japan have produced legislation to severely restrict the use of lead solders. In the European Union, the Restriction on Hazardous Substances (RoHS) Directive has barred the selling of products with leaded solders since July 1, 2006. In the European Union and Japan, directives regarding waste electrical and electronics equipment require the manufacturer to be responsible for recycling its products. These directives produce a substantial, additional cost for extracting lead solder and other hazardous materials from the discarded assembly.

Manufacturers with significant global sales are effectively required to convert their solder technology to a lead-free material due to the significant fraction of the global market represented by Europe and Japan. The nature of traditional lead-containing solders was introduced in Chapter 9 relative to the discussion of the lead–tin phase diagram. Of course, material substitutions would be expected to duplicate that nature of lead as much as possible without the use of lead in the alloy compositions. The lead-free solders should also be free of any other elements that are toxic to humans. The material substitutions are simplified if the solders and resulting joins are not covered by patents. By choosing substitute alloys with comparable material properties, the replacement solders should have minimal impact on the current manufacturing processes. Finally, acceptable alloys and the overall joining processes should both have modest costs.

As an example, a senior design team at the University of California, Davis, working in conjunction with Agilent Technologies, evaluated a variety of candidate materials for replacing a eutectic 62Sn–36Pb–2Ag solder in a 50-GHz microwave amplifier. In addition to evaluating mechanical properties by a pull test and performing microstructural examinations by optical and scanning electron microscopy, the team tested furnace temperature profiles and package assembly techniques at the Agilent Technologies facility in Santa Rosa, California. They concluded that a eutectic 90Sn–10Au alloy was a successful replacement for the lead-containing solder in this application. Although gold is effectively replacing lead, the relatively small amount of gold keeps the cost acceptably low.

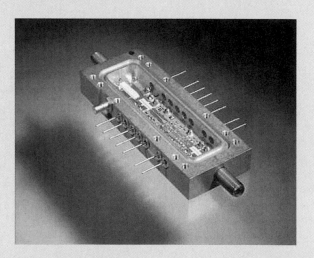

Typical microcircuit package in which a lead-containing solder is being replaced by a lead-free alloy. (Courtesy of Michael T. Powers, Agilent Technologies.)

type (container versus plate) and color (clear versus various colors). Also, glass containers with polymeric coatings are additional examples of dissimilar material combinations that make recycling more difficult. In contrast to glasses, crystalline ceramic products are, in general, not recycled.

An elaborate recycling code system is now routinely noted on a variety of polymeric products, especially food and beverage containers. As a practical matter, recycling of polymers is practical only for thermoplastic materials, such as PET, polyethylene (PE), and polypropylene (PP). The presence of fillers that are difficult to separate out presents a challenge to recycling. Thermosetting polymers are in general not amenable to recycling. Rubber, once vulcanized, behaves as a thermosetting polymer and is not generally recyclable. Rubber tires, in fact, pose a major challenge to landfills. A major benefit of thermoplastic elastomers is, as noted in Section 12.1, their recyclability.

Composite materials, by definition, are a fine-scale combination of different material components and are impractical to recycle.

The substantial demands and related expense in producing high-quality silicon wafers for the semiconductor industry have led to a substantial market for wafer reclaiming. The surface circuitry produced by the elaborate techniques described in Section 13.5 can be stripped away by mechanical and chemical means to leave behind a high-quality, albeit thinner wafer suitable for test wafer applications. In addition, semiconductor fabrication operations are scrutinized by regulatory agencies to maximize both solid and chemical waste recycling. Large volumes of polymeric packaging materials are routinely recycled at semiconductor fabrication laboratories ("fab labs").

Summary

The many properties of materials defined in the first 13 chapters of this text become the design parameters to guide engineers in the selection of materials for a given engineering design. These engineering design parameters are often dependent on the processing of the material.

In selecting structural materials, we are faced first with a competition among the five types outlined in Chapters 11 and 12. Once a given category is chosen, a specific material must be identified as an optimal choice. In general, a balance must be reached between strength and ductility. Relatively brittle ceramics and glasses can still find structural applications based on properties such as temperature resistance and chemical durability. Improved toughness is expanding their design options further. Glasses and polymers may be selected as structural materials because of optical properties. Final design issues for any structural material are environmental degradation and failure prevention.

The selection of electronic materials begins with defining the need for one of three categories: conductor, insulator, or semiconductor. The selection of semiconductors is part of a complex engineering design process that leads to increasingly complex and miniaturized electrical circuits. For structural and electronic categories, materials selection is further illustrated by specific case studies.

A wide range of environmental reactions limits the utility of engineering materials considered in this book. Oxidation is the direct chemical reaction of a metal with atmospheric oxygen. There are four mechanisms for oxidation associated with various modes of diffusion through the oxide scale. The tendency of a coating to be protective can be predicted by the Pilling–Bedworth ratio, R. Values of R between 1 and 2 are associated with a coating under moderate compressive stresses, and, as a result, it can be protective. Familiar examples of protective coatings are those on anodized aluminum and stainless steel. Unprotective coatings are susceptible to buckling and flaking off, a process known as spalling. Other atmospheric gases, including nitrogen, sulfur, and hydrogen, can lead to direct chemical attack of metals.

Corrosion is the dissolution of a metal into an aqueous environment. A galvanic cell involves two different metals with different tendencies toward ionization. The more active, or ionizable, metal is anodic and corrodes. The more noble metal is cathodic and plates out. A list of half-cell potentials showing relative corrosion tendencies is the electromotive force series. A galvanic series is a more qualitative list for commercial alloys in a given corrosive medium such as seawater. Metallic corrosion can be prevented by materials selection, design selection, protective coatings of various kinds, galvanic protection (using sacrificial anodes or impressed voltage), and chemical inhibitors. Although nonmetals are relatively inert in comparison with corrosion-sensitive metals, direct chemical attack can affect design applications. The attack of silicates by moisture and the vulcanization of rubber are examples. All materials can be selectively damaged by certain forms of radiation. Wear is the removal of surface material as a result of mechanical action, such as continuous or cyclic sliding.

As engineers, we must be increasingly aware of our responsibility to the environment. Governmental entities such as the Environmental Protection Agency oversee a vast array of regulations. DFE has become an integral part of overall professional practice. The definition of hazardous air pollutants is being broadened to include increasingly fine-scale particles, by-products of many common materials-processing technologies. Substantial energy savings and attendant reductions in pollution emissions are possible with the recycling of many engineering materials.

Key Terms

abrasive wear (522)
adhesive wear (522)
aerosol (526)
anode (516)
anodic reaction (516)
anodized aluminum (515)
Ashby chart (494)
biological material (504)
biomaterial (504)
cathode (517)
cathodic reaction (517)
corrosion (515)
corrosion fatigue (522)
corrosive wear (522)
design for the environment
  (DFE) (525)
design parameter (493)

design selection (519)
electrochemical cell (516)
electromotive force series (517)
environmental impact assessment
  (EIA) (525)
galvanic cell (516)
galvanic corrosion (515)
galvanic protection (521)
galvanic series (518)
galvanized steel (520)
green engineering (525)
half cell (517)
half-cell reaction (517)
hydrogen embrittlement (515)
impressed voltage (521)
inhibitor (521)
life-cycle assessment (LCA) (525)

light-emitting diode (LED) (511)
materials selection (519)
oxidation (514)
Pilling–Bedworth ratio (515)
PM2.5 (526)
PM10 (526)
porcelain enamel (521)
prosthesis (503)
protective coating (520)
radiation (522)
recycling (526)
sacrificial anode (521)
spalling (515)
spectral width (511)
stress-corrosion cracking (522)
surface-fatigue wear (522)
wear (522)

References

ASM Handbook, Vol. 13A: Corrosion: Fundamentals, Testing and Protection; ASM International, Materials Park, OH, 2003.

ASM Handbook, Vol. 20: Materials Selection and Design, ASM International, Materials Park, OH, 1997.

Ashby, M. F., *Materials Selection in Mechanical Design,* 3rd ed., Butterworth-Heinemann, Oxford, 2005.

Ashby, M. F., and **D. R. H. Jones,** *Engineering Materials 1—An Introduction to Their Properties, Applications, and Design,* 3rd ed., Butterworth-Heinemann, Boston, 2005.

Ashby, M. F., and **D. R. H. Jones,** *Engineering Materials 2—An Introduction to Microstructures, Processing and Design,* 2nd ed., Butterworth-Heinemann, Boston, 2006.

Dorf, R. C., *Electrical Engineering Handbook,* 3rd ed., 6-volume set, CRC Press, Boca Raton, FL, 2006.

Jones, D. A., *Principles and Prevention of Corrosion,* 2nd ed., Prentice-Hall, Upper Saddle River, NJ, 1996.

Kelly, B. T., *Irradiation Damage to Solids,* Pergamon Press, Inc., Elmsford, NY, 1966.

Rabinowicz, E., *Friction and Wear of Materials,* 2nd ed., John Wiley & Sons, Inc., New York, 1995.

Shackelford, J. F., and **W. Alexander,** *CRC Materials Science and Engineering Handbook,* 3rd ed., CRC Press, Boca Raton, FL, 2001.

Problems

At the beginning of this chapter, we noted that materials selection is a complex topic and that our introductory examples are somewhat idealistic. As in previous chapters, we shall continue to avoid subjective homework problems and instead deal with problems in the objective style. The many subjective problems dealing with materials selection will be left to more advanced courses or your own experiences as a practicing engineer.

14.1 • Material Properties—Engineering Design Parameters

[D] 14.1. A government contractor requires the use of tool steels with a tensile strength of $\geq 1{,}000$ MPa and a % elongation of $\geq 15\%$. Which alloys in Tables 14.1 and 14.2 would meet this specification?

[D] 14.2. Which additional alloys could be specified if the government contractor in Problem 14.1 would allow the selection of tool steels with a tensile strength of $\geq 1{,}000$ MPa and a % elongation of $\geq 10\%$?

[D] 14.3. In selecting a ductile cast iron for a high-strength/low-ductility application, you find the following two tables in a standard reference:

Tensile strength of ductile irons

Grade or class	T.S. (MPa)
Class C	345
Class B	379
Class A	414
65–45–12	448
80–55–06	552
100–70–03	689
120–90–02	827

Elongation of ductile irons

Grade or class	Elongation (%)
120–90–02	2
100–70–03	3
80–55–06	6
Class B	7
65–45–12	12
Class A	15
Class C	20

If specifications require a ductile iron with a tensile strength of ≥ 550 MPa and a % elongation of $\geq 5\%$, which alloys in these tables would be suitable?

[D] 14.4. In selecting a ductile cast iron for a low-strength/high-ductility application, which of the alloys in Problem 14.3 would meet the specifications of a tensile strength of ≥ 350 MPa and a % elongation of $\geq 15\%$?

[D] 14.5. In selecting a polymer for an electronic packaging application, you find the following two tables in a standard reference:

Volume resistivity of polymers

Polymer	Resistivity ($\Omega \cdot$ m)
Epoxy	1×10^5
Phenolic	1×10^9
Cellulose acetate	1×10^{10}
Polyester	1×10^{10}
Polyvinyl chloride	1×10^{12}
Nylon 6/6	5×10^{12}
Acrylic	5×10^{12}
Polyethylene	5×10^{13}
Polystyrene	2×10^{14}
Polycarbonate	2×10^{14}
Polypropylene	2×10^{15}
PTFE	2×10^{16}

Thermal conductivity of polymers

Polymer	Conductivity (J/s $\cdot$ m $\cdot$ K)
Polystyrene	0.12
Polyvinyl chloride	0.14
Polycarbonate	0.19
Polyester	0.19
Acrylic	0.21
Phenolic	0.22
PTFE	0.24
Cellulose acetate	0.26
Polyethylene	0.33
Epoxy	0.52
Polypropylene	2.2
Nylon 6/6	2.9

If specifications require a polymer with a volume resistivity of $\geq 10^{13}$ Ω m and a thermal conductivity of ≥ 0.25 J/s $\cdot$ m $\cdot$ K, which polymers in these tables would be suitable?

D 14.6. In reviewing the performance of polymers chosen in Problem 14.5, concerns over product performance lead to the establishment of more stringent specifications. Which polymers would have a volume resistivity of $\geq 10^{14}$ $\Omega \cdot$ m and a thermal conductivity of ≥ 0.35 J/s $\cdot$ m $\cdot$ K?

14.7. On a photocopy of Figure 14.3, superimpose modulus-density data for (a) 1040 carbon steel, (b) 2048 plate aluminum, and (c) Ti–5Al–2.5Sn. Take the modulus data from Table 6.1 and approximate the alloy densities by using the values for pure Fe, Al, and Ti in Appendix 1.

14.8. On a photocopy of Figure 14.3, superimpose modulus-density data for (a) the sintered alumina from Table 6.4, (b) the high-density polyethylene from Table 6.6, and (c) the E-glass/epoxy fiberglass from Table 12.7. Take the densities to be 3.8, 1.0, and 1.8 Mg/m^3, respectively.

14.2 • Selection of Structural Materials: Case Studies

14.9. Use Equation 14.4 to calculate the mass of a CFRP windsurfer mast with a length of 4.6 m, outer radius of 18 mm, shape factor of $\phi = 5$, and density of 1.5 Mg/m^3.

14.10. Repeat Problem 14.9 for a GFRP windsurfer mast with a length of 4.6 m, outer radius of 17 mm, shape factor of $\phi = 4$, and density of 1.8 Mg/m^3.

14.11. The Boeing 767 aircraft uses a Kevlar-reinforced composite for its cargo liner. The structure weighs 125 kg. **(a)** What weight savings does this weight represent compared to an aluminum structure of the same volume? (For simplicity, use the density of pure aluminum. A calculation of density for a Kevlar composite was made in Problem 12.38.) **(b)** What annual fuel savings would this one material substitution represent?

14.12. What is the annual fuel savings resulting from the Al–Li alloy substitution described in Problem 11.3?

14.13. Consider an inspection program for Ti–6Al–4V THR femoral stems. Would the ability to detect a flaw size of 1 mm be adequate to prevent fast fracture? Use the data in Table 8.3, and assume an extreme loading of five times body weight for an athletic 200-lb patient. The cross-sectional area of the stem is 650 mm^2.

14.14. Consider an inspection program for a more traditional application of the Ti–6Al–4V alloy discussed in

Problem 14.13. Is the ability to detect a 1-mm flaw size adequate in an aerospace structural member loaded to 90% of the yield strength given in Table 6.1?

14.15. Consider the loading of a simple, cantilever beam: From basic mechanics, it can be shown that the mass of the beam subjected to a deflection, δ, by a force, F, is given by

$$M = (4l^5 F/\delta)^{1/2}(\rho^2/E)^{1/2},$$

where ρ is the density, E is the elastic modulus, and the other terms are defined by the figure. Clearly, the mass of this structural member is minimized for a given loading by minimizing ρ^2/E. Given the data in Problem 12.61, which of those materials would be the optimal choice for this type of structural application? (Moduli can generally be obtained from Tables 6.1 and 12.7. The modulus of reinforced concrete can be taken as 47×10^3 MPa.)

14.16. How would the material selection in Problem 14.15 be modified if cost were included in the minimization?

14.3 • Selection of Electronic Materials: Case Studies

14.17. Verify the total owning cost calculation for the amorphous metal core in Table 14.7.

14.18. Repeat Problem 14.17 for the grain-oriented silicon steel core in Table 14.7.

D 14.19. Consider the following design considerations for a typical solid-state device, a memory cell in a FET. The cell contains a thin film of SiO_2, which serves as a small capacitor. You want the capacitor to be as small as possible, but big enough so that an α radiation particle will not cause errors with your 5-V operating signal. To ensure this result, design specifications require a capacitor with a capacitance [= (charge density $\times$ area)/voltage] of 50×10^{-15} coul/volt. How much area is required for the capacitor if the SiO_2 thickness is 1 μm? (Note that the dielectric constant of SiO_2 is 3.9, and recall the discussion of capacitance in Section 13.4.)

D 14.20. If the specifications for the device introduced in Problem 14.19 are changed so that a capacitance of 70×10^{-15} coul/volt is required, but the area is fixed by the overall circuit design, calculate the appropriate SiO_2 film thickness for the modified design.

14.21. Calculate the band gap, E_g, that corresponds to the LEDs in Table 14.8 made of (a) GaP and (b) GaAsP.

14.22. Repeat Problem 14.21 for the LEDs in Table 14.8 made of (a) GaAs and (b) InGaAs.

14.4 • Materials and Our Environment

14.23. The densities for three iron oxides are FeO (5.70 Mg/m^3), Fe$_3$O$_4$ (5.18 Mg/m^3), and Fe$_2$O$_3$ (5.24 Mg/m^3). Calculate the Pilling–Bedworth ratio for iron relative to each type of oxide, and comment on the implications for the formation of a protective coating.

14.24. Given the density of SiO$_2$ (quartz) = 2.65 Mg/m^3, calculate the Pilling–Bedworth ratio for silicon and comment on the implication for the formation of a protective coating if quartz is the oxide form.

14.25. In contrast to the assumption in Problem 14.24, silicon oxidation tends to produce a vitreous silica film with density = 2.20 Mg/m^3. Semiconductor fabrication routinely involves such vitreous films. Calculate the Pilling–Bedworth ratio for this case, and comment on the implication for the formation of a tenacious film.

14.26. Verify that the Pilling–Bedworth ratio is the ratio of oxide volume produced to metal volume consumed.

14.27. **(a)** In a simple galvanic cell consisting of Co and Cr electrodes immersed in 1-molar ionic solutions, calculate the cell potential. **(b)** Which metal would be corroded in this simple cell?

14.28. **(a)** In a simple galvanic cell consisting of Al and Mg electrodes immersed in 1-molar ionic solutions, calculate the cell potential. **(b)** Which metal would be corroded in this simple cell?

14.29. Identify the anode in the following galvanic cells, and provide a brief discussion of each answer: **(a)** copper and nickel electrodes in standard solutions of their own ions, **(b)** a two-phase microstructure of a 50:50 Pb–Sn alloy, **(c)** a lead–tin solder on a 2024 aluminum alloy in seawater, and **(d)** a brass bolt in a Hastelloy C plate, also in seawater.

D 14.30. Figure 14.19 illustrates a microstructural-scale galvanic cell. In selecting a material for the outer shell

of a marine compass, use the Cu–Zn phase diagram from Chapter 9 to specify a brass composition range that would avoid this problem.

D 14.31. In designing the hull of a new fishing vessel to ensure corrosion protection, you find that a sacrificial anode of zinc provides an average corrosion current of 2 A over the period of 1 year. What mass of zinc is required to give this protection?

D 14.32. In designing an offshore steel structure to ensure corrosion protection, you find that a sacrificial anode of magnesium provides an average corrosion current of 1.5 A over a period of 2 years. What mass of magnesium is required to give this protection?

D 14.33. The maximum corrosion-current density in a galvanized steel sheet used in the design of the new engineering laboratories on campus is found to be 5 mA/m^2. What thickness of the zinc layer is necessary to ensure at least **(a)** 1 year and **(b)** 5 years of rust resistance?

D 14.34. A galvanized steel sheet used in the design of the new chemistry laboratories on campus has a zinc coating 18 μm thick. The corrosion-current density is found to be 4 mA/m^2. What is the corresponding duration of rust resistance provided by this system?

14.35. Would you consider the following materials systems to be easy or difficult to recycle: (a) steel casters fixed to polymeric chair bases and (b) clear frozen-food bags made of low-density polyethylene? Briefly explain your answers.

14.36. Repeat Problem 14.35 for (a) polymeric trays used to carry electronic devices locally in a fabrication plant and (b) polymeric bottles with well-adhered paper labels.

14.37. Repeat Problem 14.35 for **(a)** scrap rubber erasers in a pencil factory and **(b)** scrap polypropylene tubing in a manufacturing plant.

14.38. Repeat Problem 14.35 for **(a)** a large volume of aluminum foil (well washed) from a dormitory kitchen and **(b)** scrap phenol-formaldehyde electrical parts in a manufacturing plant.

APPENDIX ONE
Physical and Chemical Data for the Elements

Atomic number	Element	Symbol	1s	2s	2p	3s	3p	3d	4s	4p	4d	4f	5s	5p	5d	5f	6s	6p	6d	7s	Atomic mass (amu)	Density of solid (at 20°C) (Mg/m³ = g/cm³)	Crystal structure (at 20°C)	Melting point (°C)	Atomic number
1	Hydrogen	H	1																		1.008			−259.34 (T.P.)	1
2	Helium	He	2																		4.003			−271.69	2
3	Lithium	Li		1																	6.941	0.533	bcc	180.6	3
4	Berylium	Be		2																	9.012	1.85	hcp	1,289	4
5	Boron	B		2	1																10.81	2.47		2,092	5
6	Carbon	C		2	2																12.01	2.27	hex.	3,826 (S.P.)	6
7	Nitrogen	N		2	3																14.01			−210.0042 (T.P.)	7
8	Oxygen	O		2	4																16.00			−218.789 (T.P.)	8
9	Flourine	F		2	5																19.00			−219.67 (T.P.)	9
10	Neon	Ne		2	6																20.18			−248.587 (T.P.)	10
11	Sodium	Na				1															22.99	0.966	bcc	97.8	11
12	Magnesium	Mg				2															24.31	1.74	hcp	650	12
13	Aluminum	Al				2	1														26.98	2.70	fcc	660.452	13
14	Silicon	Si				2	2														28.09	2.33	dia. cub.	1,414	14
15	Phosphorus	P				2	3														30.97	1.82 (white)	ortho.	44.14 (white)	15
16	Sulfur	S				2	4														32.06	2.09	ortho.	115.22	16
17	Chlorine	Cl				2	5														35.45			−100.97 (T.P.)	17
18	Argon	Ar				2	6														39.95			−189.352 (T.P.)	18
19	Potassium	K							1												39.10	0.862	bcc	63.71	19
20	Calcium	Ca							2												40.08	1.53	fcc	842	20
21	Scandium	Sc						1	2												44.96	2.99	fcc	1,541	21
22	Titanium	Ti						2	2												47.90	4.51	hcp	1,670	22
23	Vanadium	V						3	2												50.94	6.09	bcc	1,910	23
24	Chromium	Cr						5	1												52.00	7.19	bcc	1,863	24
25	Manganese	Mn						5	2												54.94	7.47	cubic	1,246	25
26	Iron	Fe						6	2												55.85	7.87	bcc	1,538	26
27	Cobalt	Co						7	2												58.93	8.8	hcp	1,495	27
28	Nickel	Ni						8	2												58.71	8.91	fcc	1,455	28
29	Copper	Cu						10	1												63.55	8.93	fcc	1,084.87	29
30	Zinc	Zn						10	2												65.38	7.13	hcp	419.58	30
31	Gallium	Ga						10	2	1											69.72	5.91	ortho.	29.7741 (T.P.)	31
32	Germanium	Ge						10	2	2											72.59	5.32	dia. cub.	938.3	32
33	Arsenic	As						10	2	3											74.92	5.78	rhomb.	603 (S.P.)	33

Electronic configuration[a]
(number of electrons in each group)

Helium core — Neon core — Argon core

Atomic number	Element	Symbol	1s	2s	2p	3s	3p	3d	4s	4p	4d	4f	5s	5p	5d	5f	6s	6p	6d	7s	Atomic mass[a] (amu)	Density of solid[b] (at 20°C) (Mg/m³ = g/cm³)	Crystal structure[c] (at 20°C)	Melting point[d] (°C)	Atomic number
34	Selenium	Se						10	2	4											78.96	4.81	hex.	221	34
35	Bromine	Br						10	2	5											79.90			−7.25 (T.P.)	35
36	Krypton	Kr						10	2	6											83.80			−157.385	36
37	Rubidium	Rb											1								85.47	1.53	bcc	39.48	37
38	Strontium	Sr											2								87.62	2.58	fcc	769	38
39	Yttrium	Y									1		2								88.91	4.48	hcp	1,522	39
40	Zirconium	Zr									2		2								91.22	6.51	hcp	1,855	40
41	Niobium	Nb									4		1								92.91	8.58	bcc	2,469	41
42	Molybdenum	Mo									5		1								95.94	10.22	bcc	2,623	42
43	Technetium	Tc									6		1								98.91	11.50	hcp	2,204	43
44	Ruthenium	Ru									7		1								101.07	12.36	hcp	2,334	44
45	Rhodium	Rh									8		1								102.91	12.42	fcc	1,963	45
46	Palladium	Pd									10										106.4	12.00	fcc	1,555	46
47	Silver	Ag									10		1								107.87	10.50	fcc	961.93	47
48	Cadmium	Cd									10		2								112.4	8.65	hcp	321.108	48
49	Indium	In									10		2	1							114.82	7.29	fct	156.634	49
50	Tin	Sn									10		2	2							118.69	7.29	bct	231.9681	50
51	Antimony	Sb									10		2	3							121.75	6.69	rhomb.	630.755	51
52	Tellurium	Te									10		2	4							127.60	6.25	hex.	449.57	52
53	Iodine	I									10		2	5							126.90	4.95	ortho.	113.6 (T.P)	53
54	Xenon	Xe									10		2	6							131.30			−111.7582 (T.P.)	54
55	Cesium	Cs															1				132.91	1.91 (−10°)	bcc	28.39	55
56	Barium	Ba															2				137.33	3.59	bcc	729	56
57	Lanthanum	La													1		2				138.91	6.17	hex.	918	57
58	Cerium	Ce										2					2				140.12	6.77	fcc	798	58
59	Praseodymium	Pr										3					2				149.91	6.78	hex.	931	59
60	Neodymium	Nd										4					2				144.24	7.00	hex.	1,021	60
61	Promethium	Pm										5					2				(145)		hex.	1,042	61
62	Samarium	Sm										6					2				150.4	7.54	rhomb.	1,074	62
63	Europium	Eu										7					2				151.96	5.25	bcc	822	63
64	Gadolinium	Gd										7			1		2				157.25	7.87	hcp	1,313	64
65	Terbium	Tb										9					2				158.93	8.27	hcp	1,356	65
66	Dysprosium	Dy										10					2				162.50	8.53	hcp	1,412	66
67	Holmium	Ho										11					2				164.93	8.80	hcp	1,474	67
68	Erbium	Er										12					2				167.26	9.04	hcp	1,529	68
69	Thulium	Tm										13					2				168.93	9.33	hcp	1,545	69

Electronic configuration[a] (number of electrons in each group)

Elements 34–54 share the Krypton core; elements 55–69 share the Xenon core.

Atomic number	Element	Symbol	1s	2s	2p	3s	3p	3d	4s	4p	4d	4f	5s	5p	5d	5f	6s	6p	6d	7s	Atomic mass (amu)	Density of solid (at 20°C) ($Mg/m^3 = g/cm^3$)	Crystal structure (at 20°C)	Melting point (°C)	Atomic number
70	Ytterbium	Yb										14					2				173.04	6.97	fcc	819	70
71	Lutetium	Lu										14			1		2				174.97	9.84	hcp	1,663	71
72	Hafnium	Hf										14			2		2				178.49	13.28	hcp	2,231	72
73	Tantalum	Ta										14			3		2				180.95	16.67	bcc	3,020	73
74	Tungsten	W										14			4		2				183.85	19.25	bcc	3,422	74
75	Rhenium	Re										14			5		2				186.2	21.02	hcp	3,186	75
76	Osmium	Os										14			6		2				190.2	22.58	hcp	3,033	76
77	Iridium	Ir										14			9						192.22	22.55	fcc	2,447	77
78	Platinum	Pt										14			9		1				195.09	21.44	fcc	1,769.0	78
79	Gold	Au										14			10		1				196.97	19.28	fcc	1,064.43	79
80	Mercury	Hg										14			10		2				200.59			-38.836	80
81	Thallium	Tl										14			10		2	1			204.37	11.87	hcp	304	81
82	Lead	Pb										14			10		2	2			207.2	11.34	fcc	327.502	82
83	Bismuth	Bi										14			10		2	3			208.98	9.80	rhomb.	271.442	83
84	Polonium	Po										14			10		2	4			(~210)	9.2	monoclinic	254	84
85	Astatine	At										14			10		2	5			(210)			≈302	85
86	Radon	Rn										14			10		2	6			(222)			-71	86
87	Francium	Fr																		1	(223)		bcc	≈27	87
88	Radium	Ra																		2	226.03		bct	700	88
89	Actinium	Ac																	1	2	(227)		fcc	1,051	89
90	Thorium	Th																	2	2	232.04	11.72	fcc	1,755	90
91	Protoactinium	Pa														2			1	2	231.04		bct	1,572	91
92	Uranium	U														3			1	2	238.03	19.05	ortho.	1,135	92
93	Neptunium	Np														4			1	2	237.05		ortho.	639	93
94	Plutonium	Pu														6				2	(244)	19.82	monoclinic	640	94
95	Americium	Am														7				2	(243)		hex.	1,176	95
96	Curium	Cm														7			1	2	(247)		hex.	1,345	96
97	Berkelium	Bk														9				2	(247)		hex.	1,050	97
98	Californium	Cf														10				2	(251)			900	98
99	Einsteinium	Es														11				2	(254)			860	99
100	Fermium	Fm														12				2	(257)			≈1,527	100
101	Mendelevium	Md														13				2	(258)			≈827	101
102	Nobelium	No														14				2	(259)			≈827	102
103	Lawrencium	Lw														14			1	2	(260)			≈1,627	103
104	Rutherfordium[e]	Rf														14			2	2	(261)				104
105	Dubnium[e]	Db														14			3	2	(262)				105
106	Seaborgium[e]	Sg														14			4	2	(266)				106

(1s–5p for atomic numbers 70–86 = Xenon core; for atomic numbers 87–106 = Radon core)

Sources: Data from

[a]*Handbook of Chemistry and Physics,* 58th ed., R. C. Weast, Ed., CRC Press, Boca Raton, FL, 1977.

[b]X-ray diffraction measurements are tabulated in B. D. Cullity, *Elements of X-Ray Diffraction,* 2nd ed., Addison-Wesley Publishing Co., Inc., Reading, MA, 1978.

[c]R. W. G. Wyckoff, *Crystal Structure,* 2nd ed., Vol. 1, Interscience Publishers, New York, 1963; and *Metals Handbook,* 9th ed., Vol. 2, American Society for Metals, Metals Park, OH, 1979.

[d]*Binary Alloy Phase Diagrams,* Vols. 1 and 2, T. B. Massalski, Ed., American Society for Metals, Metals Park, OH, 1986. T.P. = triple point. S.P. = sublimation point at atmospheric pressure.

[e]www.webelements.com

Atomic and Ionic Radii of the Elements

Atomic number	Symbol	Atomic radius (nm)	Ion	Ionic radius (nm)
1	H	0.046	H^-	0.154
2	He	—	—	—
3	Li	0.152	Li^+	0.078
4	Be	0.114	Be^{2+}	0.054
5	B	0.097	B^{3+}	0.02
6	C	0.077	C^{4+}	<0.02
7	N	0.071	N^{5+}	0.01–0.02
8	O	0.060	O^{2-}	0.132
9	F	—	F^-	0.133
10	Ne	0.160	—	—
11	Na	0.186	Na^+	0.098
12	Mg	0.160	Mg^{2+}	0.078
13	Al	0.143	Al^{3+}	0.057
14	Si	0.117	Si^{4-}	0.198
			Si^{4+}	0.039
15	P	0.109	P^{5+}	0.03–0.04
16	S	0.106	S^{2-}	0.174
			S^{6+}	0.034
17	Cl	0.107	Cl^-	0.181
18	Ar	0.192	—	—
19	K	0.231	K^+	0.133
20	Ca	0.197	Ca^{2+}	0.106
21	Sc	0.160	Sc^{2+}	0.083
22	Ti	0.147	Ti^{2+}	0.076
			Ti^{3+}	0.069
			Ti^{4+}	0.064
23	V	0.132	V^{3+}	0.065
			V^{4+}	0.061
			V^{5+}	~0.04
24	Cr	0.125	Cr^{3+}	0.064
			Cr^{6+}	0.03–0.04

Atomic number	Symbol	Atomic radius (nm)	Ion	Ionic radius (nm)
25	Mn	0.112	Mn^{2+}	0.091
			Mn^{3+}	0.070
			Mn^{4+}	0.052
26	Fe	0.124	Fe^{2+}	0.087
			Fe^{3+}	0.067
27	Co	0.125	Co^{2+}	0.082
			Co^{3+}	0.065
28	Ni	0.125	Ni^{2+}	0.078
29	Cu	0.128	Cu^{+}	0.096
			Cu^{2+}	0.072
30	Zn	0.133	Zn^{2+}	0.083
31	Ga	0.135	Ga^{3+}	0.062
32	Ge	0.122	Ge^{4+}	0.044
33	As	0.125	As^{3+}	0.069
			As^{5+}	~0.04
34	Se	0.116	Se^{2-}	0.191
			Se^{6+}	0.03–0.04
35	Br	0.119	Br^{-}	0.196
36	Kr	0.197	—	—
37	Rb	0.251	Rb^{+}	0.149
38	Sr	0.215	Sr^{2+}	0.127
39	Y	0.181	Y^{3+}	0.106
40	Zr	0.158	Zr^{4+}	0.087
41	Nb	0.143	Nb^{4+}	0.074
			Nb^{5+}	0.069
42	Mo	0.136	Mo^{4+}	0.068
			Mo^{6+}	0.065
43	Tc	—	—	—
44	Ru	0.134	Ru^{4+}	0.065
45	Rh	0.134	Rh^{3+}	0.068
			Rh^{4+}	0.065
46	Pd	0.137	Pd^{2+}	0.050
47	Ag	0.144	Ag^{+}	0.113
48	Cd	0.150	Cd^{2+}	0.103
49	In	0.157	In^{3+}	0.092
50	Sn	0.158	Sn^{4-}	0.215
			Sn^{4+}	0.074
51	Sb	0.161	Sb^{3+}	0.090
52	Te	0.143	Te^{2-}	0.211
			Te^{4+}	0.089
53	I	0.136	I^{-}	0.220
			I^{5+}	0.094
54	Xe	0.218	—	—
55	Cs	0.265	Cs^{+}	0.165
56	Ba	0.217	Ba^{2+}	0.143
57	La	0.187	La^{3+}	0.122
58	Ce	0.182	Ce^{3+}	0.118
			Ce^{4+}	0.102

Atomic number	Symbol	Atomic radius (nm)	Ion	Ionic radius (nm)
59	Pr	0.183	Pr^{3+}	0.116
			Pr^{4+}	0.100
60	Nd	0.182	Nd^{3+}	0.115
61	Pm	—	Pm^{3+}	0.106
62	Sm	0.181	Sm^{3+}	0.113
63	Eu	0.204	Eu^{3+}	0.113
64	Gd	0.180	Gd^{3+}	0.111
65	Tb	0.177	Tb^{3+}	0.109
			Tb^{4+}	0.089
66	Dy	0.177	Dy^{3+}	0.107
67	Ho	0.176	Ho^{3+}	0.105
68	Er	0.175	Er^{3+}	0.104
69	Tm	0.174	Tm^{3+}	0.104
70	Yb	0.193	Yb^{3+}	0.100
71	Lu	0.173	Lu^{3+}	0.099
72	Hf	0.159	Hf^{4+}	0.084
73	Ta	0.147	Ta^{5+}	0.068
74	W	0.137	W^{4+}	0.068
			W^{6+}	0.065
75	Re	0.138	Re^{4+}	0.072
76	Os	0.135	Os^{4+}	0.067
77	Ir	0.135	Ir^{4+}	0.066
78	Pt	0.138	Pt^{2+}	0.052
			Pt^{4+}	0.055
79	Au	0.144	Au^{+}	0.137
80	Hg	0.150	Hg^{2+}	0.112
81	Tl	0.171	Tl^{+}	0.149
			Tl^{3+}	0.106
82	Pb	0.175	Pb^{4-}	0.215
			Pb^{2+}	0.132
			Pb^{4+}	0.084
83	Bi	0.182	Bi^{3+}	0.120
84	Po	0.140	Po^{6+}	0.067
85	At	—	At^{7+}	0.062
86	Rn	—	—	—
87	Fr	—	Fr^{+}	0.180
88	Ra	—	Ra^{+}	0.152
89	Ac	—	Ac^{3+}	0.118
90	Th	0.180	Th^{4+}	0.110
91	Pa	—	—	—
92	U	0.138	U^{4+}	0.105

Source: After a tabulation by R. A. Flinn and P. K. Trojan, *Engineering Materials and Their Applications*, Houghton Mifflin Company, Boston, 1975. The ionic radii are based on the calculations of V. M. Goldschmidt, who assigned radii based on known interatomic distances in various ionic crystals.

Constants and Conversion Factors

Constants

Avogadro's number, N_A	0.6023×10^{24} mol^{-1}
Atomic mass unit (amu)	1.661×10^{-24} g
Electric permittivity of a vacuum, ϵ_0	8.854×10^{-12} C/(V $\cdot$ m)
Electron mass	0.9110×10^{-27} g
Elementary charge, e	0.1602×10^{-18} C
Gas constant, R	8.314 J/(mol $\cdot$ K)
	1.987 cal/(mol $\cdot$ K)
Boltzmann's constant, k	13.81×10^{-24} J/K
	86.20×10^{-6} eV/K
Planck's constant, h	0.6626×10^{-33} J $\cdot$ s
Speed of light (in vacuum), c	0.2998×10^{9} m/s
Bohr magneton, μ_B	9.274×10^{-24} A $\cdot$ m^2
Faraday's constant, F	$96{,}500$ C/mol

SI prefixes

giga, G	10^9
mega, M	10^6
kilo, k	10^3
milli, m	10^{-3}
micro, μ	10^{-6}
nano, n	10^{-9}
pico, p	10^{-12}

Conversion factors

Length	1 meter $= 10^{10}$ Å $= 10^9$ nm
	$= 3.281$ ft
	$= 39.37$ in.
Mass	1 kilogram $= 2.205$ lb$_m$
Force	1 newton $= 0.2248$ lb$_f$
Pressure	1 pascal $= 1$ N/m^2
	$= 0.1019 \times 10^{-6}$ kg$_f$/mm^2
	$= 9.869 \times 10^{-6}$ atm
	$= 0.1450 \times 10^{-3}$ lb$_f$/in.^{2}
Viscosity	1 Pa $\cdot$ s $= 10$ poise
Energy	1 joule $= 1$ W $\cdot$ s
	$= 1$ N $\cdot$ m
	$= 1$ V $\cdot$ C
	$= 0.2389$ cal
	$= 6.242 \times 10^{18}$ eV
	$= 0.7377$ ft lb$_f$
Temperature	°C $=$ K $- 273$
	$= (°F - 32)/1.8$
Current	1 ampere $= 1$ C/s
	$= 1$ V/Ω

APPENDIX FOUR
Properties of the Structural Materials

The related Web site provides numerous links to data sources. Many of the professional society Web sites are especially useful in this regard.

The following tables are provided as convenient compilations of the key properties of the structural materials covered in Chapters 11 and 12 of the book. Reference is frequently made to specific tables located throughout the book.

Appendix 4A

Physical Properties of Selected Materials

Material	Density [Mg/m^3]	Melting temperature [°C]	Glass transition temperature [°C]
Metals[a]			
Aluminum	2.70	660	
Copper	8.93	1,085	
Gold	19.28	1,064	
Iron	7.87	1,538	
Lead	11.34	328	
Nickel	8.91	1,455	
Silver	10.50	962	
Titanium	4.51	1,670	
Tungsten	19.25	3,422	
Ceramics and glasses[b]			
Al$_2$O$_3$	3.97	2,054	
MgO	3.58	2,800	
SiO$_2$	2.26–2.66	1,726	
Mullite (3Al$_2$O$_3$·2SiO$_2$)	3.16	1,890	
ZrO$_2$	5.89	2,700	
Silica glass	2.2		1,100
Soda–lime–silica glass	2.5		450
Polymers[c]			
Epoxy (mineral filled)	1.22		400
Nylon 66	1.13–1.15		150
Phenolic	1.32–1.46		375
Polyethylene (high-density)	0.94–0.97		
Polypropylene	0.90–0.91		
Polytetrafluoroethylene (PTFE)	2.1–2.3		

Source: Data from [a]Appendix 1; [b]D. R. Lide, *Handbook of Chemistry and Physics*, 71st ed., CRC Press, Boca Raton, FL, 1990, ceramic phase diagrams in Chapter 9, and the Corning Glass Works; and [c]J. F. Shackelford, W. Alexander, and J. S. Park, *The CRC Materials Science and Engineering Handbook*, 2nd ed., CRC Press, Boca Raton, FL, 1994, and Figure 6.52.

Appendix 4B

Tensile and Bend Test Data for Selected Engineering Materials

Material	E [GPa (psi)]	E_{flex} [MPa (ksi)]	E_{Dyn} [MPa (ksi)]	Y.S. [MPa (ksi)]	T.S. [MPa (ksi)]	Flexural strength [MPa (ksi)]	Compressive strength [MPa (ksi)]	Percent elongation at failure
Metal alloys[a]								
1040 carbon steel	$200(29 \times 10^6)$			600(87)	750(109)			17
8630 low-alloy steel				680(99)	800(116)			22
304 stainless steel	$193(28 \times 10^6)$			205(30)	515(75)			40
410 stainless steel	$200(29 \times 10^6)$			700(102)	800(116)			22
L2 tool steel				1,380(200)	1,550(225)			12
Ferrous superalloy (410)	$200(29 \times 10^6)$			700(102)	800(116)			22
Ductile iron, quench	$165(24 \times 10^6)$			580(84)	750(108)			9.4
Ductile iron, 60-40-18	$169(24.5 \times 10^6)$			329(48)	461(67)			15
3003-H14 aluminum	$70(10.2 \times 10^6)$			145(21)	150(22)			8-16
2048, plate aluminum	$70.3(10.2 \times 10^6)$			416(60)	457(66)			8
AZ31B magnesium	$45(6.5 \times 10^6)$			220(32)	290(42)			15
AM100A casting magnesium	$45(6.5 \times 10^6)$			83(12)	150(22)			2
Ti-5Al-2.5Sn	$107-110(15.5-16 \times 10^6)$			827(120)	862(125)			15
Ti-6Al-4V	$110(16 \times 10^6)$			825(120)	895(130)			10
Aluminum bronze, 9% (copper alloy)	$110(16.1 \times 10^6)$			320(46.4)	652(94.5)			34
Monel 400 (nickel alloy)	$179(26 \times 10^6)$			283(41)	579(84)			39.5
AC41A zinc	$97(14 \times 10^6)$				328(47.6)			7
50:50 solder (lead alloy)				33(4.8)	42(6.0)			60
Nb-1 Zr (refractory metal)	$68.9(10 \times 10^6)$			138(20)	241(35)			20
Dental gold alloy (precious metal)					310-380(45-55)			20-35
Ceramics and glasses[b]								
Mullite (aluminosilicate) porcelain	$69(10 \times 10^6)$					69(10)		
Steatite (magnesia aluminosilicate) porcelain	$60(10 \times 10^6)$					140(20)		
Superduty fireclay (aluminosilicate) brick	$97(14 \times 10^6)$					5.2(0.75)		
Alumina (Al_2O_3) crystals	$380(55 \times 10^6)$					340-1,000(49-145)		
Sintered alumina (~5% porosity)	$370(54 \times 10^6)$					210-340(30-49)		
Alumina porcelain (90-95% alumina)	$370(54 \times 10^6)$					340(49)		
Sintered magnesia (~5% porosity)	$210(30 \times 10^6)$					100(15)		
Magnesite (magnesia) brick	$170(25 \times 10^6)$					28(4)		
Sintered spinel (magnesia aluminate) (~5% porosity)	$238(35 \times 10^6)$					90(13)		
Sintered stabilized zirconia (~5% porosity)	$150(22 \times 10^6)$					83(12)		
Sintered beryllia (~5% porosity)	$310(45 \times 10^6)$					140-280(20-41)		
Dense silicon carbide (~5% porosity)	$470(68 \times 10^6)$					170(25)		
Bonded silicon carbide (~20% porosity)	$340(49 \times 10^6)$					14(2)		
Hot-pressed boron carbide (~5% porosity)	$290(42 \times 10^6)$					340(49)		
Hot-pressed boron nitride (~5% porosity)	$83(12 \times 10^6)$					48-100(7-15)		
Silica glass	$72.4(10.5 \times 10^6)$					107(16)		
Borosilicate glass	$69(10 \times 10^6)$					69(10)		

Appendix 4B

Continued

Material	E [GPa (psi)]	E_{flex} [MPa (ksi)]	E_{Dyn} [MPa (ksi)]	Y.S. [MPa (ksi)]	T.S. [MPa (ksi)]	Flexural strength [MPa (ksi)]	Compressive strength [MPa (ksi)]	Percent elongation at failure
Polymers[c]								
Polyethylene								
High-density	$0.830(0.12 \times 10^6)$				28(4)			15–100
Low-density	$0.170(0.025 \times 10^6)$				14(2)			90–800
Polyvinylchloride	$2.80(0.40 \times 10^6)$				41(6)			2–30
Polypropylene	$1.40(0.20 \times 10^6)$				34(5)			10–700
Polystyrene	$3.10(.045 \times 10^6)$				48(7)			1–2
Polyesters		8,960(1,230)			158(22.9)			2.7
Acrylics (Lucite)	$2.90(0.42 \times 10^6)$				55(8)			5
Polyamides (nylon 66)	$2.80(0.41 \times 10^6)$	2,830(410)			82.7(12.0)			60
Cellulosics	$3.40–28.0(0.50–4.0 \times 10^6)$				14–55(2–8)			5–40
ABS	$2.10(0.30 \times 10^6)$				28–48(4–7)			20–80
Polycarbonates	$2.40(0.35 \times 10^6)$				62(9)			110
Acetals	$3.10(0.45 \times 10^6)$	2,830(410)			69(10)			50
Polytetrafluoroethylene (Teflon)	$0.41(0.060 \times 10^6)$				17(2.5)			100–350
Polyester-type thermoplastic elastomers		585(85)			46(6.7)			400
Phenolics (phenolformaldehyde)	$6.90(1.0 \times 10^6)$				52(7.5)			0
Urethanes					34(5)			—
Urea-melamine	$10.0(1.5 \times 10^6)$				48(7)			0
Polyesters	$6.90(1.0 \times 10^6)$				28(4)			0
Epoxies	$6.90(1.0 \times 10^6)$				69(10)			0
Polybutadiene/polystyrene copolymer								
Vulcanized	$0.0016(0.23 \times 10^3)$		0.8(0.12)		1.4–3.0(0.20–0.44)			440–600
Vulcanized with 33% carbon black	$0.003–0.006(0.4–0.9 \times 10^3)$		8.7(1.3)		17–28(2.5–4.1)			400–600
Polyisoprene								
Vulcanized	$0.0013(0.19 \times 10^3)$		0.4(0.06)		17–25(2.5–3.6)			750–850
Vulcanized with 33% carbon black	$0.003–0.008(0.44–1.2 \times 10^3)$		6.2(0.90)		25–35(3.6–5.1)			550–650
Polychloroprene								
Vulcanized	$0.0016(0.23 \times 10^3)$		0.7(0.10)		25–38(3.6–5.5)			800–1,000
Vulcanized with 33% carbon black	$0.003–0.005(0.4–0.7 \times 10^3)$		2.8(0.41)		21–30(3.0–4.4)			500–600
Polyisobutene/polyisoprene copolymer								
Vulcanized	$0.0010(0.15 \times 10^3)$		0.40(0.06)		18–21(2.6–3.0)			750–950
Vulcanized with 33% carbon black	$0.003–0.004(0.4–0.6 \times 10^3)$		3.6(0.52)		18–21(2.6–3.0)			650–850
Silicones					7(1)			4,000
Vinylidene fluoride/hexafluoropropylene					12.4(1.8)			

Appendix 4B

Continued

Material	E [GPa (psi)]	E_{flex} [MPa (ksi)]	E_{Dyn} [MPa (ksi)]	Y.S. [MPa (ksi)]	T.S. [MPa (ksi)]	Flexural strength [MPa (ksi)]	Compressive strength [MPa (ksi)]	Percent elongation at failure
Composites[d]								
E-glass (73.3 vol %) in epoxy (parallel loading of continuous fibers)	56(8.1 × 10⁶)				1,640(238)	—	—	2.9
Al₂O₃ whiskers (14 vol %) in epoxy	41(6 × 10⁶)				779(113)	—	—	—
C (67 vol %) in epoxy (parallel loading)	221(32 × 10⁶)				1,206(175)	—	—	—
Kevlar (82 vol %) in epoxy (parallel loading)	86(12 × 10⁶)				1,517(220)	—	—	—
B (70 vol %) in epoxy (parallel loading of continuous filaments)	210 – 280(30 – 40 × 10⁶)				1,400–2,100(200–300)	—	—	—
Al₂O₃ (10 vol %) dispersion-strengthened aluminum	—				330(48)	—	—	—
W (50 vol %) in copper (parallel loading of continuous filaments)	260(38 × 10⁶)				1,100(160)	—	—	—
W particles (50 vol %) in copper	190(27 × 10⁶)				380(55)	—	—	—
SiC whiskers in Al₂O₃	—				—	800(116)	—	—
SiC fibers in SiC	—				—	750(109)	—	—
SiC whiskers in reaction-bonded Si₃N₄	—				—	900(131)	—	—

[a]From Table 6.1
[b]From Table 6.4
[c]From Tables 6.6 and 6.7
[d]From Table 12.7

Appendix 4C

Miscellaneous Mechanical Properties Data for Selected Engineering Materials

	Poisson's ratio ν	Brinell hardness number	Rockwell hardness R scale	Charpy impact energy [J(ft·lb)]	Izod impact energy [J(ft·lb)]	K_{IC} (MPa $\sqrt{m}$)	Fatigue limit [MPa (ksi)]
Metals and alloys[a]							
1040 carbon steel	0.30	235		180(133)			280(41)
Mild steel						140	
Medium-carbon steel						51	400(58)
8630 low-alloy steel	0.30	220		51(41)			170(25)
304 stainless steel	0.29	250		34(25)			
410 stainless steel				26(19)			
L2 tool steel							
Rotor steels (A533; Discalloy)						204–214	
Pressure-vessel steels (HY130)						170	
High-strength steels (HSS)						50–154	
Ductile iron	0.29	167		9(7)			
Cast iron						6–20	
Pure ductile metals (e.g., Cu, Ni, Ag, Al)						100–350	
Be (brittle hcp metal)						4	
3003-H14 aluminum	0.33	40					62(9)
2048, plate aluminum				10.3(7.6)			
Aluminum alloys (high strength-low strength)						23–45	
AZ31B magnesium	0.35	73		4.3(3.2)			
AM100A casting magnesium	0.35	53		0.8(0.6)			
Ti-5Al-2.5Sn	0.35	335		23(17)			69(10)
Ti-6Al-4V	0.33						410(59)
Titanium alloys						55–115	
Aluminum bronze, 9% (copper alloy)	0.33	165		48(35)			200(29)
Monel 400 (nickel alloy)	0.32	110–150		298(220)			290(42)
AC41A zinc		91					
50:50 solder (lead alloy)		14.5		21.6(15.9)			56(8)
Nb-1 Zr (refractory metal)				174(128)			
Dental gold alloy (precious metal)		80–90					

Continued

	Poisson's ratio ν	Brinell hardness number	Rockwell hardness R scale	Charpy impact energy [J(ft·lb)]	Izod impact energy [J(ft·lb)]	K_{IC} (MPa $\sqrt{m}$)	Fatigue limit [MPa (ksi)]
Ceramics and glasses[b]							
Al$_2$O$_3$	0.26					3–5	
BeO	0.26						
CeO$_2$	0.27–0.31						
MgO						3	
Cordierite (2MgO·2Al$_2$O$_3$·5SiO$_2$)	0.31						
Mullite (3Al$_2$O$_3$·2SiO$_2$)	0.25						
SiC	0.19					3	
Si$_3$N$_4$	0.24					4–5	
TaC	0.24						
TiC	0.19						
TiO$_2$	0.28						
Partially stabilized ZrO$_2$	0.23					9	
Fully stabilized ZrO$_2$	0.23–0.32						
Glass-ceramic (MgO–Al$_2$O$_3$–SiO$_2$)	0.24						
Electrical porcelain						1	
Cement/concrete, unreinforced						0.2	
Soda glass (Na$_2$O–iO$_2$)						0.7–0.8	
Borosilicate glass	0.20						
Glass from cordierite	0.26						
Polymers[c]							
Polyethylene							
High-density			40		1.4–16(1–12)	2	
Low-density			10		22(16)	1	
Polyvinylchloride			110		1.4(1)	—	
Polypropylene			90		1.4–15(1–11)	3	
Polystyrene			75		0.4(0.3)	2	
Polyesters			120		1.4(1)	0.5	40.7(5.9)
Acrylics (Lucite)			130		0.7(0.5)	—	
Polyamides (nylon 66)	0.41		121		1.4(1)	3	
Cellulosics			50 to 115		3–11(2–8)	—	
ABS			95		1.4–14(1–10)	4	
Polycarbonates			118		19(14)	1.0–2.6	
Acetals	0.35		120		3(2)	—	31(4.5)
Polytetrafluoroethylene (Teflon)			70		5(4)	—	
Polyester-type thermoplastic elastomers					—	—	
Phenolics (phenolformaldehyde)			125		0.4(0.3)	—	
Urethanes			—		—	—	
Urea-melamine			115		0.4(0.3)	—	
Polyesters			100		0.5(0.4)	—	
Epoxies			90		1.1(0.8)	0.3–0.5	

Appendix 4C

Continued

	Poisson's ratio ν	Brinell hardness number	Rockwell hardness R scale	Charpy impact energy [J(ft-lb)]	Izod impact energy [J(ft-lb)]	K_{IC} (MPa $\sqrt{m}$)	Fatigue limit [MPa (ksi)]
Composites[c]							
E-glass (73.3 vol %) in epoxy (parallel loading of continuous fibers)						42–60	
B (70 vol %) in epoxy (parallel loading of continuous filaments)						46	
SiC whiskers in Al_2O_3						8.7	
SiC fibers in SiC						25.0	
SiC whiskers in reaction-bonded Si_3N_4						20.0	

[a]From Tables 6.3, 6.10, 8.1, 8.3, and 8.4
[b]From Tables 6.5 and 8.3
[c]From Tables 6.6, 6.11, 8.2, and 8.3
[d]From Table 12.7

Appendix 4D

Thermal Properties Data for Selected Materials

	Specific heat[a] c_p [J/kg · K]	Linear coefficient of thermal expansion[b] α [mm/(mm · °C)] $\times 10^6$			Thermal conductivity[c] k [J/(s · m · K)]			
		27°C	527°C	0–1000°C	27°C	100°C	527°C	1000°C
Metals[a]								
Aluminum	900	23.2	33.8		237		220	
Copper	385	16.8	20.0		398		371	
Gold	129	14.1	16.5		315		292	
Iron (α)	444				80		43	
Lead	159							
Nickel	444	12.7	16.8		91		67	
Silver	237	19.2	23.4		427		389	
Titanium	523				22		20	
Tungsten	133	4.5	4.8		178		128	
Ceramics and glasses[b]								
Mullite ($3Al_2O_3 \cdot 2SiO_2$)				5.3		5.9		3.8
Porcelain				6.0		1.7		1.9
Fireclay refractory				5.5		1.1		1.5
Al_2O_3	160			8.8		30		6.3
Spinel ($MgO \cdot Al_2O_3$)				7.6		15		5.9
MgO	457			13.5		38		7.1
UO_2				10.0				
ZrO_2 (stabilized)				10.0		2.0		2.3
SiC	344			4.7				
TiC						25		5.9
Carbon (diamond)	519							
Carbon (graphite)	711							
Silica glass				0.5		2.0		2.5
Soda–lime–silica glass				9.0		1.7		
Polymers[c]								
Nylon 66	1,260–2,090	30–31			2.9			
Phenolic	1,460–1,670	30–45			0.17–0.52			
Polyethylene (high-density)	1,920–2,300	149–301			0.33			
Polypropylene	1,880	68–104			2.1–2.4			
Polytetrafluoroethylene (PTFE)	1,050	99			0.24			

[a]From Table 7.1
[b]From Table 7.2
[c]From Table 7.4

APPENDIX FIVE
Properties of the Electronic Materials

The related Web site provides numerous links to data sources. Many of the professional society Web sites are especially useful in this regard.

Appendix 5A

Electrical Conductivities of Selected Materials at Room Temperature

Material	Conductivity, σ ($\Omega^{-1} \cdot m^{-1}$)
Metals and alloys[a]	
Aluminum (annealed)	35.36×10^6
Copper (annealed standard)	58.00×10^6
Gold	41.0×10^6
Iron (99.9+%)	10.3×10^6
Lead (99.73+%)	4.84×10^6
Magnesium (99.80%)	22.4×10^6
Mercury	1.04×10^6
Nickel (99.95% + Co)	14.6×10^6
Nichrome (66% Ni + Cr and Fe)	1.00×10^6
Platinum (99.99%)	9.43×10^6
Silver (99.78%)	62.9×10^6
Steel (wire)	$5.71 - 9.35 \times 10^6$
Tungsten	18.1×10^6
Zinc	16.90×10^6
Semiconductors[b]	
Silicon (high purity)	0.40×10^{-3}
Germanium (high purity)	2.0
Gallium arsenide (high purity)	0.17×10^{-6}
Indium antimonide (high purity)	17×10^3
Lead sulfide (high purity)	38.4
Ceramics, glasses, and polymers[c]	
Aluminum oxide	$10^{-10} - 10^{-12}$
Borosilicate glass	10^{-13}
Polyethylene	$10^{-13} - 10^{-15}$
Nylon 66	$10^{-12} - 10^{-13}$

[a]From Tables 13.1 and 13.2
[b]From Tables 13.1 and 13.5
[c]From Table 13.1

Appendix 5B

Properties of Some Common Semiconductors at Room Temperature

Material	Energy gap, E_g (eV)	Electron mobility, μ_e[m²/(V·s)]	Hole mobility, μ_h[m²/(V·s)]	Carrier density, $n_e (= n_h)$ (m⁻³)
Elements[a]				
Si	1.107	0.140	0.038	14×10^{15}
Ge	0.66	0.364	0.190	23×10^{18}
III–V Compounds[b]				
AlSb	1.60	0.090	0.040	—
GaP	2.25	0.030	0.015	—
GaAs	1.47	0.720	0.020	1.4×10^{12}
GaSb	0.68	0.500	0.100	—
InP	1.27	0.460	0.010	—
InAs	0.36	3.300	0.045	—
InSb	0.17	8.000	0.045	13.5×10^{21}
II–VI Compounds[b]				
ZnSe	2.67	0.053	0.002	—
ZnTe	2.26	0.053	0.090	—
CdS	2.59	0.034	0.002	—
CdTe	1.50	0.070	0.007	—
HgTe	0.025	2.200	0.016	—

[a]From Table 13.5
[b]From Table 13.9

Appendix 5C

Dielectric Constant and Dielectric Strength for Selected Insulators

Material	Dielectric constant,[a] κ	Dielectric strength,[a] (kV/mm)
Al_2O_3 (99.9%)	10.1	9.1
Al_2O_3 (99.5%)	9.8	9.5
BeO (99.5%)	6.7	10.2
Cordierite	4.1–5.3	2.4–7.9
Nylon 66—reinforced with 33% glass fibers (dry-as-molded)	3.7	20.5
Nylon 66—reinforced with 33% glass fibers (50% relative humidity)	7.8	17.3
Acetal (50% relative humidity)	3.7	19.7
Polyester	3.6	21.7

[a]From Table 13.4

APPENDIX SIX
Glossary

Following are definitions of the key words that appeared at the end of each chapter.

abrasive wear Wear occurring when a rough, hard surface slides on a softer surface. Grooves and wear particles form on the softer surface.

acceptor level Energy level near the valence band in a *p*-type semiconductor. (See Figure 13.31.)

activation energy Energy barrier that must be overcome by atoms in a given process or reaction.

addition polymerization See *chain growth.*

adhesive A thermosetting polymer that serves to join two solids by secondary forces similar to those between molecular chains in thermoplastics.

adhesive wear Wear occurring when two smooth surfaces slide over each other. Fragments are pulled off one surface and adhere to the other.

advanced composites Synthetic fiber-reinforced composites with relatively high modulus fibers. The fiber modulus is generally higher than that of E-glass.

aerosol Airborne particle.

age hardening See *precipitation hardening.*

aggregate composite Material reinforced with a dispersed particulate (rather than fibrous) phase.

alloy Metal composed of more than one element.

amorphous metal Metal lacking long-range crystalline structure.

amorphous semiconductor Semiconductor lacking long-range crystalline structure.

amplifier Electronic device for increasing current.

anion Negatively charged ion.

anisotropic Having properties that vary with direction.

annealing Heat treatment for the general purpose of softening or stress-relieving a material.

annealing point Temperature at which a glass has a viscosity of $10^{13.4}$ P and at which internal stresses can be relieved in about 15 minutes.

anode The electrode in an electrochemical cell that oxidizes, giving up electrons to an external circuit.

anodic reaction The oxidation reaction occurring at the anode in an electrochemical cell.

anodized aluminum Aluminum alloy with a protective layer of Al_2O_3.

Arrhenius equation General expression for a thermally activated process, such as diffusion. (See Equation 5.1.)

Arrhenius plot A semilog plot of ln (rate) versus the reciprocal of absolute temperature $(1/T)$. The slope of the straight-line plot gives the activation energy of the rate mechanism.

Ashby chart Plot such as Figure 14.3 in which pairs of materials properties are plotted against each other, with various categories of structural materials tending to cluster together.

atactic Irregular alteration of side groups along a polymeric molecule. (See Figure 12.10.)

atomic mass Mass of an individual atom expressed in atomic mass units.

atomic mass unit (amu) Equal to 1.66×10^{-24} g. Approximately equal to the mass of a proton or neutron.

atomic number Number of protons in the nucleus of an atom.

atomic packing factor (APF) Fraction of unit-cell volume occupied by atoms.

atomic radius Distance from atomic nucleus to outermost electron orbital.

austempering Heat treatment of a steel that involves holding just above the martensitic transformation range long enough to completely form bainite. (See Figure 10.20.)

austenite Face-centered cubic (γ) phase of iron or steel.

austenitic stainless steel Corrosion-resistant ferrous alloy with a predominant face-centered cubic (γ) phase.

Avogadro's number Equal to the number of atomic mass units per gram of material (0.6023×10^{24}).

bainite Extremely fine needlelike microstructure of ferrite and cementite. (See Figure 10.9.)

base Intermediate region between emitter and collector in a transistor.

Bernal model Representation of the atomic structure of an amorphous metal as a connected set of polyhedra. (See Figure 4.22.)

bifunctional Polymer with two reaction sites for each mer, resulting in a linear molecular structure.

binary diagram Two-component phase diagram.

biological material A naturally occurring structural material, such as bone.

biomaterial An engineered material created for a biological or medical application.

biomimetic processing A ceramic fabrication technique that imitates natural processes, such as seashell formation.

bipolar junction transistor (BJT) A "sandwich" configuration such as the *p–n–p* transistor shown in Figure 13.39.

blend Molecular-scale polymeric mixture. (See Figure 12.4.)

block copolymer Combination of polymeric components in "blocks" along a single molecular chain. (See Figure 12.3.)

blow molding Processing technique for thermoplastic polymers.

body-centered cubic Common atomic arrangement for metals, as illustrated in Figure 3.4.

bond angle Angle formed by three adjacent, directionally bonded atoms.

bond energy Net energy of attraction (or repulsion) as a function of separation distance between two atoms or ions.

bond force Net force of attraction (or repulsion) as a function of separation distance between two atoms or ions.

bond length Center-to-center separation distance between two adjacent, bonded atoms or ions.

borosilicate glass High-durability, commercial glassware primarily composed of silica, with a significant component of B_2O_3.

Bragg angle Angle relative to a crystal plane from which x-ray diffraction occurs. (See Figure 3.31.)

Bragg equation See *Bragg's law.*

Bragg's law The relationship defining the condition for x-ray diffraction by a given crystal plane (Equation 3.5).

branching The addition of a polymeric molecule to the side of a main molecular chain. (See Figure 12.11.)

brass The most common copper alloy having zinc as a predominant substitutional solute.

Bravais lattice See *crystal lattice.*

brazing Joining of metal parts in which the braze metal melts, but the parts being joined may not.

Brinell hardness number (BHN) Parameter obtained from an indentation test, as defined in Table 6.9.

brittle Lacking in deformability.

brittle fracture Failure of a material following mechanical deformation with the absence of significant ductility.

bronze Copper alloy involving elements such as tin, aluminum, silicon, and nickel providing both high corrosion resistance and strength.

Bronze Age The period of time from roughly 2000 B.C. to 1000 B.C. representing the foundation of metallurgy.

buckminsterfullerene Carbon molecule named in honor of R. Buckminster Fuller, inventor of the geodesic dome, which resembled the initial fullerene, C_{60}.

buckyball Nickname for the buckminsterfullerene molecule, C_{60}.

buckytube Cylindrical buckminsterfullerene molecule composed of hexagonal carbon rings.

bulk amorphous alloy Noncrystalline metal with three large dimensions resulting from the rapid cooling of a material with multiple alloying elements.

bulk diffusion See *volume diffusion.*

Burgers vector Displacement vector necessary to close a stepwise loop around a dislocation.

capacitor Electrical device involving two electrodes separated by a dielectric.

carbon–carbon composite An advanced composite system with especially high modulus and strength.

carbon steel Ferrous alloy with nominal impurity level and carbon as the primary compositional variable.

carburization The diffusion of carbon atoms into the surface of steel for the purpose of hardening the alloy.

casting Material-processing technique involving pouring of molten liquid into a mold, followed by solidification of the liquid.

cast iron Ferrous alloy with greater than 2 wt % carbon.

cathode The electrode in an electrochemical cell that accepts electrons from an external circuit.

cathodic reaction The reduction reaction occurring at the cathode in an electrochemical cell.

cation Positively charged ion.

ceramic Nonmetallic, inorganic engineering material.

ceramic–matrix composite Composite material in which the reinforcing phase is dispersed in a ceramic.

cermet A ceramic–metal composite.

cesium chloride Simple compound crystal structure, as illustrated in Figure 3.8.

chain growth Polymerization process involving a rapid chain reaction of chemically activated monomers.

charge carrier Atomic-scale species by which electricity is conducted in materials.

charge density Number of charge carriers per unit volume.

Charpy test Method for measuring impact energy, as illustrated by Figure 8.1.

chemically strengthened glass A fracture-resistant glass produced by the compressive stressing of the silicate network as a result of the chemical exchange of larger radius K^+ ions for the Na^+ ions in the glass surface.

chemical vapor deposition The production of thin layers of materials in the fabrication of integrated circuits by means of specific chemical reactions.

chip A thin slice of crystalline semiconductor upon which electrical circuitry is produced by controlled diffusion.

clay Fine-grained soil, composed chiefly of hydrous aluminosilicate minerals.

coherent interface Interface between a matrix and precipitate at which the crystallographic structures maintain registry.

cold work   Mechanical deformation of a metal at relatively low temperatures.

collector   Region in a transistor that receives charge carriers.

colorant   Additive for a polymer for the purpose of providing color.

complete solid solution   Binary phase diagram representing two components that can dissolve in all proportions. (See Figure 9.5.)

complex failure   Material fracture involving the sequential operation of two distinct mechanisms.

component   Distinct chemical substance (e.g., Al or Al_2O_3).

composite   Material composed of a microscopic-scale combination of individual materials from the categories of metals, ceramics (and glasses), and polymers.

compound semiconductor   Semiconductor consisting of a chemical compound rather than a single element.

compression molding   Processing technique for thermosetting polymers.

concentration gradient   Change in concentration of a given diffusing species with distance.

condensation polymerization   See *step growth*.

conduction band   A range of electron energies in a solid associated with an unoccupied energy level in an isolated atom. An electron in a semiconductor becomes a charge carrier upon promotion to this band.

conduction electron   A negative charge carrier in a semiconductor. (See also *conduction band*.)

conductivity   Reciprocal of electrical resistivity.

conductor   Material with a substantial level of electrical conduction (e.g., a conductivity greater than 10^{+4} $\Omega^{-1} \cdot$ m^{-1}).

congruent melting   The case in which the liquid formed upon melting has the same composition as the solid from which it was formed.

continuous cooling transformation (CCT) diagram   A plot of percentage phase transformation under non-isothermal conditions using axes of temperature and time.

continuous fiber   Composite reinforcing fiber without a break within the dimensions of the matrix.

controlled devitrification   Processing technique for glass-ceramics in which a glass is transformed to a fine-grained, crystalline ceramic.

coordination number (CN)   Number of adjacent ions (or atoms) surrounding a reference ion (or atom).

copolymer   Alloylike result of polymerization of an intimate solution of different types of monomers.

corrosion   The dissolution of a metal in an aqueous environment.

corrosion-fatigue failure   Metal fracture due to the combined actions of a cyclic stress and a corrosive environment.

corrosive wear   Wear that takes place with sliding in a corrosive environment.

corundum   Compound crystal structure, as illustrated in Figure 3.13.

coulombic attraction   Tendency toward bonding between oppositely charged species.

covalent bond   Primary, chemical bond involving electron sharing between atoms.

creep   Plastic (permanent) deformation occurring at a relatively high temperature under constant load over a long time period.

creep curve   Characteristic plot of strain versus time for a material undergoing creep deformation. (See Figure 6.32.)

creep-rupture failure   Material fracture following plastic deformation at a relatively high temperature (under constant load over a long time period).

cristobalite   Compound crystal structure, as illustrated in Figure 3.11.

critical current density   Current flow at which a material stops being superconducting.

critical resolved shear stress   Stress operating on a slip system and great enough to produce slip by dislocation motion. (See Equation 6.15.)

critical shear stress   Theoretical stress level for the shearing of a crystal in the absence of dislocations.

cross-linking   Joining of adjacent, linear polymeric molecules by chemical bonding. See, for example, the vulcanization of rubber in Figure 6.47.

crystal (Bravais) lattice   The 14 possible arrangements of points (with equivalent environments) in three-dimensional space.

crystalline   Having constituent atoms stacked together in a regular, repeating pattern.

crystalline ceramic   A ceramic material with a predominantly crystalline atomic structure.

crystal system   The seven, unique unit-cell shapes that can be stacked together to fill three-dimensional space.

cubic   Simplest of the seven crystal systems. (See Table 3.1.)

cubic close packed (CCP)   See *face-centered cubic*.

current   Flow of charge carriers in an electrical circuit.

Debye temperature   The temperature above which the value of the heat capacity at constant volume, C_v, levels off at approximately $3R$, where R is the universal gas constant.

degree of polymerization (DP)   Average number of mers in a polymeric molecule.

degrees of freedom   Number of independent variables available in specifying an equilibrium microstructure.

delocalized electron   An electron equally probable to be associated with any of a large number of adjacent atoms.

design for the environment (DFE)   The inclusion of environmental concerns in the process of engineering design.

design parameter Material property that serves as the basis for selecting a given engineering material for a given application.

design selection Method for corrosion prevention (e.g., avoid a small-area anode next to a large-area cathode).

device A functional electronic design (e.g., the transistor).

diamond cubic Important crystal structure for covalently bonded, elemental solids, as illustrated in Figure 3.20.

dielectric Electrically insulating material.

dielectric constant The factor by which the capacitance of a parallel-plate capacitor is increased by inserting a dielectric in place of a vacuum.

dielectric strength The voltage gradient at which a dielectric "breaks down" and becomes conductive.

diffraction The directional scattering of radiation by a regular array of scattering centers whose spacing is about the same as the wavelength of the radiation. (See also *x-ray diffraction*.)

diffraction angle Twice the Bragg angle. (See Figure 3.32.)

diffractometer An electromechanical scanning device for obtaining an x-ray diffraction pattern of a powder sample.

diffusion The movement of atoms or molecules from an area of higher concentration to an area of lower concentration.

diffusional transformation Phase transformation with strong time dependence due to a mechanism of atomic diffusion.

diffusionless transformation Phase transformation that is essentially time independent due to the absence of a diffusional mechanism.

diffusion coefficient (diffusivity) Proportionality constant in the relationship between flux and concentration gradient, as defined in Equation 5.8.

diffusivity See *diffusion coefficient*.

diode A simple electronic device that limits current flow to an application of positive voltage (forward bias).

dipole Asymmetrical distribution of positive and negative charge associated with secondary bonding.

dipole moment Product of charge and separation distance between centers of positive and negative charge in a dipole.

discrete (chopped) fiber Composite reinforcing fiber broken into segments.

dislocation Linear defect in a crystalline solid.

dislocation climb A mechanism for creep deformation, in which the dislocation moves to an adjacent slip plane by diffusion.

dispersion-strengthened metal An aggregate-type composite in which the metal contains less than 15 vol % oxide particles (0.01 to 0.1 micron in diameter).

donor level Energy level near the conduction band in an *n*-type semiconductor. (See Figure 13.26.)

dopant Purposeful impurity addition in an extrinsic semiconductor.

double bond Covalent sharing of two pairs of valence electrons.

drain Region in a field-effect transistor that receives charge carriers.

drift velocity Average velocity of the charge carrier in an electrically conducting material.

ductile fracture Metal failure occurring after the material is taken beyond its elastic limit.

ductile iron A form of cast iron that is relatively ductile due to spheroidal graphite precipitates rather than flakes.

ductile-to-brittle transition temperature Narrow temperature region in which the fracture of bcc alloys changes from brittle (at lower temperatures) to ductile (at higher temperatures).

ductility Deformability. (Percent elongation at failure is a quantitative measure.)

dye Soluble, organic colorant for polymers.

dynamic modulus of elasticity Parameter representing the stiffness of a polymeric material under an oscillating load.

edge dislocation Linear defect with the Burgers vector perpendicular to the dislocation line.

effusion cell Resistance-heated furnace providing a source of atoms (or molecules) for a deposition process.

E-glass Most generally used glass-fiber composition for composite applications. (See Table 12.1.)

elastic deformation Temporary deformation associated with the stretching of atomic bonds. (See Figure 6.18.)

elastomer Polymer with a pronounced rubbery plateau in its plot of modulus of elasticity versus temperature.

electrical conduction Having a measurable conductivity due to the movement of any type of charge carrier.

electrical field strength Voltage per unit distance.

electric permittivity Proportionality constant in the relationship between charge density and electrical field strength, as defined in Equation 13.11.

electrochemical cell System providing for connected anodic and cathodic electrode reactions.

electromotive force (emf) series Systematic listing of half-cell reaction voltages, as summarized in Table 14.10.

electron Negatively charged subatomic particle located in an orbital about a positively charged nucleus.

electron density Concentration of negative charge in an electron orbital.

electronegativity The ability of an atom to attract electrons to itself.

electron hole Missing electron in an electron cloud. A charge carrier with an effective positive charge.

electron-hole pair Two charge carriers produced when an electron is promoted to the conduction band, leaving behind an electron hole in the valence band.

electronic ceramic Ceramic material with an engineering application predominantly based on its electronic properties.

electronic conduction Having a measurable conductivity due specifically to the movement of electrons.

electron orbital Location of negative charge about a positive nucleus in an atom.

emitter Region in a transistor that serves as the source of charge carriers.

enamel Glass coating on a metal substrate.

energy band A range of electron energies in a solid associated with an energy level in an isolated atom.

energy band gap Range of electron energies above the valence band and below the conduction band.

energy level Fixed binding energy between an electron and its nucleus.

energy trough See *energy well*.

energy well The region around the energy minimum in a bonding energy curve such as that shown in Figure 2.18.

engineering polymer Polymer with sufficient strength and stiffness to be a candidate for a structural application previously reserved for a metal alloy.

engineering strain Increase in sample length at a given load divided by the original (stress-free) length.

engineering stress Load on a sample divided by the original (stress-free) area.

environmental impact assessment (EIA) An evaluation of the potential impact of emissions associated with an engineering process for the purpose of identifying major problems.

epitaxy A vapor deposition technique involving the buildup of thin-film layers of one semiconductor on another while maintaining some particular crystallographic relationship between the layer and the substrate.

eutectic composition Composition associated with the minimum temperature at which a binary system is fully melted. (See Figure 9.11.)

eutectic diagram Binary phase diagram with the characteristic eutectic reaction.

eutectic reaction The transformation of a liquid to two solid phases upon cooling, as summarized by Equation 9.3.

eutectic temperature Minimum temperature at which a binary system is fully melted. (See Figure 9.11.)

eutectoid diagram Binary phase diagram with the eutectoid reaction (Equation 9.4). (See Figure 9.18.)

exhaustion range Temperature range over which conductivity in an *n*-type semiconductor is relatively constant due to the fact that impurity-donated electrons have all been promoted to the conduction band.

extended length Length of a polymeric molecule that is extended as straight as possible. (See Figure 12.9.)

extrinsic gettering The capture of oxygen in a silicon device by using mechanical damage to produce dislocations as "gettering sites."

extrinsic semiconduction Semiconducting behavior due to an impurity addition.

extrusion molding Processing technique for thermoplastic polymers.

face-centered cubic (fcc) Common atomic arrangement for metals, as illustrated in Figure 3.5.

failure analysis A systematic methodology for characterizing the failure of engineering materials.

failure prevention The application of knowledge obtained by failure analysis to prevent future catastrophes.

family of directions A set of structurally equivalent crystallographic directions.

family of planes A set of structurally equivalent crystallographic planes.

fast fracture See *flaw-induced fracture*.

fatigue The general phenomenon of material failure after several cycles of loading to a stress level below the ultimate tensile stress.

fatigue curve Characteristic plot of stress versus number of cycles to failure. (See Figure 8.10.)

fatigue failure The fracture of a metal part by a mechanism of slow crack growth, with a resulting "clamshell" fracture surface.

fatigue strength (endurance limit) Lower limit of the applied stress at which a ferrous alloy will fail by cyclic loading.

Fermi function Temperature-dependent function that indicates the extent to which a given electron energy level is filled. (See Equation 13.7.)

Fermi level Energy of an electron in the highest filled state in the valence energy band at 0 K.

ferritic stainless steel Corrosion-resistant ferrous alloy with a predominant body-centered cubic (α) phase.

ferroelectric Material that exhibits spontaneous polarization under an applied electrical field.

ferrous alloy Metal alloy composed of predominantly iron.

fiberglass Composite system composed of a polymeric matrix reinforced with glass fibers.

fiber-reinforced composite Material reinforced with a fibrous phase.

Fick's first and second laws The basic mathematical descriptions of diffusional flow. (See Equations 5.8 and 5.9.)

field-assisted sintering technique A sintering technique in which the sample is heated by the direct application of an external electrical current. As a result, the process can generally be carried out at substantially reduced temperatures and times compared with conventional sintering.

field-effect transistor (FET) Solid-state amplifier, as illustrated in Figure 13.40.

filler Relatively inert additive for a polymer, providing dimensional stability and reduced cost.

firing The processing of a ceramic by heating raw materials to a high temperature, typically above 1,000°C.

flame retardant Additive used to reduce the inherent combustibility of certain polymers.

flaw-induced fracture The rapid failure of a material due to the stress concentration associated with a pre-existing flaw.

flexural modulus Stiffness of a material, as measured in bending. (See Equation 6.12.)

flexural strength (F.S.) Failure stress of a material, as measured in bending. (See Equation 6.10.)

fluorite Compound crystal structure, as illustrated in Figure 3.10.

forward bias Orientation of electrical potential to provide a significant flow of charge carriers in a rectifier.

Fourier's law Relationship between rate of heat transfer and temperature gradient, as expressed in Equation 7.5.

fracture mechanics Analysis of failure of structural materials with preexisting flaws.

fracture toughness Critical value of the stress-intensity factor at a crack tip necessary to produce catastrophic failure.

free electron Conducting electron in a metal.

Frenkel defect Vacancy-interstitialcy defect combination, as illustrated in Figure 4.9.

fullerene See *buckminsterfullerene*.

fusion casting Ceramic processing technique similar to metal casting.

gage length Region of minimum cross-sectional area in a mechanical test specimen.

galvanic cell Electrochemical cell in which the corrosion and associated electrical current are due to the contact of two dissimilar metals.

galvanic corrosion Corrosion produced by the electromotive force associated with two dissimilar metals.

galvanic protection Design configuration in which the structural component to be protected is made to be the cathode and is, thereby, protected from corrosion.

galvanic series Systematic listing of relative corrosion behavior of metal alloys in an aqueous environment, such as seawater.

galvanized steel Steel with a zinc coating for the purpose of corrosion protection.

gate Intermediate region in a field-effect transistor. (See Figure 13.40.)

Gaussian error function Mathematical function based on the integration of the "bell-shaped" curve. It appears in the solution to many diffusion-related problems.

general diagram A binary phase diagram containing more than one of the simple types of reactions described in Section 9.2.

general yielding The slow "failure" of a material due to plastic deformation occurring at the yield strength.

gettering A process used to capture oxygen and remove it from a region of a silicon wafer where device circuitry is developed.

Gibbs phase rule General relationship between microstructure and state variables, as expressed in Equation 9.1.

glass Noncrystalline solid, unless otherwise noted, with a chemical composition comparable to a crystalline ceramic.

glass-ceramic Fine-grained, crystalline ceramic produced by the controlled devitrification of a glass.

glass container Common household product composed of approximately 15 wt % Na_2O, 10 wt % CaO, and 75 wt % SiO_2.

glass forming Processing technique for a glass.

glass transition temperature The temperature range, above which a glass becomes a supercooled liquid, and below which it is a true, rigid solid.

glaze Glass coating applied to a ceramic such as clayware pottery.

graft copolymer Combination of polymeric components in which one or more components is grafted onto a main polymeric chain.

grain Individual crystallite in a polycrystalline microstructure.

grain boundary Region of mismatch between two adjacent grains in a polycrystalline microstructure.

grain-boundary diffusion Enhanced atomic flow along the relatively open structure of the grain-boundary region.

grain-boundary dislocation (GBD) Linear defect within a grain boundary, separating regions of good correspondence.

grain growth Increase in average grain size of a polycrystalline microstructure due to solid-state diffusion.

grain-size number Index for characterizing the average grain size in a microstructure, as defined by Equation 4.1.

gram-atom Avogadro's number of atoms of a given element.

gray cast iron A form of cast iron containing sharp graphite flakes that contribute to a characteristic brittleness.

green engineering Environmentally sensitive engineering design.

Griffith crack model Prediction of stress intensification at the tip of a crack in a brittle material.

group Chemical elements in a vertical column of the periodic table.

Guinier–Preston (G.P.) zone Structure developed in the early stages of precipitation of an Al–Cu alloy. (See Figure 10.28.)

half cell The anodic or cathodic half of an electrochemical cell.

half-cell reaction   Chemical reaction associated with either the anodic or the cathodic half of an electrochemical cell.

hardenability   Relative ability of a steel to be hardened by quenching.

hardness   Resistance of a material to indentation.

hard sphere   Atomic (or ionic) model of an atom as a spherical particle with a fixed radius.

heat capacity   The amount of heat necessary to raise the temperature of one gram-atom (for elements) or one mole (for compounds) by 1 K ($= 1°C$). (See also *specific heat*.)

heat treatment   Temperature versus time history necessary to generate a desired microstructure.

heteroepitaxy   Deposition of a thin film of a composition significantly different from the substrate.

heterogeneous nucleation   The precipitation of a new phase occurs at some structural imperfection such as a foreign surface.

hexagonal   One of the seven crystal systems, as illustrated in Table 3.1.

hexagonal close packed (hcp)   Common atomic arrangement for metals, as illustrated in Figure 3.6.

high-alloy steel   Ferrous alloy with more than 5 wt % noncarbon additions.

high-strength, low-alloy (HSLA) steel   Steel with relatively high strength, but significantly less than 5 wt % noncarbon additions.

Hirth–Pound model   Atomistic model of the ledge-like structure of the surface of a crystalline material, as illustrated in Figure 4.17.

homoepitaxy   Deposition of a thin film of essentially the same material as the substrate.

homogeneous nucleation   The precipitation of a new phase occurs within a completely homogeneous medium.

Hooke's law   The linear relationship between stress and strain during elastic deformation. (See Equation 6.3.)

hot isostatic pressing (HIP)   Powder metallurgical technique combining high temperature and isostatic forming pressure.

Hume–Rothery rules   Four criteria for complete miscibility in metallic solid solutions.

Hund's rule   Electron pairing in a given orbital tends to be delayed until all levels of a given energy have a single electron.

hybrid   Woven fabric with two or more types of reinforcing fibers for use in a single composite.

hybridization   Formation of four equivalent electron energy levels (sp^3-type) from initially different levels (*s*-type and *p*-type).

hydrogen bridge   Secondary bond formed between two permanent dipoles in adjacent water molecules.

hydrogen embrittlement   Form of environmental degradation in which hydrogen gas permeates into a metal and forms brittle hydride compounds.

hypereutectic composition   Composition greater than that of the eutectic.

hypereutectoid composition   Composition greater than that of the eutectoid.

hypoeutectic composition   Composition less than that of the eutectic.

hypoeutectoid composition   Composition less than that of the eutectoid.

impact energy   Energy necessary to fracture a standard test piece with an impact load.

impressed voltage   A method of corrosion protection in which an external voltage is used to oppose the one due to an electrochemical reaction.

incongruent melting   The case in which the liquid formed upon melting has a different composition than the solid from which it was formed.

inhibitor   A substance, used in small concentrations, that decreases the rate of corrosion in a given environment.

initiator   Chemical species that triggers a chain-growth polymerization mechanism.

injection molding   Processing technique for thermoplastic polymers.

insulator   Material with a low level of electrical conduction (e.g., a conductivity less than $10^{-4}\ \Omega^{-1}\text{m}^{-1}$).

integrated circuit (IC)   A sophisticated electrical circuit produced by the application of precise patterns of diffusable *n*-type and *p*-type dopants to give numerous elements within a single-crystal chip.

interfacial strength   Strength of the bonding between a composite matrix and its reinforcing phase.

intermediate   An oxide whose structural role in a glass is between that of a network former and a network modifier.

intermediate compound   A chemical compound formed between two components in a binary system.

interplanar spacing   Distance between the centers of atoms in two adjacent crystal planes.

interstitialcy   Atom occupying an interstitial site not normally occupied by an atom in the perfect crystal structure or an extra atom inserted into the perfect crystal such that two atoms occupy positions close to a singly occupied atomic site in the perfect structure.

interstitial solid solution   An atomic-scale combination of more than one kind of atom, with a solute atom located in an interstice of the solvent crystal structure.

intrinsic gettering   The capture of oxygen in a silicon device by heat-treating the silicon wafer to form SiO_2 precipitates below the surface region where the integrated circuit is being developed.

intrinsic semiconduction   Semiconducting behavior independent of any impurity additions.

invariant point   Point in a phase diagram that has zero degrees of freedom.

ion   Charged species due to an electron(s) added to or removed from a neutral atom.

ionic bond   Primary, chemical bond involving electron transfer between atoms.

ionic packing factor (IPF)   The fraction of the unit-cell volume for a ceramic occupied by the various cations and anions.

ionic radius   See *atomic radius*. (Ionic radius is associated, of course, with an ion rather than a neutral atom.)

Iron Age   The period of time from roughly 1000 B.C. to 1 B.C. during which iron alloys largely replaced bronze for tool and weapon making in Europe.

isostrain   Loading condition for a composite in which the strain on the matrix and dispersed phase is the same.

isostress   Loading condition for a composite in which the stress on the matrix and dispersed phase is the same.

isotactic   Polymeric structure in which side groups are along one side of the molecule. (See Figure 12.10.)

isothermal transformation diagram   See *TTT diagram*.

isotope   Any of two or more forms of a chemical element with the same number of protons, but different numbers of neutrons.

Izod test   Impact test typically used for polymers.

Jominy end-quench test   Standardized experiment for comparing the hardenability of different steels.

kaolinite   Silicate crystal structure, as illustrated in Figure 3.14.

kevlar   DuPont trade name for a para-aramid fiber widely used in advanced composite systems.

kinetics   Science of time-dependent phase transformations.

Knudsen cell   See *effusion cell*.

laminate   Fiber-reinforced composite structure in which woven fabric is layered with the matrix.

lattice constant   Length of a unit-cell edge and/or angle between crystallographic axes.

lattice direction   Direction in a crystallographic lattice. (See Figure 3.24 for standard notation.)

lattice parameter   See *lattice constant*.

lattice plane   Plane in a crystallographic lattice. (See Figure 3.26 for standard notation.)

lattice point   One of a set of theoretical points that are distributed in a periodic fashion in three-dimensional space.

lattice position   Standard notation for a point in a crystallographic lattice, as illustrated in Figure 3.22.

lattice translation   Vector connecting equivalent positions in adjacent unit cells.

lever rule   Mechanical analog for the mass balance with which one can calculate the amount of each phase present in a two-phase microstructure. (See Equations 9.9 and 9.10.)

life-cycle assessment (LCA)   A "cradle-to-grave" evaluation of environmental and energy impacts of a given product design.

light-emitting diode (LED)   An electro-optical device illustrated in Figure 14.14.

linear coefficient of thermal expansion   Material parameter indicating dimensional change as a function of increasing temperature. (See Equation 7.4.)

linear defect   One-dimensional disorder in a crystalline structure, associated primarily with mechanical deformation. (See also *dislocation*.)

linear density   The number of atoms per unit length along a given direction in a crystal structure.

linear molecular structure   Polymeric structure associated with a bifunctional mer and illustrated by Figure 2.15.

liquid-erosion failure   A special form of wear damage in which a liquid is responsible for the removal of material.

liquid-metal embrittlement   A form of degradation in which a material loses some ductility or fractures below its yield stress in conjunction with surface wetting by a lower-melting-point liquid metal.

liquidus   In a phase diagram, the line above which a single liquid phase will be present.

lithography   Print-making technique applied to the processing of integrated circuits.

long-range order (LRO)   A structural characteristic of crystals (and not glasses).

low-alloy steel   Ferrous alloy with less than 5 wt % noncarbon additions.

lower-yield point   The onset of general plastic deformation in a low-carbon steel. (See Figure 6.10.)

malleable iron   A traditional form of cast iron with modest ductility. It is first cast as white iron and then heat-treated to produce nodular graphite precipitates.

martempering   Heat treatment of a steel involving a slow cool through the martensitic transformation range to reduce stresses associated with that crystallographic change.

martensite   Iron-carbon solid-solution phase with an acicular, or needlelike, microstructure produced by a diffusionless transformation associated with the quenching of austenite.

martensitic stainless steel   Corrosion-resistant ferrous alloy with a predominant martensitic phase.

martensitic transformation   Diffusionless transformation most commonly associated with the formation of martensite by the quenching of austenite.

mass balance   Method for calculating the relative amounts of the two phases in a binary microstructure. (See also *lever rule*.)

materials selection   Decision that is a critical component of the overall engineering design process.

matrix   The portion of a composite material in which a reinforcing, dispersed phase is embedded.

Maxwell–Boltzmann distribution   Description of the relative distribution of molecular energies in a gas.

mean free path   The average distance that an electron wave can travel without deflection.

medium-range order   Structural ordering occurring over the range of a few nanometers in an otherwise noncrystalline material.

melting point   Temperature at which a solid-to-liquid transformation occurs upon heating.

melting range   Temperature range over which the viscosity of a glass is between 50 and 500 P.

mer   Building block of a long-chain or network (polymeric) molecule.

metallic   Material having various properties such as ductility and electrical conductivity characteristic of metals.

metallic bond   Primary, chemical bond involving the nondirectional sharing of delocalized electrons.

metal–matrix composite   Composite material in which the reinforcing phase is dispersed in a metal.

metastable   A state that is stable with time, although it does not represent true equilibrium.

microcircuit   Microscopic-scale electrical circuit produced on a semiconductor substrate by controlled diffusion.

microstructural development   Changes in the composition and distribution of phases in a material's microstructure as a result of thermal history.

Miller–Bravais indices   Four-digit set of integers used to characterize a crystalline plane in the hexagonal system.

Miller indices   Set of integers used to characterize a crystalline plane.

mixed-bond character   Having more than one type of atomic bonding (e.g., covalent and secondary bonding in polyethylene).

mixed dislocation   Dislocation with both edge and screw character. (See Figure 4.13.)

modulus of elasticity   Slope of the stress-strain curve in the elastic region.

modulus of elasticity in bending   See *flexural modulus*.

modulus of rigidity   See *shear modulus*.

modulus of rupture (MOR)   See *flexural strength*.

mole   Avogadro's number of atoms or ions in the compositional unit of a compound (e.g., a mole of Al_2O_3 contains 2 moles of Al^{3+} ions and 3 moles of O^{2-} ions).

molecular-beam epitaxy (MBE)   A highly controlled ultrahigh-vacuum deposition process.

monomer   Individual molecule that combines with similar molecules to form a polymeric molecule.

Moore's law   During much of the history of the integrated circuit, the number of transistors produced in a single chip has roughly doubled every 2 years. (See Figure 13.45.)

near net-shape processing   Material processing with the goal of minimizing any final shaping operation.

net-shape processing   Material processing that does not require a subsequent shaping operation.

network copolymer   Alloylike combination of polymers with an overall network, rather than linear, structure.

network former   Oxides that form oxide polyhedra, leading to network structure formation in a glass.

network modifier   Oxides that do not form oxide polyhedra and, therefore, break up the network structure in a glass.

network molecular structure   Polymeric structure associated with a polyfunctional mer and illustrated in Figure 12.7.

neutron   Subatomic particle without a net charge and located in the atomic nucleus.

noncrystalline   Atomic arrangement lacking in long-range order.

noncrystalline solid   Solid lacking in long-range structural order.

nondestructive testing   The evaluation of engineering materials without impairing their usefulness.

nonferrous alloy   Metal alloy composed predominantly of an element(s) other than iron.

nonmetallic   Material such as a ceramic or polymer that is not metallic in nature.

nonoxide ceramic   Ceramic material composed predominantly of a compound(s) other than an oxide.

nonprimitive unit cell   Crystal structure having atoms at unit-cell positions in addition to the unit-cell corners.

nonsilicate glass   Glass composed predominantly of a compound(s) other than silica.

nonsilicate oxide ceramic   Ceramic material composed predominantly of an oxide compound(s) other than silica.

nonstoichiometric compound   Chemical compound in which variations in ionic charge lead to variations in the ratio of chemical elements, (e.g., $Fe_{1-x}O$).

n-type semiconductor   Extrinsic semiconductor in which the electrical conductivity is dominated by negative charge carriers.

nuclear ceramic   Ceramic material with a primary engineering application in the nuclear industry.

nucleation   First stage of a phase transformation, such as precipitation. (See Figure 10.2.)

nucleus   Central core of atomic structure, about which electrons orbit.

nylon   An important engineering polymer.

Ohm's law   Relationship between voltage, current, and resistance in an electrical circuit. (See Equation 13.1.)

1–2–3 superconductor   The material $YBa_2Cu_3O_7$, which is the most commonly studied ceramic superconductor and whose name is derived from the three metal ion subscripts.

optical fiber   A small-diameter glass fiber in which digital light pulses can be transmitted with low losses. (See Figure 1.10.)

orbital shell   Set of electrons in a given orbital.

ordered solid solution   Solid solution in which the solute atoms are arranged in a regular pattern. (See Figure 4.3.)

overaging   Continuation of the age-hardening process until the precipitates coalesce into a coarse dispersion, becoming a less effective dislocation barrier and leading to a drop in hardness.

oxidation   Reaction of a metal with atmospheric oxygen.

oxide glass   Noncrystalline solid in which one or more oxides are the predominant components.

partially stabilized zirconia (PSZ) ZrO_2 ceramic with a modest second component addition (e.g., CaO) producing a two-phase microstructure. Retention of some ZrO_2-rich phase in PSZ allows for the mechanism of transformation toughening.

particulate composite   Composite material with relatively large dispersed particles (at least several microns in diameter) and in a concentration greater than 25 vol %.

Pauli exclusion principle   Quantum-mechanical concept that no two electrons can occupy precisely the same state.

pearlite   Two-phase eutectoid microstructure of iron and iron carbide. (See Figure 9.2.)

periodic table   Systematic graphical arrangement of the elements indicating chemically similar groups. (See Figure 2.2.)

peritectic diagram   Binary phase diagram with the peritectic reaction (Equation 9.5). (See Figure 9.22.)

peritectic reaction   The transformation of a solid to a liquid and a solid of a different composition upon heating, as summarized by Equation 9.5.

phase   Chemically homogeneous portion of a microstructure.

phase diagram   Graphical representation of the state variables associated with microstructures.

phase field   Region of a phase diagram that corresponds to the existence of a given phase.

photonic material   Optical material in which signal transmission is by photons rather than by the electrons of electronic materials.

photoresist   Polymeric material used in the lithography process.

piezoelectricity   An electrical response to mechanical pressure application.

pigment   Insoluble, colored additive for polymers.

Pilling–Bedworth ratio   Ratio of oxide volume produced to the metal volume consumed in oxidation. (See Equation 14.13.)

planar defect   Two-dimensional disorder in a crystalline structure (e.g., a grain boundary).

planar density   The number of atoms per unit area in a given plane in a crystal structure.

plastic   See *polymer*.

plastic deformation   Permanent deformation associated with the distortion and reformation of atomic bonds.

plasticizer   Additive for the purpose of softening a polymer.

PM2.5   Airborne particulate matter less than 2.5 μm in diameter.

PM10   Airborne particulate matter less than 10 μm in diameter.

p–n junction   Boundary between adjacent regions of p-type and n-type material in a solid-state electronic device.

point defect   Zero-dimensional disorder in a crystalline structure, associated primarily with solid-state diffusion.

point lattice   See *crystal (Bravais) lattice*.

Poisson's ratio   Mechanical property indicating the contraction perpendicular to the extension caused by a tensile stress. (See Equation 6.5.)

polar molecule   Molecule with a permanent dipole moment.

polyethylene (PE)   The most widely used polymeric material. See Figure 3.18 for the unit-cell structure.

polyfunctional   Polymer with more than two reaction sites for each mer, resulting in a network molecular structure.

polymer   Engineering material composed of long-chain or network molecules.

polymeric molecule   A long-chain or network molecule composed of many building blocks (mers).

polymerization   Chemical process in which individual molecules (monomers) are converted to large molecular-weight molecules (polymers).

polymer–matrix composite   Composite material in which the reinforcing phase is dispersed in a polymer.

porcelain enamel   Silicate glass coating on a metal substrate for the purpose of shielding the metal from a corrosive environment.

powder metallurgy   Processing technique for metals involving the solid-state bonding of a fine-grained powder into a polycrystalline product.

precious metal   Generally corrosion-resistant metal or alloy, such as gold, platinum, and their alloys.

precipitation-hardened stainless steel   Corrosion-resistant ferrous alloy that has been strengthened by precipitation hardening.

precipitation hardening   Development of obstacles to dislocation motion (and thus to increased hardness) by the controlled precipitation of a second phase.

preexponential constant   Temperature-independent term that appears before the exponential term of the Arrhenius equation. (See Equation 5.1.)

primary bond   Relatively strong bond between adjacent atoms resulting from the transfer or sharing of outer orbital electrons.

primitive unit cell   Crystal structure that has atoms located at unit-cell corners only.

processing   Production of a material into a form convenient for engineering applications.

proeutectic   Phase that forms by precipitation in a temperature range above the eutectic temperature.

proeutectoid   Phase that forms by solid-state precipitation in a temperature range above the eutectoid temperature.

property averaging   Determination of the overall property (e.g., elastic modulus) of a composite material as the geometrical average of the properties of the individual phases.

prosthesis   Device for replacing a missing body part.

protective coating   A barrier between a metal and its corrosive environment.

proton   Positively charged subatomic particle located in the atomic nucleus.

p-type semiconductor   Extrinsic semiconductor in which the electrical conductivity is dominated by positive charge carriers.

pure oxide   Ceramic compound with a relatively low impurity level (typically less than 1 wt %).

quantum dot   Semiconducting material with three thin dimensions of the scale associated with a quantum well.

quantum well   Thin layer of semiconducting material in which the wavelike electrons are confined within the layer thickness.

quantum wire   Semiconducting material with two thin dimensions of the scale associated with a quantum well.

radiation   Various photons and atomic-scale particles that can be a source of environmental damage to materials.

radius ratio   The radius of a smaller ion divided by the radius of a larger one. This ratio establishes the number of larger ions that can be adjacent to the smaller one.

random network theory   Statement that a simple oxide glass can be described as the random linkage of "building blocks" (e.g., the silica tetrahedron).

random solid solution   Solid solution in which the solute atoms are arranged in an irregular fashion. (See Figure 4.3.)

random walk   Atomistic migration in which the direction of each step is randomly selected from among all possible orientations. (See Figure 5.6.)

rapidly solidified alloy   Metal alloy formed by a rapid solidification process.

rate-limiting step   Slowest step in a process involving sequential steps. The overall process rate is thereby established by that one mechanism.

recovery   Initial state of annealing in which atomic mobility is sufficient to allow some softening of the material without a significant microstructural change.

recrystallization   Nucleation and growth of a new stress-free microstructure from a cold-worked microstructure. (See Figures 10.30a–d.)

recrystallization temperature   The temperature at which atomic mobility is sufficient to affect mechanical properties as a result of recrystallization. The temperature is approximately one-third to one-half times the absolute melting point.

rectifier   See *diode*.

recycling   The reprocessing of a relatively inert engineering material.

reflection rules   Summary of which crystal planes in a given structure cause x-ray diffraction. (See Table 3.4.)

refractory   High-temperature-resistant material, such as many of the common ceramic oxides.

refractory metal   Metals and alloys (e.g., molybdenum) that are resistant to high temperatures.

reinforcement   Additive (e.g., glass fibers) providing a polymer with increased strength and stiffness.

relaxation time   The time necessary for the stress on a polymer to fall to 0.37 (= $1/e$) of the initial applied stress.

repulsive force   Force due to the like-charge repulsion of both (negative) electron orbitals and (positive) nuclei of adjacent atoms.

residual stress   The stress remaining within a structural material after all applied loads are removed.

resistance   Property of a material by which it opposes the flow of an electrical current. (See Equation 13.1.)

resistivity   The material property for electrical resistance normalized for sample geometry.

resolved shear stress   Stress operating on a slip system. (See Equation 6.14.)

reverse bias   Orientation of electrical potential to provide a minimal flow of charge carriers in a rectifier.

Rockwell hardness   Common mechanical parameter, as defined in Table 6.9.

root-mean-square length   Separation distance between the ends of a randomly coiled polymeric molecule. (See Equation 12.4 and Figure 12.8.)

sacrificial anode   Use of a less noble material to protect a structural metal from corrosion. (See Figure 14.22.)

saturation range   Temperature range over which conductivity in a p-type semiconductor is relatively constant due to the fact that all acceptor levels are "saturated" with electrons.

Schottky defect   A pair of oppositely charged ion vacancies. (See Figure 4.9.)

screw dislocation   Linear defect with the Burgers vector parallel to the dislocation line.

secondary bond   Atomic bond without electron transfer or sharing.

Seebeck effect   The development of an induced voltage in a simple electrical circuit as the result of a temperature differential.

Seebeck potential   Induced voltage due to a dissimilar metal thermocouple between two different temperatures. (See Figure 13.14.)

segregation coefficient   Ratio of the saturation impurity concentrations for the solid and liquid solution phases, as defined by Equation 13.22.

self-diffusion Atomic-scale migration of a species in its own phase.

self-propagating high-temperature synthesis (SHS) Material processing involving the heat evolved by certain chemical reactions to sustain the reaction and produce the final product.

semiconductor Material with a level of electrical conductivity intermediate between that for an insulator and a conductor (e.g., a conductivity between $10^{-4}\ \Omega^{-1} \cdot m^{-1}$ and $10^{+4}\ \Omega^{-1} \cdot m^{-1}$).

shear modulus Elastic modulus under pure shear loading. (See Equation 6.8.)

shear strain Elastic displacement produced by pure shear loading. (See Equation 6.7.)

shear stress Load per unit area (parallel to the applied load). See Equation 6.6.

short-range order (SRO) Local "building block" structure of a glass (comparable with the structural unit in a crystal of the same composition).

silica Silicon dioxide. One of various stable crystal structures, as illustrated in Figure 3.12.

silicate Ceramic compound with SiO_2 as a major constituent.

silicate glass Noncrystalline solid with SiO_2 as a major constituent.

sintering Bonding of powder particles by solid-state diffusion.

slip casting Processing technique for ceramics in which a powder-water mixture (slip) is poured into a porous mold.

slip system A combination of families of crystallographic planes and directions that corresponds to dislocation motion.

soda–lime–silica glass Noncrystalline solid composed of sodium, calcium, and silicon oxides. The majority of windows and glass containers are in this category.

sodium chloride Simple compound crystal structure, as illustrated in Figure 3.9.

softening point Temperature at which a glass has a viscosity of $10^{7.6}$ P, corresponding to the lower end of the working range.

softening temperature The temperature at which a material in a thermal expansion experiment can no longer support the weight of the length-monitoring probe.

soft sphere Atomic (or ionic) model that acknowledges that the outer orbital electron density does not terminate at a fixed radius.

soldering Joining of metal parts by the adhesion of melted solder to the surface of each part, without the melting of those parts.

sol–gel processing Technique for forming ceramics and glasses of high density at a relatively low temperature by means of an organometallic solution.

solid solution Atomic-scale intermixing of more than one atomic species in the solid state.

solidus In a phase diagram, the line below which only one or more solid phases are present.

solute Species dissolving in a solvent to form a solution.

solution hardening Mechanical strengthening of a material associated with the restriction of plastic deformation due to solid-solution formation.

solution treatment Heating of a two-phase microstructure to a single-phase region.

solvent Species into which a solute is dissolved in order to form a solution.

source Region in a field-effect transistor that provides charge carriers.

spalling Buckling and flaking off of an oxide coating on a metal due to large compressive stresses.

spark plasma sintering (SPS) See *field-assisted sintering technique (FAST)*.

specific heat The amount of heat necessary to raise the temperature of one unit mass of material by 1 K ($= 1°C$). (See also *heat capacity*.)

specific strength Strength per unit density.

spectral width A range of wavelengths.

stabilizer Additive for a polymer for the purpose of reducing degradation.

state Condition for a material, typically defined in terms of a specific temperature and composition.

state point A pair of temperature and composition values that defines a given state.

state variables Material properties, such as temperature and composition, that are used to define a state.

static fatigue For certain ceramics and glasses, a degradation in strength that occurs without cyclic loading.

steady-state diffusion Mass transport that is unchanging with time.

steel A ferrous alloy with up to approximately 2.0 wt % carbon.

step growth Polymerization process involving individual chemical reactions between pairs of reactive monomers.

Stone Age The time when our human ancestors, or hominids, chipped stones to form weapons for hunting. This age has been traced back as far as 2.5 million years ago.

strain hardening The strengthening of a metal alloy by deformation (due to the increasing difficulty for dislocation motion through the increasingly dense array of dislocations).

strain-hardening exponent The slope of a log-log plot of true stress versus true strain between the onset of plastic deformation of a metal alloy and the onset of necking. This parameter is an indicator of the alloy's ability to be deformed.

strength-to-weight ratio See *specific strength*.

stress-corrosion cracking (SCC) A combined mechanical and chemical failure mechanism in which a noncyclic tensile stress (below the yield strength) leads to the initiation and propagation of fracture in a relatively mild chemical environment.

stress intensity factor Parameter indicating the greater degree of mechanical stress at the tip of a pre-existing crack in a material under a mechanical load.

stress relaxation Mechanical phenomenon in certain polymers in which the stress on the material drops exponentially with time under a constant strain. (See Equation 6.17.)

stress-rupture failure See *creep-rupture failure.*

substitutional solid solution An atomic-scale combination of more than one kind of atom, with a solute atom substituting for a solvent atom at an atomic lattice site.

superalloy Broad class of metals with especially high strength at elevated temperatures.

superconductor Material that is generally a poor conductor at elevated temperatures, but that, upon cooling below a critical temperature, has zero resistivity.

superplastic forming Technique for forming complex-shaped metal parts from certain fine-grained alloys at elevated temperatures.

surface diffusion Enhanced atomic flow along the relatively open structure of a material's surface.

surface fatigue wear Wear occurring during repeated sliding or rolling of a material over a track.

syndiotactic Regular alteration of side groups along a polymeric molecule. (See Figure 12.10.)

temperature coefficient of resistivity The coefficient that indicates the dependence of a metal's resistivity on temperature. (See Equation 13.9.)

tempered glass A strengthened glass involving a heat treatment that serves to place the exterior surface in a residual compressive state.

tempered martensite An $\alpha + Fe_3C$ microstructure produced by heating the more brittle martensite phase.

tempering A thermal history for steel, as illustrated in Figure 10.17, in which martensite is reheated.

tensile strength (T.S.) The maximum engineering stress experienced by a material during a tensile test.

terminator Chemical species that ends a chain-growth polymerization mechanism.

thermal activation Atomic-scale process in which an energy barrier is overcome by thermal energy.

thermal conductivity Proportionality constant in the relationship between heat-transfer rate and temperature gradient, as defined in Equation 7.5.

thermal shock The fracture (partial or complete) of a material as the result of a temperature change (usually a sudden cooling).

thermal vibration Periodic oscillation of atoms in a solid at a temperature above absolute zero.

thermocouple Simple electrical circuit for the purpose of temperature measurement. (See Figure 13.14.)

thermoplastic elastomer Compositelike polymer with rigid, elastomeric domains in a relatively soft matrix of a crystalline, thermoplastic polymer.

thermoplastic polymer Polymer that becomes soft and deformable upon heating.

thermosetting polymer Polymer that becomes hard and rigid upon heating.

III–V compound Chemical compound between a metallic element in group III and a nonmetallic element in group V of the periodic table. Many of these compounds are semiconducting.

tie line Horizontal line (corresponding to constant temperature) that connects two phase compositions at the boundaries of a two-phase region of a phase diagram. (See Figure 9.6.)

tilt boundary Grain boundary associated with the tilting of a common crystallographic direction in two adjacent grains. (See Figure 4.19.)

tool steel Ferrous alloy used for cutting, forming, or otherwise shaping another material.

toughness Total area under the stress-strain curve.

transducer A device for converting one form of energy to another form.

transfer molding Processing technique for thermosetting polymers.

transformation toughening Mechanism for enhanced fracture toughness in a partially stabilized zirconia ceramic involving a stress-induced phase transformation of tetragonal grains to the monoclinic structure. (See Figure 8.7.)

transistor A solid-state amplifier.

TTT diagram A plot of the time necessary to reach a given percent transformation at a given temperature. (See Figure 10.6.)

twin boundary A planar defect separating two crystalline regions that are, structurally, mirror images of each other. (See Figure 4.15.)

II–VI compound Chemical compound between a metallic element in group II and a nonmetallic element in group VI of the periodic table. Many of these compounds are semiconducting.

ultrasonic testing A type of nondestructive testing in which defects are detected using high-frequency acoustical waves.

unit cell Structural unit that is repeated by translation in forming a crystalline structure.

upper yield point Distinct break from the elastic region in the stress-strain curve for a low-carbon steel. (See Figure 6.10.)

vacancy Unoccupied atom site in a crystal structure.

vacancy migration Movement of vacancies in the course of atomic diffusion without significant crystal-structure distortion. (See Figure 5.5.)

valence Electronic charge of an ion.

valence band A range of electron energies in a solid associated with the valence electrons of an isolated atom.

valence electron Outer orbital electron that takes part in atomic bonding. In a semiconductor, an electron in the valence band.

van der Waals bond See *secondary bond*.

vapor deposition Processing technique for semiconductor devices involving the buildup of material from the vapor phase.

viscoelastic deformation Mechanical behavior involving both fluidlike (viscous) and solidlike (elastic) characteristics.

viscosity Proportionality constant in the relationship between shearing force and velocity gradient, as defined in Equation 6.19.

viscous deformation Liquidlike mechanical behavior associated with glasses and polymers above their glass transition temperatures.

vitreous silica Commercial glass that is nearly pure SiO_2.

voltage A difference in electrical potential.

volume (bulk) diffusion Atomic flow within a material's crystal structure by means of some defect mechanism.

vulcanization The transformation of a polymer with a linear structure into one with a network structure by means of cross-linking.

wafer Thin slice from a cylindrical single crystal of high-purity material, usually silicon.

wear Removal of surface material as a result of mechanical action.

wear failure Surface-related damage phenomena, such as wear debris on sliding contact surfaces.

welding Joining of metal parts by local melting in the vicinity of the join.

whisker Small, single-crystal fiber with a nearly perfect crystalline structure that serves as a high-strength composite reinforcing phase.

white cast iron A hard, brittle form of cast iron with a characteristic white crystalline fracture surface.

whiteware Commercial fired ceramic with a typically white and fine-grained microstructure. Examples include tile, china, and pottery.

window glass Common structural material composed of approximately 15 wt % Na_2O, 10 wt % CaO, and 75 wt % SiO_2.

working range Temperature range in which glass product shapes are formed (corresponding to a viscosity range of 10^4 to 10^8 P).

woven fabric Composite reinforcing fiber configuration, as illustrated in Figure 12.14.

wrought alloy Metal alloy that has been rolled or forged into a final, relatively simple shape following an initial casting operation.

wrought process Rolling or forging of an alloy into a final, relatively simple shape following an initial casting step.

x-radiation The portion of the electromagnetic spectrum with a wavelength on the order of 1 nanometer. X-ray photons are produced by inner orbital electron transitions.

x-radiography A type of nondestructive testing in which defects are inspected using the attenuation of x-rays.

x-ray diffraction The reinforced scattering of x-ray photons by an atomic structure. Bragg's law indicates the structural information available from this phenomenon.

yield point See *upper yield point*.

yield strength (Y.S.) The strength of a material associated with the approximate upper limit of Hooke's law behavior, as illustrated in Figure 6.4.

Young's modulus See *modulus of elasticity*.

Zachariasen model Visual definition of the random network theory, as illustrated in Figure 4.21b.

zinc blende Compound crystal structure, as illustrated in Figure 3.21.

zone refining Technique for purifying materials by passing an induction coil along a bar of the material and using principles of phase equilibria, as illustrated in the feature box in Chapter 9.

Answers to Practice Problems (PP) and Odd-Numbered Problems

 Note that the solutions to all Practice Problems are provided on the related Web site.

Chapter 2

PP 2.1 (a) 3.38×10^{10} atoms, (b) 2.59×10^{10} atoms

PP 2.2 3.60 kg

PP 2.3 (a) 19.23 mm, (b) 26.34 mm

PP 2.4 (b) Mg: $1s^2 2s^2 2p^6 3s^2$, Mg^{2+}: $1s^2 2s^2 2p^6$; O: $1s^2 2s^2 2p^4$, O^{2-}: $1s^2 2s^2 2p^6$ (c) Ne for both cases of Mg^{2+} and O^{2-}

PP 2.5 $F_c = 20.9 \times 10^{-9}$ N, $F_R = -F_c = -20.9 \times 10^{-9}$ N

PP 2.6 (a) $\sin(109.5°/2) = R/(r + R)$ or $r/R = 0.225$
(b) $\sin 45° = R/(r + R)$ or $r/R = 0.414$

PP 2.7 CN = 6 for both cases

PP 2.8

H   H   H   H   H   H   H   H
| | | | | | | |
C = C   and ··· — C — C — C — C — C — C — ···, where $R = CH_3$
| | | | | | | |
H   R   H   R   H   R   H   R

PP 2.9 Same as for Practice Problem 2.8, except $R = C_6H_5$

PP 2.10 (a) 60 kJ/mol, (b) 60 kJ/mol

PP 2.11 794

PP 2.12 A greater degree of covalency in the Si–Si bond provides even stronger directionality and a lower coordination number.

2.1 8.33×10^{21} atoms

2.3 2.21×10^{15} atoms Si and 4.41×10^{15} atoms O

2.5 1.11×10^{12} atoms Al and 1.66×10^{12} atoms O

2.7 (a) 0.310×10^{24} molecules O_2, (b) 0.514 mol O_2

2.9 (a) 1.41 g, (b) 6.50×10^{-4} g

2.11 4.47 nm

2.15 -1.49×10^{-9} N

2.17 -8.13×10^{-9} N

2.19 -10.4×10^{-9} N

2.23 Pink

2.25 22.1×10^{-9} N

2.27 (b) 1.645

2.29 229 kJ

2.31 (b) 60 kJ/mol, (c) 678 kJ

2.33 335 kJ/mol

2.35 50,060 amu

2.37 8.99 kJ

2.39 (b) 60 kJ/mol, (c) 32,020 amu

2.49 392 m^2/kg

2.51 3.05×10^{23} atoms/(m^3 atm)

Chapter 3

PP 3.2 (a) $a = (4/\sqrt{3})r$, (b) $a = 2r$

PP 3.3 7.90 g/cm^3

PP 3.4 (a) 0.542, (b) 0.590, (c) 0.627

PP 3.5 3.45 g/cm^3

PP 3.6 5.70×10^{24}

PP 3.7 Directionality of covalent bonding dominates over efficient packing of spheres.

PP 3.8 5.39 g/cm^3

PP 3.9 (a) Body-centered position: $\frac{1}{2}\frac{1}{2}\frac{1}{2}$ (b) same, (c) same

PP 3.10 (a) 000, $\frac{1}{2}\frac{1}{2}\frac{1}{2}$, 111 (b) same, (c) same

PP 3.12 (a) $\langle 100 \rangle = [100]$, [010], [001]
$[\bar{1}00]$, $[0\bar{1}0]$, $[00\bar{1}]$

PP 3.13 (a) 45°, (b) 54.7°

PP 3.16 (a) 4.03 atoms/nm, (b) 1.63 atoms/nm

PP 3.17 **(a)** 7.04 atoms/nm^2, **(b)** 18.5 atoms/nm^2

PP 3.18 (1.21 Ca^{2+} + 1.21 O^{2-})/nm

PP 3.19 10.2 (Ca^{2+} or O^{2-})/nm^2

PP 3.20 2.05 atoms/nm

PP 3.21 7.27 atoms/nm^2

PP 3.22 78.5°, 82.8°, 99.5°, 113°, 117°

3.7 1.74 g/cm^3

3.13 Diameter of opening in center of unit cell = 0.21 nm

3.15 10.9 g/cm^3

3.17 0.317

3.19 1.24 eV

3.21 0.12

3.23 3.09 g/cm^3

3.27 **(b)** 26.6°, **(c)** 26.6°

3.29 Tetrahedron

3.31 $[0\bar{1}1]$

3.33 $[\bar{1}11]$, $[1\bar{1}1]$, $[1\bar{1}\bar{1}]$, and $[\bar{1}1\bar{1}]$

3.35 $[\bar{1}10]$, $[1\bar{1}0]$, $[101]$, $[\bar{1}0\bar{1}]$, $[011]$, and $[0\bar{1}\bar{1}]$

3.37 $[111]$, $[\bar{1}\bar{1}\bar{1}]$, $[1\bar{1}\bar{1}]$, and $[\bar{1}11\bar{1}]$

3.43 **(a)** (100), (010), ($\bar{1}$00), and (0$\bar{1}$0), **(b)** (100) and ($\bar{1}$00)

3.45 **(a)** 000, 112, 224

3.47 **(a)** 000, $\frac{1}{2}\frac{1}{2}1$, 112

3.53 000, $\frac{1}{2}\frac{1}{2}0$, $\frac{1}{2}0\frac{1}{2}$, $0\frac{1}{2}\frac{1}{2}$

3.55 000, $\frac{1}{2}\frac{1}{2}\frac{1}{2}$

3.63 Cl^- at 000, $\frac{1}{2}\frac{1}{2}0$, $\frac{1}{2}0\frac{1}{2}$, $0\frac{1}{2}\frac{1}{2}$, and Na^+ at $00\frac{1}{2}$, $\frac{1}{2}\frac{1}{2}\frac{1}{2}$, $\frac{1}{2}01$, $0\frac{1}{2}1$

3.67 6.56(Zn^{2+} or S^{2-})/nm^2

3.69 Zn^{2+} at 000, $\frac{1}{2}\frac{1}{2}0$, $\frac{1}{2}0\frac{1}{2}$, $0\frac{1}{2}\frac{1}{2}$ and S^{2-} at $\frac{1}{4}\frac{1}{4}\frac{1}{4}$, $\frac{3}{4}\frac{3}{4}\frac{1}{4}$, $\frac{3}{4}\frac{1}{4}\frac{3}{4}$, $\frac{1}{4}\frac{3}{4}\frac{3}{4}$

3.71 44.8°, 65.3°, 82.5°

3.73 (220), (310), (222)

3.77 (100), (002), (101)

3.79 48.8°, 52.5°, 55.9°

3.81 56.5°, 66.1°, 101°, 129°, 142°

3.83 Cr

3.85 38.44°, 44.83°, 65.19°

3.87 69.05°, 106.37°, 156.20°

Chapter 4

PP 4.1 No, % radius difference >15%

PP 4.2 **(b)** Roughly 50% too large

PP 4.3 5.31×10^{25} m^{-3}

PP 4.4 0.320 nm

PP 4.5 **(a)** 16.4 nm, **(b)** 3.28 nm

PP 4.6 G $\cong$ 2

PP 4.7 0.337

4.1 Rule number 3 (electronegativities differ by 27%), rule number 4 (valences are different)

4.3 Rule number 2 (different crystal structures), rule number 3 (electronegativities differ by 19%), possibly rule number 4 (same valences shown in Appendix 2, although Cu^+ is also stable)

4.9 0%

4.11 0.6023×10^{24} Mg^{2+} vacancies

4.13 **(a)** 10.0×10^{-6} at %, **(b)** 9.60×10^{-6} wt %

4.15 **(a)** 10.0×10^{-6} at %, **(b)** 11.0×10^{-6} wt %

4.17 5.00×10^{21} m^{-3}

4.19 0.269×10^{24} m^{-3}

4.21 **(a)** 2.67, **(b)** 1.33

4.23 **(a)** 3.00, **(c)** 2.67

4.25 $\approx$5.6

4.27 170 μm, 41.3 μm

Chapter 5

PP 5.1 0.572 kg/(m$^4 \cdot$ s)

PP 5.2 **(a)** 1.25×10^{-4}, **(b)** 9.00×10^{-8}, **(c)** 1.59×10^{-12}

PP 5.3 6.52×10^{19} atoms/(m$^2 \cdot$ s)

PP 5.4 **(a)** 0.79 wt % C, **(b)** 0.34 wt % C

PP 5.5 **(a)** 0.79 wt % C, **(b)** 0.34 wt % C

PP 5.6 970°C

PP 5.7 2.88×10^{-3} kg/h

PP 5.8 21.9 mm

5.1 500 kJ/mol

5.3 290 kJ/mol

5.15 179 kJ/mol

5.17 4.10×10^{-15} m^2/s

5.19 1.65×10^{-13} m^2/s

5.21 264 kJ/mol

5.23 Al^{3+} diffusion controlled ($Q = 477$ kJ/mol)

5.25 0.324×10^{-3} kg/m$^2 \cdot$ h

5.27 12.0×10^{12} atoms/s

5.29 86.6 kJ/mol

Chapter 6

PP 6.1 **(d)** 193×10^3 MPa (28.0×10^6 psi), **(e)** 275 MPa (39.9 ksi), **(f)** 550 MPa (79.8 ksi), **(g)** 30%

PP 6.2 **(a)** 1.46×10^{-3}, **(b)** 2.84×10^{-3}

PP 6.3 9.9932 mm

PP 6.4 **(a)** 80 MPa, **(b)** 25 MPa

PP 6.5 80.1 MPa

PP 6.6 3.57×10^{-5}

PP 6.7 **(a)** 0.2864 nm, **(b)** 0.2887 nm

PP 6.8 0.345 MPa (50.0 psi)

PP 6.9 4.08 mm

PP 6.10 **(a)** 1.47×10^{-3}% per hour, **(b)** 1.48×10^{-2}% per hour, **(c)** 9.98×10^{-2}% per hour

PP 6.11 **(a)** $\approx 550°C$, **(b)** $\approx 615°C$

PP 6.12 **(a)** 83.2 days, **(b)** 56.1 days

PP 6.13 511 to 537°C

6.1 108 GPa

6.3 **(a)** 4.00 GPa

6.5 3.46×10^5 N (7.77×10^4 lb$_f$)

6.7 **(a)** 2.96 MPa, **(b)** 2.39×10^{-5}

6.9 **(b)** Ti $-$ 5Al $-$ 2.5Sn, **(c)** 1040 carbon steel

6.13 **(a)** 78.5°, **(b)** 31.3 GPa

6.15 **(a)** 28.4 MPa, **(b)** 4.96×10^4 N

6.17 4.69 μm

6.19 8,970 MPa

6.21 **(a)** 1.11 GPa, **(b)** 1.54 GPa

6.23 97.5 MPa

6.25 $\left(\dfrac{dF}{da}\right)_{a_0} = -42\dfrac{K_A}{a_0^8} + 156\dfrac{K_R}{a_0^{14}}$

6.27 0.136 MPa

6.29 20.4 MPa

6.31 $(1\bar{1}1)[110]$, $(\bar{1}11)[110]$, $(11\bar{1})[101]$, $(\bar{1}11)[101]$, $(11\bar{1})[011]$, $(1\bar{1}1)[011]$

6.37 $(211)[\bar{1}11]$, $(121)[1\bar{1}1]$, $(112)[11\bar{1}]$
$(\bar{2}11)[111]$, $(1\bar{2}1)[111]$, $(11\bar{2})[111]$
$(2\bar{1}1)[11\bar{1}]$, $(\bar{1}21)[1\bar{1}1]$, $(\bar{1}12)[1\bar{1}1]$
$(21\bar{1})[1\bar{1}1]$, $(12\bar{1})[\bar{1}11]$, $(1\bar{1}2)[\bar{1}11]$

6.39 400 ± 140 MPa

6.41 26.7 MPa

6.43 Rockwell B99

6.45 Y.S. = 400 MPa, T.S. $\cong$ 550 MPa

6.47 347 BHN

6.49 **(a)** 252 kJ/mol, **(b)** 1.75×10^{-5}% per hour

6.51 1.79 hours

6.53 **(a)** 30.3 hr, **(b)** 303 hr

6.55 **(a)** 8.13×10^{-8} mm/mm/hr, **(b)** 14 years

6.57 **(a)** 247 days, **(b)** **(i)** 0.612 MPa, **(ii)** 0.333 MPa, **(iii)** 0.171 MPa

6.59 1.37 days

6.61 **(a)** 405 kJ/mol, **(b)** 759 to $1,010°C$, **(c)** 1,120 to $1,218°C$

6.63 15°C

Chapter 7

PP 7.1 392 J/kg·K $\approx$ 385 J/kg·K

PP 7.2 0.517 mm

PP 7.3 1.86×10^6 J/m^2·s

PP 7.4 670°C

PP 7.5 $\approx 250°C$ to $1,000°C$ ($\approx 700°C$ midrange)

7.1 **(a)** 66.6 kJ, **(b)** 107 kJ, **(c)** 282 kJ

7.3 2,940

7.5 10.07 mm

7.7 49.4°C

7.9 -11.7 kW

7.11 -43.7 kW

7.13 357 MPa (compressive)

7.15 35.3 MPa (compressive)

7.17 277 MPa

7.19 No

7.21 -1.12 kW

7.23 155 times

Chapter 8

PP 8.1 $\leq 0.20\%$

PP 8.2 **(a)** 12.9 mm, **(b)** 2.55 mm

PP 8.3 **(a)** 169 MPa, **(b)** 508 MPa

PP 8.4 77.5 MPa

PP 8.5 **(a)** 213 s, **(b)** 11.6 s

PP 8.6 **(a)** 0.289, **(b)** 1.19×10^{-27}

8.1 1040 carbon steel, 8630 low-alloy steel, L2 tool steel, ductile iron, Nb–1Zr

8.3 1.8 wt % Mn

8.5 ≥ 11.3 mm

8.9 Yes

8.11 57 μm $\leq a \leq$ 88 μm

8.13 286 μm

8.19 226 MPa

8.21 18.3 MPa

8.23 52.8%

8.25 91°C

8.27 (a) 1.75×10^6 per year, (b) 17.5×10^6 over 10 years

8.29 32.4 hr

8.31 6.67 MPa

8.33 $+0.14$ mm, -0.13 mm

8.35 0.43 mm to 0.68 mm

8.37 0.046%

Chapter 9

PP 9.1 (a) 2, (b) 1, (c) 0

PP 9.2 The first solid to precipitate is β; at the peritectic temperature, the remaining liquid solidifies, leaving a two-phase microstructure of solid solutions β and γ.

PP 9.3 $m_L = 952$ g, $m_{SS} = 48$ g

PP 9.4 $m_\alpha = 831$ g, $m_{Fe3C} = 169$ g

PP 9.5 38.5 mol % monoclinic and 61.5 mol % cubic

PP 9.6 (a) 667 g, (b) 0.50

PP 9.7 60.8 g

PP 9.9 (a) $\approx 680°C$, (b) solid solution β with a composition of ≈ 100 wt % Si, (c) 577°C, (d) 84.7 g, (e) 13.0 g Si in eutectic α, 102.0 g Si in eutectic β, 85.0 g Si in proeutectic β

PP 9.11 (a) For 200°C, (i) liquid only, (ii) L is 60 wt % Sn, (iii) 100 wt % L; for 100°C, (i) α and β, (ii) α is ≈ 5 wt % Sn, β is ≈ 99 wt % Sn, (iii) 41.5 wt % α, 58.5 wt % β
(b) For 200°C, (i) α and liquid, (ii) α is ≈ 18 wt % Sn, L is ≈ 54 wt % Sn, (iii) 38.9 wt % α, 61.1 wt % L; for 100°C, (i) α and β, (ii) α is ≈ 5 wt % Sn, β is ≈ 99 wt % Sn, (iii) 62.8 wt % α, 37.2 wt % β

PP 9.12 0 mol % Al_2O_3 = 0 wt % Al_2O_3, 60 mol % Al_2O_3 = 71.8 wt % Al_2O_3, 44.5 mol % SiO_2 = 36.1 wt % SiO_2, 55.5 mol % mullite = 63.9 wt % mullite

9.3 (a) 2, (b) 1, (c) 2

9.17 (a) $m_L = 1$ kg, $m_\alpha = 0$ kg; (b) $m_L = 667$ g, $m_\alpha = 333$ g; (c) $m_L = 0$ kg, $m_\alpha = 1$ kg

9.19 (a) $m_L = 1$ kg; (b) $m_L = 611$ g, $m_{\alpha-Pb} = 389$ g; (c) $m_{\alpha-Pb} = 628$ g, $m_{\beta-Sn} = 372$ g; (d) $m_{\alpha-Pb} = 606$ g, $m_{\alpha-Sn} = 394$ g

9.21 (a) $m_L = 50$ kg; (b) $m_\alpha = 20.9$ kg, $m_\beta = 29.1$ kg; (c) $m_\alpha = 41.8$ kg, $m_\beta = 8.2$ kg; (d) $m_\alpha = 50$ kg; (e) $m_\alpha = 39.9$ kg, $m_\beta = 10.1$ kg; (f) $m_\alpha = 33.2$ kg, $m_\beta = 16.8$ kg

9.23 760 g

9.25 3 kg

9.27 58.3 mol % SiO_2 and 41.7 mol % mullite

9.29 (a) Periclase (ss) and spinel (ss), periclase is ≈ 100 mol % MgO and spinel is ≈ 50 mol % alumina, 1 wt % periclase and 99 wt % spinel; (b) spinel (ss) with 50.3 mol % alumina, 100 wt % spinel (ss)

9.31 6.6 wt % CaO

9.33 18.6 wt % SiO_2

9.35 (a) β, (b) ≈ 60 % B

9.39 0.858

9.41 8.75 wt % Si

9.53 2.3 kg

9.55 (a) $\approx 910°C$, (b) α with a composition of ≈ 30 wt % Zn, (c) $= 903°C$, (d) $\approx 750°C$ to $\approx 220°C$

9.61 $m_\alpha = 58.2$ g, $m_\beta = 58.8$ g

Chapter 10

PP 10.1 582°C

PP 10.2 (a) ≈ 1 s (at 600°C), ≈ 80 s (at 300°C); (b) ≈ 7 s (at 600°C), $\approx 1,500$ s = 25 min (at 300°C)

PP 10.3 (a) ≈ 10% fine pearlite + 90% γ, (b) ≈ 10% fine pearlite + 90% bainite, (c) ≈ 10% fine pearlite + 90% martensite (including a small amount of retained γ)

PP 10.4 >90% for 0.5 wt % C, ≈ 20% for 0.77 wt % C, 0% for 1.13 wt % C

PP 10.5 (a) ≈ 15 s, (b) $\approx 2\frac{1}{2}$ min. (c) ≈ 1 hour

PP 10.6 (a) $\approx 7°C/s$ (at 700°C), (b) $\approx 2.5°C/s$ (at 700°C)

PP 10.7 (a) Rockwell C38, (b) Rockwell C25, (c) Rockwell C21.5

PP 10.8 (a) 7.55%, (b) 7.55%

PP 10.10 62 wt %

10.1 8.43×10^{12}

10.3 32.9 kJ/mol

10.5 $-2\sigma/\Delta G_v$

10.9 (a) 100% bainite, (b) austempering

10.11 (a) 100% bainite, (b) >90% martensite, balance retained austenite

10.13 (b) $\approx 710°C$, (c) coarse pearlite

10.15 (b) $\approx 225°C$, (c) martensite

10.19 100% fine pearlite

10.23 (a) $> \approx 10°C/s$ (at 700°C), (b) $< \approx 10°C/s$ (at 700°C)

10.25 84.6% increase

10.27 Rockwell C53

10.31 Alloys 4340 and 9840

10.33 $\approx 400°C$

10.35 93.8 kJ/mol

10.37 No

10.39 36% CW

10.41 209 kJ/mol

10.43 (a) 503°C, (b) 448°C, (c) 521°C

10.45 475 kJ/mol

10.47 59.9 kJ/mol

Chapter 11

PP 11.1 2.75 Mg/m^3 (3003 Al), 2.91 Mg/m^3 (2048 Al)

PP 11.2 14.8 wt % Na_2O, 9.4 wt % CaO, 1.1 wt % Al_2O_3, 74.7 wt % SiO_2

PP 11.3 (a) 34 < % Ni < 79, (b) 34% Ni alloy

PP 11.4 (a) 63%, (b) 86%

PP 11.5 845 kg

PP 11.6 876°C to 934°C

11.1 (a) 7.85 Mg/m^3, (b) 99.7%

11.3 (a) 5.94%, (b) 4,455 kg

11.5 332 kg

11.7 10.1 Mg/m^3

11.9 72.5 wt % Al_2O_3, 27.5 wt % SiO_2

11.11 36.1 wt % SiO_2, 63.9 wt % mullite

11.13 0.462 kg

11.15 13.7 wt % Na_2O, 9.9 wt % CaO, 76.4 wt % SiO_2

11.17 84.4 kg

11.19 1.30 nm

11.21 0.0425 μm

11.23 4.0 vol %

11.25 0.100 μm

11.27 (a) 450 MPa, (b) 8%

11.31 1,587°C

11.33 520°C

11.35 22.6 g

11.37 sintered beryllia

11.39 60.3 g

11.41 75.0 g

Chapter 12

PP 12.1 (a) 0.243 wt %, (b) 0.121 wt %

PP 12.2 833

PP 12.3 (a) 15.5%, (b) 33.3%

PP 12.4 3.37 g

PP 12.5 1.49 Mg/m^3

PP 12.6 (a) 1.82 Mg/m^3, (b) 2.18 Mg/m^3

PP 12.7 14.1 Mg/m^3

PP 12.8 39.7 × 10^3 MPa

PP 12.9 0.57 W/(m·K)

PP 12.10 3.2 − 38% error

PP 12.11 6.40 × 10^6 mm (epoxy), 77.8 × 10^6 mm (composite)

12.1 21,040 amu

12.3 1.35 g

12.5 19.8 wt %

12.7 34,060 amu

12.9 612

12.11 0.690

12.13 80.5 nm

12.15 152 nm

12.17 (a) 50,010 amu, (b) 4.87 nm, (c) 126 nm

12.19 393

12.21 296 kg

12.23 4.52 × 10^{25} amu

12.25 0.0426

12.27 (a) 43.5 wt %, (b) 1.46 Mg/m^3

12.29 45.9%

12.31 0.131 mm

12.33 23.0 wt %

12.35 1.50 Mg/m^3

12.37 119,100 amu

12.39 330 kg

12.41 9.58 Mg/m^3

12.43 289 × 10^3 MPa

12.45 286 × 10^3 MPa

12.49 49.2 × 10^3 MPa

12.53 0.38% error

12.55 4.1 − 16% error

12.57 (69.3 to 104) ×10^6 mm

12.59 7.96 × 10^6 mm

12.63 12.8 × 10^6 kJ

12.65 acrylics, polyamides, polycarbonates

12.67 0.341 ≤ v_{clay} ≤ 0.504

Chapter 13

PP 13.1 (a) 1.73 V, (b) 108 mV

PP 13.2 8.17 × 10^{23} electrons

PP 13.3 6.65×10^{23} atoms

PP 13.4 1.59 hr

PP 13.5 2.11×10^{-44}

PP 13.6 2.32×10^{-9}

PP 13.7 **(a)** $34.0 \times 10^6 \ \Omega^{-1} \cdot m^{-1}$, **(b)** $10.0 \times 10^6 \ \Omega^{-1} \cdot m^{-1}$

PP 13.8 $43.2 \times 10^{-9} \ \Omega \cdot m$

PP 13.9 7 mV

PP 13.10 1.30×10^{-8} A/m^2

PP 13.11 **(a)** $3.99 \times 10^{-4} \ \Omega^{-1} \cdot m^{-1}$, **(b)** $2.50 \times 10^3 \ \Omega \cdot m$

PP 13.12 **(a)** $24.8 \ \Omega^{-1} \cdot m^{-1}$

PP 13.13 $474 \ \Omega^{-1} \cdot m^{-1}$

PP 13.14 5.2×10^{21} atoms/m^3 ($\ll 20 \times 10^{24}$ atoms/m^3)

PP 13.15 8.44×10^{-8}

PP 13.16 **(a)** $135 \ \Omega^{-1} \cdot m^{-1}$

PP 13.17 **(a)** 16.6 ppb, **(b)** 135°C, **(c)** $55.6 \ \Omega^{-1} \cdot m^{-1}$

PP 13.18 4.07×10^{21} atoms/m^3

PP 13.19 $2.20 \times 10^4 \ \Omega^{-1} \cdot m^{-1}$

PP 13.20 0.9944 (electrons), 0.0056 (holes)

PP 13.21 1.31 parts per billion (ppb) Al

13.1 **(a)** 55.0 A, **(b)** 3.14×10^{-9} A, **(c)** 7.85×10^{-19} A

13.3 **(a)** 40.0 m/s, **(b)** 12.5 μs

13.5 5.66 Ω

13.7 734 m

13.9 994°C

13.11 **(a)** 0.0353, **(b)** 0.0451

13.13 3.75×10^{-13} (for GaP), 1.79×10^{-6} (for GaSb)

13.15 29.0 mV

13.17 24.6 mV

13.19 $\approx 500°C$

13.21 8.80 m

13.23 6.29 kW

13.25 31.9×10^{-9} W

13.27 $\approx 2{,}210$

13.31 4.34×10^{-5} C/m^2

13.33 1.20×10^{-4} C/m^2

13.35 5.2×10^{-10}

13.37 **(a)** 0.657 (electrons), 0.343 (holes); **(b)** 0.950 (electrons), 0.050 (holes)

13.39 $1.74 \times 10^4 \ \Omega^{-1} \cdot m^{-1}$

13.43 4.77%

13.45 **(a)** 9.07×10^{-6} mol %, **(b)** 4.54×10^{21} atoms/m^3 ($<< 1{,}000 \times 10^{24}$ atoms/m^3)

13.47 **(a)** Electron hole, **(b)** 3.29×10^{18} m^{-3}, **(c)** 7.6 m/s

13.49 $15.8 \ \Omega^{-1} \cdot m^{-1}$

13.51 285°C

13.53 $5.41 \ \Omega^{-1} \cdot m^{-1}$

13.55 $2.07 \ \Omega^{-1} \cdot m^{-1}$

13.57 317 K (= 44°C), taking room temperature as 300 K

13.59 **(a)** 1.11×10^{-10} K, **(b)** 9.31×10^{-7} K, **(c)** 289 K

13.65 7.5 K

13.67 0.1°C

13.69 0.973 (electrons), 0.027 (holes)

13.71 2.76 ppm

13.73 **(a)** 55.1 ppm Pb, **(b)** 4.1 ppm Pb

13.75 **(a)** 10^3 m/s, **(b)** 7.14×10^3 V/m, **(c)** 5.14 gigahertz

13.79 **(a)** 10^3 m/s, **(b)** 7.14×10^3 V/m, **(c)** 10.3 gigahertz

13.81 $38.1 \times 10^6 \ \Omega^{-1} \cdot m^{-1}$

Chapter 14

PP 14.1 S7 fan cooled from 940°C and single tempered at 540°C; S7 fan cooled from 940°C and single tempered at 425°C

PP 14.2 15.2×10^6 l

PP 14.3 697 nm

PP 14.4 **(a)** 1.75, **(b)** yes

PP 14.5 **(a)** Zinc corroded, **(b)** 1.100 V

14.1 S5 oil-quenched from 870°C and single-tempered at 650°C
L2 oil-quenched from 855°C and single-tempered at 540°C

14.3 80–55–06

14.5 Polyethylene and polypropylene

14.9 2.53 kg

14.11 **(a)** 120 kg, **(b)** 9.96×10^4 l (per aircraft)

14.13 Adequate

14.15 Carbon fiber-reinforced polymer

14.19 1.45×10^{-9} m^2

14.21 **(a)** 2.19 eV, **(b)** 1.91 to 2.10 eV

14.23 1.78 (FeO), 2.10 (Fe$_3$O$_4$), 2.15 (Fe$_2$O$_3$)

14.25 2.27

14.27 **(a)** 0.467 V, **(b)** Cr

14.29 **(a)** nickel, **(b)** tin-rich phase, **(c)** 2024 aluminum, **(d)** brass

14.31 21.8 kg

14.33 **(a)** 7.50 μm, **(b)** 37.5 μm

14.35 **(a)** difficult, **(b)** easy

14.37 **(a)** difficult, **(b)** easy

Index

Glass artifacts, 2
Glass-ceramics, 348, 363–365
Glass-forming process, 372
Glass transition temperature, 193
Glazes, 362
Gold, 358
Golden Gate Bridge, 3, 4
Graft copolymer, 392
Grain boundary, 114
Grain boundary diffusion, 146
Grain-boundary dislocations, 115
Grain growth, 304, 332–334, 334–335, 338
Grains, 114, 115
Grain-size number, 116
Gram-atom, 23
Graphite, 71
Gray cast iron, 289, 293, 352, 353
Green engineering, 525
Griffith, Alan Arnold, 167n
Griffith crack model, 167
Grove, Andrew, 484, 485
Grove, William, 145
Guinier, Andre, 329n
Guinier-Preston zones, 329, 330

Half cell, 517
Half-cell reactions, 517
Hardenability, 304, 324–327
Hardness, 181–185
Hard-sphere, 32
Hastelloy, 357
Heat capacity, 210–213
Heat treatment, 304, 363
Heteroepitaxy, 473
Heterogeneous nucleation, 306
Hexagonal close-packed (hcp) structure, 63, 64, 65
High-alloy steels, 349
High-density polyethylene (HDPE), 392
High-strength, low-alloy steels, 349
Hip-joint replacement materials, 503–506
Hirth, John Price, 114n
Hirth-Pound model, 114
Homoepitaxy, 473
Homogeneous nucleation, 306
Hooke, Robert, 154n
Hooke's law, 154
Hot isostatic pressing, 366
H_2O, 260

Human scale, 17, 18
Hume-Rothery, William, 103n
Hume-Rothery rules, 103–104
Hund's rule, 432
Hybridization, 25
Hybrids (woven fabrics), 400
Hydrogen bridge, 48, 49
Hydrogen embrittlement, 252, 515
Hydroxyapatite coatings, 505
Hypereutectic composition, 287
Hypereutectoid composition, 288, 316
Hypoeutectic composition, 287
Hypoeutectoid composition, 289
Hysteresis, 200

Impressed voltage, 521
Inconel, 357
Incongruent melting, 272
Induced dipole, 48
Ingots, 77
Inhibitor, 521
Initiator, 382
Injection molding, 416
Insulators, 449–451
    dielectric constant and dielectric strength for selected, 552
    electrical conductivities of, 485
    electrical conductivities of, at room temperature, 428
Integrated circuits, 76, 479
Intel, 484–485
Interdiffusion, 132–134
Interfacial strength, 409
Intermediate compound, 275
Intermediates, 362
Interplanar spacing, 93
Interstitial diffusion, 135
Interstitial (interstitialcy), 108
Interstitial solid solution, 104
Intrinsic gettering, 472
Intrinsic semiconduction, 451
Invariant point, 263–264
Ionic bond, 28–39
    coordination number, 32–34
    coulombic attraction and, 29
    nondirectional characteristic of, 28
Ionic packing factor (IPF), 66
Ionic radius, 32
Ions, 28
Iridium, 358

PHYSICAL AND CHEMICAL DATA FOR SELECTED ELEMENTS [a]

Atomic number	Element	Symbol	Atomic mass (amu)	Density of solid (at 20°C) ($Mg/m^3 = g/cm^3$)	Crystal structure (at 20°C)	Melting point (°C)	Atomic number
1	Hydrogen	H	1.008			−259.34 (T.P.)	1
2	Helium	He	4.003			−271.69	2
3	Lithium	Li	6.941	0.533	bcc	180.6	3
4	Beryllium	Be	9.012	1.85	hcp	1,289	4
5	Boron	B	10.81	2.47		2,092	5
6	Carbon	C	12.01	2.27	hex.	3,826 (S.P.)	6
7	Nitrogen	N	14.01			−210.0042 (T.P.)	7
8	Oxygen	O	16.00			−218.789 (T.P.)	8
9	Fluorine	F	19.00			−219.67 (T.P.)	9
10	Neon	Ne	20.18			−248.587 (T.P.)	10
11	Sodium	Na	22.99	0.966	bcc	97.8	11
12	Magnesium	Mg	24.31	1.74	hcp	650	12
13	Aluminum	Al	26.98	2.70	fcc	660.452	13
14	Silicon	Si	28.09	2.33	dia. cub.	1,414	14
15	Phosphorus	P	30.97	1.82 (white)	ortho.	44.14 (white)	15
16	Sulfur	S	32.06	2.09	ortho.	115.22	16
17	Chlorine	Cl	35.45			−100.97 (T.P.)	17
18	Argon	Ar	39.95			−189.352 (T.P.)	18
19	Potassium	K	39.10	0.862	bcc	63.71	19
20	Calcium	Ca	40.08	1.53	fcc	842	20
21	Scandium	Sc	44.96	2.99	fcc	1,541	21
22	Titanium	Ti	47.90	4.51	hcp	1,670	22
23	Vanadium	V	50.94	6.09	bcc	1,910	23
24	Chromium	Cr	52.00	7.19	bcc	1,863	24
25	Manganese	Mn	54.94	7.47	cubic	1,246	25
26	Iron	Fe	55.85	7.87	bcc	1,538	26
27	Cobalt	Co	58.93	8.8	hcp	1,495	27
28	Nickel	Ni	58.71	8.91	fcc	1,455	28
29	Copper	Cu	63.55	8.93	fcc	1,084.87	29
30	Zinc	Zn	65.38	7.13	hcp	419.58	30
31	Gallium	Ga	69.72	5.91	ortho.	29.7741 (T.P.)	31
32	Germanium	Ge	72.59	5.32	dia. cub.	938.3	32
33	Arsenic	As	74.92	5.78	rhomb.	603 (S.P.)	33
34	Selenium	Se	78.96	4.81	hex.	221	34
35	Bromine	Br	79.90			−7.25 (T.P.)	35
36	Krypton	Kr	83.80			−157.385	36
37	Rubidium	Rb	85.47	1.53	bcc	39.48	37
38	Strontium	Sr	87.62	2.58	fcc	769	38
39	Yttrium	Y	88.91	4.48	hcp	1,522	39
40	Zirconium	Zr	91.22	6.51	hcp	1,855	40
41	Niobium	Nb	92.91	8.58	bcc	2,469	41
42	Molybdenum	Mo	95.94	10.22	bcc	2,623	42

[a] For a complete listing, see Appendix 1.